Seasonal Snowpacks

Processes of Compositional Change

NATO ASI Series

Advanced Science Institutes Series

A series presenting the results of activities sponsored by the NATO Science Committee, which aims at the dissemination of advanced scientific and technological knowledge, with a view to strengthening links between scientific communities.

The Series is published by an international board of publishers in conjunction with the NATO Scientific Affairs Division

A Life Sciences B Physics	Plenum Publishing Corporation London and New York
C Mathematical and Physical Sciences D Behavioural and Social Sciences E Applied Sciences	Kluwer Academic Publishers Dordrecht, Boston and London
F Computer and Systems Sciences G Ecological Sciences H Cell Biology I Global Environmental Change	Springer-Verlag Berlin Heidelberg New York London Paris Tokyo Hong Kong Barcelona Budapest

NATO-PCO DATABASE

The electronic index to the NATO ASI Series provides full bibliographical references (with keywords and/or abstracts) to more than 30000 contributions from international scientists published in all sections of the NATO ASI Series. Access to the NATO-PCO DATABASE compiled by the NATO Publication Coordination Office is possible in two ways:

- via online FILE 128 (NATO-PCO DATABASE) hosted by ESRIN, Via Galileo Galilei, I-00044 Frascati, Italy.
- via CD-ROM "NATO-PCO DATABASE" with user-friendly retrieval software in English, French and German (© WTV GmbH and DATAWARE Technologies Inc. 1989).

The CD-ROM can be ordered through any member of the Board of Publishers or through NATO-PCO, Overijse, Belgium.

Series G: Ecological Sciences Vol. 28

Seasonal Snowpacks

Processes of Compositional Change

Edited by

T. D. Davies

University of East Anglia
School of Environmental Sciences
Norwich NR4 7TJ
United Kingdom

M. Tranter

University of Southampton
Department of Oceanography
Southampton SO9 5NH
United Kingdom

H. G. Jones

Université du Québec
Institut National de la Recherche Scientifique
2700 rue Einstein
Sainte-Foy, Québec G1V 4C7
Canada

Springer-Verlag
Berlin Heidelberg New York London Paris Tokyo
Hong Kong Barcelona Budapest
Published in cooperation with NATO Scientific Affairs Division

Proceedings of the NATO Advanced Research Workshop on Processes of Chemical Change in Snowpacks held in Maratea, Italy, July 23–27, 1990.

Library of Congress Cataloging-in-Publication Data
NATO Advanced Research Workshop on Processes of Chemical Change in Snowpacks (1990 : Maratea, Italy) Seasonal snowpacks : processes of compositional change / edited by T. D. Davies, M. Tranter, H. G. Jones.— (NATO ASI series. Series G, Ecological sciences ; vol. 28) "Published in cooperation with NATO Scientific Affairs Division." "Proceedings of the NATO Advanced Research Workshop on Processes of Chemical Change in Snowpacks held in Maratea, Italy, July 23–27, 1990"—T. p. verso. Includes bibliographical references and index.
ISBN-13: 978-3-642-75114-1 e-ISBN-13: 978-3-642-75112-7
DOI: 10.1007/ 978-3-642-75112-7
1. Snow—Congresses. I. Davies, Trevor D. II. Tranter, M. (Martyn) III. Jones, H. G. (H. Gerald), 1936– . IV. Title. V. Series: NATO ASI series. Series G, Ecological sciences ; no. 28. GB2601.2.N36 1990 551.57'84—dc20
91-26115

Softcover reprint of the hardcover 1st edition 1991

Typesetting: Camera-ready by authors
31/3140-543210 – Printed on acid-free paper

TABLE OF CONTENTS

PREFACE

The NATO Advanced Research Workshop (ARW) was held in Maratea, Italy, July 23-27, 1990. The principal theme of the ARW was the study of the processes which control the compositional evolution of seasonal snowcover, both at the snow-atmosphere interface and within the snowpack. The purpose of this Preface is to outline the reasons why the ARW was convened, and to provide a brief introduction to the contents of this proceedings volume.

The ARW represented a logical succession to a previous NATO Advanced Study Institute (ASI) (Jones and Orville-Thomas, 1987). The ASI considered the physics and chemistry of snow and meltwaters, and the hydrogeochemistry of meltwater pathways at the catchment-scale. It was an interdisciplinary meeting in which scientists, in different areas and disciplines of snow science, learned from their colleagues and attempted to set the base of knowledge in the subject. It was established that seasonal snowcover can be significant in the chemical dynamics of ecosystems in many temperate, alpine and subarctic regions. A major conclusion of the ASI was that the knowledge of physical processes in seasonal snowcover was more advanced than that on the chemical composition and dynamics of snowcover. The effects of biological activity had been almost totally ignored. The ASI recognised that this imbalance in knowledge stemmed from the poor integration of physical and chemical studies. A strong recommendation was, therefore, the adoption of a more integrated approach to the study of snowcover composition; in particular, it was recommended that more effort be directed towards interdisciplinary work in the laboratory, and controlled experiments in the field. It was also resolved to convene an ARW in 1990.

The ARW had specific aims:; to consolidate and extend contacts; to assess and synthesise results from experimental research; to consider the limitations of research methodology; and review the theoretical base from which to propose future studies. The ARW was designed to encourage as much scientific exchange as possible. A number of speakers were invited to make the contributions which are reproduced in this volume. In order to maximise informed discussion, their papers were circulated beforehand to "discussants". The discussants provided feedback to the speakers before the Workshop, and also led the subsequent discussions on the presentations. Some discussants have provided a written critique, or complementary information; these contributions are also published, following the appropriate main paper. The invited papers have been peer-reviewed, by two or more ARW participants, or by other scientists. The short discussants' contributions have not been peer-reviewed.

A "Review Panel" sat throughout the ARW. Besides engaging in the scientific discussion throughout the Workshop, the panel formulated questions to be addressed on the last day of the ARW. The Panel represented a synthesising mechanism for the meeting, and directed the final sessions on the "ways forward" for research into the chemistry of seasonal snowcover.

The proceedings reflect the structure of the ARW. The initial contributions deal with the chemistry of snowfall and how the the composition of the snowcover can be changed by subsequent dry deposition. The way in which the snowpack can be modified through wind transport of snow particles is also considered. Subsequent papers discuss some of the problems in coupling models of meltwater that incorporate the hydrologic flowpaths within the snow, the solute transport processes, and in-pack transformations of chemical species during snowmelt. A case in point is the necessity to clarify the thermodynamics of the concentrated solutions of ionic species in small liquid inclusions as the snowcover starts to melt. In addition, some snowpacks may show significant changes in nutrient species (e.g. N) due to microbiological activity in interstitial pockets of meltwater. Although these problems are difficult to resolve, and have not been fully explored as distinct model components, progress is gradually being made as explicit physico-chemical models are being developed.

The degree to which different processes affect snowcover composition will vary both with the local and regional environment. Two of the papers consider the composition of snowfall and snowcover in two very different areas; high-altitude snow-fields of the Austrian Alps, and a low-lying, coastal, urban area in Japan. A third paper examines the distribution and fate of organic micropollutants in the perennial snow of the Canadian Arctic archipelago.

Perennial snows contain chemical signatures which are used to reconstruct past atmospheric conditions. However, post-depositional modification is often overlooked in the paleo-interpretation of snow/firn/ice cores and may lead to erroneous conclusions concerning past climate. Two papers are devoted to this issue. The first reviews the work on high-altitude, mid/low latitude glaciers, and examines the type of information available. The second is more specifically orientated towards high latitude samples; it assesses the reality of links between historical records and contemporaneous atmospheric conditions by considering the stability of the chemical signals from deposition to recovery and analysis.

Finally, in view of the current interest in future climatic change, a discussion on the implications of projected change on the dynamics of seasonal meltwater discharge and chemical composition is presented.

As previously mentioned, this Preface sets the scene; it does not provide a detailed summary of the individual contributions. Nor does it provide a resulting assessment of the state-of-the-science. If we had done this, it would have duplicated the contribution of members of the Review Panel, which concludes this volume. The reader will be able to judge, from their remarks, how successful was the ARW in achieving its objectives, stated above.

The editors of this volume would like to thank all the participants at the ARW, whether they were contributors, discussants or reviewers. Agreement is not often the lot of scientists, but objectivity ruled the day, and discussions were as invigorating as they were vigorous. We also acknowledge the kind financial and administrative support of NATO, which made the ARW possible.

Trevor D. Davies

Martn Tranter

H. Gerald Jones

Reference

Jones H. G. and Orville-Thomas W. J. (1987) Seasonal Snowcovers: Physics, Chemistry, Hydrology. NATO ASI Series C: Mathematical and Physical Sciences Vol. 211; D. Reidel Publishing Company, Dordrecht. 746pp.

ARW PARTICIPANTS

R. C. Bales	Department Hydrology and Water Resources University of Arizona Tucson, Arizona, USA
L. A. Barrie	Atmospheric Environment Service Downsview Ontario, Canada
P. J. Barry	Chalk River Nuclear Laboratories Chalk River Ontario, Canada
P. Brimblecombe	School of Environmental Sciences University of East Anglia Norwich, UK
S. H. Cadle	Environmental Science Department General Motor Research Laboratories Warren, Michigan, USA
M. H. Conklin	Department Hydrology and Water Resources University of Arizona Tucson, Arizona, USA
R. E. Davis	US Army CRREL Hanover New Hampshire, USA
T. D. Davies	School of Environmental Sciences University of East Anglia Norwich, UK
J. Dozier	CSL/CRSED University of California Santa Barbara, California, USA
W. Giger	EAWAG Dubendorf, Switzerland
Y. T. Gjessing	Institute of Geophysics University of Bergen Bergen, Norway
D. J. Gregor	National Water Research Institute Environment Canada Burlington, Ontario, Canada
R. W. Hoham	Department of Biology Colgate University Hamilton, New York, USA

A. J. Johannes	Department of Chemical Engineering Oklahoma State University Stillwater, Oklahoma, USA
H. G. Jones	INRS-EAU Universite du Quebec Quebec, Canada
A. Kovar	Institute for Analytical Chemistry Technical University Vienna, Austria
A. R. W. Marsh	National Power TEC Leatherhead, UK
F. Maupetit	Laboratoire de GG Environnement Domaine Universitaire St. Martin d'Heres, France
E. M. Morris	British Antarctic Survey Cambridge, UK
A. Neftel	Physics Institute University of Berne Berne, Switzerland
N. E. Peters	USGS-WRD Doraville, Georgia, USA
J. W. Pomeroy	National Hydrology Research Institute Saskatoon Saskatchewan, Canada
H. Puxbaum	Institute for Analytical Chemistry Technical University Vienna, Austria
D. Shooter	Department of Chemistry University of Auckland Auckland, New Zealand
K. Suzuki	Department of Geography Metropolitan University Tokyo, Japan
M. Tranter	Department of Oceanography University of Southampton Southampton, UK
S. Tsiouris	Laboratory of Agricultural Chemistry Aristotelian University Thessaloniki, Greece

M. Williams | CSL/CRSED
University of California
Santa Barbara, California, USA

E. W. Wolff | British Antarctic Survey
Cambridge, UK

SNOW FORMATION AND PROCESSES IN THE ATMOSPHERE THAT INFLUENCE ITS CHEMICAL COMPOSITION

L.A. Barrie
Atmospheric Environment Service
4905 Dufferin Street
Downsview, Ontario, Canada M3H 5T4

1. INTRODUCTION

The chemical composition and physical form of snowfall is controlled by a range of complex physical and chemical processes occurring on spatial scales ranging from the atomic dimensions of a water molecule (10^{-10} m) to the size of precipitating cloud system (10^5 to 10^6 m) and on temporal scales ranging from the time to grow ice particles by diffusion (1 to 10^2 s) to the lifetime of a precipitating cloud system (10^4 to 10^5s). The processes occur in a turbulent atmosphere and involve three phases of water. In the last thirty years, the cloud physics research community has made major advances in our understanding of snow formation. In particular, the role of the ice phase has received considerable attention since in extra-tropical regions, it plays a key role. There are several comprehensive texts in this field (Mason, 1971; Hobbs, 1974; Pruppacher and Klett, 1980) as well as symposia proceedings(e.g. Cloud Physics, 1990) which the reader is encouraged to consult as background to this review.

At first, the role of clouds and precipitation in scavenging atmospheric chemical compounds was only considered by the cloud physics community when it pertained to processes of cloud and precipitation formation or cloud electricity. But in the past twenty years, the field of precipitation scavenging of substances from the atmosphere has received more and more attention (Precip. Scav. I, II &III; Slinn, 1984; Barrie and Schemenauer, 1989). Currently, there is considerable work being undertaken to explain the chemical composition of precipitation(rain and snow). Processes of incorporating particles and gases into ice and snow crystals are not being neglected since they play a key role in precipitation formation and pollutant removal from the atmosphere even for storms delivering rain to the ground.

2. PROPERTIES OF ATMOSPHERIC ICE RELEVANT TO CHEMICAL COMPOSITION

2.1 Ice Nucleation

When the temperature of air decreases below O°C, water suspended as cloud droplets does not immediately freeze. Supercooling can take place to temperatures as low as -40°C.

Above -40°C, the transition in the atmosphere from water to ice is effected by heterogeneous nucleation whereas below -40°C, it occurs by homogeneous nucleation. The former involves

NATO ASI Series, Vol. G 28
Seasonal Snowpacks
Edited by T. D. Davies et al.

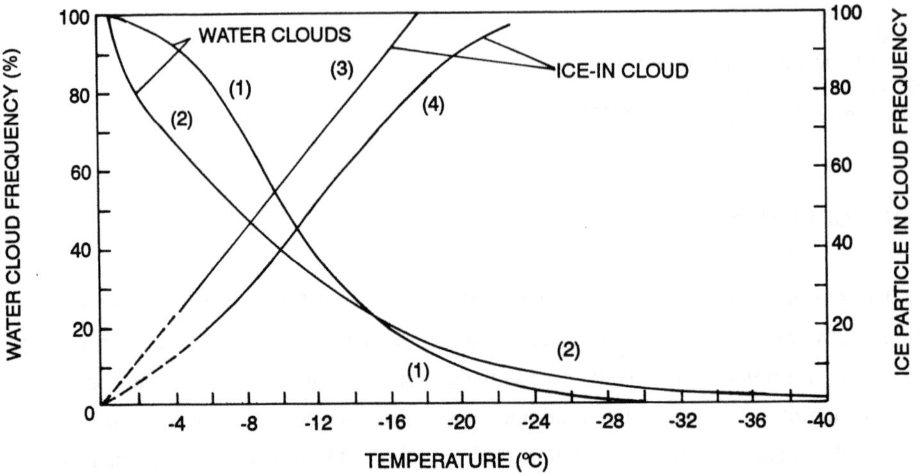

Figure 1 The frequency of supercooled clouds and of clouds containing ice crystals as a function of temperature. Curves 1 and 2 have the left ordinate, curves 3 and 4 the right one (compiled by Pruppacher and Klett, 1980).

initiation of the ice phase on the surface of insoluble particles called ice nuclei (IN). The latter occurs by the spontaneous formation of ice from clusters of water molecules.

Since the ice phase is generally present in clouds well above -40°C (Fig. 1), heterogeneous nucleation is considered the predominant pathway to the frozen phase.

There are three mechanisms of heterogeneous ice nucleation: (i) **deposition** of water vapour onto the IN (ii) **immersion-freezing nucleation** that involves the initiation of the ice phase by an IN within a supercooled cloud droplet (incorporated either in condensation nuclei or later by diffusion or inertial impaction onto a cloud drop) and (iii) **contact nucleation** whereby an IN particle initiates the ice phase in supercooled drops by collisional contact with them.

IN are "preferred" aerosol particles which are able to initiate the ice phase in the three modes described above (Pruppacher, 1986). They constitute only a small fraction of the existing particulate matter in the atmosphere. They are generally water insoluble, larger in diameter than 1 μm, exhibit surface crystallographic characteristics similar to that of ice, have surface ionic or preferred hydrogen bonds and contain "surface active sites". Their abundance in the atmosphere is highly temperature and composition dependent.

A review of the origin of IN by Vali (1985) concludes that as suspected earlier (Mason, 1971), clay minerals are important ice nuclei. They often appear at the centre of snow crystals. Some organic materials can also be efficient IN. In the mid- 1970's, Vali et al (1976) showed that decomposition of vegetation such as tree leaves produce IN in soil that were capable of initiating freezing between -4°C and -10°C. Fresch (1973) showed that a single strain of bacteria (pseudonomas syringae) in plant litter act as IN. Many other studies have shown the efficiency of particular bacteria as IN (Levin and Yankofsky, 1983; Yankofsky et al 1981).

Production of ice particles in clouds by splintering during the riming process as proposed by Hallet and Mossop (1974) still remains an important but not sufficient source of ice crystals in clouds. Mossop (1985) has concluded that the droplet size distribution in a cloud is a critical factor in this process. The criteria for predicting the presence of ice in a cloud remain unclear. This is not surprising considering the large variability in IN in the atmosphere and the complexity of ice multiplication. The latter seems to be a limiting factor in determining presence of ice. Recent observations suggest that ice can form in clouds with much warmer tops than predicted by the Hallet-Mossop criteria (Rangno and Hobbs, 1988).

An assessment of the state of knowledge of ice particle formation in clouds by Hobbs (1990) leads to the hypothesis that contact nucleation of drizzle-sized drops followed by ice nucleation in pockets of high supersaturation with respect to water that may occur locally near colliding drops, freezing drizzle drops and near drops accreting on graupel is the mechanism for ice formation in clouds.

2.2 Ice Crystal Nomenclature and Habit

At this stage it is necessary to introduce terminology commonly used in atmospheric ice physics (Pruppacher and Klett, 1980). At temperatures where supercooled water droplets coexist with ice particles, the latter grow at the expense of the former. This may occur by water vapor diffusion to the ice particles because the saturation vapor pressure of water is higher over water than over ice or by colliding and freezing of droplets on ice crystals. The first mechanism is called **deposition** and the second **riming**. Ice particles that have been formed by deposition of water vapour are called **ice or snow crystals.**

If they clump together by collision, they are called **snowflakes.** In the initial stages of riming as long as the original form of the crystal is apparent, the ice particles are called **rimed snow crystals.** When riming proceeds to the point that the original crystal shape is obliterated, the ice particle is called **graupel, soft hail or a snow pellet**. Further riming produces **small hail.** Particles are called **sleet** if they are hard transparent or irregular ice particles of frozen drops or partially melted and refrozen snow crystals.

The form of ice crystals in the atmosphere has been investigated extensively since the 1950's both in the laboratory and in clouds. The picture that has emerged (Magono and Lee,1966) is summarized in Figure 2. The form depends on degree of supersaturation of water vapor and temperature. The crystallographic structure of an ice crystal at atmospheric temperatures is generally hexagonal. It has 2 basal planes parallel to the conventional "a" axis and 6 prism planes parallel to the "c" axis. For crystals growing by vapor deposition, the dependence of crystal habit on temperature is regulated by the availability of water. At supersaturations only somewhat greater than that over ice, habit varies little with temperature. It is a thick hexagonal plate with a height to diameter ratio of 0.81. In contrast, at greater supersaturations near that over water, there is a marked variation of crystal habit with temperature. As temperature decreases from 0°C,

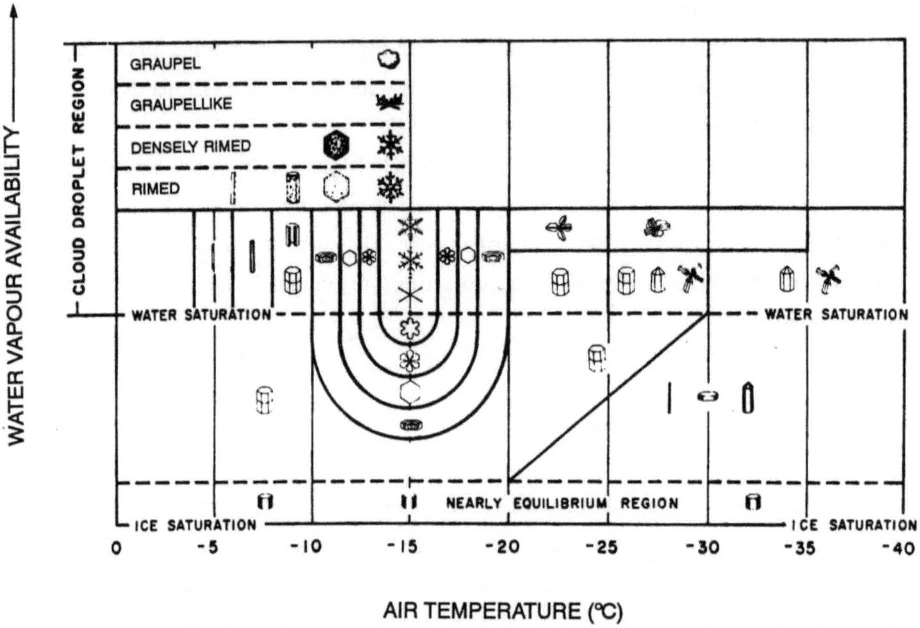

Figure 2 Temperature and humidity dependence of the growth of natural snow crystals of various types (adapted from Magono ad Lee, 1966).

a cycle from plate to column to plate to column occurs at transition temperatures of -4°C, -9°C and -22°C. Finally, at even higher supersaturations where supercooled cloud droplets are very abundant which occurs generally for temperatures above -15°C, rimed plates or columns are found. Note that for a given temperature, the humidity strongly influences the crystal shape. For example, near -15°C as vapor supply is increased, the snow crystal habit varies from a thick plate to a thin plate to a sector plate to a dendrite. Near -5°C, it varies from a short, solid column to a needle.

The shape of a snow crystal controls its interaction with atmospheric particles and gases. As an air parcel bearing ice crystals moves through different temperature and moisture regimes in a cloud, the growth behaviour depicted in Figure 2 can lead to extremely complex shapes . This complexity has been captured beautifully in photographic surveys of real snow crystals (Bentley and Humphreys, 1962; Magono and Lee, 1966).

2.3 Snowflake Formation And Sedimentation

Snowflakes form by the collision and sticking of ice crystals. Field observations show that the frequency of occurrence of this process is dependent on temperature and crystal habit. The largest snow flake diameters occur near -1°C with a secondary maximum between -12 and -17°C (Fig. 3). Upon contact, ice crystals stick to each other by freezing together and mechanically

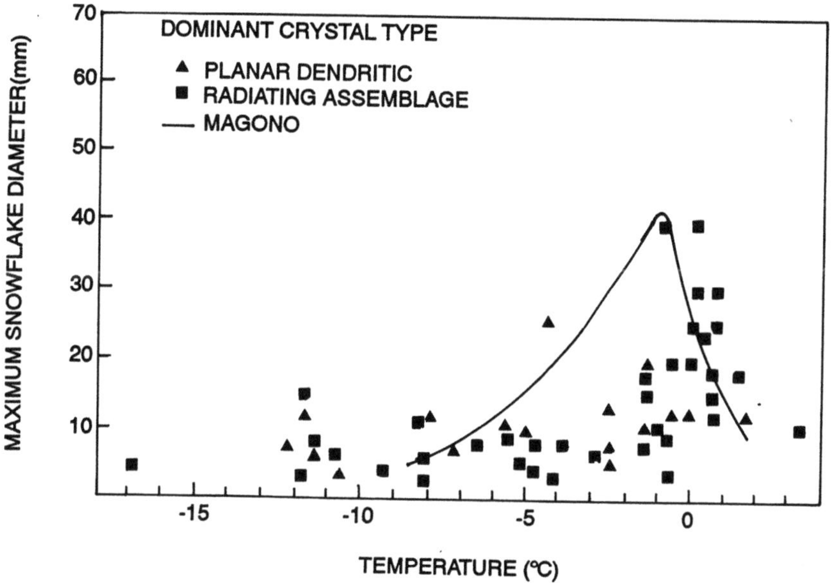

Figure 3 Maximum observed snowflake dimensions as a function of air temperature for two types of snowflake composition (adapted from Rodgers, 1974)

interlocking. The first effect is most pronounced near 0°C where a substantial pseudo-liquid film on the ice surface is present to promote the formation of an ice-neck between the crystals. The secondary maximum in clumping at around -15°C is related to formation of the classical snow crystal shape of dendritic plates whose complex shape favor clumping (Fig. 2). The formation of snowflakes has an effect on the chemical composition of snowfall since ice particle shape strongly affects interactions with gases, particles and supercooled water droplets. It also enhances the removal of water and chemical substances from a precipitating system by increasing the sedimentation rate of ice bearing particles.

The sedimentation velocity of single ice particles is fairly well known. It varies strongly with crystal habit as illustrated by the results for plate-like crystals in Fig. 4 where typical ranges are 10 to 70 cm s^{-1}. Columnar crystals settle at rates close to that calculated from the drag on equivalent cylinders. Equations for calculating sedimentation velocities can be found in Pruppacher and Klett (1980). Because of the variety and complexity of snowflake shapes, it is difficult to characterize settling velocities in a simple way (generally for maximum dimensions greater than 1 mm they are in the range 70 to 150 cm s^{-1}) and the complex nature of snowfall is beyond our quantitative understanding of ice processes in clouds (Beard, 1987).

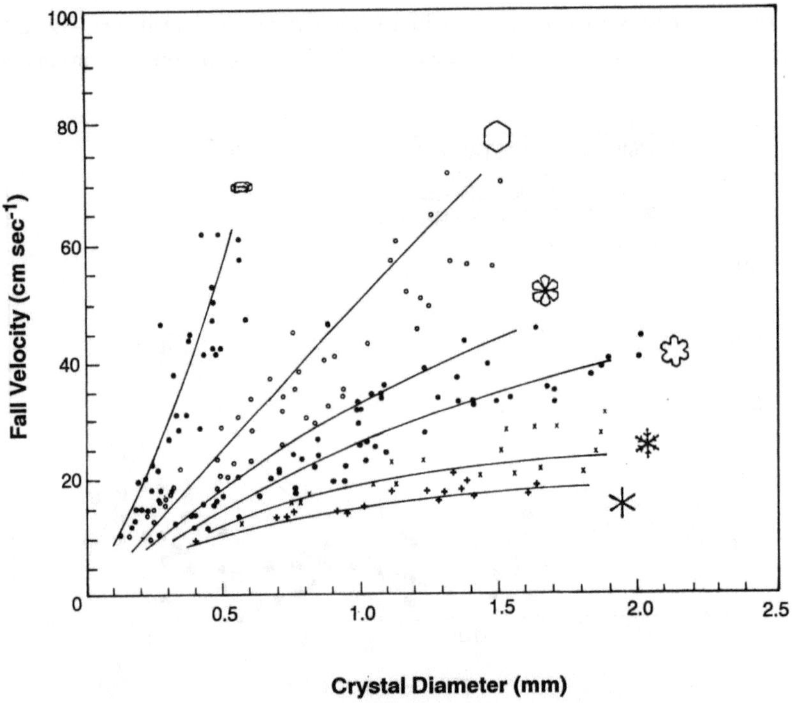

Figure 4 Variation with size of the observed and computed terminal fall velocity of ice crystals of various shapes, at 10°C, 1000mb (adapted from Kajikawa, 1972).

3. MACROSCOPIC PROPERTIES OF PRECIPITATION SYSTEMS

Precipitation at mid-latitudes is often associated with extra-tropical cyclones (Stewart, 1990) which have cloud patterns similar to that shown by the satellite photo of North America in Figure 5. These low pressure systems typically involve the intrusion of a tongue of warm southern air northward to override colder northern air. This warm air is, in turn, undercut from the west-northwest by the colder northern air flowing south. A warm front marks the boundary between advancing warm air and the colder air over which it is rising. A cold front lies between the advancing cold air in the northwest and the warm air. Different precipitation scavenging situations arise depending upon location relative to these fronts in the cyclone (Fig. 6). In the vicinity of the cold front and in the warm sector ahead of the cold front convective clouds of 2 to 20 km diameter form in clusters (Fig. 6 top). Each cloud draws air from the atmospheric boundary layer below. North of the surface warm front, the situation is quite different. Clouds

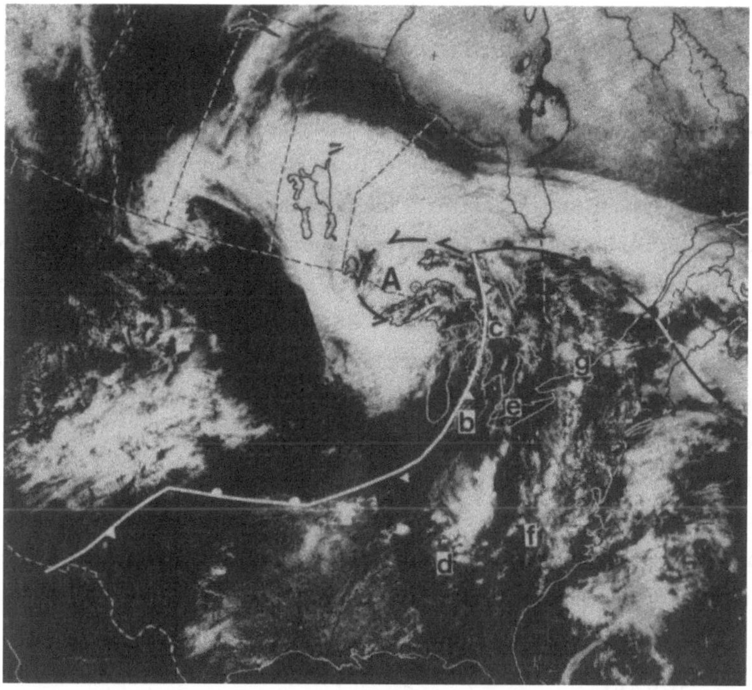

Figure 5 A satellite photo in the visible waveband illustrating the typical cloud and frontal structure associated with an extratropical cyclone occurring May 12, 1985. Triangles mark a cold front and semi-circles a warm front. Convective cloud bands are visible in the warm sector between b and c, e and d, and between g and f. Warm frontal cloud is present in central Quebec and northwestern Ontario.

form over large areas as a result of widespread uplift (Fig. 6 middle). The air in which they form originates from the near surface layer several hundreds of kilometres to the south. Precipitation falls from the overriding air into the layer below.

The nature of precipitating clouds in extra-tropical cyclones has been extensively studied using satellite photography, radar and aircraft (see the series of papers referenced in Hertzmann and Hobbs, 1988; also review by Stewart, 1990). Precipitation tends to occur from organized bands of convective clouds(see example in the warm air sector of Fig. 5). They also occur imbedded in widespread layer cloud north of the surface warm front.

The nature of a precipitation event at a ground level location that is associated with the passage of an extra-tropical cyclone depends on the intensity of the cyclone, its speed and the location of the site relative to it. A climatology of storm features from this perspective done for the eastern United States (Thorp and Scott, 1982) shows that, in general, there is a marked difference in storm types between winter and summer. In winter, storms move twice as fast across the region (15 m s^{-1} versus 7 m s^{-1}) but their extent is such that the average duration is much longer (26 hr) than in summer (2.5 hr). Precipitation intensity is lower in winter than in summer (0.9 versus 2.5 mm hr^{-1}). The fraction of storms that are isolated convective bands is higher in summer than in winter.

The above detailed discussion of a type of cloud system prevalent at mid-latitudes serves to illustrate the complexity of macro-scale features of storm systems in the atmosphere. Of course, other important cloud systems and other areas of the globe should not be overlooked. In the polar regions for instance, cirrus-like clouds of ice crystals forming in air that is cooling radiatively and precipitating to the ground in apparently clear skies (i.e. diamond dust) is a very prevalent form of precipitation throughout the troposphere (Hoff, 1988). During winter, it delivers snow bearing chemicals to northern glaciers that are important sources of historical records of temperature and chemical composition of the atmosphere. Orographically induced precipitation systems (Fig. 6 bottom) so prevalent in mountainous regions are also important sources of snow. It is well known that because the atmosphere holds less water vapour at low temperatures, cloud formation and snow showers are initiated by a much lower slope of land in winter than in summer. This coupled with abundant sources of water vapour from unfrozen lake and ocean surfaces explain the snow belts that occur downwind of such water on adjacent land areas.

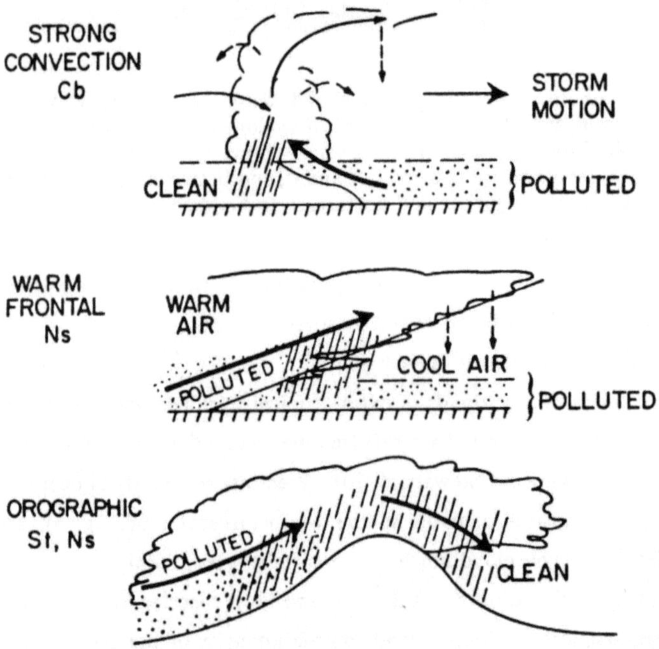

Figure 6 A schematic of different precipitation scavenging situations.

4. WITHIN-CLOUD PROCESSES OF SNOW FORMATION & CHEMICAL INTERACTIONS

Notable references on cloud microphysical processes are Pruppacher and Klett (1980), Slinn (1984), Beard (1987) and Barrie and Schemenauer (1989). Pollutants are scavenged from the atmosphere via the sequence of processes depicted in Figure 7. In order to understand the composition of snow, it is important to understand the condensation process leading to the formation of super-cooled cloud water. The condensation of water occurring generally on hygroscopic aerosols (cloud condensation nuclei CCN) entering a cloud as well as the dissolution of soluble gases (step 1) leads to supercooled cloud drops that contain inorganic ions, insoluble particulate matter and organic constituents. This is known as 'nucleation scavenging'. It is a major pathway of chemicals from the air into snow. Typically in precipitating clouds, condensation takes place on particles of diameter greater than 0.2 μm (Pruppacher and Klett, 1980). However, not all particles larger than 0.2 μm are active CCN. Their ability to serve as

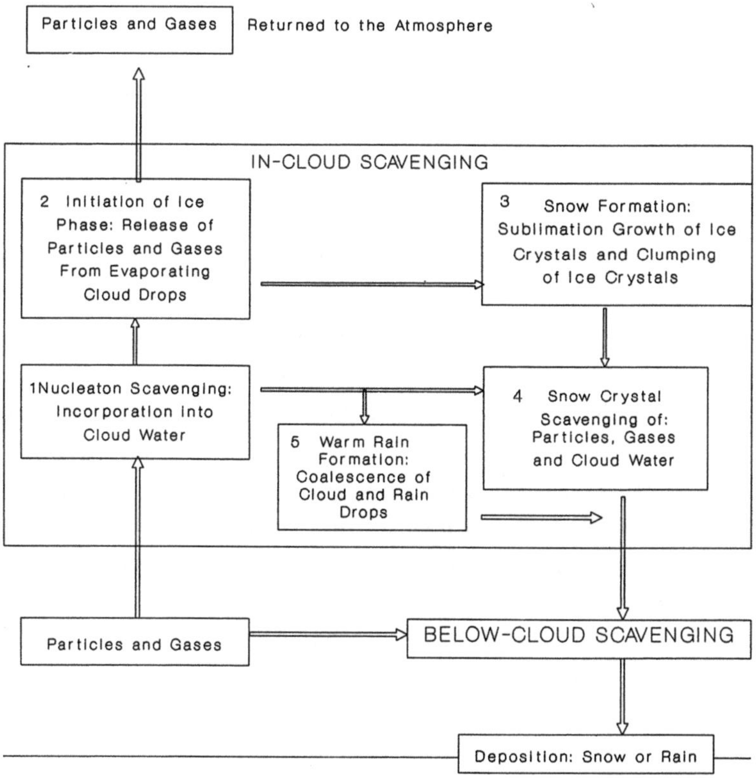

Figure 7 The precipitation scavenging process.

an active nucleus is influenced strongly by their soluble salt content. Those containing no salts (e.g. silicate minerals from soil) may be hydrophobic and inactive CCN. However, these particles often become active after some time in the atmosphere as they are coated with soluble salts from gas to particle conversion on their surfaces and from coagulation with other aerosols. Generally, the greater the supersaturation in a cloud the lower the minimum particle size of active CCN that actually grow into cloud drops (Jensen and Charlson, 1984).

In precipitating clouds, the nucleation scavenging efficiency for the total particle mass is generally in the range 50 to 100%. Daum et al (1984) observed that for precipitating stratus, the mass of aerosol in the air between the droplets is generally low but not negligible compared to the mass in cloud water. Many studies of the nucleation scavenging efficiency of $SO_4^=$ and NO_3^- (particles plus HNO_3) mass have been undertaken by a variety of investigators in a variety of clouds (e.g. Hegg et al ,1982; Hegg and Hobbs, 1988; Leaitch et al, 1988). The general picture is that scavenging is highly variable ranging between 30 and 95% depending upon cloud supersaturation. In other words, much of the particle and soluble gas mass enter cloud water during the condensation process.

During and after water droplet formation by condensation, gaseous compounds may dissolve in the liquid cloud water and may react with other dissolved substances such as 0_3, H_20_2, O_2 and soluble free-radicals to form soluble compounds. The generation of soluble trace gases in the cloud interstitial air by chemical reactions can also affect cloud water composition (Leaitch et al, 1988; Strapp et al, 1988; Hegg, 1989; Hegg and Larson, 1989; Lelieveld and Crutzen, 1990).

Two things can happen to cloud water that strongly influence the formation and chemical composition of snow: (i) if ice is present in-cloud, the droplets can evaporate feeding ice growth. Water vapour, driven by the difference in saturation vapour pressure of water and ice, diffuses from the drops to the ice crystals (steps 2 and 3). The gaseous and particulate matter in cloud water is then returned to the atmosphere in a form that is not necessarily the same as when it entered the water; (ii) cloud water can be incorporated into precipitation either by being captured by large (diameter > 200 μm) snow crystals (steps 1-4) by riming or by coalescence of rain and cloud drops (i.e. the warm rain formation process steps 1-5). The warm rain process is of little relevance in snow formation. Riming is the collection by snow of supercooled cloud drops 2 to 50 μm diameter by inertial impaction and interception as the heavier snow crystals or flakes fall past the small drops. The cloud drops freeze upon contact. Pollutants trapped in the cloud water thereby enter precipitation and fall from the cloud. Feng and Grant (1982) observed that in mid-latitude snowfalls, half the mass of snowflakes can be collected cloud water. However, this observation is highly dependent on location, cloud type and other factors.

Particles and gases that do not enter cloud water during condensation (step 1) or that are released back into the air during vapor growth of the ice phase (step 2) can be incorporated into ice crystals (steps 3 and 4). For particles, this occurs by Brownian diffusion, electrical attraction, inertial impaction and phoretic processes. Gases may react with the ice surfaces or supercooled water drops.

Precipitation leaving cloud base can consist of rimed ice, unrimed ice or rain drops. Scott (1981) has found that snow storms with mostly unrimed snow crystals (i.e. scavenging steps 1-2-3-4) remove $SO_4^=$ from the atmosphere much less efficiently than storms with rimed snow crystals remove them (scavenging steps 1-2-3-4 and 1-4). This conclusion is probably also valid for most atmospheric compounds that are in active CCN or in soluble or water reactive gases.

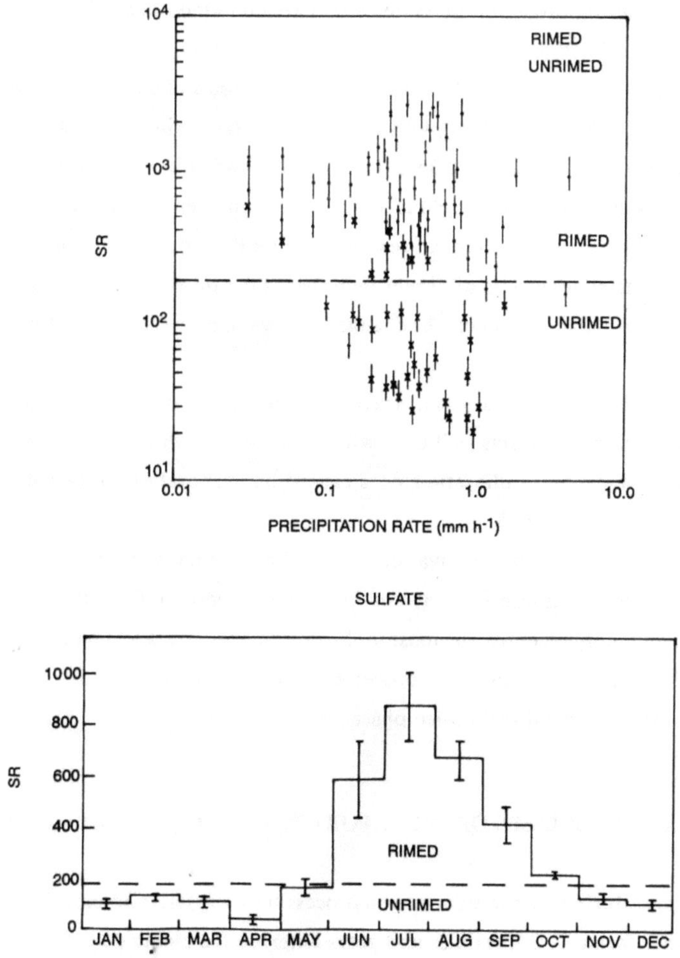

Figure 8 Top The sulphate mass scavenging ratio (concentration in snow to that in air) averaged over several hours as a function of precipitation rate for rimed and unrimed snowfalls on the east side of Lake Michigan (Scott, 1981).
Bottom Annual variation of the monthly mean sulphate mass scavenging ratio deduced for Greenland from observations by Davidson et al (1989).

The scavenging ratio (SR) defined as the concentration of a substance in precipitation falling from a storm divided by the concentration of a substance in air entering the storm is dependent on precipitation rate and snow formation process. For $SO_4^=$ particles (Fig. 8 top), it is a factor of 3 to 20 higher for rimed storms than for unrimed storms. Thus, whether the concentration of a substance in precipitation reflects that in cloud water of the storm from which it has fallen depends on the snow formation process. Differences are least for that involving the warm rain process (step 1-5) intermediate for snow formation by riming (step 1-2-3-4 and 1-4) and most for that involving unrimed snow crystal formation (step 1-2-3-4). In other words, the scavenging efficiency of substances by precipitation systems that are contained in active CCN or in soluble gases increases from unrimed to rimed to warm rain storms. For $SO_4^=$, this ratio is approximately 1:10:100, respectively (Hogan, 1982). Observations in the Arctic and on Greenland show a strong seasonality in SR of $SO_4^=$ (Fig. 8 bottom) with a strong summer maximum and a winter minimum (Davidson et al, 1989). This probably reflects less rimed snowfall in winter than in summer.

Particles and gases that enter the cloud but are not removed by precipitation exit at the evaporating edges. As a result of chemical reactions inside the cloud, the chemical composition of air at the outflowing edges can be different than that at the inflow regions. Furthermore, the size distribution of particulate matter can be changed as a result of in-cloud chemical and physical transformations.

Snow falling below the cloud base, scavenges particles and reactive gases from the atmosphere by the same mechanisms as it does within the cloud with the exception of the riming process and with the addition of phoretic forces caused by evaporation in the sub-saturated air beneath the cloud (see section 5.3).

Numerical simulations of the removal of chemical constituents from the atmosphere by a precipitation system that takes into account cloud dynamics, microphysics and chemistry are only in their infancy. Until only recently, the most sophisticated ones (e.g. Flossmann and Pruppacher, 1988) that contain detailed microphysical processes have been limited to only two dimensional dynamics of a convective cloud and no ice phase.

5. SMALL SCALE PROCESSES OF INCORPORATION OF CHEMICALS INTO SNOW

At several stages during the snow formation process snow crystals and snowflakes accumulate chemicals that were formerly in particles and gases entering the cloud: (i) during the nucleation of ice (ii) during the process of ice growth when water vapour is diffusing to ice crystals as a result of higher water vapour pressure gradients over supercooled water than over ice (iii) during the subsequent fall of snow crystals and snowflakes through the cloud and (iv) during their fall from cloud base to ground (i.e. below-cloud scavenging). All of these processes contribute to the morphology of the ice particle and distribution of chemicals throughout it. These features may have an influence on chemical movement in a snowpack subsequent to deposition.

5.1 During Ice Nucleation

If ice crystals originate from vapour deposition, that is, ice nucleation on an insoluble IN, chemicals characteristic of active IN (see section 2.1) enter snow at this stage and are in the center of a crystal. On the other hand, if they form by immersion freezing or contact nucleation of supercooled cloud drops, not only are the ice nuclei materials incorporated into the crystal but so also are cloud water constituents. During freezing, soluble gases may or may not be retained in the frozen cloud drop (see section 5.6). The mechanism of ice nucleation can be identified microscopically. Pruppacher and Klett (1978) reviewed the existing observations. In general, snow crystals with frozen droplets at their centers are quite abundant ranging from 19% of all crystals at -9 to -10°C to 48% at -15 to 16°C to 23% at -21 to -22°C. The diameter of the frozen centre drop ranged from 2.5 to 25 μm.

5.2 During Diffusional Growth of Ice

When ice crystals grow by diffusion at the expense of supercooled cloud water, there is a flux of vapor from water to ice and a flux of heat from ice to water. Because the difference between water vapour pressure over water and that over ice has a relative maximum at -12°C, diffusional growth is a maximum at -15°C (Pruppacher and Klett, 1980). Let us first consider particles in this setting. In the air, they undergo diffusiophoretic and thermophoretic forces associated with the moisture and heat flux, respectively. Particles less than 3 μm diameter are dominated by thermophoretic forces that keeps them away from the growing ice crystal. However, particles bigger than 3 μm are dominated by diffusiophoretic forces and are incorporated into the ice crystal (Martin et al, 1980). If the particle is hygroscopic, it tends to be captured by the ice crystal at diameters much lower than 3 μm. This is apparently due to the growth of the particle in humidity gradients and deposition as a larger particle (Prodi, 1983). Thus during the diffusional ice growth stage, cloud drops evaporate leaving residues of involatile materials in particles in the interstitial air. Ice crystals capture large particles that are present either originally in interstitial air or that are released through cloud water evaporation. During this process, the insoluble fraction of particles will be incorporated into the ice rather than on its surface. Whether the soluble fraction of particles remains on the surface or is distributed throughout the ice depends on the type of ion. Some ions such as those of the halogens or ammonium are incorporated into ice more readily than others such as sulphate (Workman, 1954).

Next let us consider the fate of gases during diffusional growth of ice at the expense of cloud water. Initially, they are present either in interstitial air (if they have a low solubility or reactivity) or, if they are volatile, released from evaporating drops. They interact with snow crystals by physical dissolution and chemical reaction. Little is known about this process for most gases. Gaseous uptake by ice apparently depends on solubility and reactivity. Gases like SO_2 can be incorporated into growing ice at this stage (see section 5.5).

5.3 During The Fall Of Snow Through The Cloud

As snow crystals fall through a cloud they interact with other ice crystals to form snow flakes. Because of the particle size dependence of the terminal velocity of cloud elements in the atmosphere, the two types of ice then can capture slower falling aerosol particles or supercooled cloud droplets. The interaction between ice crystals and particles is a complex one resulting in particles coating the crystal's surface with little attendant alteration of crystal form. Even in the relatively simple case of a plate-like or columnar ice crystal scavenging particles, the collection efficiency depends on crystal and aerosol size distributions, on the degree of charging, on evaporation, atmospheric pressure and electrostatic charge (see 100% relative humidity curves in Figure 9). Add to this an extremely variable ice crystal shape and one is left with a complicated situation to handle theoretically, experimentally or observationally. Inside a cloud, relative humidities are so close to 100% that evaporation is relatively unimportant. The exception could be near the cloud edges where entrainment of outside air is taking place. In general, plate-like crystals are more efficient scavengers than columnar crystals (compare Figs. 9 a and b). This combined with the generally greater abundance of platelike crystals means that columnar crystal scavenging is often an insignificant scavenging pathway. A recent laboratory investigation by Mitra et al (1990a) has shown that large dendritic snowflakes capture particles in the 0.2 to 0.4 μm diameter range not by inertial impaction as is seen for single ice crystals (Fig. 9) but by a filtering effect as air moves through the crystal. It also shows that snow flakes are considerably more efficient in capturing these particles than single crystals.

The process of riming of snow crystals and flakes also brings to the snow the chemicals that are in cloud water. In contrast to aerosol capture, riming causes not only a chemical alteration of the crystal but also physical alteration. A comparison of Figures 9 and 10 shows that the collection efficiency of cloud drops whose main mass is generally in drop sizes larger than 10 μm diameter by falling snow crystals is often higher than it is for aerosols. The difference is likely even greater for snow flakes due to the filtering effect discovered by Mitra et al(1990a). This fact combined with the relatively small size and low concentration of chemical mass in aerosols in cloud air compared to the chemical mass in cloud water from nucleation scavenging means that, if it occurs, the riming process will dominate the delivery of chemicals to snow. The effectiveness of riming of relatively clean snow formed in the upper reaches of a cloud by more acidic cloud water to yield snow at the ground that has an acid content between that of the initial snow and the cloud water has been demonstrated clearly by aircraft cloud chemistry observations (Isaac and Daum, 1987).

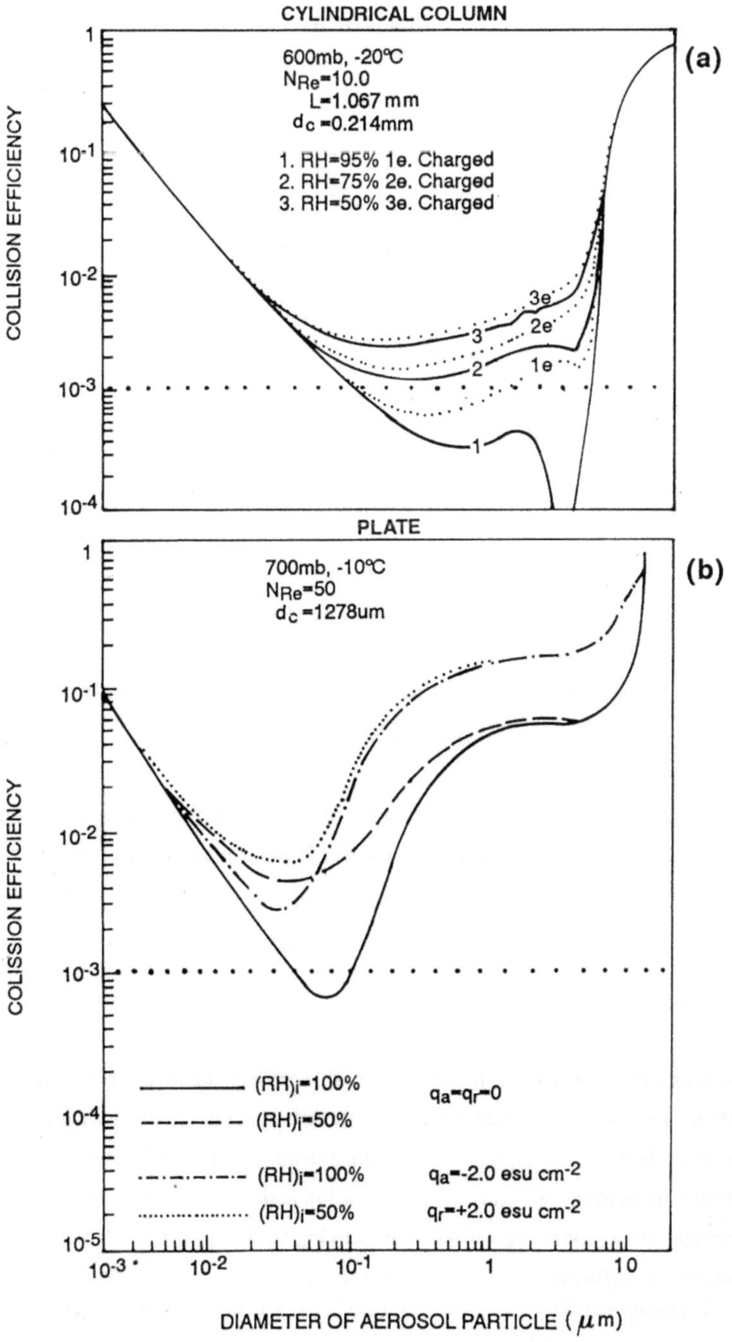

Figure 9 A comparison of the collision efficiency of a columnar (a) and plate-like (b) ice crystal with aerosols as a function of aerosol size, ambient relative humidity and crystal charge. From theoretical simulations of Miller and Wang (1989) and Martin et al (1980), respectively.

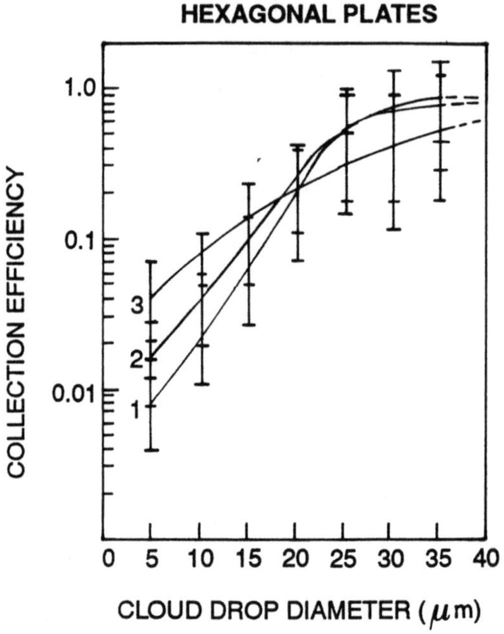

Figure 10 Observed collection efficiency of natural snow crystal (hexagonal plate) for supercooled droplets for crystal diameters of different sizes: (1) 800 to 1000 μm, (2) 600 to 800 μm and (3) 400 to 600 μm (adapted from Kajikawa, 1974).

During the riming process, the fate of gases dissolved in the droplets depends on their solubility. If very soluble they are trapped. However if not too soluble (e.g. SO_2), they can be at least partly excluded from the crystal (see section 5.5).

5.4 Below-Cloud Scavenging

As snow falls from the cloud, it interacts with aerosols much as it would while falling through the cloud. The only difference is that the interaction occurs in an environment of relative humidity less than 100% and of higher aerosol concentrations. The effects of relative humidity on aerosol capture are evident in Figure 9. The situation is the reverse of that described in section 5.2. Snow particles are evaporating rather than growing so that particles of diameter less than 3 μm whose capture is influenced by thermophoretic forces are driven down the temperature gradient into the crystal surface. Hence the capture is enhanced. Interactions of gases with snow below cloud are discussed in section 5.5.

It is generally agreed that in a below-cloud situation, on the basis of equal mass of precipitation, snow is a more effective scavenger of particles than rain in a below-cloud situation (Georgii, 1965, Graedel and Franey, 1975, Schuman, 1989). This is consistent with the observations of Murukamai et al (1983) that below-cloud scavenging by snow contributed on average more to the total aerosol scavenged (in-cloud and below-cloud) (50%) than rain (30%).

5.5 Incorporation Of Soluble Or Reactive Gases Into Ice

In two instances during snow formation, sudden freezing of a supercooled cloud droplet occurs: (i) during immersion freezing or contact nucleation of ice and (ii) during riming of snow particles. These process generally produce so much latent heat compared to that being removed from the droplet by convective diffusion that as the ice phase grows into the liquid phase, the latter is warmed to 0°C. Furthermore, ice growth preferentially excludes certain ions from its lattice forcing them to concentrate in the rapidly shrinking liquid phase (Workman, 1954).

Experiments have been done to test the retention of gases (R) during this process. Thus Iribarne and Pyshnov (1990a, b) found that retention was 100% for the very soluble gases HCl, HNO_3, NH_3 and H_2O_2 but not for the less soluble gas SO_2 (water pH of 2.3 to 2.6). Since snow in Greenland is more rimed in summer than in winter (Davidson et al, 1989; also Fig. 8 bottom), this phenomenon may amplify the summer maximum in H_2O_2 concentrations in Greenland glacial snow and ice reported by Neftel et al (1984, 1986) which reflects the seasonal variation in atmospheric H_2O_2 concentrations. For SO_2, R was 62% which is considerably higher than the results of Lamb and Blumenstein (1987) who found R less than 12%. One would expect that for droplet pH's typical of real clouds that are higher than those used by these investigators, the increased solubility of SO_2 as it dissociates reversibly into HSO_3^- ion will result in higher R values.

Another process of incorporating soluble and reactive gases into ice during diffusional growth of ice crystals has been reported by Valdez et al. (1989). They found that at -15°C, SO_2 is incorporated readily into bulk ice growing by vapor diffusion but not into evaporating ice. When ice was growing, the gas was captured at concentrations in ice equivalent to its equilibrium concentration with water at 0°C. This implicates, a pseudo-liquid layer of water on the ice surface as the vehicle for incorporation. The gases ozone and formaldehyde appeared to inhibit rather than enhance the process. This work has been extended to SO_2 uptake by real snow crystals rather than bulk ice (Mitra et al, 1990b). The results of Valdez et al (1989) were confirmed. In addition, the striking observation was made that H_2O_2 dramatically enhanced the uptake during ice growth. Little is known about the temperature dependence of this process for SO_2 uptake. Furthermore, the uptake of other gases has not been investigated in the laboratory. However, field observations indicate that, below-cloud scavenging of HNO_3 by snow is an important process of incorporation of nitrate into snow reaching the ground (Isaac and Daum, 1987).

6. CONCLUSION

The process of snow formation and incorporation of atmospheric chemicals into snow is a complex one that is not fully understood. Some aspects are not even qualitatively within our grasp. Our ability to predict the incorporation of aerosols into snow crystals, although qualitatively better than for gases, is in practice made intractable by our inability to predict the distributions and types of hydrometeors in a storm. For that matter if we could predict them, current computer capacity limits our understanding of how they interact in real storms. Much more research is needed on gas uptake by snowing clouds. Recent advances in knowledge of SO_2 uptake by ice during diffusional growth and cloud droplet freezing need to be put into the larger picture of scavenging by clouds using dynamic cloud chemistry models.

ACKNOWLEDGEMENT

The author thanks Prof. H. Pruppacher and Dr. G. Isaac for their helpful comments.

REFERENCES

Barrie, L.A. and R.S. Schemenauer, 1989, Wet deposition of heavy metals, in Control and Fate of Atmospheric Metals, ed. Pacync and Ottar, NATO ASI Series, 203-231, Kluwer Academic Pub. Beard, K.V., 1987, Cloud and precipitation physics research 1983-1986, Rev. of Geophysics, 25, 357-370.

Bentley, W.A. and W.J. Humphreys, 1962, Snow Crystals, original 1931 McGraw Hill, Dover Publication Reprint, 1962.

Cloud Physics, 1990, Conference On Cloud Physics, Proc. Meeting 23-27 July San Francisco, Cal., American Met. Soc., 45 Beacon St. Boston Mass. 02108.

Davidson, C.I., J.R. Harrington, M.J. Stephenson, M.J. Small, F.P. Boscoe and R.E. Gandley, 1989, Seasonal variations in sulfate, nitrate and chloride in the Greenland ice sheet relation to atmospheric concentrations, Atmos. Envir., 23, 2483-2495.

Daum, P.H., S.E. Schwartz and L. Newman, 1984, 'Acidic and related constituents in liquid water stratiform clouds', J. Geophys. Res., 89, 1147-1158.

Feng, D. and L.O. Grant, 1982, Correlation of snow crystal habits, number flux and snowfall intensity from ground observations, Ppts. Cloud Physics Conf., Am. Met. Soc. Boston, 485-587.

Flossmann, A.I. and H.R. Pruppacher, 1988, A theoretical study of the wet removal of atmospheric pollutants. Part III: the uptake redistribution and deposition of $(NH_4)_2$ SO_4 particles by a convective cloud using a two dimensional cloud dynamics model, J. Atmos. Sci., 45, 1857-1871.

Fresch, R.W., 1973, Res. Rept. AR-106, Dept. of Atmos. Resources, University of Wyoming, Laramie.

Georgii, H. W., 1965, Untersuchungen uber Ausregnen und Auswaschen atmosphaerischer Spurenstoffe durch Wolken und Niederschlag, Berichte Deutscher Wetterdienst 14, No. 100.

Graedel, T.E. and J.P. Franey, 1975, Field measurements of sub-micron aerosol washout by snow, Geophys. Res. Lett., 2, 325-328.

Hallett, J. and S.C. Mossop, 1974, Production of secondary particles during the riming process, Nature, 249, 26-28.

Hobbs, P.V., 1990, Ice in clouds, Proc. 1990 Cloud Physics Conference, American Met. Soc., 600-606.

Hegg, D.A., P.V. Hobbs and L.F. Radke, 1989, Measurements of the scavenging of sulphate and nitrate in clouds, Atmos. Envir., 18, 1939-1946.

Hegg, D.A., 1989, The relative importance of major aqueous sulphate formation reactions in the atmosphere, Atmos. Res., 11, 323-333.

Hegg. D.A. and T.V. Larson, 1989, The effects of microphysical parameterization on model predictions of sulphate production in clouds, Proc. Symp. Clouds in Atmospheric Chemistry and Global Climate, Anaheim. Amer. Met. Soc., Boston, Mass.

Hegg, D.A. and P.V. Hobbs, 1988, Comparisons of sulphate and nitrate production in cloud on mid-Atlantic and Pacific N.W. Coasts of the United States, J. Atmos. Chem., 7, 325-333.

Hertzman, O. and P.V. Hobbs, 1988, The mesoscale and microscale structure and organization of clouds and precipitation in midlatitude cyclones: Part XIV: three dimensional airflows and vorticity budget of rainbands in a warm occlusion, J. Atmos. Sci., 45, 893-930.

Hobbs, P.V., 1974, Ice Physics, Oxford University Press, Oxford.

Hoff, R.M., 1988, Vertical structure of Arctic haze observed by Lidar J. App. Met., 27, 125-139.

Hogan, A.W., 1982, Estimation of sulphate deposition, J. App. Met., 21, 1933-1936.

Iribarne, J.V. and T Pyshnov, 1990, The effect of freezing on the composition of supercooled droplets-I. Retention of HNO3, HCl, NH3, and H2O2, Atmos. Envir. 24a, 383-388.

Iribarne, J.V., T Pyshnov and B. Naik, 1990, The effect of freezing on the composition of super-cooled drops-II. Retention of S(IV), Atmos. Envir., 24A, 389-398.

Isaac, G. A. and P.H. Daum, 1987, A winter study of air, cloud and precipitation chemistry in Ontario, Canada, Atmos. Envir., 21, 1587-1600.

Lamb, D. and R. Blumenstein, 1987, Measurement of the entrapment of sulphur dioxide by rime ice, Atmos. Envir., 21, 1765-1772.

Leaitch, W.R., J.W. Strapp, G.A. Isaac and J.G. Hudson, 1986, Cloud scavenging of aerosol sulphate in polluted atmospheres, Tellus 38B, 328-344.

Levin, Z. and S.A. Yankofsky, 1983, Contact versus immersion freezing of freely suspended droplets by bacterial ice nucleii, J. Cli. App. Met., 22, 1964-1966.

Kajikawa, M., 1972, J. Meteor. Soc. Japan, 50, 577.

Kajikawa, M., , 1974, J. Meteor. Soc. Japan, 52, 328.

Magono, C. and C.W. Lee, 1966, J. Fac. Sci., Hokkaido University, Ser. 7, 2 & 4.

Mason, R.J., 1971, The Physics of Clouds, 2nd Ed. Oxford Univ. Press, London.

Martin, J.J., P.K. Wang and H.R. Pruppacher, 1980, 'A theoretical study of the effect of electric charges on the efficiency with which aerosol particles are collected by ice crystal plates', J. of Colloid and Interface Sci., 78, 44-55.

Mitra, S.K., S. Barth and H.R. Pruppacher, 1990a, A laboratory study of the efficiency with which aerosol particles are scavenged by snow flakes, Atmos. Envir., 24A, 1247-1254.

Mitra, S.K., S. Barth and H.R. Pruppacher, 1990b, A laboratory study on the scavenging of SO_2 by snow crystals, Atmos. Envir., 24A, 2307-2313.

Mossop, S.C., 1985, Microphysical properties of supercooled cumulus clouds in which an ice particle multiplication operated, Quart. J. Roy. Met. Soc., 111, 183-198.

Murakami, M., T. Kimura, C. Magono and K. Kikuchi, 1983, Observations of precipitation scavenging for water soluble particles, J. Met Soc. Japan, 61, 346-357.

Neftel, A., P. Jacob and D. Klockow, 1984, Measurements of hydrogen peroxide in polar ice samples, Nature, 311(5981), 43-45.

Neftel, A., P. Jacob and D. Klockow, 1986, Long term record of H_2O_2 in polar ice cores, Tellus, 38B, 262-270.

Precip. Scavenging I, 1970, Proc. Symposium held at Richland, Washington, eds. Englemann and Slinn, NTIS Rep. # CONF-700601.

Precip. Scavenging II, 1977, Proc. Symposium held at Champaign, Illinois, 1974, eds. Semonin and Beadle, NTIS Rep. # CONF-741003.

Precip. Scavenging III, 1983, Proc. of Symposium held in Santa Monica, Cal., 1982, eds. Pruppacher, Semonin and Slinn, Precipitation Scavenging Dry Deposition And Resuspension, Elsevier, 1462 pp.

Pruppacher, H.R. and J.D. Klett, 1980, Microphysics of clouds and precipitation, Reidel Co., Dordrecht, Boston, London, 714 pp.

Pruppacher, H.R., 1986, The Role of cloud physics in atmospheric multiphase systems: ten basic statements, in Chemistry of Multiphase Atmospheric Systems, ed. W. Jaeschke, NATO ASI 66, Springer 133-190.

Rangno, A.L. and P.V. Hobbs, 1983, Production of ice particles in clouds due to aircraft penetrations, J. Clim. App. Met., 22, 214-232.

Rangno, A.L. and P.V. Hobbs, 1984, Further observations of the production of ice particles in clouds by aircraft, J. Clim. App. Met., 23, 985-987.

Rangno, A.L. and P.V. Hobbs, 1988, Criteria for the onset of significant concentrations of ice particles in cumulus clouds, Atmos. Res., 22, 1-13.

Rodgers, D.C., 1974, Res. Rep. June 1974, No. AR 110, Dept of Atmos. Resources, University of Wyoming, Laramie.

Schumann, T., 1989, Large discrepancies between theoretical and field determined scavenging coefficients, J. Aerosol Sci., 20, 1159-1162.

Scott B.C., 1981, Sulphate washout ratios in winter storms, J. App. Met., 20, 619-625.

Slinn, W.G.N., 1984, Precipitation Scavenging, Chapt. 11, in Atmospheric Science And Power Production. D. Randerson, U.S. DOE/TIC-27601, Report. 466-524.

Stewart, R.E., 1990, Extra-tropical cyclones and precipitation, Proc. 1990 Cloud Physics Conference, American Met. Soc., 566-570.

Strapp, J. W., W.R. Leaitch, K.G. Anlauf, J.W. Bottenheim, P. Joe, R.S. Schemenauer, H.A. Weibe, G.A. Isaac, T.J. Kelly and P.H. Daum, 1988, Winter cloud water and air composition in central Ontario, J. Geophys. Res. 93, D4, 3760-3772.

Thorp, J.M. and B.C. Scott, 1982, Preliminary calculations of average storm duration and seasonal precipitation rates for the N.E. section of the United States, Atmos. Envir., 16, 1763-1794.

Valdez, M.P., G.A. Dawson, and R.C. Bales, 1989, Sulphur dioxide incorporation into ice depositing from the vapor, J. Geophys. Res., 94D, 1095-1103.

Vali, G., 1985, Atmospheric ice nucleation - a review, J. Rech.Atmos., 19, 105-115.

Vali, G., M. Christensen, R.W. Fresch, E.L. Galyen, L.R. Malu and R.C. Schnell, 1987, J. Atmos. Sci., 33, 1565.

Workman, S.E., 1954, On geochemical effects of freezing, Science, 119, 73.

Yankofsky, S.A., Z. Levin, T. Bertold and N. Sandlermon, 1981, Some basic characteristics of bacterial freezing nucleii, J. Appl. Meteor., 20, 1013-1019.

DRY DEPOSITION TO SNOWPACKS

Steven H. Cadle
Environmental Science Department
General Motors Research Laboratories
Warren, MI., U.S.A 48090-9055

INTRODUCTION

Dry deposition refers to the transfer of gases and particles to ground-based surfaces, where they are removed. The two main reasons for studying dry deposition are: (1) The determination of the quantity of a species entering the environment. Interest usually centers on species such as SO_4^{-2} or toxic trace metals that may harm components of aquatic or terrestrial ecosystems. (2) The determination of removal rates of a species from the atmosphere so that its atmospheric concentration and lifetime can be determined. Over the last several years, much of the dry deposition research has focused on acid deposition issues.

Dry deposition is a complex process that involves turbulent transfer through the atmosphere, transfer across a quasi-laminar layer at the surface, and uptake at the surface. These are complex processes that vary with atmospheric conditions, as well as the nature of both the depositing species and the deposition surface. Because of this complexity, the measurement of dry deposition rates is a much more complex and demanding task than the measurement of wet deposition. Indeed, there are no reference methods for the measurement of dry deposition, and little likelihood that a routine method for measuring the deposition rates of all the species of interest will be developed in the near term. Because of these problems, there is much less information on dry deposition rates than on wet deposition rates.

NATO ASI Series, Vol. G 28
Seasonal Snowpacks
Edited by T. D. Davies et al.

This chapter will start with a general discussion of the processes that control dry deposition and the methods for measuring and calculating dry deposition rates. This discussion applies to dry deposition to any surface, although information of special relevance to snowpacks will be highlighted. This will be followed by a specific discussion of the state-of-knowledge regarding deposition velocities to snowpacks, and the relative importance of wet and dry deposition to the chemical composition of snowpacks.

THE DRY DEPOSITION PROCESS

Deposition velocity

The deposition velocity is a convenient parameter that relates the atmospheric concentration of the depositing species to its flux via equation (1).

$$F = v_d C \tag{1}$$

Deposition velocity is defined as being positive when there is a downward flux. Since meteorologists define an upward flux as positive, the left side of equation (1) is sometimes written as -F. The flux is specified as being to the projected ground area, rather than the actual surface area. Four points should be noted about this relationship. First, it is only valid when the concentration of the depositing species at the surface is zero. A concentration greater at the surface implies that the flux will be upward when the atmospheric concentration is less than the surface concentration. Second, no significant atmospheric sources or sinks should be present between the ground and the point at which the concentration is measured. Third, since the concentration of a depositing species increases with height, and the flux is independent of height, v_d must decrease with height. Thus, it is generally recommended that a reference height be specified whenever v_d is given. Historically, reference heights have been 1-1.5 m, although heights up to 10 m are becoming more common. Fourth, dry deposition is directly related to atmospheric concentration. Thus, dry deposition rates will be highest near sources, and will rapidly decrease as the pollutants are dispersed in the atmosphere. This simple relationship between the deposition rate and the atmospheric concentration is very useful, and is the focus of much of the current dry deposition research.

Resistance model

Figure 1 shows a simple resistance model of the dry deposition process. R_a is the resistance to transport through the atmosphere, R_b is resistance to transport across the boundary layer (a very thin layer in contact with the surface), and R_s is resistance to uptake at the surface. This latter term is frequently written as R_c when a canopy is present or R_t when the resistance is defined as a transfer resistance. This simple, three component model should be adequate for most situations involving a uniform snowpack in an open area. In instances where there is reason to believe that more than one type of adsorption site is present at the surface, a second, parallel resistance, R_{s1}, can be added to the model. In more complex situations, such as a snowpack in a wooded area, at least two additional resistances (R_{b2} and R_{s2} in Fig. 1) would be added in parallel to the original resistances to account for deposition to the canopy. If emissions are expected from the surface, an upward flux term, E, is sometimes added (Albritton et al., 1987) to emphasize this complication.

The resistances in this model have units of s cm^{-1}. The three resistances in the simplest from of the model are related to v_d via equation (2).

$$v_d = (R_a + R_b + R_s)^{-1} \tag{2}$$

Additional resistances in parallel and in series are treated mathematically in the same manner as an electrical circuit. The model provides a very convenient separation of the major factors controlling dry deposition, both from a conceptual

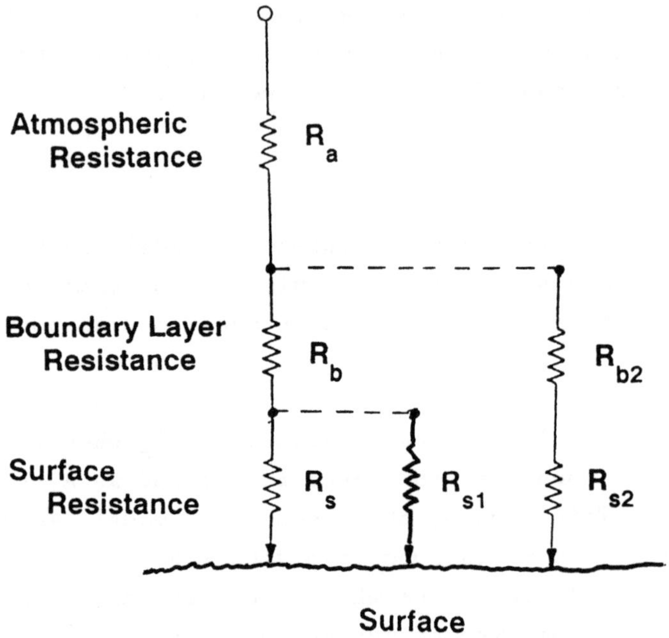

Figure 1. Resistance model for the dry deposition of gases.

and a mathematical standpoint. The three resistances are discussed in sequence, below.

Aerodynamic resistance. The mixed layer, which is also referred to as the planetary boundary layer, is the region of the atmosphere where the wind structure is influenced by the earth's surface. This layer should not be confused with the surface boundary layer, which is discussed in the next section. The height of the mixed layer is variable, with typical daytime values of roughly 500 m. Turbulence in the mixed layer is strongly influenced by the atmospheric temperature profile. The dry, adiabatic temperature profile is used as the reference temperature profile. This profile is calculated for a parcel of air that is allowed to slowly expand with no gain or loss of heat when it is displaced upwards. This parcel of air is in neutral equilibrium in the atmosphere since it has no tendency to either rise or settle. If the actual atmospheric temperature decreases faster with height than this reference profile, then the atmosphere is unstable, since a parcel of air displaced upwards or downwards continues to move in that direction. Alternately, if the actual temperature profile decreases slower than the reference profile, the atmosphere is stable, since air parcels are inhibited from movement. Atmospheric stability plays an important role in determining the atmospheric resistance.

Transport through the mixed layer is caused by turbulence. Diffusion is generally slow by comparison and can be neglected when calculating dry deposition rates. Turbulence is induced when heating of the surface results in increased buoyancy of the air parcel adjacent to the surface, and when interaction of the wind with surface elements causes wind shear. Overall, turbulence increases with wind speed, surface roughness, and atmospheric instability. R_a, the term describing resistance to transport in the atmospheric mixed layer, decreases with increasing turbulence.

Two meteorological parameters used to describe the wind profile above a surface are the friction velocity, u_*, and the surface roughness, z_o (Sehmel, 1980). These are related to the measured wind speed, u, at a height, z, above ground level, the von Karman's constant, k, which is approximately 0.4, and the zero-displacement plane, d, via equation (3).

$$u = u_* k^{-1} \ln(z - d + z_o)/z_o \tag{3}$$

The d term is only used for surfaces with high physical roughness, and thus is zero for a snowpack. The terms d and z_o are obtained from graphical procedures, and thus have no physical meaning, although d is about 75 percent of the canopy height and z_o is about 0.15 times the physical roughness height. The friction velocity is typically a few percent of the average wind speed.

The aerodynamic resistance is calculated from equation (4),

$$R_a = [\ln(z/z_o) - \Psi]/ku_* \tag{4}$$

where Ψ is an atmospheric stability correction function. For more detailed discussions of aerodynamic resistance and resistance models, the reader is referred to Hosker and Lindberg (1982), Fowler (1984), Seinfeld (1986), Hicks et al. (1987), and Hicks and Matt (1988).

<u>Quasi-laminar boundary layer resistance.</u> Turbulent velocities decrease to almost zero very close to the surface. For the ideal case of air flow over a smooth surface, the flow in the layer next to the surface is laminar. The thickness of this layer depends on wind speed, but is typically on

the order of 1 mm. A similar layer forms over nonideal surfaces. However, the roughness elements of the surface are such that the layer is probably highly variable, with some roughness elements extending through the layer. This layer is referred to as the quasi-laminar layer (Hicks, 1984) to reflect the fact that it is probably turbulent at times. Equation (5)

$$R_b = Au_*Sc^{\alpha} \tag{5}$$

shows that the boundary layer resistance for a gas or small particle is a function of the friction velocity, the Schmidt number (Sc), and two experimentally determined parameters, A and α. The Schmidt number is the viscosity divided by the pollutant diffusivity. The values of A and α are uncertain, but are frequently taken to be 5 and 2/3, respectively (Hicks, 1984;, Voldner et al., 1986).

Exchange across the quasi-laminar boundary layer is limited primarily by diffusion for gases and by Brownian motion for small particles. However, as particle size increases, inertial impaction and gravitational settling become significant factors. For very large particles the deposition rate is controlled primarily by gravitational settling.

Stefan flow and phoretic forces, which are not accounted for in equation (5), may also affect the transport across the quasi-laminar boundary layer. Stefan flow results from the injection of a gas into the atmosphere from the surface. As Hicks (1984) points out, sublimation or evaporation of 0.2 $g\ m^{-2}\ s^{-1}$ of water (0.07 cm of water per hour) from the surface will result in a Stefan flow of 0.2 $mm\ s^{-1}$ away from the surface. This flow rate is equivalent to the predicted deposition rate of small particles to smooth surfaces. Phoretic effects include thermophoresis, diffusiophoresis, and electrophoresis. Thermophoresis depends on local temperature gradients and tends to drive particles away from hot surfaces. Diffusiophoresis is caused by water evaporation or sublimation, but is a different effect than Stefan flow. If the concentration of water is higher on the downward facing side of a particle than on the upward facing side, the downward facing side will be impacted by more water molecules. Since water has a lower molecular weight than the other air constituents, there will be a net downward force. Electrophoresis is caused by electrical forces acting on charged particles. The effects of thermophoresis and diffusiophoresis are generally thought to be small. Electrophoretic forces, however, may be important in some cases.

An additional complication has been suggested for hygroscopic particles such as sulfuric acid. These particles grow rapidly at high relative humidities. In the situation where water is evaporating or subliming from the surface, a strong gradient in the relative humidity will occur near ground level. A hygroscopic particle entering this layer may grow rapidly, thereby changing its dry deposition rate.

<u>Surface resistance.</u> Uptake of gases at the surface depends on the state of the surface and the reactivity of the depositing species. There may be some species that have zero surface resistance, and clearly many for which the resistance will be infinite. For snow, a major factor can be the presence or absence of a liquid layer at the surface. Water soluble gases such as HNO_3, HCl, SO_2, and H_2O_2 may deposit much more rapidly to liquid water or a quasi-liquid water layer, than to ice. The pH of a water layer may also have an effect on the deposition rate of some gases. SO_2, for example, is much more soluble at high pH's than at low ones. Also, reaction of one depositing species with a previously deposited or co-depositing species may affect deposition

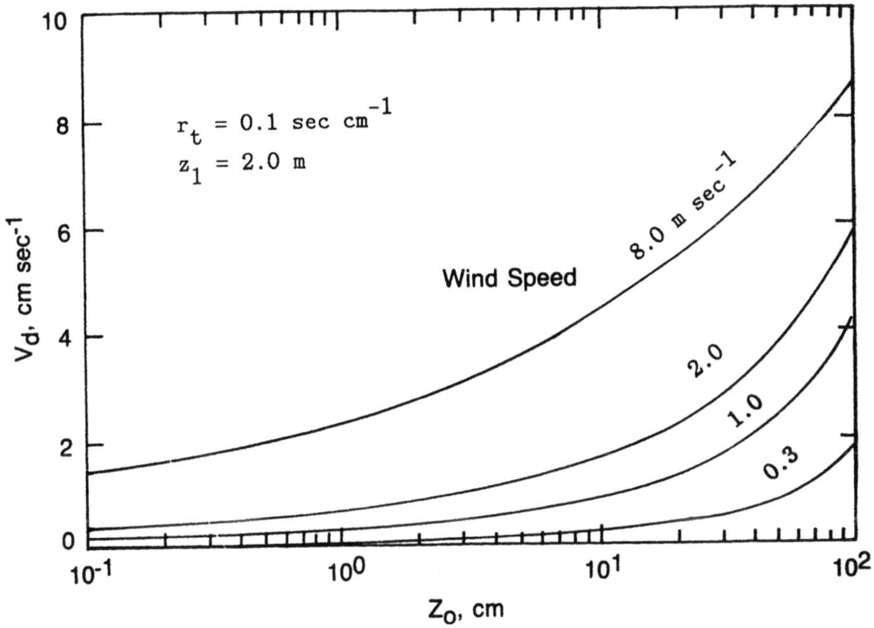

Figure 2. Calculated deposition velocities, v_d, for a gas as a function of wind speed and surface roughness, z_o. Adapted from Seinfeld (1986).

rates. An example of this effect is the reaction of SO_2 with H_2O_2.

Most workers assume that any particle that reaches a surface is retained, i.e., that surface resistance is zero. This assumption ignores the possibility of particle bounce and re-entrainment. Both of these processes are thought to be localized, and thus to have little net effect. Surface effects will be discussed in more detail in a later section that addresses deposition rates of individual species.

Predicted deposition velocities

Deposition velocities can be calculated using the resistance model discussed above. These calculations give a good feel for the impact of variables on deposition rates. Figure 2 shows deposition velocities given by Seinfeld (1986) as a function of wind speed and surface roughness for a gas at a reference height of 2 m and a surface resistance of 0.1 s cm^{-1}. Examination of Figure 2 shows that dry deposition is a strong function of wind speed. The combination of low wind speeds at night and the presence of a nighttime inversion (stable atmosphere) frequently leads to a strong diurnal variation in the dry deposition velocity. This variation will be strongest when R_s is zero. The other point to note from Figure 2 is that deposition rates increase with increasing surface roughness.

Table 1 gives the surface roughness for a variety of surfaces as summarized by Sehmel (1980) and Voldner et al. (1986). A level snowpack has a very low surface roughness. Thus, the dry deposition rates of gases and small particles to a snow surface are expected to be considerably lower than to many other surfaces. In an area with seasonal snowpacks, the deposition velocity of many species is likely to be a minimum in the winter. It should be kept in mind that the aerodynamic resistance for the situation where the snowpack is below a forest canopy or is in an urban area will be determined by the interactions of wind with the canopy or structures, rather than by the snowpack. The situation in a forest can be quite complex, since the wind speed will be a function of height within the canopy. Thus, deposition rates to snow retained on branches and leaves in the canopy could be significantly higher than the deposition rate to the snowpack. Detailed models have been developed for dry deposition to a canopy (Meyers and Baldocchi, 1988).

Table 1. Surface roughness, z_o (cm)

Snow	0.01
Level desert	0.03
Grass, 3 cm height	0.7
Tall, thin grass	10
Shrubs	20
Corn	30
Forest	90
Urban	100

Figure 3 shows particle deposition velocities calculated by Sehmel (1980) as a function of particle size and roughness height, at a reference height of 1 m and a friction velocity of 20 cm s^{-1}. The deposition velocities of very small particles (0.03 μm diameter) are controlled by Brownian diffusion. As the particles increase in size, the Brownian diffusion rate drops, resulting in deposition velocities as low as 0.002 cm s^{-1} for 0.2 - 0.3 μm particles. At larger particle sizes, the deposition velocity increases due to the increased role of impaction and gravitational settling. Deposition of particles larger than 30 μm is controlled almost entirely by gravitational settling. Such particles can have deposition rates greater than 10 cm s^{-1}.

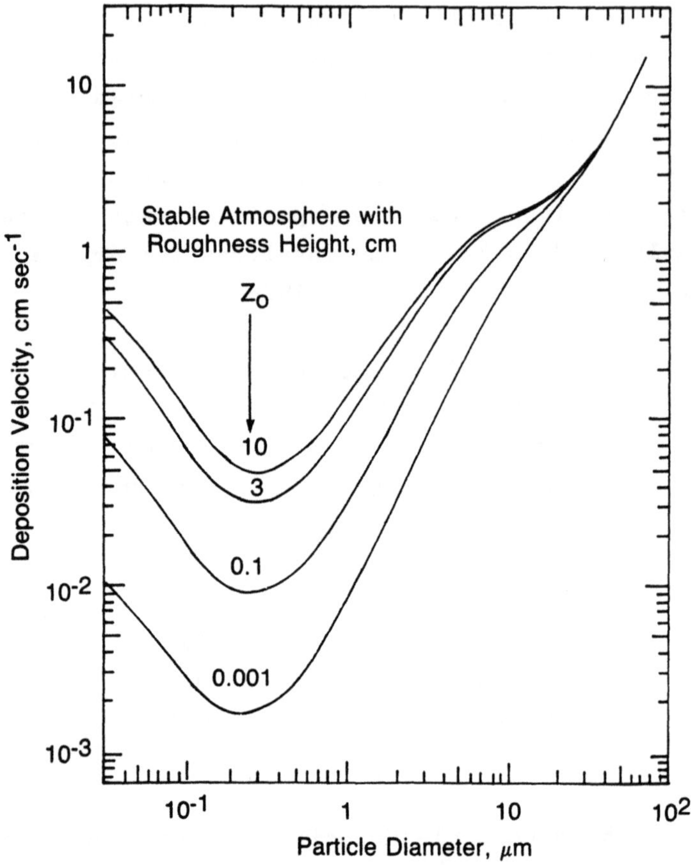

Figure 3. Calculated particle deposition velocities as a function of surface roughness and particle diameter. The particle density is unity. Adapted from Sehmel (1980).

Figure 4, which was adapted from Whitby (1978), shows typical particle number, surface area, and volume distributions for atmospheric particles in an urban area. The particles are generally classified into three ranges. The nuclei mode particles are in the 0.005 to 0.03 μm size range. These particles dominate the number distribution and can have fairly large deposition velocities. However, there is usually very little mass associated with these particles. Therefore, they do not contribute significantly to the total dry deposition flux. These particles rapidly coagulate. Particles in the 0.03 to 3.0 μm range are the accumulation mode particles. They have very low deposition velocities, and tend to remain in the atmosphere for long periods of time. These particles can comprise most of the particle mass in aged air masses in remote areas, and thus frequently comprise the particles of interest in dry deposition studies. Particles greater than 1 μm are referred to as coarse particles. They have high deposition velocities, and consequently have short atmospheric lifetimes, but can be a very important contributor to the dry deposition flux. Noll and Fang (1989) have experimentally examined dry deposition rates of large particles and developed a dry deposition model for coarse particles that includes an effective inertial coefficient.

It is important to note that deposition velocities measured for particulate species such as SO_4^{-2} or NO_3^- are almost always for particles in a fairly broad range of particle sizes. Thus, the measured deposition velocity is really an integration of deposition velocities across the particle size range of that species. Since the deposition velocity of particles larger than 1 μm increases rapidly, it is frequently found that a few larger particles are the dominant source of the deposited material. For example, Coe and Lindberg (1987) used electron microscopy to examine particles dry deposited on leaves and inert surfaces. They

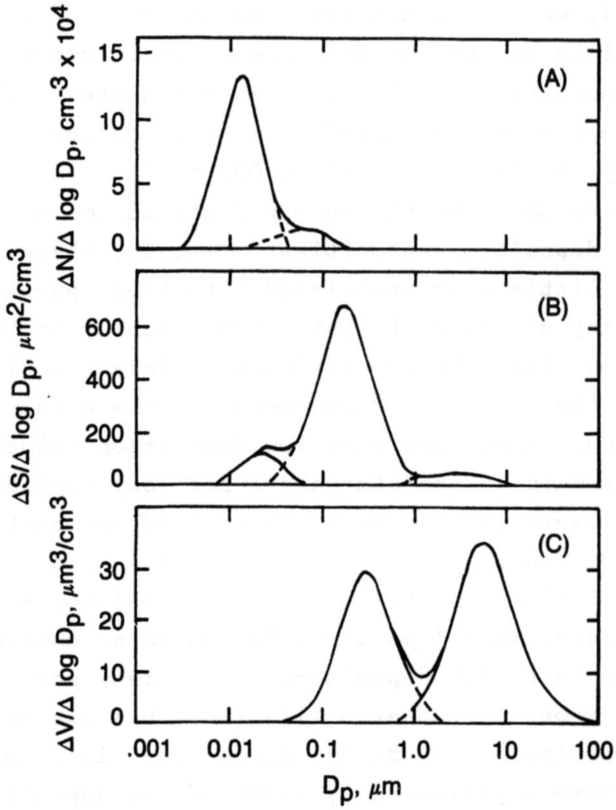

Figure 4. Number, surface area, and volume particle size distributions for typical urban aerosol. Adapted from Whitby (1978).

found that particle mass was dominated by particles larger than 10 μm in diameter. When reporting particle deposition velocities, or when calculating deposition rates from atmospheric concentrations of particulate matter, it is important that size distribution information be available. Speciated particle size distributions are commonly determined from cascade impactor samples. Unfortunately, most impactors are not designed to efficiently collect very large particles. This can be a serious problem. For example, Dulac et al. (1989) estimate that errors in impactor data of mineral and sea-salt particles could lead to an order of magnitude underestimation of their dry deposition rate in the Mediterranean area.

MEASUREMENT METHODS

A wide variety of methods have been used, or have been proposed for use, to measure dry deposition rates. These methods can be separated into the following categories; micrometeorological methods, mass balance methods, surrogate surfaces, snowpack sampling, and long-term monitoring. Not all of these methods have been used for dry deposition studies to snowpacks, although they all could be used. The strengths and weaknesses of these method are discussed briefly below. A brief discussion of throughfall measurements is also presented. Throughfall can transfer material to the snowpack that was dry deposited to a canopy. Thus, while it does not measure direct inputs to the snowpack, it is important in determining snowpack composition. Recent reviews of these methods have been published by Hicks (1984), Hicks et al. (1986), Hicks (1986), Businger (1986), and Nicholson (1988).

Micrometeorological methods

The micrometeorological methods use measurements of atmospheric concentrations and meteorological variables to either directly or indirectly measure the deposition flux. These measurements are generally done on the time scale of a few hours or less.

Eddy correlation. The eddy correlation method directly measures flux by continuously monitoring variations in vertical wind speed and concentration. The measurements are generally made at several meters above the surface on a tower. However, recent developments of very fast sensors have enabled aircraft measurements as well. Sensor response time for tower based measurements must be 1 s or faster. This is a severe limitation to the method, since analyzers with that response time are not available for very many species. Heat and water vapor fluxes must also be measured so that corrections involving these parameters can be incorporated (Baldocchi et al., 1988). As with all micrometeorological methods, siting requirements must be carefully met. The surface must be uniform for a distance of at least 100 times the height of the measurement. Care must be taken to minimize effects from the instruments on the tower, and to ensure that their orientations are proper. Concentrations should be constant during the measurement period, which is typically 15 to 60 minutes.

Another important restriction is that there must be no immediate sources or sinks of the species being monitored. Complicating factors are not always obvious. For example, a ground-based source of NH_3 can remove HNO_3 by forming particulate NH_4NO_3 (Huebert and Luke, 1988; Brost and Delany, 1988). This process would appear as a net flux, although the HNO_3 is not depositing on the surface. A similar problem could arise in the measurement of O_3, if a ground-based source of NO reacted rapidly with the O_3 to form NO_2. Strong humidity gradients could cause problems with the measurement of the deposition rate of hygroscopic particles. The upward flux of hygroscopic particles could appear to be larger than the downward flux, due to their growth in the more humid air. Clearly, this method must be conducted with care, and is not suitable for routine monitoring.

Eddy accumulation. In this method, the fast-response analyzers used for eddy correlation are replaced with two fast-response samplers controlled by the vertical velocity signal. One sampler is for updrafts and the other for downdrafts. This approach is used to overcome the lack of fast-response analyzers for species of interest. The analytical precision for the collected pollutant mass must be 1% for a flux accuracy of 20%. This method remains in the development stage.

Vertical gradient measurement. This method uses integrated measurements of concentration at two or more heights above the surface to determine the concentration gradient. The concentration gradient is multiplied by the eddy diffusivity to obtain a flux. The advantage of the method is that fast response sensors are not required. Thus, sampling can be performed with instruments with response times slower than those required for the eddy correlation method. Also, samplers such as filter packs and diffusion denuders that collect integrated samples can be used. Precision requirements of the method are severe. A precision of 1 to 2% in the concentration measured at each level is required to obtain a 10 to 20% precision in the gradient. Since analytical instruments and samplers are frequently more precise than accurate, it is recommended that concentrations be measured at both levels with the same instrument whenever possible. The method is limited to rapidly depositing species only, since the concentration gradient of slowly depositing species is too small. Hicks et al.(1986) estimate the accuracy of the method to be 30 to 50%.

Variance method. It has been shown that the flux in the form of an eddy correlation is equal to the product of the variances of the correlating quantities (vertical wind speed and concentration) with a correlation coefficient. Since variances can be measured faster and more accurately than the eddy correlation, this approach has some advantages (Businger, 1986). Some information on the correlation coefficient is available, but more work needs to be done. A disadvantage of the method is that it does not give information on the direction of the flux.

Mass balance methods

Atmospheric mass balance methods measure the horizontal flux into and out of a volume of air. Fluxes through the top of the volume and any other source and sink terms must also be known. This approach requires extensive sampling over distances of a 100 km or more with aircraft measurements, and thus is not generally practical. A more practical approach that has been demonstrated successfully is to release a species of interest along with a non-depositing tracer over a horizontally uniform surface. Only ground-based concentration measurements are required. Deposition rates must be high enough to produce changes in concentration that can be accurately measured. Thus, the method is most suitable for large particles or rapidly depositing gases. To date, the method has only been applied under stable atmospheric conditions over uniform surfaces. Extension to other conditions will require further development and exploratory studies (Hicks et al., 1986).

Ecosystem mass balance methods are similar in concept to the atmospheric mass balance approach. All sources and sinks of the species of interest other than dry deposition must be accurately known. This technique is difficult, but may produce seasonal average deposition fluxes for well controlled situations, such as snowpacks. Factors affecting snowpacks are discussed in more detail below.

Surrogate surfaces

The possibility of collecting dry deposition on an inert surface protected from rain events is attractive since such collectors can be located anywhere for long periods of time, and then brought back into a central laboratory for analysis. Dustfall collectors were widely used in the 50's and 60's to collect radioactive particles. These particles were primarily in the coarse particle mode whose dry deposition rates are determined by gravitational settling. The dustfall collectors do a good job of collecting such particles. When interest grew in wet and dry acidic deposition, networks of wet and dry bucket samplers were setup. While the dry deposition buckets do a reasonable job of collecting coarse particles, they cannot simulate the various surface parameters that determine the dry deposition rates of accumulation mode particles. In addition, they disrupt the airflow in their immediate vicinity. Finally, the reactivity of their surfaces is such that they do not simulate the collection of gases by most natural surfaces. Therefore, the use of dry deposition buckets has greatly diminished.

A variety of flat plate collection surfaces, pans, and buckets have been used to collect dry deposition. Deposition rates vary significantly between these collectors (Dasch, 1982). Recently, an aerodynamically shaped, inert collection surface has been designed to overcome the problem of sampler induced turbulence (Hicks et al., 1986). This collector can be used for the collection of particles. Presumably the surface could be constructed from a wide variety of materials in an effort to simulate natural surface roughnesses. To date, Teflon has been the most widely used material since its inertness prevents the collection of unknown amounts of reactive gases. Still, significant differences could exist between this collector and real surfaces for coarse particles if bounce off rates are different. Large differences might also occur between this collector and real surfaces for accumulation mode particles due to differences in the boundary layer. More work needs to be done relating the collection efficiency of surrogate surfaces to that of real surfaces.

Snowpack sampling

Snowpacks provide the opportunity of directly measuring dry deposition rates. This can be done by monitoring the change in concentration of a given species in the snowpack, or a well defined surface layer of the snowpack, during periods with no wet deposition. Alternately, corrections for wet deposition can be made. If simultaneous measurements are made of atmospheric concentrations, time-averaged deposition velocities can be obtained. Spatial variability in the snowpack composition and depth can be a major problem with this technique. Consideration must be given to other snowpack processes that might affect the concentrations of the species being considered. Losses due to melting are one obvious problem. Delmas and Jones (1986) note that care must be taken in situations where the surface snow is redistributed by wind. Also, it has been shown that both biological activity in the snowpack and leaching of detritus can have a major impact on the composition of snowpacks in forested areas during the melt period (Jones, 1988).

This technique is not applicable to reactive species such as H_2O_2 and O_3 since they will not be quantitatively recovered from the snowpack. In addition, the increase in snowpack concentrations of species such as SO_4^{-2}, NO_3^-, and

Cl^- can be due to the deposition of both particles and gases. Thus, it can be difficult to determine accurate deposition velocities from these measurements.

Chamber experiments of dry deposition to snowpacks are also very useful, since they avoid some of the problems associated with direct snowpack sampling. Two approaches have been used. In one approach, changes in the ambient concentration in a chamber placed over the snowpack in the field are used to determine deposition rates that are primarily related to the surface resistance (Gravenhorst and Bottger, 1983). The second approach involves laboratory chamber experiments in which both the ambient concentration of the depositing species and the changes in the snowpack loading of the species can be monitored (Valdez et al., 1987). This approach permits the careful manipulation of important variables such as temperature, concentration of co-depositing species, and, in the case of artificial snowpacks, the concentration of other species in the snow. Presumably, this approach could be used to determine if the metamorphic changes in a snowpack can affect surface resistances.

Another approach is to use shaved ice placed in a petri dish at the snowpack surface as a surrogate surface for dry deposition collection (Molesky and DeWalle, 1988). This approach avoids the problem of the natural variability in concentrations in snowpacks, and also permits the measurement of small changes in concentration, since there is a negligible background concentration in the ice. Possible disadvantages are the unrealistic morphology of shaved ice, and the lack of realistic concentrations of contaminants in the artificial snow.

Long term monitoring

None of the methods for measuring dry deposition rates discussed above are suitable for long term monitoring programs of all the species of interest. Therefore, long term monitoring programs have taken alternative approaches. One approach, termed the concentration method, is to calculate deposition rates from measurements of the atmospheric concentrations of the species of interest and previously determined deposition velocities that are appropriate for the general atmospheric and surface conditions at the site. The approach currently taken by the National Dry Deposition Network (Meyers et al., 1988) is to measure average atmospheric concentrations of depositing species as well as real time meteorological parameters. R_a is calculated from the standard deviation of the wind direction, σ_θ , using equations (6) and (7)

under stable and neutral conditions $$R_a = 4/(u\sigma_\theta)^2 \qquad (6)$$

under unstable conditions $$R_a = 9/(u\sigma_\theta)^2 \qquad (7)$$

Hicks et al. 1985; Hicks et al. 1987). Conditions are determined to be unstable if net radiation is positive and σ_θ exceeds ten degrees. R_b is determined from equation (5) by estimating u_* from the approximation $R_a = uu_*^{-2}$. R_a and R_b are combined with information on the state of the surface in a resistance model to calculate the dry deposition velocity. Intensive studies with other dry deposition methods at the same sites are performed to refine the model, and to determine how to extend these results to broader geographical areas. This approach has been termed the inferential method to distinguish it from the simpler concentration method.

Throughfall

When a snowpack is present under a forest canopy, much of the dry deposition probably occurs to the canopy, rather than to the snowpack surface (Wesley et al., 1983; Molesky and DeWalle, 1988). Interaction of snow with the canopy may remove some of the deposited species. Comparing concentrations in wet deposition collected above and below (throughfall) the canopy is the traditional way of examining this issue for rain events. However, little work has been done for snow, probably because retention of snow in the canopy and the pickup of detritus from the canopy are major complicating factors. Mollitor and Raynal (1983) collected snowfall and snow throughfall samples and found large increases in the concentrations of some species due to canopy interactions. However, it is not known how much of the increase was due to dry deposition since the effects of detrital inputs was not investigated. Zajac and Grodzinska (1982) found that trace metal concentrations in the snowpack downwind from a steel mill were higher in the snowpack under the canopy than in the open. Differences were attributed to removal of dry-deposited particles from the canopy by the snow. Cadle et al. (1984) found significant differences for some species in the snowpack between open and wooded areas, but again no attempt was made to distinguish between detrital inputs and dry deposition. At this time, the significance of throughfall remains an open question.

SURVEY OF DEPOSITION VELOCITIES TO SNOWPACKS

Dry deposition rates and deposition velocities for a variety of gaseous and particulate species have been reported for snowpacks. Table 2 presents a summary of these measurements. Recent reviews of deposition velocities to all surfaces have been published elsewhere (McMahon and Denison, 1979; Sehmel, 1980; Hicks, 1984; Voldner et al., 1986).

Table 2. Deposition velocities to snowpacks

Species	Deposition velocity, cm s^{-1}	Method	Reference
SO_2	0.02 - 0.14	Laboratory study/ model	Bales et al. (1987)
SO_2	0.25±0.2	Snowpack sampling	Barrie and Walmsley (1978)
SO_2	0.06 - 0.15	Snow surface	Cadle et al. (1985)
HNO_3	1.4		
Ca (MMD = 4.4)	2.1		
Mg (MMD = 1.5)	1.5		
Na (MMD = 1.8)	0.44		
K (MMD = 0.9)	0.51		
NH_4^+ (MMD = 0.46)	0.1		
Elemental carbon (MMD = 0.4)	<0.2	Snowpack sampling	Cadle and Dasch (1988)
O_3	0.08	Gradient	Colbeck and Harrison (1985)
SO_2	0.082	Snow surface	Dasch and Cadle (1986)
HNO_3	2.0		
NH_4^+ (MMD = 0.43)	0.083		

Ca (MMD = 5.4)	2.0		
Cl^- (MMD = 3.8)	4.3		
Elemental carbon	0.55	Snowpack sampling	Dasch and Cadle (1989)
Organic carbon	0.28		
Crustal elements	3.4	Snowpack sampling	Davidson et al. (1981)
Enriched elements	0.12		
Crustal elements	0.2	Snowpack sampling	Davidson et al. (1985)
SO_4^{-2}	0.03±0.01		
Pb	0.16	Snowpack sampling	Dovland and Eliassen (1976)
SO_2	0.1		
Particulate (0.15 - 0.3 μm)	0.034±.014	Eddy correlation (partially snow covered field)	Duan et al. (1988)
Particulate (0.5 - 1.0 μm)	0.021±0.005		
SO_2	0.031	Eddy Correlation	Edwards and Ogram (1986)
NO_2	-0.015		
SO_2	0.1	Chamber	Granat and Johansson (1983)
NO	<0.03		
NO_2	<0.03		
NO_2	<0.015	Chamber	Gravenhorst and Bottger (1983)

Table 2. (Continued)

Species	Deposition velocity, $cm\ s^{-1}$	Method	Reference
SO_4^{-2} (0.7 μm)	0.039, 0.096	Snow sampling, isotopic labeled	Ibrahim et al. (1983)
SO_4^{-2} (7.0 μm)	0.16, 0.096		
HNO_3	<0.02 - 0.57	Chamber	Johansson and Granat (1986)
Particulate (0.05 - 1.0 μm)	0.01-1.0	Gradient	Kikuchi and Yoshida (1987)
SO_2	0.22±0.07	Gradient	Nicholson and Davies (1988)
Particulate (0.15 - 0.3 μm)	<0.1	Gradient	Sievering and Pueschel (1982)
NO_2	0.005 - 0.012	Chamber study	Valdez et al. (1987)
O_3	0.013 - 0.082	Eddy correlation	Wesley et al. (1981)
Particulate (<1.0 μm)	-0.1	Eddy correlation	Wesley and Hicks (1979)
SO_2 unstable	1.6	Gradient	Whelpdale and Shaw (1974)
neutral	0.52		
stable	0.05		

Sulfur dioxide

Sulfur dioxide is the most frequently studied species, undoubtedly due to its role in the acid deposition process. Table 2 lists eight studies that have measured SO_2 deposition velocities. The values range from 0.02 to 0.22 cm s^{-1}. The variability can be attributed to both differences in atmospheric conditions and differences in the surface. The effect of surface condition was noted in three field studies (Granat and Johansson, 1983; Cadle et al., 1985, Dasch and Cadle, 1986) that found deposition velocity increased as the temperature approached 0 °C. The increase was attributed to the greater solubility of SO_2 in either the quasi-liquid layer or in melt water as compared to ice. It is reasonable to expect that the pH of the quasi-liquid layer or melt water would affect the deposition rate since it is well known that SO_2 solubility decreases with decreasing pH. This effect has not been studied in the field. Since SO_3^{-2} is rarely found in snowpack samples, it is believed that most dry deposited SO_2 is oxidized to SO_4^{-2}. Thus, the presence of oxidants in the snowpack may also affect the deposition rate. This was confirmed by Lee and Kinsley (1986) who studied the deposition of SO_2 in an ice-coated annular denuder. They found very little deposition of SO_2 to the ice in the absence of other pollutants. However, when H_2O_2 was simultaneously added to the air stream, deposition of SO_2 increased from <1% of the diffusion-controlled rate to 45% of the diffusion controlled rate. H_2O_2, which is present in trace quantities in the atmosphere, is very effective at oxidizing SO_2 to SO_4^{-2} in solution. The most thorough study of SO_2 dry deposition to snow was conducted in the laboratory (Bales et al., 1987a; Valdez et al., 1987; Bales et al., 1987b). In these experiments, it was found that SO_2 penetrated up to 7 cm into the snow. Deposition rates to fresh snow were double the rate to metamorphized snow. Ozone, which can oxidize SO_2 to SO_4^{-2} in solution, had little effect on the deposition velocity.

Overall, it was concluded that the main controlling factor was the liquid water-to-air ratio in the snowpack. The incorporation of SO_2 into ice deposited from the vapor phase (Valdez et al., 1989) has also been studied. It was found that SO_2 was readily incorporated into the ice, and that O_3 and formaldehyde, which complexes SO_2, had a small inhibiting role.

Gaseous nitrogen compounds

There are several trace gaseous nitrogen compounds commonly found in the atmosphere that could contribute to the total nitrogen load of a snowpack. A review of nitrogen compounds, their dry deposition, and their atmospheric chemistry is given by McRae and Russell (1984). The most abundant compounds are NO, NO_2, HNO_3, HNO_2, PAN (peroxy-acetylnitrate), and NH_3. Of these, only NO, NO_2, and HNO_3 have been studied. Granat and Johansson (1983) performed the only study on NO, a chamber study in which no deposition of NO could be detected ($v_d < 0.03$ cm s^{-1}). The chamber studies of Granat and Johansson (1983) and Gravenhorst and Bottger (1983) both found negligible NO_2 deposition. Similarly, the eddy correlation study by Edwards and Ogram (1986) found a small flux away from the snowpack, corresponding to a deposition velocity of -0.015 cm s^{-1}. Only Valdez et al. (1987) have reported uptake of NO_2. Their laboratory study yielded dry deposition velocities of 0.005 cm s^{-1} in the dark and 0.012 cm s^{-1} in the sunlight.

The deposition rate of HNO_3 is much less certain. Cadle et al. (1985) and Dasch and Cadle (1985) used snow samples to estimate the dry deposition rate at 1.4 and 2.0 cm s^{-1}, respectively, after correcting for particulate NO_3^- deposition. Lee and Kinsley (1986) conducted annular denuder experiments in which they found that HNO_3 adsorption on ice was diffusion limited. Huebert et al. (1983) measured snow scavenging ratios for HNO_3 and found that it was rapidly adsorbed on snow. Although this is not a study of deposition to a snowpack, the apparent high accommodation coefficient suggests that HNO_3 deposition rates to a snowpack should be very high. Other workers (Chang, 1984) generally assume that HNO_3 has a high affinity for adsorption on ice and snow. On the other hand, Johansson and Granat (1986) have made the only direct measurements of HNO_3 deposition to a snowpack. They observed a strong temperature effect on the surface resistance, which was negligible only when the temperature exceeded -3 °C. The deposition velocities ranged from <0.02 to 0.57 cm s^{-1} over the temperature range of -18 to -2 °C. An additional complication has been reported by Neubauer and Heumann (1988), who found that the NO_3^- concentrations in Antarctica are highest in new snow, lower in old snow, and lowest in firm core samples. They concluded that HNO_3 must either evaporate or decompose photochemically.

Ozone

Two studies of ozone deposition to snowpacks are given in Table 2. Colbeck and Harrison (1985) conducted a gradient study in which they measured an ozone deposition velocity of 0.08 cm s^{-1}. Wesley et al. (1981) measured ozone deposition velocities of 0.013 to 0.082 cm s^{-1} using the eddy correlation technique. The low deposition velocities indicate that surface resistance to ozone deposition on snowpacks is high. These results are in agreement with a box study of surface resistances performed by Galbally and Roy (1980). They found that the surface resistance to snow (10 - 20 s cm^{-1}) is an order of magnitude higher than the surface resistance to soil or grass.

Particulate

Ten studies of the dry deposition of particles to snow are listed in Table 2. Deposition velocities range from 0.02 to 4.3 cm s^{-1}. This range is not surprising, since the deposition velocity is highly dependent on particle size. Duan et al. (1988) measured deposition velocities for 0.15 to 0.3 μm and 0.5 to 1.0 μm size particles using eddy correlation. The deposition velocities were 0.034 ± 0.014 and 0.021 ± 0.005, respectively. Sievering and Pueschel (1982) used a gradient method to determine that the deposition velocity of 0.15 to 0.3 μm particles was <0.1 cm s^{-1}. Ibrahim et al. (1983) measured deposition velocities of 0.039 and 0.096 cm s^{-1} for 0.7 μm particles under stable and unstable conditions, respectively. Cadle et al. (1985) and Cadle and Dasch (1985) reported deposition velocities of 0.1 and 0.08 cm s^{-1} for NH_4^+ particles with mass median diameters (MMD) of 0.46 and 0.4 μm. Since most SO_4^{-2} is present in the atmosphere as ammonium sulfate particles, it is reasonable to expect that SO_4^{-2} particles had a similar deposition velocity. Cadle and Dasch (1988) also report a deposition velocity of <0.2 cm s^{-1} for elemental carbon particles with a MMD of 0.4 μm. Davidson et al. (1985) report a deposition velocity of 0.03 ± 0.01 for SO_4^{-2} particles to the Greenland ice sheet. A particle size distribution was not given, but it can be expected to be in the accumulation mode. Crustal elements in the same study had deposition velocities of 0.2 cm s^{-1}. Overall, there is good agreement that particles in the accumulation mode have low deposition velocities, ranging from 0.02 to 0.1 cm s^{-1}. The larger deposition velocities reported in Table 2 are for particles that probably had a significant coarse particle mode. For example, Cadle et al. (1985) measured a deposition velocity of 2.1 cm s^{-1} for Ca particles with a MMD of 4.4 μm. Davidson (1981) reported an average deposition velocity for crustal elements of 3.4 cm s^{-1} when sampling snow near bare soil in the Himalayas, but a value of 0.2 cm s^{-1} (Davidson et al. , 1985) for the Greenland ice sheet. In the latter case the aerosol would have traveled much further, thus giving more time for the removal of coarse particles. The highest deposition velocity reported, 4.3 cm s^{-1} was for Cl^- containing particles in an urban area that had a MMD of 3.8 μm. These particles were generated from local, salted roads. It is likely that a few very large particles contributed to the observed deposition velocity.

RELATIVE CONTRIBUTION OF WET AND DRY DEPOSITION

The relative contributions of wet and dry deposition to snowpack compositions are examined so that the importance of studying these processes can be determined. A large number of factors are involved. (1) Dry deposition is a continuous process whereas wet deposition is much more variable. The frequency and magnitude of wet deposition events will be a major factor in determining the relative importance of the two processes. (2) Dry deposition rates are directly related to the atmospheric concentration of the depositing species. Therefore, dry deposition rates of pollutants will be higher near urban areas (Foland and Gjessing, 1975) and point sources than in rural and remote regions. The composition of wet deposition is more uniform geographically. (3) For particulate species there is the additional confounding factor of changes in particle size with time in the atmosphere. Since large particles are rapidly removed, the particle size distributions of some species will shift to smaller particle sizes with increasing residence time. This effect will be strongest for crustal elements and mechanically generated particles since they usually have a significant coarse particle component. Clearly, the dry deposition of crustal particles can be expected to decrease with distance from the nearest areas lacking a snowcover. (4) Ambient temperature will affect the deposition rates of SO_2 and HNO_3 as well as other water-soluble species since the surface resistance for these species appears to be controlled by the amount of water in the snowpack. Thus, climate will affect the relative wet and dry deposition rates of these species. (5) The presence of a canopy may significantly affect the dry deposition rates to the snowpack since much of the material will be deposited in the canopy. It is uncertain how much of this material is transferred to the snowpack by throughfall. Clearly, the relative contributions of wet and dry deposition to the composition of the snowpack will vary considerably on a regional scale. Significant variations on a smaller scale may occur due to changes in the canopy, non-uniform heating of the snowpack, or non-uniform atmospheric turbulence in areas of complex terrain.

A number of studies that have addressed the relative contributions of wet and dry deposition on a seasonal or longer time basis are summarized below. Studies that calculate dry deposition rates from ambient concentrations and deposition velocities are included, since only a limited number of studies have measured both the wet and dry deposition rates over extended periods of time.

Dasch and Cadle (1986) measured wet and dry deposition rates in an urban area during one winter. Dry deposition was measured using exposed snow surfaces. Dry deposition contributed 61, 42, 48, 58, and 80% of the total deposition of Ca, NH_4^+, NO_3^-, SO_4^{-2}, and Cl^-, respectively. In a separate study at the same location, Dasch and Cadle (1989) determined that dry deposition was responsible for 48 and 75% of the particulate organic carbon and elemental carbon, respectively, deposited during two winters.

As expected, studies in rural areas show lower dry deposition contributions. Cadle et al. (1986) found that dry deposition accounted for less than 15% of the SO_4^{-2}, NO_3^-, and NH_4^+ and approximately 25% of the Ca, Mg, Na, and K during an average precipitation year in rural northern Michigan. Dry deposited elemental carbon comprised 23% of the total elemental carbon deposited during two winters at the same site (Cadle and Dasch, 1988). Molesky and DeWalle (1988) found similar dry deposition rates of Ca, Mg, SO_4^{-2}, and NO_3^- to surrogate snow samples in Pennsylvania, suggesting that these rates may be typical for a wide region. Semkin and Jeffries (1988) have measured wet and dry deposition at the Turkey Lakes watershed in Ontario. They report that dry deposition accounted for 10.5, 5.3, 12.1, 4.8, and 27.7% of the total deposition in the snowpack for NH_4^+, Ca, SO_4^{-2}, NO_3^-, and Cl^-, respectively. In an interesting comparison, Sirois and Vet (1988) calculated the dry deposition contribution to the Turkey Lakes watershed using monthly mean deposition velocities derived from a resistance model. Deposition velocities for SO_2, SO_4^{-2}, HNO_3, and NO_3^- averaged 0.06, 0.08, 3.1, and 0.15, cm s^{-1}, respectively. They concluded that 10% and 27% of the SO_4^{-2} and NO_3^-, respectively, is dry deposited. While there is good agreement between the two studies for SO_4^{-2}, the calculated NO_3^- dry deposition was much larger than the amount measured. Semkin and Jeffries (1988) point out that Johansson and Granat (1986) measured much lower HNO_3 dry deposition velocities to snowpacks than were used in this study, and that this likely accounts for the difference between predicted and measured NO_3^- deposition. An alternate hypothesis is that the high HNO_3 deposition velocity is appropriate for the forested areas of the watershed, but that much of the dry deposited HNO_3 remains in the canopy until a rain event.

Johansson and Granat (1986) estimated that at temperatures below -2 °C, HNO_3 dry deposition would only amount to

4% of the measured wet deposition at their site. However, if temperatures were consistently ≥2 °C, then HNO_3 deposition would constitute 40% of the wet deposited NO_3^-.

Barrie et al. (1984), and Barrie and Sirois (1986) have used ambient concentrations and deposition velocities derived from a resistance model (Voldner and Sirois, 1986) to calculate dry deposition rates of SO_4^{-2} and NO_3^- at six sites in eastern Canada. The wintertime HNO_3 deposition velocity was approximately 1.5 cm s^{-1}. The fraction of the total SO_4^{-2} and NO_3^- dry deposited in the winter varies from site-to-site. At the ELA (Barrie and Sirois, 1986) site it constitutes 20 to 25% of the total SO_4^{-2} and 55 to 70% of the NO_3^- deposition. At Chalk River the dry deposition of SO_4^{-2} was 40 to 50% of the total during January and February. No distinction is made between deposition to a canopy as opposed to deposition to the snowpack. Ragland and Wilkening (1983) modeled the wet and dry deposition contributions of SO_2 and SO_4^{-2} from regional sources to the Rainy Lake watershed in Ontario. They found that dry deposition of SO_2 dominates the total deposited sulfur in the winter from regional sources. However, they did not include SO_2 and SO_4^{-2} transported long distances into the region. Inclusion of these sources would greatly impact their total deposition quantities.

Davidson et al. (1981) estimated the relative contributions of wet and dry deposition of trace elements to the Greenland ice sheet by assuming a particle deposition velocity of 0.1 cm s^{-1}. It was concluded that dry deposition contributes less than 25% of the trace metals with an anthropogenic origin. However, it was noted that crustal elements may have larger average particle sizes, and thus may have a greater dry deposition contribution. Similarly, Davidson et al. (1987) estimated that SO_4^{-2} has an average deposition velocity between 0.02 and 0.05 cm s^{-1} at Dye 3 in the Arctic, and that SO_4^{-2} dry deposition contributes approximately 10-30% of the total SO_4^{-2} deposited at this site. However, it is noted that during portions of the Arctic winter there is less precipitation and lower particle scavenging ratios due to the presence of unrimed snow. At these times dry deposition may be the dominant input. Barrie (1985) has also considered the relative roles of wet and dry deposition for remote glaciers and snow fields. He concludes that wet deposition dominates except in cases with (1) low annual accumulations of snowfall, (2) the presence of large particles emitted from volcanoes, or (3) when wet deposition scavenging is minimized by the presence of unrimed snow.

As expected, these results show that the relative importance of wet and dry deposition inputs to a snowpack are variable, due to the variety of factors that have been discussed above. Dry deposition can be a major contributor to the composition of snowpacks in urban areas and, presumably, near point sources. In rural and remote areas dry deposition is rarely the dominant source of a species present in a snowpack, and frequently contributes 25% or less of the total loading.

CONCLUSIONS AND RECOMMENDATIONS

Dry deposition is a complex process that depends on meteorological conditions, the nature of the deposition surface, and the nature of the depositing species. No technique is suitable for measuring the deposition rates of all the species of interest, and few methods are suitable for the routine monitoring of dry deposition. Because of this problem, there is relatively little information about dry deposition rates to snowpacks. Deposition rates can be modeled using resistance models as long as meteorological and surface information is available. However, detailed information about surface resistances is not available for many important species. For example, no information appears to be available regarding surface resistances of snowpacks for PAN, HNO_2, or NH_3. It is commonly assumed that HNO_3 surface resistance is low, yet one study finds that it is a strong function of temperature, and yet another study suggests that HNO_3 can be revolatilized from snowpacks. H_2O_2 may be an important oxidant for SO_2 absorbed on snow, thereby affecting its deposition rate, yet no dry deposition rates for H_2O_2 are available. This may be due, in part, to the lack of specific, convenient measurement methods for ambient H_2O_2. There is interest in organic compounds in snowpacks (Gregor and Gummer, 1989) as well. Yet, the relative importance of wet and dry deposition for these species is not known.

Resistance models can do a good job of modeling dry deposition for simple situations, such as a continuous, flat snowpack in the absence of a canopy. However, modeling is much more difficult for complex terrains (Hicks and Meyers, 1988) since advective effects can significantly affect atmospheric transport. Little work has been done on the effects of varying surface elements, or the edge effects that occur at forest boundaries. Also, very little is known about dry deposition rates to forest canopy elements in the winter, and the removal of dry deposited material by snow throughfall. All of these areas need attention before reliable modeling can be performed.

It must be remembered that particle deposition rates are highly dependent on their size. Therefore, studies in which particle deposition rates are measured should include particle size measurements. Care should be taken that the samplers chosen for both total particulate and for particle size distributions collect large particles. Unfortunately, the trend in the US has been to measure only particles smaller than 10 μm because of the PM-10 particulate standard.

Researchers have a major advantage when studying the dry deposition of some species to snowpacks as opposed to many other surfaces - namely the possibility of conducting field and laboratory studies in which the snow is quantitatively sampled. It is recommended that more fundamental laboratory studies on surface resistance be conducted. Also, additional careful snowpack sampling should be carried out to better define the relative contributions of wet and dry deposition.

ACKNOWLEDGMENT

Helpful discussions with J. M. Dasch are gratefully acknowledged.

REFERENCES

Albritton D, Fehsenfeld F, Hicks B, Miller J, Liu S, Hales J, Shannon J, Durham J, and Patrinos A (1987) NAPAP Interim Assessment Volume III, Chapter 4. Atmospheric Processes.

Bales R C, Valdez M P, and Dawson G A (1987) Gaseous deposition to snow 2. Physical-chemical model for SO_2 deposition, J Geophys. Res. 92, D8, 9789-9799.

Bales R C, Valdez M P, Dawson G A, and Stanley D A (1987) Physical and chemical factors controlling gaseous deposition of SO_2 to snow, In: Seasonal Snowcovers: Physics, Chemistry, Hydrology, H G Jones and W J Orville-Thomas, eds., D. Reidel Publishing.

Baldocchi, D D, Hicks B B, and Meyers T P (1988) Measuring biosphere-atmosphere exchanges of biologically related gases with micrometeorological methods, Ecology 69, 1331-1340.

Barrie L A (1985) Atmospheric particles: their physical/ chemical characteristics and deposition processes relevant to the chemical composition of glaciers, Ann. Glaciol. 7, 100-108.

Barrie L A, Anlauf K, Wiebe H A, and Fellin P (1984) Acidic pollutants in air and precipitation at selected rural locations in Canada, In: Deposition Both Wet and Dry. B Hicks, ed. Acid Precipitation Series - Vol. 4. Butterworth Publishers, Boston-London. 15 -35.

Barrie L A and Sirois A (1986) Wet and dry deposition of sulphates and nitrates in eastern Canada: 1979-1982, Water, Air and Soil Pollut. 30, 303-310.

Barrie L A and Walmsley J L (1978) A study of sulphur dioxide deposition velocities to snow in northern Canada, Atmos. Environ. 12, 2321-2332.

Brost R A and Delany A C (1988) Numerical modeling of concentrations and fluxes of HNO_3, NH_3, and NH_4NO_3 near the surface, J. Geophys. Res. 93, D6, 7137-7152.

Businger J A (1986) Evaluation of the accuracy with which dry deposition can be measured with current micrometeorological techniques, J. Clim. Appl. Meteor. 25, 1100-1124.

Cadle S H and Dasch J M (1988) Wintertime concentrations and sinks of atmospheric particulate carbon at a rural location in northern Michigan, Atmos. Environ. 22, 1373-1381.

Cadle S H, Dasch J M, and Mulawa P A (1985) Atmospheric concentrations and the deposition velocity to snow of nitric acid, sulfur dioxide and various particulate species, Atmos. Environ. 19, 1819-1827.

Cadle S H, Dasch J M, and Grossnickle N E (1984) Retention and release of chemical species by a northern Michigan snowpack, Water, Air, and Soil Pollut. 22, 303-319.

Cadle S H, Dasch J M, and Vandekopple R (1986) Wintertime wet and dry deposition in northern Michigan, Atmos. Environ. 20, 1171-1178.

Chang T Y (1984) Rain and snow scavenging of HNO_3 vapor in the atmosphere, Atmos. Environ. 18, 191-197.

Coe J M and Lindberg S E (1987) The morphology and size distribution of atmospheric particles deposited on foliage and inert surfaces, J. Air Pollut. Control Assoc. 37, 237-243.

Colbeck I and Harrison R M (1985) Dry deposition of ozone: some measurements of deposition velocity and of vertical profiles to 100 metres, Atmos. Environ. 19, 1807-1818.

Dasch J M (1982) A comparison of surrogate surfaces for dry deposition collection, In: Proc. Fourth International Conference on Precipitation Scavenging, Dry Deposition, and Resuspension, Santa Monica, CA.

Dasch J M and Cadle S H (1986) Dry deposition to snow in an urban area, Water, Air, and Soil Pollut. 29, 297-308.

Dasch J M and Cadle S H (1989) Atmospheric carbon particles in the Detroit urban area: wintertime sources and sinks, Aerosol Sci. and Technol. 10, 236-248.

Davidson C I, Chu L, Grimm T C, Nasta M A, and Quamoos M P (1981) Wet and dry deposition of trace elements onto the Greenland ice sheet, Atmos. Environ. 15, 1429-1438.

Davidson C I, Honrath R E, Kadane J B, Tsay, R S, Mayewski P A, Lyons W B, and Heidam N Z (1987) The scavenging of atmospheric sulfate by arctic snow, Atmos. Environ. 21, 871-882.

Davidson C I, Santhanam S, Fortmann R C, and Olson M P (1985) Atmospheric transport and deposition of trace elements onto the Greenland ice sheet, Atmos. Environ. 19, 2065-2081.

Delmas V and Jones H G (1986) Wind as a factor in the direct measurement of the dry deposition of acid pollutants to snowcovers, NATO Advanced Institute on Seasonal Snowcovers: Physics, Chemistry, Hydrology, Les Arc, France, 321-335.

DeWalle, D R, Halvewrson H G, and Sharpe W E (1987) Snowpack dry deposition of sulfur: a four-day chronicle, Eastern Snow Conference, 43rd, 185-189.

Dovland H and Eliassen A (1976) Dry deposition on a snow surface, Atmos. Environ. 10, 783-785.

Duan B, Fairall C W, and Thomson D W (1988) Eddy correlation measurements of the dry deposition of particles in winter-time, J. Clim. Appl. Meteorol. 27, 642-652.

Dulac F, Buat-Menard P, Ezat U, Mwelki S, Bergamedtti G. (1989) Atmospheric input of trace metals to the western Mediterranean: uncertainties in modelling dry deposition from cascade impactor data, Tellus 41B, 362-378.

Edwards G C and Ogram G L (1986) Eddy correlation measurements of dry deposition fluxes using a tunable diode laser absorption spectrometer gas monitor, Water, Air, and Soil Pollut. 30, 187-194.

Forland E J and Gjessing Y T (1975) Snow contamination from washout, rainout, and dry deposition, Atmos. Environ. 9, 339-352.

Fowler D (1984) Transfer to terrestrial surfaces. Phil. Trans. R. Soc. Lond. B 305, 281-297.

Galbally I E and Roy C R (1980) Destruction of ozone at the earth's surface, Quart. J. R. Met. Soc. 106, 599-620.

Granat L and Johansson C (1983) Dry deposition of SO_2 and NO_x in winter, Atmos. Environ. 17, 191-192.

Gravenhorst G and Bottger A (1983) Field measurements of NO and NO_2 fluxes to and from the ground, In: Acid Deposition: Proceedings of the Commission of the European Communities Workshop, Reidel Publishing Company.

Gregor D J and Gummer W D (1989) Evidence of atmospheric transport and deposition of organochlorine pesticides and polychlorinated biphenyls in Canadian arctic snow, Environ. Sci. Technol. 23, 561-565.

Hicks B B (1984) Chapter A-7. Dry Deposition Processes, In: The acidic deposition phenomenon and its effects: Critical assessment review papers, Vol. I.

Hicks B B, Baldocchi D D, Hosker Jr, R P, Hutchison B A, Matt D R, McMillen R T, and Satterfield L C (1985) On the use of monitored air concentrations to infer dry deposition, NOAA Technical Memorandum ERL ARL-141.

Hicks B B, Baldocchi D D, Meyers T P, Hosker Jr, R P, and Matt D R (1987) A preliminary multiple resistance routine for deriving dry deposition velocities from measured quantities, Water, Air, and Soil Pollut. 36, 311-330.

Hicks B B and Matt D R (1988) Combining biology, chemistry, and meteorology in modeling and measuring dry deposition, J. Atmos. Chem. 6, 117-131.

Hicks B B and Meyers T P (1988) Measuring and modelling dry deposition in mountainous areas, In: Acid Deposition at High Elevation Sites, M H Unsworth and D Fowler eds, Kluwer Acadenis Publishers.

Hicks B B (1986) Measuring dry deposition: A re-assessment of the state of the art, Water, Air, and Soil Pollut. 30, 75-90.

Hicks B B, Wesely M L, Lindberg S E, and Bromberg S M (1986) Proceedings of the NAPAP Workshop on Dry Deposition, March 25-27, Harpers Ferry, WV

Hosker R P Jr. and Lindberg S E (1982) Review: Atmospheric deposition and plant assimilation of gases and particles, Atmos. Environ. 16, 889-910.

Huebert B J, Fehsenfeld F C, Norton R B, and Albritton D (1983) The scavenging of nitric acid vapor by snow, In: Precipitation Scavenging, Dry Deposition, and Resuspension, Pruppacher et al, eds, Elsevier Science Publishing Co, Inc, 293-300.

Huebert B J and Luke W T (1988) Measurements of concentration and dry surface fluxes of atmospheric nitrates in the presence of ammonia, J. Geophys. Res. 93, D6, 7127-7136.

Ibrahim M, Barrie L A, and Fanak F (1983) An experimental and theoretical investigation of the dry deposition of particles to snow, pine trees and artificial collectors, Atmos. Environ. 17, 781-788.

Johansson C and Granat L (1986) An experimental study of the dry deposition of gaseous nitric acid to snow, Atmos. Environ. 20, 1165-1170.

Jones H G (1988) Nitrate, sulfate and hydrogen fluxes through a boreal forest snowpack during the spring melt period, Internationale Vereinigung fur Theoretische und Angewandte Limnologie. 23, 2286-2290.

Jones H G and Deblois C (1986) Chemical dynamics of N-containing ionic species in a boreal forest snowcover during the spring melt period, Hydrological Processes 1, 271-282.

Kikuchi K and Yoshida Y (1987) Vertical distributions of the concentrations of submicron aerosol particles near the ground surface, J. Meteor. Soc. of Japan 65, 963-971.

Lee L N and Kinsley M T (1986) Adsorption of nitric acid, hydrogen peroxide, and sulfur dioxide on ice, EOS Trans. AGU 67, 898.

McMahon T A and Denison P J (1979) Empirical atmospheric deposition parameters - a survey, Atmos. Environ. 13, 571-585.

McRae G J and Russel A G (1984) Dry deposition of nitrogen-containing species, In: Deposition Both Wet and Dry. B Hicks ed, Acid Precipitation Series - Vol. 4, 153-193, Butterworth Publishers, Boston-London.

Meyers T P and Baldocchi D D (1988) A comparison of models for deriving dry deposition fluxes of O_3 and SO_2 to a forest canopy, Tellus 40B, 270-284.

Meyers T P, Matt D R, Baldocchi D D, and Hicks B B (1988) On the use of measurements and models of dry deposition, APCA Paper 88-119.2, presented at the 81st Annual Meeting of APCA, Dallas TX.

Molesky E W and DeWalle D R (1988) Use of surrogate snow samples to measure dry deposition on snowpacks in forests, Proceedings of the 1988 Eastern Snow Conference.

Mollitor A V and Raynal D J (1983) Atmospheric deposition and ionic input in Adirondack forests, APCA Paper 83-34.1, presented at the 76th Annual Meeting of APCA, Atlanta GA.

Neubauer J and Heumann K G (1988) Nitrate trace determinations in snow and firm core samples of ice shelves at the Weddell Sea, Antarctica, Atmos. Environ. 22, 537-545.

Nicholson K W (1988) The dry deposition of small particles: a review of experimental measurements, Atmos. Environ. 22, 2653-2666.

Nicholson K W and Davies T D (1988) The dry deposition of sulphur dioxide at a rural site, Atmos. Environ. 22, 2885-2889.

Noll K E and Fang K Y P (1989) Development of a dry deposition model for atmospheric coarse particles, Atmos. Environ. 23, 585-594.

Ragland K W and Wilkening K E (1983) Intermediate-range grid model for atmospheric sulfur dioxide and sulfate concentrations and depositions, Atmos. Environ. 17, 935-947.

Sehmel G A (1980) Particle and gas dry deposition: a review. Atmos.Environ. 14, 983-1011.

Seinfeld J H (1986) Atmospheric Chemistry and Physics of Air Pollution, John Wiley and Sons.

Semkin R G and Jeffries D S (1988) Chemistry of atmospheric deposition, the snowpack, and snowmelt in the Turkey Lakes watershed, Can. J. Fish. Aqua. Sci. 45, 38-46.

Sirois A and Vet R (1988) Detailed sulfate and nitrate atmospheric deposition estimates at the Turkey Lakes watershed, Can. J. Fish. Aqua. Sci. 45, 14-25.

Sievering H and Pueschel R (1982) Impact of particle characterization in confounding reported particle deposition velocities, Atmos. Environ. 16, 359-361.

Valdez M P, Bales R C, Stanley D A, and Dawson G A (1987) Gaseous deposition to snow 1. Experimental study of SO_2 and NO_2 deposition, J. Geophys. Res. 92, D8, 9779-9787.

Valdez M P, Dawson G A, and Bales R C (1989) Sulfur dioxide incorporation into ice depositing from the vapor, J. Geophys. Res. 94, D1, 1095-1103.

Voldner E C, Barrie L A, and Sirois A (1986) A literature review of dry deposition of oxides of sulphur and nitrogen with emphasis on long-range transport modelling in North America, Atmos. Environ. 20, 2101-2123.

Voldner E C and Sirois A (1986) Monthly mean spatial variation of dry deposition velocities of oxides of sulphur and nitrogen, Water, Air and Soil Pollut. 30, 179-186.

Wesely M L, Cook D R, and Hart R L (1983) Fluxes of gases and particles above a deciduous forest in wintertime, Bound. Layer Meteorol. 27, 237-255.

Wesely M L, Cook D R, and Williams R M (1981) Field measurement of small ozone fluxes to snow, wet bare soil, and lake water, Bound. Layer Meteor. 20, 459-471.

Wesely M L and Hicks B B (1979) Dry deposition and emission of small particles at the surface of the earth, Proceedings of the 4th Symposium on Turbulence, Diffusion, and Air Quality, Reno, Nevada, 510-513.

Whelpdale D M and Shaw R W (1974) Sulphur dioxide removal by turbulent transfer over grass, snow, and water surfaces, Tellus XXVI 196-205.

Whitby K T (1978) The physical characteristics of sulfur aerosols, Atmos. Environ. 12, 135-159.

Zajac P K and Grodzinska K (1982) Snow contamination by heavy metals and sulfur in Cracow agglomeration (southern Poland), Water, Air, Soil Pollut. 17, 269-280.

DISCUSSION OF "DRY DEPOSITION TO SNOWPACKS"

Martha H. Conklin
Department of Hydrology and Water Resources
University of Arizona
Tucson, AZ 85718, U.S.A.

Dry deposition is part of the overall atmosphere-surface exchange process and is the result of a net balance between the atmosphere-to-surface fluxes and the surface-to-atmospheric fluxes. In the overall process, gasses can be reversibly absorbed or adsorbed deposited and particulate matter can be resuspended. Desorption and resuspension can be very important processes in snowpacks, as the physical properties of snowpacks can change on the time scales of hours. Understanding overall deposition processes to snow and ice is crucial to relating concentrations of chemical species measured in snowpacks to atmospheric concentrations.

The data collected in temperate regions with high annual precipitation indicate that wet deposition is probably more important than dry deposition, at least for the species that have been studied. In areas where there is limited precipitation and snowfall is unrimed, dry deposition may be the more significant process (Davidson, 1989). Dry deposition may play an important role in polar regions, where glacial records are being used to reconstruct past atmospheric conditions.

The process of dry deposition is usually separated into three steps corresponding to (i) atmospheric transport to a quasi-laminar sublayer above the surface; (ii) transport through the quasi-laminar sublayer and (iii) deposition to the surface. These processes are often modeled as a series of resistances, r_a, r_b, and r_s, repectively. An additional process, due to gravitational effects, can be important for particulate matter deposition and a fourth resistance, gravitational resistance r_g, is considered to be in parallel with the sum of series resistances. When this resistance is included the expression for deposition velocity, v_D, becomes:

$$v_D = \frac{1}{(r_a + r_b + r_s)} + \frac{1}{r_g}$$

The importance of this fourth resistance is illustrated by Davidson (1989) in calculations of deposition velocities for a range of particle diameters and densities.

Of these processes, atmospheric transport is the best understood (for uniform terrain), using expressions for the eddy diffusion developed using Monin-Obukhov similarity theory. These equations have been well parameterized for momentum flux over uniform snow and ice surfaces (Andreas, 1986). The atmospheric resistance is highly dependent on atmospheric stability and surface roughness, with lower resistivity for unstable conditions. For very smooth surfaces

NATO ASI Series, Vol. G 28
Seasonal Snowpacks
Edited by T. D. Davies et al.

surface roughness ranges from 10^{-3} cm to 10^{-1} cm for ice and snow, respectively. Measurements for atmospheric resistances range from 4 to 0.02 $cm^{-1}sec$ for stable to unstable atmospheric conditions for a highly soluble gas, HNO_3. These were made using vertical gradient measurements over grasslands (Huebert, 1983; Huebert and Robert, 1985) and from collected dew (Pierson et al., 1986). Atmospheric transport is not as well understood over complex (e.g. mountainous) terrain. Currently, the U.S. Forest Service in Fort Collins, CO is conducting experiments using eddy correlation measurement in uneven terrain. Dry deposition of gaseous species to snow-covered forest canopies are being measured by Earth and Planetary Sciences at Harvard University. The appropriateness of using micrometeorological techniques to measure v_D over nonuniform surfaces has not yet been determined.

The importance of the other two resistances depends on their relative magnitude compared to the atmospheric-transport resistance. Transport across the quasi-laminar layer can occur by diffusion, interception, inertial motion, and sedimentation. Most studies of dry deposition mechanisms have been conducted for vegetated surfaces, which have indicated that the transport through the quasi-laminar layer is not significant. Snow surfaces are significantly different from plant surfaces, with processes such as hygroscopic growth in the boundary layer being important. In plant studies, the main resistance through the quasi-laminar layer has been identified as Brownian-diffusion resistance. Barry (1967) has reported a correlation between surface resistance and diffusivity based on experiments made over a snow surface:

$$r_b = Sc^{0.8}\overline{U}_{30}^{-1}C_f^{-1}$$

Where Sc is the Schmidt number, $\overline{U}_{30}$ is average wind speed at 30 cm and C_f is the slip correction factor.

Transport through the quasi-laminar layer also depends on other properties of the depositing species as well as the physical properties of the snowpack (e.g. smoothness of surface, relative humidity). Ibrahaim et al. (1983) conducted a tracer test of particulate matter over a snow surface to develop an understanding of the important parameters for modeling deposition velocity. They developed expressions for sedimentation surface resistance, interception of particles by snow grains, and hygroscopic growth. Hygroscopic growth was calculated with the assumption that there is over 90 percent humidity above the snow surface when surface temperatures exceed -10°C. Davidson and Wu (1989) have identified other mechanisms that might be important for transport through the quasi-laminar layer: electrostatic forces, thermophoresis and diffusiophoresis.

Davidson (1989) used the Ibrahaim model to the conditions measure at Dye 3, Greenland (0 percent relative humidity). Important parameters for the model were particle density, roughness height and atmospheric stability. High deposition velocities are predicted to occur with unstable atmospheric conditions, high windspeeds, and rough surfaces.

The importance of the third resistance, surface interactions, is highly dependent on the substance depositing and the condition of the surface. For highly soluble gasses, the liquid-like character of the disordered region at the ice-air interface (quasi-liquid layer) can be important in contributing to lower surface resistances. To estimate the importance of surface resistances for the highly soluble and reactive gas SO_2, Bales et al. (1987) developed a mathematical model for the deposition velocity due to diffusion of SO_2 into a snowpack over a time $\bar{t}$ assuming that SO_2 could dissolve in this quasi-liquid layer or liquid water at 0°C:

$$v_{Ds} = \frac{1}{r_s} = \frac{K_H X_m \rho_B}{\alpha_0} \left\{ \frac{4D}{\pi \bar{t}} \left[1 + \frac{K_H r_{wa}}{\alpha_o} \right]^{-1} \right\}^{0.5}$$

Where v_{Ds} is deposition velocity due to surface deposition, K_H is Henry's constant, X_m is liquid water mass fraction, ρ_B is bulk density of the snow, α_o is the first ionization fraction of dissolved SO_2, D is the diffusivity, r_{wa} is the liquid-water-to-air ratio. This model is for the simplest case with surface resistance due to SO_2 dissolution and dissociation, with no oxidation reactions. Calculated deposition velocities indicate that the penetration of SO_2 in the snowpack is greatly affected by the partitioning into the "aqueous" phase. The important parameters were the liquid water content of the snow and the concentration of chemical species (e.g. H^+) in the disordered region. Experimental results on the same system yielded the same trends (Valdez et al., 1987). Measured surface resistances ranged from 59 to 2 $cm^{-1}s$. The lowest r_s were measured in wet snow conditions, considerably larger than those observed for atmospheric resistances. These are comparable to the measured series resistances for SO_2 in field studies, 32 to 0.6 $cm^{-1}s$ (Cadle, this volume).

The results of the work on SO_2 can be extrapolated to other species of concern. Gaseous deposition and uptake of highly soluble species such as HNO_3, HCl and H_2O_2 should be limited by atmospheric transport as the surface resistance will be very low. Diffusion into the snowpack should be minimal for these gasses, except under extremely cold conditions (when the disordered region is no longer significant). For less-soluble, reactive gasses such as O_2, O_3, organic acids and aldehydes surface resistance will depend on the aqueous-phase reactions on the snow grains. Aqueous-phase reactions that must be considered include changes in pH as gasses dissociate (e.g. SO_2 and NH_3) as well as pH changes due to any reactions (e.g. the oxidation of SO_2). For low solubility, less reactive gasses such as NO and NO_2, surface resistance might be considerably higher. Valdez et al. (1987) found the surface resistance for NO_2 to be approximately one to two orders of magnitude higher than SO_2.

For most of the field measurements of v_D, the relative contributions of different resistances are not separated. There is a need for more field experiments to be conducted that monitor the changes in the condition of the snowpack (i.e. temperature changes and gradients, liquid water content, snow movement) during the measurement period. Atmospheric conditions (stability and wind speed) should be monitored simultaneously. In deposition studies of particulate matter, size distributions, density, relative humidity should be measured. For gaseous deposition, the presence of any reactive species in the snowpack should be determined. Gasses with different solubilities and reactivities should be released to develop models that can be used to predict deposition velocities of different species. Two techniques for deposition velocity measurements: isotopic tracer studies and micrometeorological measurements have been underused. In summary, more careful measurements on dry deposition are needed to relate concentrations of species accumulated in the snow and ice to atmospheric concentrations.

REFERENCES

Andreas EL (1986) *A theory for the scalar roughness and scalar transfer coefficients over snow and sea ice.* CRREL Report 86-9, U.S. Army Cold Regions Research and Engineering Laboratory, New Hampshire.

Bales RC, Valdez MP, and Dawson GA (1987) Gaseous deposition to snow: II. Physical-chemical model for SO_2 deposition. *Journal of Geophysical Research* 92:9789-9799 .

Barry PJ (1967) The use of radioactive tracer gasses to study the rate of exchange of water vapour between air and natural surfaces. In: GE Stout (ed) *Isotope Techniques in the Hydrologic Cycle*, Geophysical Monograph Series, Number 11, American Geophysical Union, Washington, D. C., pp. 69-76.

Davidson CI (1989) Mechanisms of wet and dry deposition of atmospheric contaminants to snow surfaces. In: H Oeschger and CC Langway, Jr (ed) *The Environmental Record in Glaciers and Ice Sheets*, John Wiley & Sons, New York, pp. 29-51.

Davidson C and Wu YL (1989) Dry deposition of particles and vapors. In: DC Adriano (ed) *Acid Precipitation*, Springer-Verlag, New York.

Huebert BJ (1983) Meaurements of the dry-deposition of nitric acid to grasslands and forest. In: HR Pruppacher, RG Semonin and WGN Slinn (ed) *Precipitation Scavenging, Dry Deposition and Resuspension*, Elsevier-North Holland, New York, pp. 785-794.

Huebert BJ and Robert CH (1985) The dry deposition of nitric acid to grass. *Journal of Geophysical Research* 90:2085-2090 .

Ibrahaim M, Barrie LA, and Fanaki F (1983) *Atmospheric Environment* 17:781-788 .

Pierson WR, Brachaczek WW, Gorse RA Jr, Japar SM, and Norbeck JM (1986) On the acidity of dew. *Journal of Geophysical Research* 91:4083-4096 .

Valdez MP, Bales RC, Stanley DA, and Dawson GA (1987) Gaseous deposition to snow: I. Experimental study of SO_2 and NO_2 deposition. *Journal of Geophysical Research* 92:9779-9789 .

THE IMPACT OF BLOWING SNOW ON SNOW CHEMISTRY

J.W. Pomeroy[1], T.D. Davies[2] and M. Tranter[3]

INTRODUCTION

During and after deposition to open and windswept areas, snow is redistributed by the wind to surfaces that are relatively sheltered from erosive aerodynamic forces. This redistribution impacts the chemistry of snow and snowmelt by transporting snow and its contaminants across the surface and by forming snowpacks whose physical characteristics such as depth, density, particle bonding, crystal habit and porosity vary widely over a basin. The variation of these physical characteristics in windswept conditions has been well documented (for example Steppuhn and Dyck, 1974; Tabler, 1975; Gray, 1978; Langham, 1981; McKay and Gray, 1981; Tabler et al., 1990). Likely in association with this physical variability, a chemical variability has been observed (Davies et al., 1987; 1988; Tranter et al., 1987). Delmas and Jones (1987) measured a reduction of ion concentrations in surface snow after a drifting snow episode on a lake snowcover in eastern Canada, they suggested the nature of

---

[1] National Hydrology Research Institute,
Environment Canada,
11 Innovation Blvd.,
Saskatoon,
S7N 3H5
Canada

[2]School of Environmental Sciences,
University of East Anglia
Norwich
NR4 7TJ
U.K.

[3]Department of Oceanography
The University
Southampton
SO9 5NH
U.K.

NATO ASI Series, Vol. G 28
Seasonal Snowpacks
Edited by T. D. Davies et al.

chemical changes during blowing snow is complex and should be considered when calculating dry deposition to snowpacks.

This paper reviews the processes of snow transport and chemical change during transport, presents recent measurements of changes in snow chemistry during wind transport and compares the measurements to a model of snow chemistry changes caused by sublimation of blowing snow.

DEFINITION OF THE PROBLEM

Changes to snow chemistry associated with wind transport of snow may be considered a result of the transport of chemical species between sources and sinks and the transformation of these species along the flow paths. The contaminant transport mechanisms during blowing snow are turbulent transport of chemicals bonded to blowing snow particles, aerosols and as free vapour. Transformations of chemicals in blowing snow are not necessarily conservative with respect to the snow particle. During wind transport snow particles may scavenge or release contaminants as aerosols and vapour and water vapour may sublimate from the particle. The release of contaminants from sources and the absorption by sinks are affected by changes to the snow surface associated with the flux of blowing and falling snow. Using the frame of reference of a control volume of blowing particles above a layer of surface snow, the changes in the concentration of ions in these snows may be found as the result of ion and snow fluxes to the control volume and transformations of ions and ice resident in snow particles within the volume. Figure 1 is such a control volume; aerosols, gases and blowing snow enter from the upwind boundary and exit through the downwind boundary, an exchange of gas and aerosols occurs at the top of the boundary as does wet deposition of ions in snowfall. The bottom boundary is the snow surface where snow and aerosols may be deposited or re-entrained and gases absorbed or released by the surface snow. The system of ion exchange in Figure 1 though complex, is largely controlled by blowing snow transport processes and ion exchange with blowing snow particles.

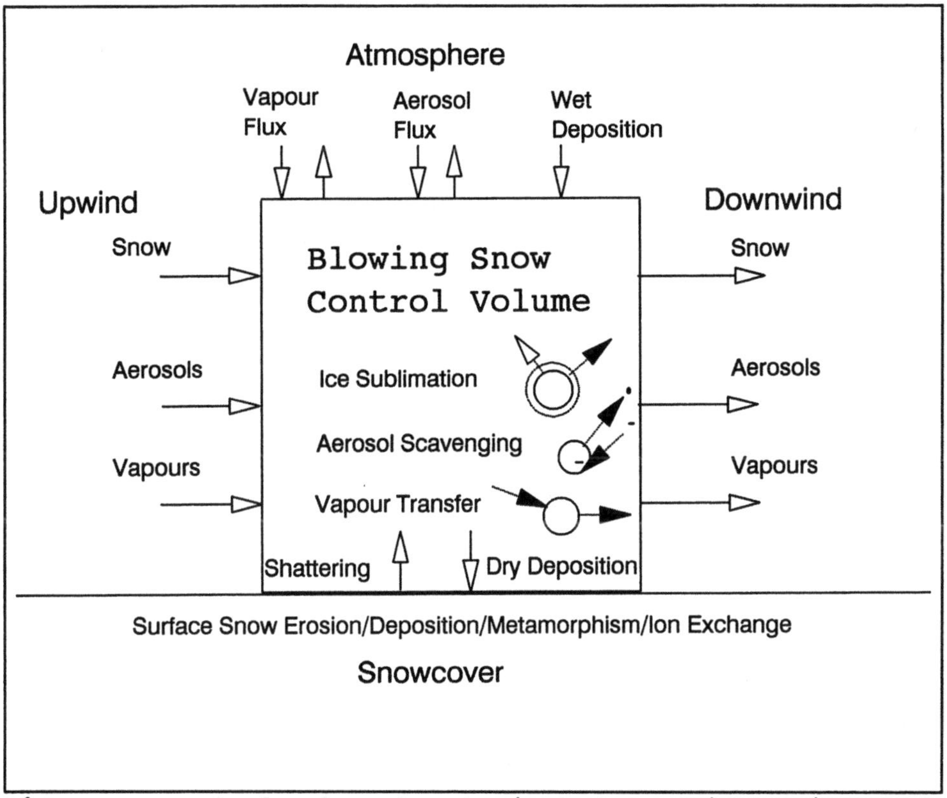

Figure 1. Control volume for chemical changes in blowing snow.

Neglecting chemical processes in snowfall genesis (considered by Barrie in this volume), the next section reviews blowing snow physical and chemical properties that may impact snow chemistry.

BLOWING SNOW PROPERTIES THAT IMPACT CHEMISTRY

Erosion of Surface Snow

Erosion of surface snow proceeds when the shear force transferred from the atmosphere to the snow surface exceeds inter-particle cohesive forces and particle inertia sufficiently to support the weight of newly eroded particles (Schmidt, 1980; Schmidt, 1986; Pomeroy and Gray, 1990). This shear force is most effectively transferred by the impact of saltating particles.

Anderson and Haff (1988) modelled erosion of non-uniform, non-cohesive particles showing that the impact of large particles with higher energy trajectories usually "splash" out a number of small particles with lower energy trajectories. This finding may apply to erosion of cohesive snow particles. The resistance to snow particle ejection provided by cohesion increases with the strength and number of snow surface bonds, hence poorly bonded crystals are more likely to undergo ejection from the surface. The measurements of Martinelli and Ozment (1985) suggest a great spatial variability to surface bond strength, often with location on snow dunes and topographic features. The ratio of exposed surface area to particle mass increases dramatically when snow is eroded, as does the ventilation rate of the exposed surfaces.

Blowing Snow Particles

Lamb et al. (1986) demonstrated that contaminant concentrations in falling snow particles are stratified with regard to particle size. Blowing snow particles however, do not usually resemble the original precipitation or snow surface crystals, having been fragmented during ejection and rebound along the snow surface and sublimated when lifted above the surface. Schmidt (1981; 1984) and Pomeroy and Male (1988) noted blowing snow particles are relatively amorphous; a slightly oblong spheroid shape is often observed. Particle density is typically 900 kg/m^3 and size decreases with height from a mean radius (of a sphere of equivalent volume) of 100 μm just above the surface to 40 μm at about 2 m height (Budd, 1966; Schmidt, 1982a). The size distribution of blowing-snow particles is skewed towards smaller radii and has been fitted to the two-parameter gamma distribution for suspended (Budd, 1966) and saltating (Schmidt, 1981) blowing snow. The form of the gamma distribution is

$$f(r) = \frac{r^{(\alpha-1)} e^{(-\frac{r}{\beta})}}{\beta^{\alpha}\Gamma(\alpha)} \qquad (1)$$

where f(r) is the relative frequency of a particle of radius r, α is the distribution shape parameter (dimensionless), β is the scale parameter (m) and Γ denotes a gamma function. The mean snow particle radius is equal to the product of α and β and the variance of radii is equal to the product of α and β^2. Whilst a more comprehensive expression using turbulence, snowfall rate and surface condition parameter is desirable, Pomeroy (1988) fitted approximate expressions for the variation of mean particle radius (of a sphere of equivalent volume), r_m (m), with height, z (m) and α parameter with height to measurements Schmidt (1982a) made over a range of heights from 0.05 to 1 m above the surface. The expressions are,

$$r_m = 4.6 \times 10^{-5} z^{-0.258} \tag{2}$$

and

$$\alpha = 4.08 + 12.6 z \tag{3}$$

where the coefficient in Eq. 2 has units $m^{1.258}$ and the coefficient in Eq. 3 has units m^{-1}. Figure 2 shows modelled distributions of blowing snow particle radii for various heights above the surface.

Given a gamma distribution, the mean blowing snow particle mass, m_m, may be found from α and the mean particle radius as,

$$m_m = \frac{4}{3} \pi \rho_p r_m^3 (1 + \frac{3}{\alpha} + \frac{2}{\alpha^2}) \tag{4}$$

The difference between the mean blowing snow particle mass and the mass of a blowing snow particle of mean radius is notable, reaching a maximum near the surface where the mean particle mass is 70% greater than the mass of particle with mean radius.

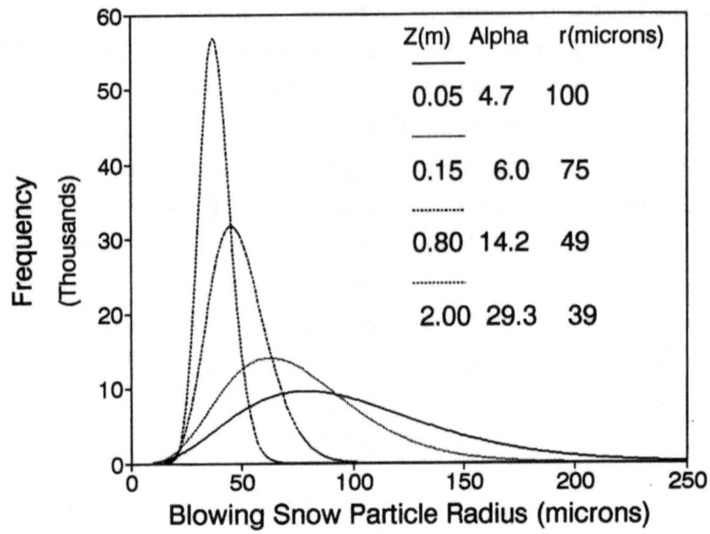

Figure 2. Two-parameter gamma distributions of blowing snow particle radii typically found at four heights above the snow surface.

Ventilation of, or wind velocity relative to, blowing snow particles is an important parameter to calculation of heat and vapour transfer to the particle and ion scavenging rates. For suspended snow, i.e. above a height of about 0.1 m, the horizontal speed of blowing snow particles is equivalent to the mean horizontal wind speed (Schmidt, 1982a). Assuming blowing snow particles fall with a terminal mean vertical velocity and that inertia restricts their ability to follow the paths of inertia-less fluid points in the turbulent atmosphere, the particle ventilation rate v_r (m/s) may be modelled as consisting of a steady vertical and a fluctuating three-dimensional component. The steady component is found using an expression Pomeroy and Male (1986) developed from Carrier's drag law that relates blowing snow terminal fall velocity to the particle radius. For simplicity, the fluctuating component is found using an empirical approximation of Lee's (1975) calculation of the root mean square turbulent velocity relative to a typical blowing snow particle over a continuous snowcover. Adding the steady (left side of right hand) and fluctuating (right side of right hand) components of ventilation velocity yields:

$$V_r - 1.1 \times 10^7 \, r^{1.8} + 0.01061 \, u_z^{1.36} \tag{5}$$

where u_z is the mean wind speed (m/s) at height z and the units of the coefficients are $m^{-0.8}s^{-1}$ and dimensionless respectively. Noting the decrease with height of snow particle radius as shown in Equation 2, the mean component of the ventilation rate decreases with height whilst given a logarithmic mean wind speed profile the turbulent component increases slightly with height and increases with upper-level wind speed.

The particle impact process and sharp wind speed gradients at the snow surface in saltating snow have been hypothesized to generate an electrostatic charge in blowing snow. The phenomenon may interfere with polar radio communication during blizzards and Wishart (1970) reported measurements of a negative electrostatic charge for Antarctic blowing snow particles that are derived from surface snow.

Wind Transport

A major impact of snow redistribution on snow chemistry is the transport of ions carried by snow within and between basins. This transport may occur during or after initial surface deposition. The two primary modes of snow transport are saltation and suspension. Saltation occurs nearest the surface and is the dominant mode at low wind speeds whilst suspension occurs to heights of several tens of metres and dominates at high wind speeds.

-Saltation

Saltation of snow refers to the transport of snow particles bounding along the surface in parabolic trajectories roughly one or two centimetres high and twenty centimetres long. Particles are shattered and abraded during rebound and ejection from the surface, especially if the original crystal is intricate. This shattering may release large aerosols previously contained in snow crystals.

Snow is eroded to and deposited from the saltation layer; these fluxes and the weight of snow supported by the layer are controlled by the balance of atmospheric shear stresses and the snow flux to/from the suspended transport layer above. When the total atmospheric shear stress exceeds that exerted on the non-erodible surface and that required to shatter bonds and expel surface snow particles, saltation may be maintained (Pomeroy and Gray, 1990). The condition at which saltation is just maintained is termed the "threshold" and the shear stress at this point (if all the surface is erodible) is the "threshold shear stress". Shear stress applied to the erodible surface in excess of the threshold level "powers" saltation transport. The balance of shear stresses implied in this argument is related to saltation transport rate Q_{salt} (kg/(m s)) as follows,

$$\frac{Q_{salt}\, g}{u_p} = e\,(\tau - \tau_n - \tau_t) \qquad \textbf{(6)}$$

where g is the gravitational constant (m/s^2), u_p is the mean horizontal velocity of saltating particles - the saltation velocity, e is the efficiency of saltation, τ is the shear stress, n referring to that exerted on the non-erodible surface and t to that exerted at the threshold of transport.

The measurements of Pomeroy and Gray (1990) indicated an inverse relationship between efficiency of saltation transport and wind speed; possibly explained by greater kinetic friction resulting from particle impact and ejection at higher wind speeds. As well, Schmidt (1986) and Pomeroy and Gray (1990) suggested that the saltation velocity is proportional to a wind speed within the saltation layer. Over a complete snowcover, saltation layer wind speed is relatively independent of height and of the wind speed above the saltation layer; evidence for this is the increase in the apparent aerodynamic roughness height with shear stress. The aerodynamic roughness height, z_0, has been postulated to increase in proportion to the height of saltating particle trajectories, the saltating particles themselves acting as roughness elements (Owen, 1964). Assuming snow particles are ejected at some vertical velocity proportional to the friction velocity, u*, their maximum trajectory height should be proportional to $u*^2$. Following this reasoning Tabler (1980), Tabler and Schmidt (1986) and Pomeroy and Gray (1990) have fitted measurements of aerodynamic roughness height and friction velocity during blowing snow to the following equation,

$$z_0 = c \frac{u*^2}{2g} \tag{7}$$

where u* is the friction velocity (m/s). Empirical values for the dimensionless coefficient c are shown in Table 1. Because variable z_0 during blowing snow, the relationship between friction velocity and upper-level wind speed is not linear, the degree of non-linearity depends upon the value of c. Schmidt (1986) noted that for short time periods, the coefficient c varies even more widely with snow conditions than is shown by mean values as reported in Table 1.

A saltation transport expression Pomeroy and Gray (1990) derived from the shear stress apportionment theory, observations of wind speed reductions near the saltation layer and measurements of the saltation mass flux is:

$$Q_{salt} = \frac{0.68\,\rho}{u^*\,g}\,[u_t^*\,u^{*2} - u_t^*\,u_n^{*2} - u_t^{*3}] \tag{8}$$

Table 1. Values of blowing snow aerodynamic roughness height coefficient c derived from wind speed profile measurements during blowing snow.

Investigator	Conditions	c
Tabler (1980)	70% snowcovered lake	.026
Tabler and Schmidt (1986)	Snowcovered Wyoming Short-grass Plateau	.063
Pomeroy and Gray (1990)	Snowcovered Canadian Prairie grainfield	.120
Pomeroy and others (Unpublished)	Snow and ice covered, boulder-strewn Scottish Highland Plateau	.089

where the coefficient, 0.68 (m/s), reflects the saltation velocity and efficiency, ρ is atmospheric density (kg/m^3), and the subscripts t and n to friction velocities refer to the apportionment of shear stress applied to the erodible surface at the threshold and that applied to the non-erodible surface. Typical threshold friction velocities are 0.2 m/s for fresh dry snow and 0.35 m/s for wind-hardened snow (Kind, 1981). The non-erodible friction velocity is usually small for continuous snowcovers. The expression predicts that for a fresh, loose snowcover, saltation transport will begin at low wind speeds and increase only moderately with wind speed, whilst for hardened snowcovers transport begins at higher wind speeds and increases sharply with wind speed. In both cases the relationship between saltation transport and upper-level wind speed is approximately linear.

Implications of this saltation theory are that topography, snow surface structure and wind speed determine whether the threshold condition for saltation is met. The quantity of snow transported in saltation increases approximately linearly with wind speed and the mass concentrations of saltating snow in the atmosphere vary from 0.4 to 0.9 kg/m^3. Blowing snow must pass through the saltation layer while being eroded or deposited and saltation shattering may release aerosols from snow crystals, a phenomenon possibly observed by Delmas and Jones (1987).

-Suspension

Suspended snow is that prevented by the upward drag force of turbulent eddies from regularly contacting the surface. These forces are incapable of entraining surface snow (Schmidt, 1982b; Anderson and Hallet, 1986; Pomeroy, 1988) so the source from which turbulent diffusion of suspended snow proceeds is the saltation layer. Hence, without concurrent precipitation, suspended snow can not occur unless the threshold condition for saltation is exceeded. Suspension favours the lifting of smaller particles, leading to a sorting of particle size with height as demonstrated by Eq.s 2 and 3. Snow may be suspended to several tens of metres in height, though the mass concentration of blowing snow in the atmosphere declines sharply with height.

The transport rate of suspended snow, Q_{susp} (kg/(m s)) is found by integrating mass concentration, η_z (kg/m^3), over height, z, with reference to the concentration, η_r, set to 0.8 kg/m^3 at height z_r, the interface between saltation and suspension, where

$$Q_{susp} = \frac{u^*}{k} \int_{z_r}^{z_b} \eta_z \ln\left(\frac{z}{z_0}\right) dz \qquad (9)$$

and k is von Kàrmàn's constant (0.4), z_0 is the aerodynamic roughness height and z_b is the height of the top of the boundary layer. The vertical gradient of mass concentration is extremely difficult to describe and model for conditions of irregular terrain, incomplete boundary-layer development, intermittent saltation transport, a highly stratified atmosphere and flow around obstacles. For tractability the following discussion assumes a fully-developed atmospheric boundary-layer for which diffusion of snow is in a steady-state, i.e. the momentum transfer variables and the mass concentration of blowing snow at some height are unchanging with time. In practice these conditions are typically found in the lowest metres of the atmosphere over snow covered surfaces with several hundred metres of relatively level and uniform upwind fetch.

For these fully-developed conditions, the gradient of mass concentration can be modelled by analogy to one-dimensional laminar diffusion, modified for continuity of mass of sublimating snow particles undergoing gravitational settling in a turbulent atmosphere where:

$$\frac{\partial\eta}{\partial t} = \frac{-\partial(w\eta - \mathbf{K}_s\frac{\partial\eta}{\partial z})}{\partial z} - [dm_m/m_m dt]\,\eta \qquad \textbf{(10)}$$

and w is the mean vertical snow particle velocity, κ_s is the turbulent diffusivity of blowing snow at height z and $dm_m/(m_m\ dt)$ is the rate coefficient for sublimation of snow particles, indexed to the snow particle of mean mass. Integrating Eq. 10 for steady-state conditions ($d\eta/dt = 0$) provides an equation similar in form to that developed by Shiotani and Arai (1953) and confirmed by Budd et al. (1966), except that in the following development, w* is height dependent,

$$\eta_I = \eta_i\ [\frac{z_I}{z_i}]^{w^*} \qquad \textbf{(11)}$$

The dimensionless diffusion parameter, w^* is a ratio between the average gravitational settling and upward turbulent velocity of blowing snow particles that is difficult to calculate because of the uncertainty in the degree of "slip" of blowing snow particles against inertia-less fluid points in the atmosphere and the possibility of "turbulent burst" mechanisms for upward diffusion. However, measurements in fully developed blowing snow (Pomeroy, 1988) provide an empirical solution for w* as a function of height,

$$w^* = -0.8412\ z^{-0.544} \tag{12}$$

where the coefficient has units of $m^{1.544}\ s^{-1}$. This relationship suggests that as a result of selective diffusion of particles, atmospheric turbulence and particle fall velocity reach a steady balance that is independent of wind speed and snow surface conditions.

The reference height for turbulent diffusion of snow from a mass concentration of 0.8 kg/m^3 (estimated for the saltation layer) increases linearly with friction velocity (Pomeroy, 1989)

$$z_r = 0.05628\ u* \tag{13}$$

where the coefficient has units of seconds. The resulting mass concentration of suspended snow decreases logarithmically with height, the rate of decrease diminishing with wind speed. For instance, at heights of 0.1 and 1.0 m the mass concentrations of suspended snow given a 6 m/s 10-m wind speed are 0.04 and less than 0.001 g/m^3 respectively, for a 20 m/s wind speed the concentrations are 200 and 2 g/m^3.

-Mass Flux

Using numerical techniques based on the preceding development (Pomeroy, 1988) to model saltation and suspension transport, the vertical profile of horizontal mass flux to a height of 10 metres may be calculated for various wind speeds (Fig 3). Mass flux in the saltation layer declines with wind speed and that in the suspended layer increases rapidly with wind speed; hence the height at which the maximum mass flux occurs increases from near the surface at threshold wind speeds to approximately 0.1 to 0.15 m at higher wind speeds (15 to 20 m/s). As a result, the suspended transport rate increases rapidly with wind speed and becomes the dominant mode of transport for wind

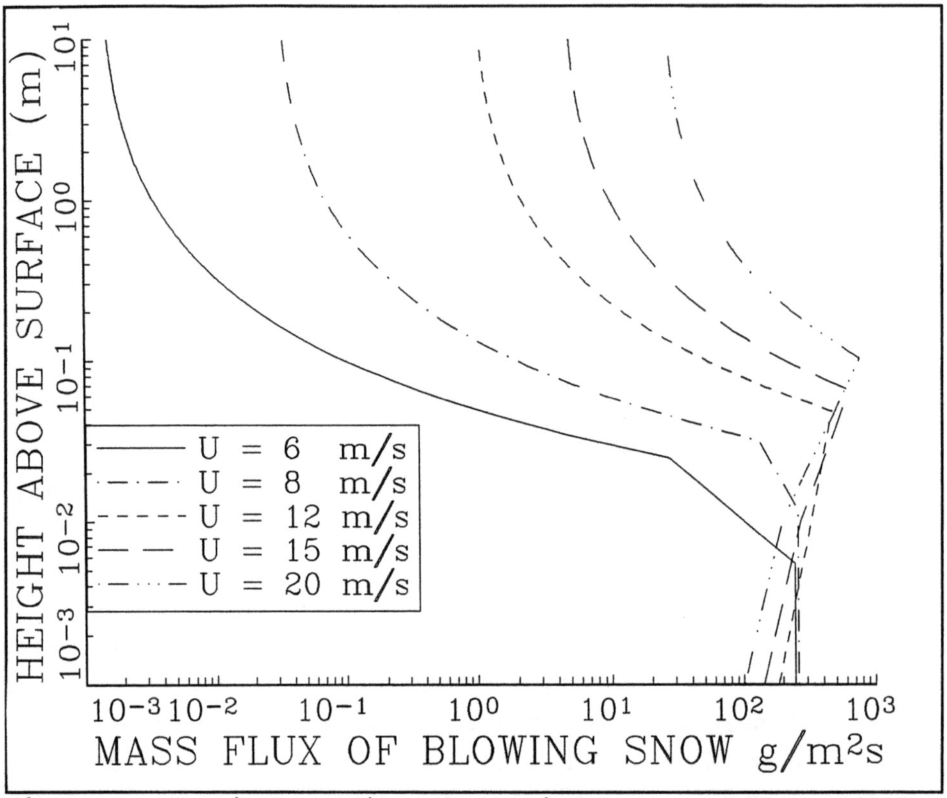

Figure 3. Vertical profiles of blowing snow mass flux modelled for fully-developed flow over a continuous snowcover. The wind speed U is that at a 10-m height. The excessively sharp changes in gradients are a modelling artifact.

speeds above the near-threshold condition. In example, for near-threshold wind speeds saltation comprises almost all transport but at u_{10} = 8 m/s suspension comprises 75% of the total and more than 90% at u_{10} greater than about 17 m/s. Substituting the 10-m wind speed, u_{10} for friction velocity and integrating the mass flux to a height of 5-m provides the total blowing snow transport rate Q (kg/(m s) as a function of u_{10}. An approximation of this relationship, valid over the range of u_{10} from 6.5 (the threshold condition) to 25 m/s, is

$$Q = \frac{u_{10}^{4.04}}{458800} \tag{14}$$

with the coefficient having units kg^{-1} m$^{5.04}$ s$^{3.04}$. Note that this approximation is derived only for completely snowcovered and unobstructed upwind fetches of length greater than 500 m. Under such conditions strong winds, even of short duration, can transport vast quantities of snow.

Sublimation of Blowing Snow

Sublimation of ice from blowing snow particles can increase the concentration of contaminants if contaminants are not removed during the sublimation process. Delmas and Jones (1987) suggested that sublimation of blowing snow may release significant numbers of contaminant aerosols from snow particles. However, noting that sublimation of blowing snow is a gradual mass transfer that proceeds from an intermediary quasi-liquid layer surrounding the particle and that thermo and diffusio-phoretic forces during sublimation favour aerosol scavenging rather than release, the more likely tendency is for aerosols to remain incorporated in a blowing snow particle during sublimation. Gaseous transfer from the quasi-liquid particle surface could be significant, depending on the relative concentrations in the atmosphere and in the snow particle.

The dispersed particulate nature of blowing snow means that sublimation cannot be considered as from a "simple" snow surface but from a dispersion of non-uniform particles in a turbulent atmosphere. This discussion therefore starts at the single spherical particle scale and may be integrated up to the snow-storm scale using parameters noted in previous sections.

Sublimation of ice from a blowing snow particle is governed by the laws of conservation of mass and energy. The theoretical equation for evaporation of a sphere in turbulent air is

$$\left(\frac{dm_m}{dt}\right) = 2\,\pi\,D\,r_m\,Sh\,(\rho - \rho_m) \qquad \textbf{(15)}$$

where r_m = radius of the snow particle possessing mean mass, m_m; D = diffusivity of water vapour in the atmosphere; Sh = Sherwood number - a turbulent mass transfer parameter, ρ = water vapour density in the atmosphere, ρ_m = water vapour density at the particle surface. Convective heat transfer to a sublimating spherical particle is expressed as

$$L_s \left(\frac{dm_m}{dt}\right) = 2\,\pi\,\lambda_T\,r_m\,Nu\,(T_m - T) \tag{16}$$

where, L_s = latent heat of sublimation; λ_T = thermal conductivity of the atmosphere; Nu = Nusselt number - a turbulent heat transfer parameter; T = ambient atmospheric temperature; T_m = atmospheric temperature at the particle surface. Using the Clausius-Clapeyron equation for the water vapour density difference, Thorpe and Mason (1966) combined the above mass and energy balances to provide a sublimation rate expression for ice spheres. Schmidt (1972) added the term Q_r, equal to the radiant energy received by the particle, to the right-hand-side of Eq. 16. By assuming that sublimating blowing snow particles are in thermo-dynamic equilibrium, Schmidt expressed the mass and energy balance for a particle of mass m_m in the following form,

$$\frac{dm_m}{dt} = \frac{2\,\pi\,r_m\,\sigma - \dfrac{Q_r}{\lambda_T\,T\,Nu}\left[\dfrac{L_s\,M}{R\,T} - 1\right]}{\dfrac{L_s}{\lambda_T\,T\,Nu}\left[\dfrac{L_s\,M}{R\,T} - 1\right] + \dfrac{1}{D\,\rho_s\,Sh}} \tag{17}$$

where σ = ambient atmospheric undersaturation of water vapour with respect to ice; M = molecular weight of water; R = universal gas constant; ρ_s = saturation density of water vapour at T. Lee (1975) confirmed Schmidt's model, with modifications for turbulence; Tabler (1975) and Pomeroy et al. (1990a) confirmed application of the sublimation calculation using field measurements of the mass balance of wind-blown snow and operational procedures for calculating sublimation. The procedure for determining the environmental variables in Eq. 17 is involved and is fully described by Pomeroy (1988).

Enhancement of ion concentration due to sublimation of blowing snow may be calculated using Eq. 17 in the following framework. The normalized concentration $C_x(t)$ of an ion (x) in a mass of snow where the mixture changes over time may be expressed as:

$$C_x(t) = \frac{\left[\frac{m_x(t)}{m_m(t)}\right]}{\left[\frac{m_x(i)}{m_m(i)}\right]} \tag{18}$$

where m_x is the mass of ion, m_m is the mean mass of a blowing snow particle, i represents the initial condition and t is the time elapsed. If m_x is constant over time, the blowing snow sublimation rate provides the change in m_m and hence $C_x(t)$.

The concentration $C_x(t)$ is plotted against time in Fig. 4 for conditions of 70% relative humidity, -5 °C temperature, 10 m/s wind speed and 120 J/(m^2 s) incoming solar radiation and a variety of initial snow particle radii. The curve for a radius of 40 μm is representative of the concentration enhancement of blowing snow at heights near 2-m, whilst that for the radius of 100 μm is representative of the concentration enhancement of saltating snow under an identical ambient atmosphere. Figure 4 suggests that a vertical gradient in ion concentration enhancement will develop, a result of the gradient in snow particle size. To show the sensitivity of concentration enhancement to ambient atmospheric conditions, $C_x(t)$ for a transport time of 30 sec. is plotted against air temperature for three different relative humidities in Fig. 5. There is a sharp decline in normalized concentration as relative humidity increases and temperature decreases. This suggests a gradient of increasing concentration with height will develop because

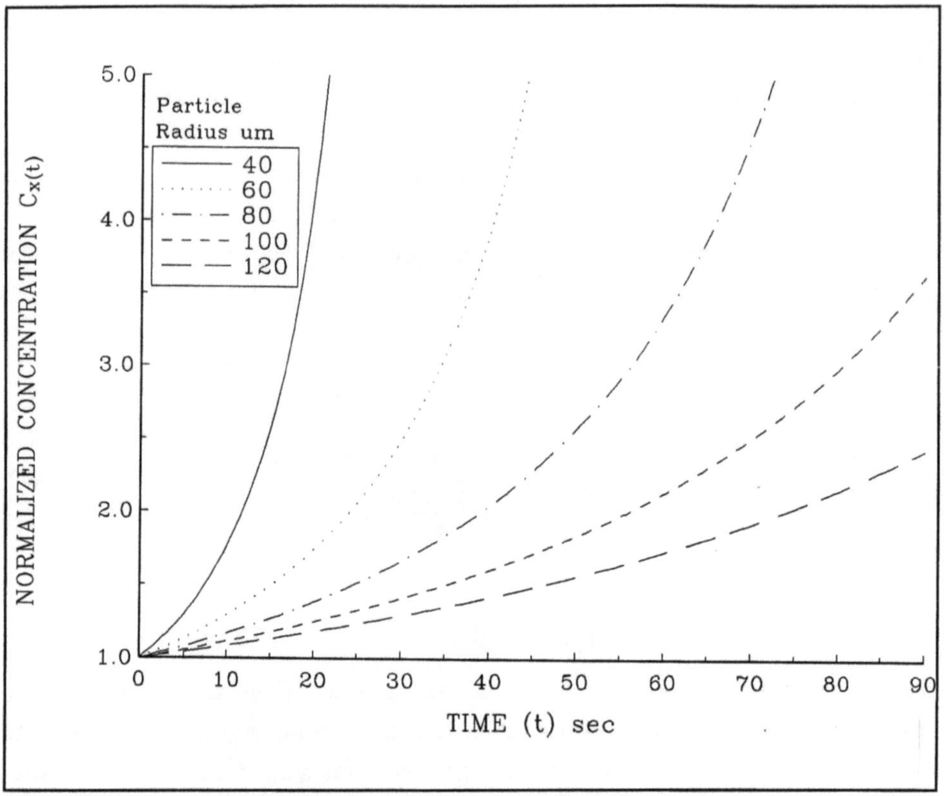

Figure 4. Concentration enhancement due to blowing snow sublimation as a function of time. Sublimation is calculated for u = 10 m/s, T = -5 °C, RH = 70% and Q_r = 120 J/(m^2 s).

relative humidity is usually higher and wind speed and temperature lower near the surface during blowing snow. Vertical gradients of particle size, wind speed, air temperature and humidity should cause a notable increase in sublimation rate with height and hence enhancement of contaminant concentrations should increase with the height of the blowing snow particle. For a column of blowing snow, a dramatic enhancement of contaminant concentration may occur as blowing snow sublimates in relatively "mild" and "arid" winter conditions.

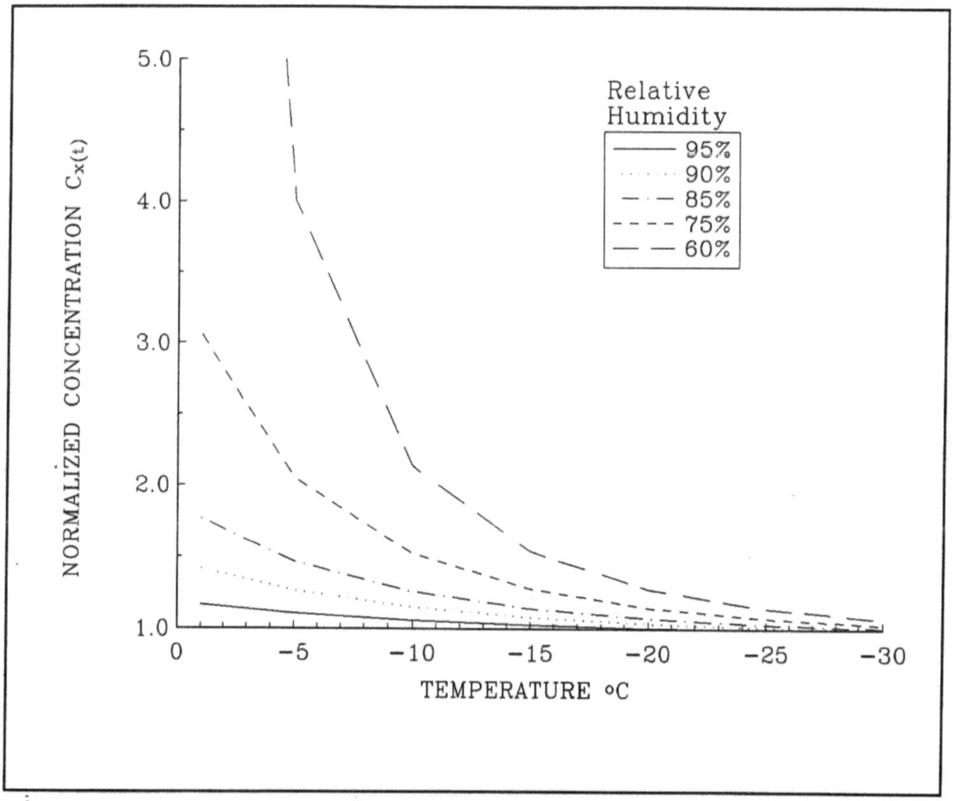

Figure 5. Concentration enhancement due to blowing snow sublimation as a function of temperature. Sublimation is calculated for $u = 10$ m/s, $Q_r = 120$ J/(m^2 s), $r_m = 60$ μm and $t = 60$ s.

Snow Accumulation/Depletion

The accumulation/depletion of surface snow is the lower boundary of a mass-balance of blowing and falling snow in the atmospheric boundary-layer (Pomeroy, 1989). As shown in Fig. 6, the mass-balance may be applied to a control volume of blowing snow: a column of the atmosphere extending 5-metres in height above a unit area of surface. In this control volume the surface accumulation is the sum of the snowfall and the difference between the snow blowing into and out of the volume, less that snow lost to sublimation within the volume, where:

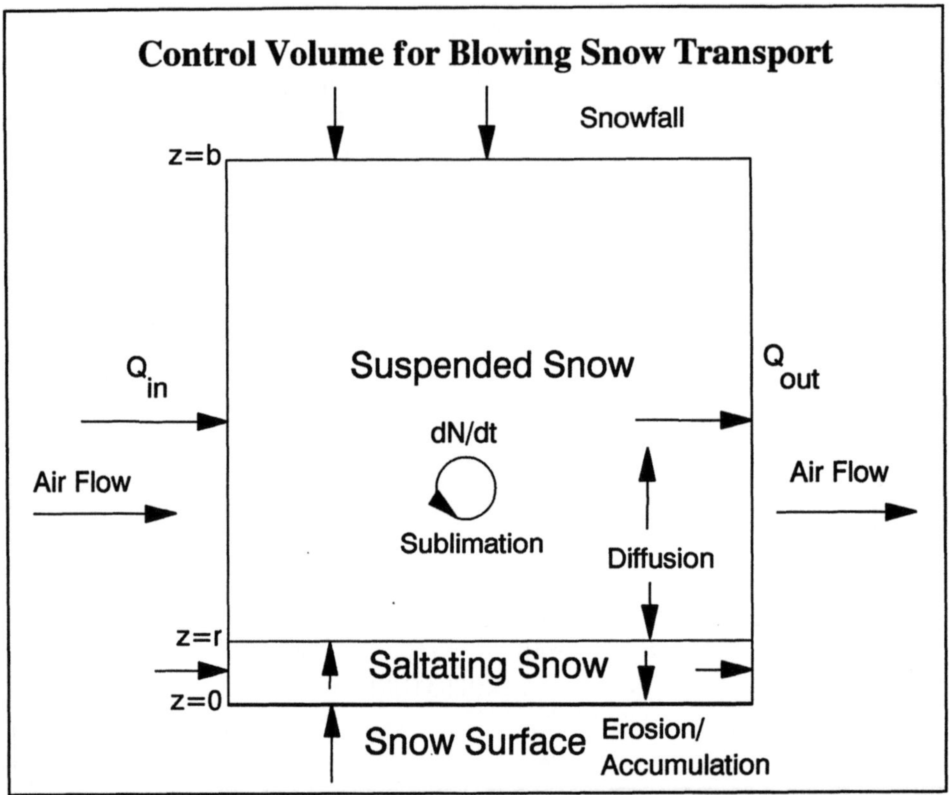

Figure 6. Control volume for wind transport of snow. z is height, b is the top of the boundary-layer, r is the interface between saltation and suspension, Q is horizontal snow flux.

$$Accum = Q_{in} - Q_{out} - \int_0^5 \frac{d\eta_z}{dt} dz + S \tag{19}$$

and Accum is the rate of surface snow mass accumulation per unit area, Q is the horizontal transport rate of blowing snow into and out of the control volume, $d\eta/dt$ is the sublimation rate for a unit volume of atmosphere and S is the downward flux of falling snow.

The measurements of Takeuchi (1980) and considerations of atmospheric boundary-layer development theory (Hunt and Weber, 1979) suggest that the boundary-layer condition where on average the blowing snow entering a 5-m high volume equals the blowing snow leaving it, requires a level, snowcovered, uniform upwind fetch of greater than 400-500 m. These fetches occur on northern plains, upland plateaus and ice sheets, hence for many locations in these regions the average snow accumulation in excess of vegetation/ roughness element storage is a balance between the quantity of snowfall and losses to blowing snow sublimation. In more irregular terrain and for incomplete snowcovers, the balance of blowing snow entering and leaving a control volume exerts the primary control on snow accumulation. For irregular terrain situations the transport rate algorithms presented must be modified for incomplete snowcover and short fetches. Suitable modifications have not yet been specified, though Tabler and Schmidt (1986) presented methods for distributing the snow drift accumulation behind snowfences and in topographic depressions.

Noting that ions are carried by or situated in the snow considered in the mass-balance, it is clear that a simple wet deposition velocity describes accumulation inadequately and a control volume mass-balance may be constructed for "wet" ion transport. Such a mass balance is,

$$Accum_x = Q_{x_{in}} - Q_{x_{out}} - \int_0^5 \frac{dM_{x_z}}{dt}\, dz + S_x \qquad \textbf{(20)}$$

where for ion x carried by snow, $Accum_x$ is the rate of surface accumulation of ion x, Q_x is the horizontal transport rate of ion x into and out of the control volume, the third term represents the rate of scavenging of the ion x by snow and S_x is the vertical flux of ion carried by snowfall into the control volume. Because of strong turbulence, it may be assumed that dry deposition by aerosols will be small compared to the rates of wet deposition during snow redistribution. If the concentrations of ions in snow, blowing snow transport and sublimation rates, snowfall rates and ion scavenging rates are known then the accumulation of ions at the snow surface may be solved for. It is important to consider that whilst in a steady-state scenario sublimation does not affect the ion transport rate, over a blowing snow event or season where Accum is controlled by topographic or vegetation capacity, the concentration of ions in accumulating snow and hence sublimation must be considered to find the long-term $Accum_x$, the "snow-ion holding" capacity of a catchment.

Snow deposited by wind transport has different properties from surface snow that has not blown. Saltation impact shattering and sublimation of ice reduces blowing snow particle size from the original crystal. In-situ surface snow is subjected to the shattering action of particle impact and in-pack sublimation due to relatively high ventilation associated with the high winds during blowing snow (Langham, 1981). Wind-hardening and densification result. Bonds between deposited snow particles form rapidly from the time of contact. Jellinek (1957) reported that the work to disaggregate deposited particles doubles within one day and triples within three days. Fresh snow deposited with light winds has a typical density of 100 kg/m^3. However snow deposited by drifting has densities typically three times this value. As snow is buried by overlying drifts its density increases by compaction from the applied pressure. Tabler (1985) provided an expression based on extensive measurements, for the density, ρ_d (kg/m$^{3)}$ of dry, drifted snow:

$$\rho_s = 522 - \frac{304}{1.485y}\,(1 - e^{-1.48y}) \tag{21}$$

where y is the depth of snow (m) and the constant 522 has units kg m^{-3}. The maximum density which can be achieved is just over 500 kg/m^3. Narita's (1971) observations of snowpack structure showed a decrease in the total surface area of snow particles per unit mass; given a density of 100 kg/m^3 the total surface area is 25.4 m^2/kg, whilst for a density of 300 kg/m^3 the area is 21.7 m^2/kg. Perla (1985) found a more dramatic change in total surface area with density, suggesting about 45 m^2/kg for a density of 100 kg/m^3 and concurring with Narita's (1971) observations for the 300 kg/m^3 value.

Scavenging During Transport

Snow particles may be blown for distances of 3 to 5 km, resulting in atmospheric residence times of up to 10 min. (Tabler, et al., 1990). In the lowest metres of the atmosphere during blowing snow, water vapour density is normally undersaturated because of strong turbulent mixing. Hence, scavenging of ions by blowing snow is not accomplished by riming and most resembles the scavenging processes of below-cloud precipitation; i.e. it is dominated by vapour transfer. Below-cloud scavenging may significantly contaminate snow particles. For instance, Zinder et al. (1988) suggested up to 80% of contaminants in snow falling by a mountainside were scavenged below the cloud in humid but sub-saturated conditions. Beard et al. (1983) listed several mechanisms of aerosol scavenging by evaporating drops of 60-90 μm radius, that also apply to sublimating blowing snow particles. They are:

1) inertia
2) interception
3) gravitation
4) thermophoresis
5) diffusiophoresis
6) Brownian motion
7) electrostatic attraction/repulsion.

Diffusion of soluble gases such as HNO_3, H_2O_2, HCl, H_2SO_4 and NH_3 to the thin quasi-liquid layers on the surface of snow particles may be added to this list (Dawson and Brimblecombe, 1983; Huebert et al., 1983).

Scavenging by Brownian motion dominates for aerosols of radii less than 0.01 μm; inertial impact scavenging is added to other modes for radii greater than 1 or 2 μm. Na^+, often carried by larger aerosols, may be subject to inertial impact and scavenged more rapidly than the other ions carried by aerosols (Zinder et al, 1988). SO_4^{2-} is carried by aerosols of 0.1 to 1

μm radius; without condensation, aerosols of this size are scavenged primarily by particle interception, phoretic effects and electrostatic attraction (Hegg and Hobbs, 1983). The ventilation velocity (Eq. 5) may be multiplied by the snow particle projected area to calculate the volume rate of atmospheric interception for a blowing snow particle falling through a turbulent atmosphere containing aerosols. Sophisticated convective diffusion models of aerosol transport to columnar ice crystals such as that of Miller and Wang (1989) can be combined with turbulent mass and energy transfer models for blowing snow particles (Eq.s 15 and 16). Sublimation creates gradients of temperature and water vapour pressure very near the surface of blowing snow particles that may favour thermophoretic and diffusiophoretic scavenging of aerosols. For evaporating, falling snow particles, convective diffusion models suggest collection efficiencies for aerosols from 0.1 to 1.0 μm radius may increase an order of magnitude with a relative humidity decrease from 95% to 50% (Miller and Wang, 1989). Murakami et al. (1985) and Sauter and Wang (1989) note that collection efficiency decreases with increasing size and compactness of a snow crystal. The field measurements of Lamb et al. (1986) provide strong evidence that small snow particles are much more efficient than large particles in phoretic capture of aerosols. The collection efficiency of blowing snow particles due to phoresis should increase with height above the surface in response to declining particle size.

Electrostatic charge may increase the aerosol collection efficiency of snow crystals by two orders of magnitude (Wang and Sauter, 1985). When particles are highly charged, electrostatic attraction of aerosols can be an important scavenging mechanism (Sauter and Wang, 1989). The measurements of Murakami and Magono (1983) suggest positively charged snow particles have twice the collection efficiency of negatively charged snow. They note an inverse relationship between the increase in collection efficiency due to electrostatic effects and the snow particle ventilation velocity. Strong electrostatic charges in snow blowing near to the snow surface might result in relatively high aerosol collection efficiencies.

Zinder et al. (1988) measured increases of 50% and 150% in snow particle concentrations of NO_3^- and NH_4^+ respectively as snow fell 1200 m in humid conditions along a mountain slope. These increases are much greater than for ions carried primarily by aerosols. Hence, they suggested that gaseous scavenging operates much faster than aerosol scavenging to falling snow particles. Huebert et al. (1983) noted the scavenging coefficient for HNO_3 vapour is five times larger than that for NO_3^- in aerosols, suggesting gaseous transfer is the dominant mode for this ion. SO_2 gas may also be scavenged and oxidised to SO_4^{2-} from highly polluted atmospheres (Hegg and Hobbs, 1983). Dawson and Brimblecombe (1983) suggest the scavenging of gases should be partitioned by their solubility. Given the scavenging coefficient of 25 x 10^{-5} for HNO_3 vapour (Huebert et al., 1983), after 30 sec. of falling snow the atmospheric concentration is reduced by less than 1%, though after 10 min. a 16% reduction occurs.

These results should be applied to blowing snow with caution. It would seem for blowing snow that scavenging of gaseous NO_3^- and NH_4^+ would proceed the fastest, followed by SO_4^{2-} in polluted atmospheres and by aerosol Na^+. It is possible that variations in blowing snow particle electrostatic charge could strongly favour the scavenging of other charged aerosols. Scavenging of aerosols by smaller snow particles is favoured, except when electrostatic charge works to enhance scavenging by larger snow particles near the surface. The blowing snow particle surface area/total mass ratio increases less than proportionately with decreasing mean radius. If gaseous scavenging is proportional to surface area then smaller snow particles will not have notably greater scavenging rates per unit mass than larger particles. Typical scavenging collection efficiencies are between 10^{-2} and 10^{-3} for blowing snow sized snow crystals; therefore, except in nearly-saturated water vapour or highly charged electrostatic conditions, snow particle ion concentration changes due to aerosol scavenging should be slower than those due to sublimation.

RECENT EXPERIMENTAL FINDINGS

Observations

An examination of the chemistry of blowing snow, eroding snow and accumulating snow and associated meteorological conditions during wind transport was conducted during March, 1989 in the Cairngorms, Scotland, the results of which are summarised here. The high plateaux of the Cairngorms are particularly suited for such a study as they are subject to frequent and severe blowing snow storms and the precipitation is often heavily contaminated by pollutants from industrialised Europe.

Vertical profiles of wind speed, temperature, humidity, blowing snow particle flux and blowing snow chemistry were monitored in the bottom two metres of the atmosphere over a plateau above the corrie of Ciste Mhearad. Electrostatic charge was measured from blowing snow particles at 0.25 m height, collected in a Faraday cage. When blowing snow was sampled for chemistry, surface snow was also sampled at sites of erosion and of deposition. Erosion and deposition sites were distinguished by snow surface hardness, the presence of a drift-crust and dune features. The concentration of various anions and cations was measured from filtered samples using a Dionex ion chromatograph and a Pye Unicam atomic absorption spectrophotometer. A more complete description of the study site, experimental methods, chemical analysis and instrumentation is given by Pomeroy et al. (1990b).

The measurements discussed here were collected under two distinct conditions; Condition A (22 March) developed from snow blowing over a continuous snowcover during snowfall, and Condition B (29 March) developed from snow blowing out of the snow-filled corrie of Ciste Mhearad under clear skies, and sublimating whilst travelling over a locally discontinuous snowcover (ice-covered fell-field exposed between snowpatches). Environmental measurements during the two blowing snow conditions are summarized in Table 2. Selected environmental parameters during the two sets of measurements are graphed in Fig. 7.

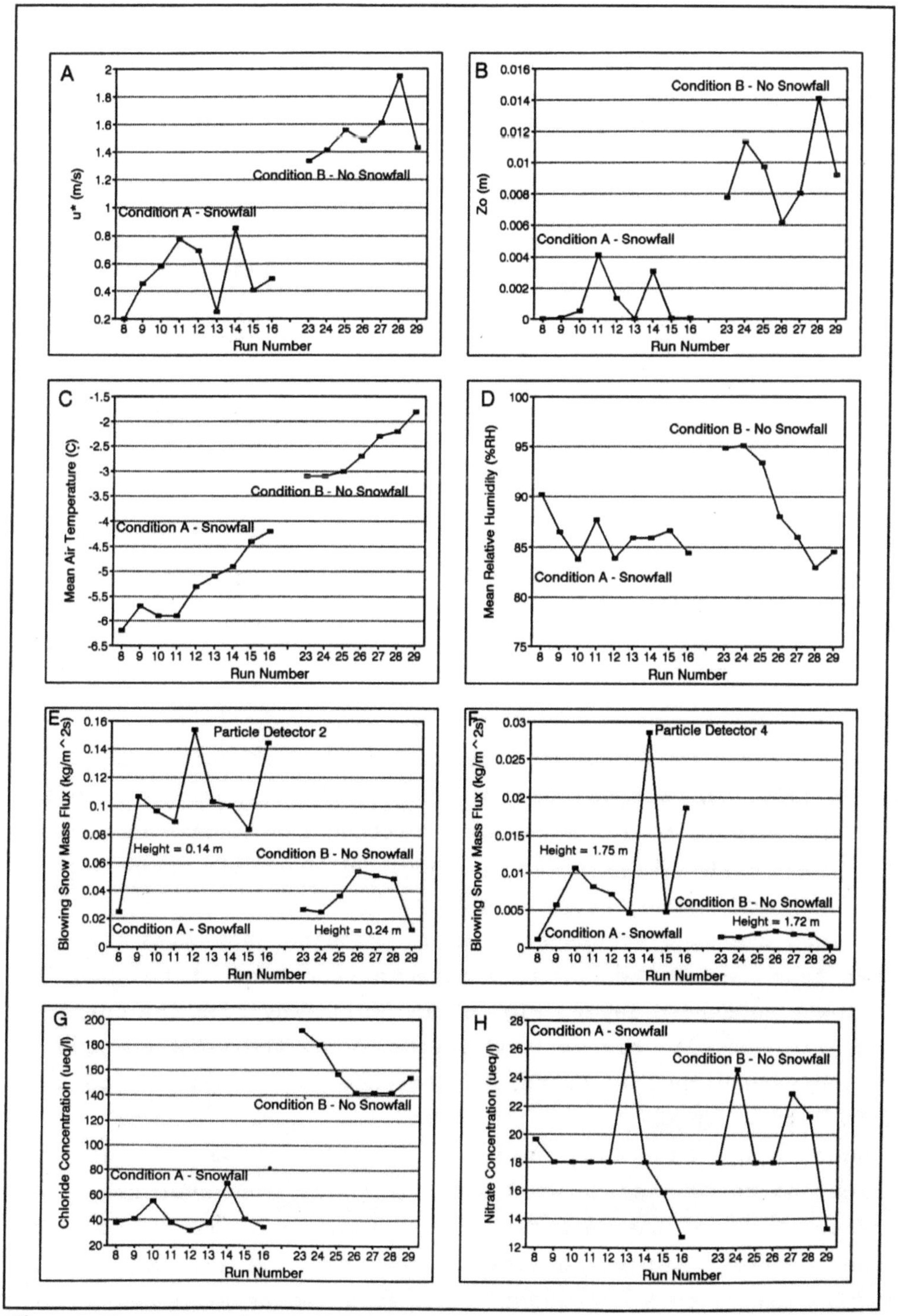

Figure 7. Examples of atmospheric parameters recorded during blowing snow chemistry measurements.

Table 2. Summary of environmental observations for blowing snow conditions.

Blowing Snow Condition	A	B
Concurrent Snowfall	Yes	No
Local Snowcover	Continuous	Discontinuous (95% ice/rock).
Source of Blowing Snow	Cairn Gorm, local & falling snow.	Ciste Mhearad, 300 m distant.
Age of Snow Surface	< 1 day	1-3 Days
Electrostatic Charge of Blowing Snow	Negative	Positive
Wind Speed at 2-m	8-14 m/s	18-24 m/s
Air Temperature	-5.5 °C	-3.0 °C
Relative Humidity	85%	89%

Friction velocity (Fig. 7a) and surface aerodynamic roughness height (Fig. 7b) are calculated from vertical profiles of mean wind speed (0.15 to 1.9 m height), fitting the logarithmic case quite well. Relative humidity (Fig. 7d) and air temperature (Fig. 7c) were measured at heights of approximately 0.3 and 0.9 m; height-averaged values are shown. Blowing snow mass flux is shown from two of the particle detectors mounted at heights in the lower suspended layer (Fig 7e) and well away from the surface (Fig 7f). Despite higher wind speeds in B, the snow transport rate appears higher in A, this difference is largely due to the limited snow supply in B. Interestingly, during snowfall (A) changes in the mass flux well away from the surface (Fig. 7f) correspond to changes in wind speed (Fig. 7a), however near the surface (Fig. 7e) the mass flux is relative unresponsive to wind speed changes. Concentrations of Cl^- (Fig. 7g) and NO_3^- (Fig. 7h) in blowing snow are shown from measurements at a height of approximately 0.1 m. Changes in NO_3^- concentration during the measurement period do not appear clearly associated with the changes in other environmental parameters. However, Cl^- concentration appear inversely correlated with blowing snow mass flux for Condition B (without snowfall) and positively correlated with mass flux for Condition A (with snowfall). The two peaks in Cl^- concentration during snowfall are associated with peaks in upper-level blowing snow flux and wind speed. There are several possible explanations for the behaviour of Cl^-. The peaks during "A" were associated with very heavy atmospheric snow loads and relatively strong turbulence; an increase in sublimation rate and resulting greater concentration of contaminants or the transport of more aerosols or more highly contaminated snow to the site are therefore possible explanations for the peaks in Cl^- concentration. During "B" wind speeds are high and snow supply is limited. If lulls in mass flux during high wind speeds are associated with restriction of snow supply, then during such lulls blowing snow particles will have, on average, undergone greater travel and sublimation than under conditions of less restricted snow supply. The inverse correlation between Cl^- and mass flux during Condition B is a possible manifestation of this effect.

Changes in Blowing Snow Chemistry

Mean ion concentrations in different snow-types (surface snow at sites of erosion and of deposition, and blowing snow at the four heights above the surface) are shown in Fig. 8(a, b). The concentrations are averages of results from approximately 8 distinct samplings (runs in Fig. 7) of snow chemistry; each sampling was conducted over from 10 to 60 min. The values shown are normalized with respect to eroding surface snow values sampled on the plateau. Measurements in Fig. 8 are distinguished between the two blowing snow conditions, concentrations in Fig. 8a refer to the blowing snow with snowfall over a continuous snowcover (condition "A"), and concentrations in Fig. 8b to blowing snow without snowfall over a discontinuous snowcover (condition "B").

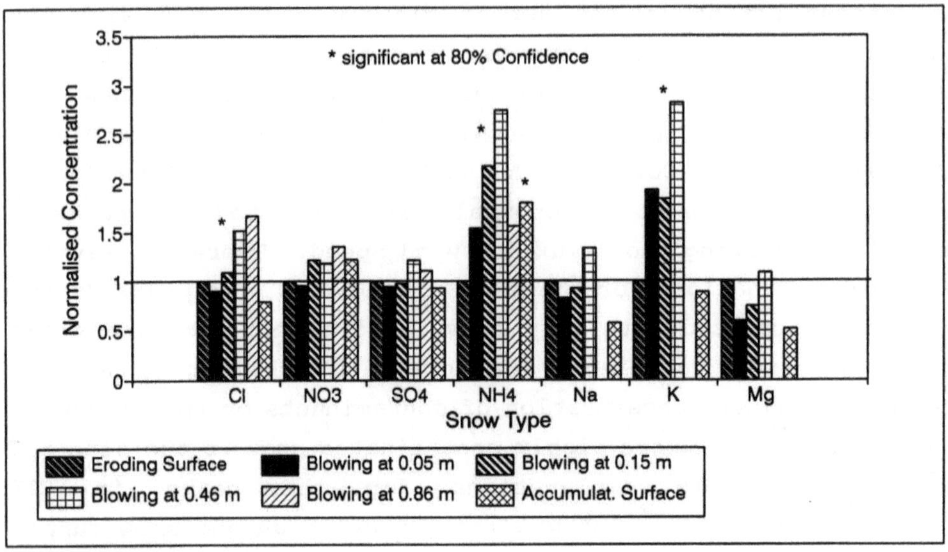

8a) Mean of 9 samplings during blowing and falling snow over a continuous snowcover. Blowing snow heights are 0.05, 0.15, 0.46 and 0.86 m.

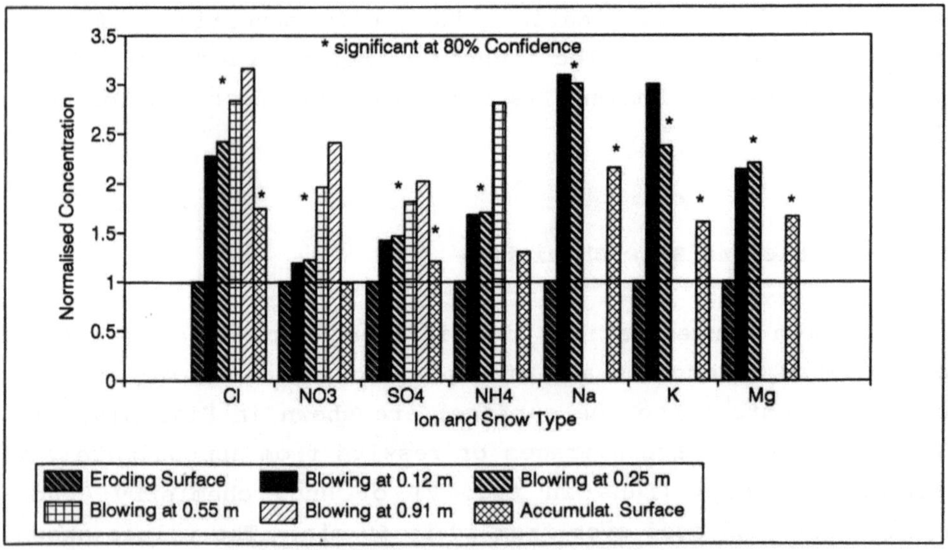

8b) Mean of 7 samplings during blowing snow over a discontinuous snowcover. Blowing snow heights are 0.12, 0.25, 0.55 and 0.91 m.

Figure 8. Mean ion concentrations in snow during drifting, normalised to eroding snow values. For each ion the sequence of snow-classes from left to right: Eroding surface snow, Blowing snow (4 heights), Accumulating surface snow.

For both conditions A and B, mean ion concentrations are greater in blowing snow than in eroding surface snow except in A for Cl^-, NO_3^-, SO_4^{2-}, Na^+ and Mg^{2+} at the lowest blowing snow height and for Na^+ and Mg^{2+} at the second lowest height.

Blowing snow ion concentrations from all four heights were combined and compared to eroding snow values using the Student's "t" test. The probability (0 - 1) that the measured concentrations reflect actual differences between eroding and blowing snow is shown for each ion and both blowing snow conditions in Table 3. The dissimilarities between the two conditions are indicated by the probabilities. For condition B there is approximately 0.99 or better probability that the ions in blowing snow are more concentrated than those in eroding snow. However, for condition A, only NH_4^+ and K^+ display this probability, the next highest being Cl^- at 0.84.

Based on an analysis of variance F-test, the probability of a difference between measured concentrations of eroding and accumulating snow is shown for each ion and both conditions in Table 4. For condition A only NH_4^+ has a probability greater than 0.8 of being more concentrated in accumulating snows, whilst Mg^+, Na^+ and Cl^- have probabilities greater than 0.7 of being more concentrated in eroding snows. For condition B the probabilities for differences between eroding and accumulating snow are greater than 0.9 for Na^+, Cl^- and Mg^+, greater than 0.8 for K^+ and SO_4^{2-} and greater than 0.75 for NH_4^+. NO_3^- never shows notably different levels between the accumulating and eroding snows and NH_4^+ displays less striking but similar tendencies. Interestingly, in condition A, NO_3^- and NH_4^+ concentrations display the greatest enhancement in accumulating snow and in condition B this enhancement is not observable.

To evaluate the significance of the change in ion concentration with height, an analysis of variance was conducted for each ion within conditions A and B with the 4 measurement levels as "treatments". In condition A the ions generally show an increase in mean concentrations with height. The increased ionic concentrations with height are significant at the 99% level of confidence for Cl^-, at the 90% level of confidence for Na^+, at the 80% level of confidence for Mg^+, NO_3^- and NH_4^+ and at only the

Table 3. Probability of a difference in ion concentration between eroding surface and blowing snows.

Blowing Snow Condition	Cl^-	NO_3^-	SO_4^{2-}	NH_4^+	Na^+	K^+	Mg^{2+}
A	.836	.713		.991	.032	.997	
B	.999	.989	.999	.996	.999	.999	.999

all significant differences are for blowing snow concentrations greater than eroding surface snow concentration.

65% level of confidence for K^+. In condition B, only the anions show a significant increase in concentration with height. Noting that the cations are available for only the two lowest heights; Na^+ and K^+ concentrations decrease with height, the decrease in K^+ being significant at the 75% level of confidence. Increased concentration with height is significant at the 99% level of confidence for SO_4^{2-}, NO_3^- and Cl^- and at the 95% level of confidence for NH_4^+. The limited range of sampling heights for the cations precludes conclusions regarding their change with height.

Processes of Change in Blowing Snow Chemistry

-Falling & Blowing Snow

Certain processes are indicated by the changes to snowpack chemistry during blowing snow with snowfall over a continuous snowcover (condition A). It is important to note that with heavy snowfall the sequence of snow particle change is not simply eroding to blowing to accumulating snow, but often falling to blowing to accumulating snow. The "dilution" of snow blown over long fetches results in blowing snow being less sublimated than in condition B and accumulating snow not necessarily being derived from eroding snow. The sequence of ionic change with snow type displayed in Fig 8a may be divided into three groups:

Table 4. Probability of a difference in ion concentration between eroding and accumulating surface snows.

Blowing Snow Condition	Cl^-	NO_3^-	SO_4^{2-}	NH_4^+	Na^+	K^+	Mg^{2+}
A	.728*	.661	.363*	.977	.758*	.305*	.777*
B	.934	.130	.807	.784	.943	.856	.892

* Denotes accumulating concentration less than eroding concentration.

1) ions scavenged as aerosols, the sea salt/cloud condensation nuclei ions, Cl^-, Na^+, K^+ and Mg^{2+},

2) ions scavenged gaseously, NH_4^+ and NO_3^-,

3) SO_4^{2-}, scavenged in both gaseous and aerosol forms.

Accumulating snow shows no enhancement of ion concentrations compared to eroding snow, except for the ions subject to gaseous scavenging, NH_4^+ and NO_3^-. SO_4^{2-} shows no difference and is intermediary in this respect. There is an increase in all ion concentrations with height, however, concentrations of sea salt ions in blowing snow near to the surface are lower (except K^+) than are eroding snow concentrations. Apart from NH_4^+, the most dramatic increases in ion concentration with height are displayed by the aerosol ions.

A combination of processes is suggested to account for the observed ion concentration profiles. The increase in concentration with height is primarily due to sublimation of ice from snow particles. The rate of sublimation increases with height as a result of the gradients of water vapour deficit, ventilation rate, temperature and ratio of particle surface area to mass. However, some scavenging of NH_3 vapour by blowing snow particles may be indicated by the notably increased concentration with height displayed by NH_4^+. Saltation particle shattering may release and resuspend relatively large aerosols within the snow crystal. In falling snow these aerosols contain the sea-salt ions that dominate the cloud condensation nuclei. Evidence of this resuspension is manifest in Fig. 5a as reduced concentrations of sea-salts near the surface in blowing snow and in accumulating snow. This does not necessarily suggest that saltation shattering of eroding surface snow particles releases similar proportions of sea-salt ions. Eroding surface snow is usually somewhat densified and rounded as it is subject to saltation abrasion and wind action before erosion. Though eroding snow is further abraded during transport, it is probably not subject to the severe shattering that a fresh snow crystal would experience upon entering the saltation layer, and hence may retain more of its aerosols. Ions contained in solution within the snow particle such as NO_3^-, NH_4^+ and SO_4^{2-} would not be dispersed so readily by particle shattering, though, because of the somewhat aerosol nature of SO_4^{2-} it may be intermediary in this respect.

-Blowing Snow over a Discontinuous Snowcover

For condition B, with no snowfall, the opportunity exists for substantial sublimation and ion scavenging because of the long transport distance, restricted snow particle replacement and high wind speeds. The transformation of snow here is less complex than in condition A, being only from eroding snow to blowing snow to accumulating snow. The contribution of ions from the exposed fell-field surface is thought to be small as an ice-layer covered much of the plateau. All ion concentrations in blowing snow are enhanced from eroding snow concentrations and with the exception of NH_4^+ and NO_3^- are enhanced in accumulating snow from eroding snow concentrations (Fig 8b). Where fully measured, blowing snow ion concentrations increase with height.

The similarity between concentrations of NH_4^+ and NO_3^- in eroding and accumulating snow may be due to vapour transfer of these ions after snow accumulation, as they return to pre-transport levels to balance their atmospheric concentrations. The slower rate of snow transport and accumulation in condition B promotes snowpack exposure whilst rapid accumulations in condition A may have not. Blowing snow eroded from older surface snow should be subject to less saltation shattering than falling snow. In concurrence with this there is no evidence for resuspension of aerosols in the concentrations of sea-salts at the lowest blowing snow heights; they all exceed the eroding snow concentrations by a large degree. A positive electrostatic charge on this day may have increased the scavenging efficiency for aerosols, contributing to high concentrations of aerosol scavenged ions at lower blowing snow heights.

-Modelling Enhancement of Concentrations: Sublimation

To examine the role of sublimation of ice in changing concentrations of ions in blowing snow, an algorithm has been devised that estimates blowing snow sublimation (Eq. 17) by calculating the sublimation that a column of initially fully-developed blowing snow would experience travelling over a fetch of uniform conditions (Pomeroy et al., 1990b). The initial mean particle mass is "sublimated" for 0.5 sec., reset for the resulting radius, sublimated, reset, etc. Normalized ion concentration at any time, t, is found using Eq.s 17 and 18 with m_p = particle mass and $m_x(i) = m_x(t)$. Time, t, is the fetch distance divided by particle speed, assumed to be equal to wind speed at the chemistry sampling height. Initial conditions are shown in Table 2 for condition B. The wind speed and initial particle mass depend on height and are set to values predicted for the sampling heights. It is assumed that similar conditions may have extended for 1000 m back to the base of Ciste Mhearad. Table 5 shows initial wind speed and particle mass and the resultant $C_x(t)$ at times representing fetch distances of 300 m and 1000 m. The two fetches represent an "envelope" of the concentration effects due to sublimation during transport from the possible surface snow sources to the measurement site.

The envelope of normalized concentration due to sublimation and the normalized concentrations of measured ions are graphed against height above the snow surface in Fig. 9. The normalized concentrations of NO_3^-, SO_4^{-2}, and NH_4^+ all fit within the envelope, however the other ions display values that exceed the envelope, particularly at the lowest height.

Table 5. Normalized concentration of ion, $C_x(t)$, due to sublimation of ice from blowing snow particles: T = -2.6 °C, RH = 89%.

Height m	U_z m/s	r_m μm	$C_{x(t)}$ (300 m)	$C_{x(t)}$ (1000 m)
0.12	9.8	98	1.15	1.66
0.25	12.6	80	1.17	1.82
0.55	15.7	60	1.25	2.45
0.91	17.6	50	1.33	3.48

The sensitivity of the sublimation rate to variation in atmospheric parameters over the fetch should be considered in assessing Fig 9. For instance, the sublimation response to a step change in relative humidity is skewed in that an increase of 5% RH reduces ion concentrations by 25% while a decrease of 5% RH raises concentrations by 41%. However, the vertical profiles of concentration for the exceeding ions suggest a process other than sublimation is responsible for at least part of their values at low heights. Furthermore their values are not likely to be sampling artifacts as they are reflected in high accumulating snow concentrations for these ions. It is possible that blowing snow particles near the surface are scavenging aerosols, and as Na^+ displays the greatest excess concentration, scavenging large aerosols. A mechanism that would favour this scavenging to snow, blowing just above the surface, is electrostatic charge which being positive should be relatively effective (Murakami and Magono, 1983), and if generated in the saltation layer is strongest near to the snow surface. Noting this supplementary role of scavenging processes, the concentration enhancement due to sublimation is the right order of magnitude to account for the higher concentrations of "conservative" blowing snow ions such as sulphate and the nitrogen compounds and, accounts for a large part of the enhancement of other ions.

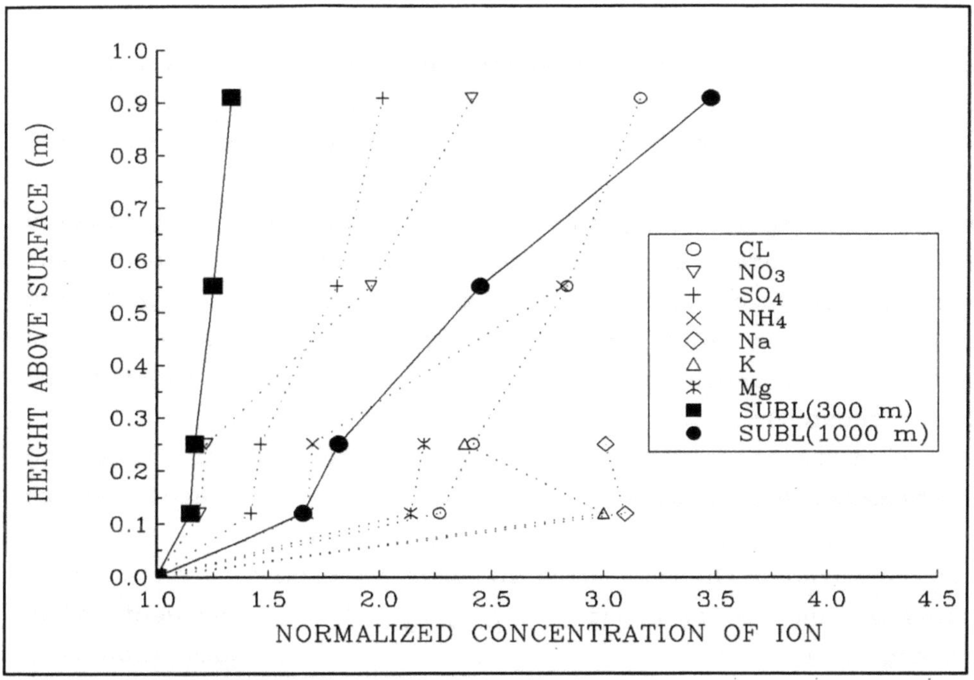

Figure 9. Normalised mean concentrations of ions in eroding surface and blowing snow (Condition B) and normalised concentrations modelled using blowing snow sublimation rates for two possible upwind fetches: 300 and 1000 m.

CONCLUSIONS

Physical processes during blowing snow may have profound impacts on snowpack chemistry, by not only transporting contaminants across the landscape but by altering their concentrations in snow. Impact and shattering of snow particles during erosion from the surface exposes and may release aerosols imbedded in the crystals; these aerosols are easily resuspended by the turbulence generated during blowing snow. An electrostatic charge develops on snow moving near the surface and strengthens the forces favouring aerosol scavenging by blowing snow particles. Atmospheric turbulence preferentially lifts smaller particles and hence sorts blowing snow particle size by height; this size gradient can lead to vertical gradients of sublimation and ion scavenging. Wind transport of snow within and between basins redistributes chemicals deposited by both wet and dry deposition over the winter season, and strongly affects the winter ionic input to small basins in windswept areas. Sublimation of ice from blowing snow proceeds at rates sufficient to enhance chemical concentrations in snow particles several fold over typical transport distances in open landscapes. The sensitivity of sublimation rates to humidity and air temperature means that snow transport distances and ion concentration enhancement are strongly dependent on the concurrent meteorological conditions. After undergoing transport and sublimation, wind-blown snow accumulates in drifts whose location and form are dependant on topographic exposure, vegetation, upwind snowcover supply, wind speed and direction.

Wind transport provides opportunities for further scavenging of atmospheric chemicals by snow. As blowing snow rarely occurs in a saturated atmosphere, scavenging proceeds at rates that are slow compared with scavenging by snow in clouds. Theoretical evidence and below-cloud scavenging studies suggest that scavenging by blowing snow is dominated by gaseous transfer, though phoretic forces due to sublimation and attraction by electrostatic charge may promote notable aerosol scavenging by blowing snow particles in certain favourable situations.

Measurements in Scotland indicate striking changes in snow chemistry during wind transport. Ion concentrations in blowing snow are found to increase with height, and at a height of 1-m can be three times the concentrations in eroding surface snow. This concentration enhancement is sometimes reflected in accumulating snow, where ion concentrations can be twice that in eroding snow. However, during blowing snow with snowfall some ions have concentrations near the surface in blowing snow at and in accumulating snow that are diminished in comparison to those in eroding snow. The degree and nature of ion concentration change in wind-blown snow varies primarily with the atmospheric form of the ion, presence of snowfall, structural change in the snow crystal and opportunity for snow particle sublimation.

During snowfall, the relatively low concentrations of sea-salt ions in blowing snow near the surface may indicate release of aerosols from falling snow as relatively intricate falling snow crystals are shattered upon entering the saltating layer of blowing snow. Concomitant low concentrations of these ions in accumulating snow suggests that sea-salt aerosols are resuspended after release from saltating snow and that falling snow particles can undergo a significant degree of chemical change in the saltation layer before deposition to the snow surface. Converse to the aerosol loss there is evidence for gaseous scavenging of NH_4^+ by blowing and falling snow, evidence supported by enhanced concentrations in accumulating snow.

Without snowfall, rather than release of aerosols, there is evidence for scavenging of sea-salt aerosols by blowing snow near the surface. A strong, positive electrostatic charge held by near-surface blowing snow particles contributes to this aerosol scavenging and the added ion concentration enhancement. Without contribution from falling snow, enhanced concentrations of ions in blowing snow are usually reflected in accumulating surface snow, however, the ions subject to gaseous transfer, NO_3^- and NH_4^+, return to eroding snow concentrations after accumulation. This chemical change upon deposition may be due to a gaseous flux of these ions from the new snow surface to the atmosphere, in response to atmospheric concentrations and volatilization in the snowpack.

Enhanced ion concentrations in blowing snow and the increase in ion concentration with height are largely due to sublimation of blowing snow particles. Predictions from a model of blowing snow sublimation that account for vertical gradients of snow particle size and wind speed in calculating the enhancement of contaminant concentrations match measurements of NO_4^-, SO_4^{2-} and NH_4^+ in blowing snow (without falling snow) quite well. Similar concentration gradients develop during blowing snow with falling snow, though addition of falling snow with its distinctive chemistry modifies the gradient somewhat. Sea-salt ions can be subject to aerosol scavenging and release during blowing snow and NO_3^- and NH_4^+ are subject to relatively rapid gaseous scavenging and volatilization, hence, amongst the ions measured in blowing snow SO_4^{2-} is the most "conservative" in it behaviour, displaying and conserving the enhancement of concentration due to blowing snow sublimation most faithfully.

ACKNOWLEDGEMENTS

Dr. E.M. Morris of the British Antarctic Survey, Cambridge and Dr. R.I. Perla of the National Hydrology Research Institute, Saskatoon made valuable comments and suggestions regarding the manuscript. The support of the North Atlantic Treaty Organization through a post-doctoral fellowship; the Royal Society, London; the USDA Forest Service, Rocky Mountain Forest and Range Experiment Station, Laramie, Wyoming, USA and the Division of Hydrology, University of Saskatchewan, Saskatoon made the research possible. The able assistance of A. Jones and M. Varly, School of Environmental Sciences, University of East Anglia, Norwich; Prof. H.G. Jones, INRS, St. Foy, Canada; Dr. Z. Zeman, Czechoslovakian Academy of Sciences, Most; and D. Pomeroy in inclement and hazardous field conditions is greatly appreciated.

REFERENCES

Anderson RS, Haff PK (1988) Simulation of aeolian saltation. Science 241:820-823

Anderson RS, Hallet B (1986) Sediment Transport by Wind: Toward a General Model. Geological Society of America Bulletin 97:523-535

Beard KV, Ochs HT, Leong KH (1983) Particle Scavenging by Evaporating Cloud Drops. In: Pruppacher et al. (eds) Precipitation Scavenging, Dry Deposition and Resuspension. Elsevier Science Publishing, Toronto, p 517-525

Budd WF (1966) The drifting of non-uniform snow particles. Studies in Antarctic Meteorology, American Geophysical Union Antarctic Research Series 9:59-70

Budd WF, Dingle WRJ, Radok U (1966) The Byrd snow drift project: Outline and basic results. Studies in Antarctic Meterology, American Geophysical Union Antarctic Research Series 9:71-134

Davies TD, Brimblecombe P, Tranter M, Tsiouris S, Vincent CE, Abrahams P, Blackwood IL (1987) The removal of soluble ions from melting snowpacks. In: Jones HG, Orville-Thomas W (eds). Seasonal Snowcovers: Physics, Chemistry, Hydrology, North Atlantic Treaty Organization Advanced Study Institute Series C-211. D. Reidel, Dordrecht, p 337-392

Davies TD, Brimblecombe P, Blackwood IL, Tranter M, Abrahams PW (1988) Chemical composition of snow in the remote Scottish Highlands. In: Unsworth MH, Fowler D (eds) Acid Deposition at High Elevation Sites, North Atlantic Treaty Organization Advanced Study Institute Series C-252. Kluwer Academic Publishers, Dordrecht, p 517-539

Dawson GA, Brimblecombe P (1983) Precipitation scavenging of highly soluble gases. In: Pruppacher et al. (eds) Precipitation Scavenging, Dry Deposition and Resuspension. Elsevier Science Publishing, Toronto, p 687-696

Delmas V, Jones HG (1987) Wind as a factor in the direct measurement of the dry deposition of acid pollutants to snowcovers. In: Jones HG, Orville-Thomas W (eds) Seasonal Snowcovers: Physics, Chemistry, Hydrology, North Atlantic Treaty Organization Advanced Study Institute Series C-211. D. Reidel, Dordrecht, p 321-336

Gray DM (1978) Snow accumulation and distribution. In: Colbeck S, Ray M (eds) Modelling of Snow Cover Runoff. US Army Cold Regions Research and Engineering Laboratory, Hanover, NH. p 3-33

Hegg DA, Hobbs PV (1983) Preliminary measurements on the scavenging of sulphate and nitrate by clouds. In: Pruppacher et al. (eds) Precipitation Scavenging, Dry Deposition and Resuspension. Elsevier Science Publishing, Toronto, p 79-89

Huebert BJ, Fehsenfeld FC, Norton RB, Albritton D (1983) The scavenging of nitric acid vapour by snow. In: Pruppacher et al. (eds) Precipitation Scavenging, Dry Deposition and Resuspension. Elsevier Science Publishing, Toronto, p 293-302

Hunt JCR, Weber AH (1979) A lagrangian statistical analysis of diffusion from a ground-level source in a turbulent boundary-layer. Quarterly Journal of the Royal Meteorology Society 105:423-443

Jellinek HHG (1957) Compressive Strength Properties of Snow, US Army Snow, Ice and Permafrost Establishment, Research Report 34

Lamb D, Mitchell D, Blumenstein R (1986) Snow chemistry in relation to precipitation growth forms. In: Pre-print Volumes, 23rd Conference on Radar Meteorology and the Conference on Cloud Physics, Snowmass, Colorado. American Meteorological Society, Boston, p 77-80

Langham EJ (1981) Physics and properties of snowcover. In: Gray DM, Male, DH (eds) Handbook of Snow, Principles, Processes, Management and Use. Pergamon Press, Toronto, p 275-337

Lee LW (1975) Sublimation of Snow in Turbulent Atmosphere, Ph.D. Thesis. University of Wyoming, Laramie, WY

Kind RJ (1981) Snow Drifting. In: Gray DM, Male, DH (eds) Handbook of Snow, Principles, Processes, Management and Use. Pergamon Press, Toronto, p 338-359

Martinelli M, Ozment A (1985) Some strength features of natural snow surfaces that affect snow drifting. Cold Regions Science and Technology 11:267-283

McKay GA, Gray DM (1981) The distribution of snowcover. In: Gray DM, Male, DH (eds) Handbook of Snow, Principles, Processes, Management and Use. Pergamon Press, Toronto, p 153-190

Miller NL, Wang PK (1989) A theoretical determination of the efficiency with which aerosol particles are collected by falling columnar ice crystals. Journal of the Atmospheric Sciences 46(12):1656-1663

Murakami M, Magono C (1983) An experimental study of electrostatic effect on aerosol scavenging by snow crystals. In: Pruppacher et al. (eds) Precipitation Scavenging, Dry Deposition and Resuspension. Elsevier Science Publishing, Toronto, p 541-549

Murakami M, Magono C, Kikuchi K (1985) Experiments on aerosol scavenging by natural snow crystals part iii, the effect of snow crystal charge on collection efficiency. Journal of the Meteorological Society of Japan 63(6):1127-1137

Narita H (1971) Specific surface of deposited snow ii (In Japanese with English summary). Low Temperature Science, Ser. A 29:76-79

Perla R (1985) Snow in strong or weak temperature gradients. Part ii: section-plane analysis. Cold Regions Science and Technology 11:181-186

Pomeroy JW (1988) Wind Transport of Snow, Ph.D. Thesis. Division of Hydrology, University of Saskatchewan, Saskatoon, Canada.

Pomeroy JW (1989) A process-based model of snow drifting. Annals of Glaciology 13:237-240

Pomeroy JW, Male DH (1986) Physical modelling of blowing snow for agricultural production. In: Steppuhn H, Nicholaichuk W (eds) Proceedings of the Symposium, Snow Management for Agriculture, Great Plains Agricultural Council Publication No 120. Water Studies Institute, Saskatoon, Canada, p 73-108

Pomeroy JW, Male DH (1987) Wind transport of seasonal snowcovers. In: Jones HG, Orville-Thomas W (eds) Seasonal Snowcovers: Physics, Chemistry, Hydrology, North Atlantic Treaty Organization Advanced Study Institute Series C-211. D. Reidel, Dordrecht, p 119-140

Pomeroy JW, Male DH (1988) Optical properties of blowing snow. Journal of Glaciology 34(116):3-10

Pomeroy JW, Gray DM (1990) Saltation of snow. Water Resources Research 26(7):1583-1594

Pomeroy JW, Nicholaichuk W, Gray DM, McConkey B, Granger RJ, Landine PG (1990a) Snow Management and Meltwater Enhancement, National Hydrology Research Institute Report, Contribution No. CS-90021. Environment Canada, Saskatoon

Pomeroy JW, Davies TD, Tranter M (1990b) Relationships between snow chemistry and blowing snow: initial findings. In: Prowse T (ed) Proceedings of the Northern Hydrology Symposium. National Hydrology Research Institute, Environment Canada, Saskatoon

Sauter DP, Wang PK (1989) An experimental study of the scavenging of atmospheric aerosols by natural snow crystals. Journal of the Atmospheric Sciences 46(12):1650-1655

Schmidt RA (1972) Sublimation of Wind-transported Snow - A Model, United States Department of Agriculture, Forest Service Research Paper RM-90. Rocky Mountain Forest and Range Research Station, Fort Collins, CO

Schmidt RA (1980) Threshold windspeeds and elastic impact in snow transport. Journal of Glaciology 26(94):453-467

Schmidt RA (1981) Estimates of threshold windspeed from particle sizes in blowing snow. Cold Regions Science and Technology 4:187-193

Schmidt RA (1982a) Vertical profiles of wind speed, snow concentration and humidity in blowing snow. Boundary-Layer Meteorology 23:223-246

Schmidt RA (1982b) Properties of blowing snow. Reviews of Geophysics and Space Physics 20:39-44

Schmidt RA (1984) Measuring particle size and snowfall intensity in drifting snow. Cold Regions Science and Technology 9:121-129

Schmidt RA (1986) Transport rate of drifting snow and the mean wind speed profile. Boundary-layer Meteorology 34:213-241

Shiotani M, Arai H (1953) A short note on the snow-storm. In: Proceedings, 2nd Japanese National Congress of Applied Mechanics, 1952, p 217-218

Steppuhn H, Dyck GE (1974) Estimating true basin snowcover. In: Advanced Concepts and Techniques in the Study of Snow and Ice Resources. National Academy of Sciences, Washington, D.C., p 314-318.

Tabler RD (1975) Estimating the transport and evaporation of blowing snow. In: Snow Management on the Great Plains, Great Plains Agricultural Council Publication No. 73. University of Nebraska, Lincoln, p 85-105

Tabler RD (1985) Ablation rates of snow fence drifts at 2300-metres elevation in Wyoming. Proceedings, Western Snow Conference 53:1-12

Tabler RD, Schmidt RA (1986) Snow erosion, transport and deposition in relation to agriculture. In: Steppuhn H, Nicholaichuk W (eds) Proceedings of the Symposium, Snow Management for Agriculture, Great Plains Agricultural Council Publication No 120. Water Studies Institute, Saskatoon, Canada, p 11-58

Tabler RD, Pomeroy JW, Santana BW (1990) Drifting snow. In: Ryan WL, Crissman RD (eds) Cold Regions Hydrology and Hydraulics, Technical Council on Cold Regions Engineering Monograph. American Society of Civil Engineers, New York p 95-145

Takeuchi M (1980) Vertical profiles and horizontal increase of drift snow transport. Journal of Glaciology 26(94):481-492

Thorpe AD, Mason BJ (1966) The evaporation of ice spheres and ice crystals. Journal of Applied Physics 17:541-548

Tranter M, Davies TD, Brimblecombe P, Vincent CE (1987) The composition of acidic meltwaters during snowmelt in the Scottish Highlands. Water, Air, and Soil Pollution 36:75-90

Wang PK, Sauter D (1985) Collection of aerosol particles by snow crystals. Proceedings of the Eastern Snow Conference, 1985, 171-176

Wishart ER 1970 Electrification of Antarctic drifting snow. In: International Symposium on Antarctic Glaciological Exploration, International Association for Scientific Hydrology Publ. No. 86:316-324

Zinder B, Schumann T, Waldvogel A (1988) Aerosol and hydrometeor concentrations and their chemical composition during winter precipitation along a mountain slope ii, enhancement of below-cloud scavenging in a stably stratified atmosphere. Atmospheric Environment 22(12):2741-2750

Links Between Snowpack Physics and Snowpack Chemistry

Robert E. Davis
Geophysical Sciences Branch
U.S. Army Cold Regions Research and Engineering Laboratory
Hanover, New Hampshire 03755 USA

1. Introduction

Seasonal snow at the Earth's surface contains a record of the chemistry in the atmosphere during snow crystal formation and subsequent snow pack accumulation. Solutes and other contaminants are incorporated into snow during crystal formation in the clouds, by scavenging and riming as the particles fall, and then by dry deposition. An established snow cover may represent the total chemical load on the environment integrated over several months. It is a widespread observation that this chemical load may be removed from the pack as an ionic pulse, with as much as 80 percent of the solutes eluting with the first 20 percent of the melt.

The location of the solutes in relation to the ice-air interface of the snow grains is one of the main factors controlling the uptake of solutes by liquid water. Another important factor is the area of the air-ice interface or pore surface. This usually decreases over time in a seasonal snow cover. Almost as soon as snow crystals are deposited they begin to change. The speed of these changes is partly because snow is close to or at the melting point of ice. The temperature gradients that are present in snowcovers control the recrystallization processes. Grain boundaries migrate and some grains grow while others are consumed. This causes the original location of solutes in or on the ice crystals to change. Some ions are relocated to the air-ice interface, while others are incorporated into the bond regions between snow grains.

Water flow in snow early in the melt season is often quite heterogeneous, flowing laterally along textural or structural discontinuities, and vertically in flow fingers or macropores. As a consequence, solutes from the near-surface layers and the pore surfaces are leeched first, and the species picked up deeper in the pack depend on the exact path taken by the melt water. There is much evidence to suggest that some chemical species are likely to be removed earlier than others. This *preferential elution* has been observed in the laboratory as well as in the field and is commonly accepted, although field observations are often confounded by nonuniform initial distributions of solutes. The proximity

NATO ASI Series, Vol. G 28
Seasonal Snowpacks
Edited by T. D. Davies et al.

of different solutes to the surface of the ice grains, and their chemical interactions with the liquid-like layer on the ice surface play important roles, but so does the sequence of solute loading to the pack.

This paper continues with two major parts. Theories and observations of dry and wet snow metamorphism are surveyed with discussions the location and migration of chemical species. This is followed by a review of observations of heterogeneous water flow and some attempts to model percolation in two modes. Next, the theory of water flow coupled to solute transport is presented for homogeneous flow in a homogeneous snow layer. A method accounting for water flow in multiple paths is described summarizing the difficulties of coupling solute flow. Both the discussion of metamorphism and water flow in snow finish up with comments on the disparity between theory and measurements, especially as it relates to the effects of stratigraphy of snow covers. The review here does not cite all of the work in this field, but summarizes what this author considers to be the important concepts and gaps in understanding the links between snow pack physics and chemistry.

2. Snow Metamorphism

2.1. Distribution and migration of solutes at the grain scale

Solutes and insoluble particles can be incorporated inside snow crystals or be deposited on the outside. Impurities are incorporated into ice crystals during growth and become defects in their crystal structure. The defects of most interest in examining the behavior of solutes in snow are gross defects, where solid or gas inclusions exist in the crystal. Because very few substances dissolve substitutionally in ice, the role of point defects is not discussed. The condensation nuclei of snow crystals are frequently solid particles, found in the central portion of the snow crystal (Kumai, 1951; 1961; 1976; 1977). The compositions of the snow crystal nuclei studied by Kumai were mostly clay particles and sodium chloride, with lesser amounts of soot and other combustion products, and hygroscopic chemical compounds. Near urban areas Kumai (1985; 1987) found a greater number of combustion products in the form of fly ash particles. While chemical species can be incorporated into snow crystals during their formation, riming is an important mechanism depositing concentrated solutes on the outsides of the crystals. During the formation of snow crystals in clouds a distillation process causes water droplets to lose mass to growing ice crystals. The resulting concentration of solutes in the droplets can be much greater than in the ice crystals. These droplets are frequently supercooled and readily freeze to snow crystals and other obstacles they collide with, such as the surface of a snowpack. In the supercooled droplets, most dissolved gases stay inside, except sulfate, which is excluded to some degree (Barrie, this volume). Thus, in general it is more likely that chloride will be found in the interior regions of a snow crystal and sulfate on the outside.

Gross defects in snow grains are not static. Solid particles in ice crystals may generate liquid inclusions if they preferentially absorb solar radiation, or if they are solutes and their eutectic temperatures are exceeded. Vapor inclusions within the ice may also form from gross defects. The internal melting of ice crystals produces interesting forms called Tyndall figures, which migrate toward the warm part of a crystal and change in shape (Nakaya, 1956). Brine pockets will also migrate through a crystal of ice toward the warm side of a temperature gradient depending on the temperature and the temperature gradient (Hoekstra et al., 1965). The diffusion rates of vapor, liquid, and solid gross defects in ice crystals is slow relative to the migration of grain boundaries owing to metamorphism in seasonal snow (e.g. Hoekstra et al., 1965; Kuroiwa, 1975). In addition, the direction of diffusion of these inclusions is typically toward the side of the ice grain that is growing, so it seems unlikely that solutes can reach the grain surface by diffusion alone. However, because most seasonal snowcovers undergo extensive recrystallization during the course of metamorphism, it can be expected that most solutes will be redistributed to the outsides of the ice crystals when the grain boundary reaches the location of the defect.

Evidence has long existed for the presence of a liquid-like film on ice surfaces at temperatures below the melting point of ice. This disordered region has been observed to migrate on snow grains in response to temperature gradients and differences in curvature (Kuroiwa, 1962; 1975). The film appears to migrate from warm to cold and from convexities to concavities. The thickness of this layer and its bulk properties are not well known (Nenow, 1984), but models and measurements of its electrical properties suggest that the ionic strengths can be quite high. The charge buildup near the ice surface in the quasi-liquid layer is much different than within the ice crystal, and can attract ionic defects near the surface. The diffusion coefficient of water molecules and impurities on the ice surface greatly exceeds the diffusion inside the lattice (Glen, 1974), so that ions located on the ice-air interface are much more mobile. Takagi (1989) suggests that the thickness of the film can be estimated as a function of temperature and the radius of curvature of the ice particle. Colbeck (1978) suggests that the film thickness may control diffusion of chemical species when the film is heterogeneous and the size of the film is on the order of ionic diameters, thus proposing a mechanism for preferential diffusion and incorporation of ions. Ion diffusion within the layer increases with increasing temperature and increasing solute content. Kuroiwa (1975) observed the migration of the quasi-liquid layer on the grain surfaces to be so pronounced during metamorphism as to cause the movement of bubbles. This behavior has been enhanced with a solute (KCl), increasing the thickness of the quasi-liquid layer. Chemical species can hence be transported in the process of surface diffusion away from convex regions of a snow grain and toward concavities such as the neck/bond areas between grains. A more detailed discussion of the character of the quasi-liquid layer can be found in Bales *(this volume).*

2.2. Dry snow metamorphism

An ice particle in a snow cover undergoes continual metamorphism because it is not in equilibrium with its own vapor. Depending on the snow density and porosity, the temperature and the temperature gradient, the presence and amount of liquid water, and the type and concentration of impurities, different crystal forms evolve at different rates. Except when subjected to strong temperature gradients, snow grains reduce their surface free energy by reducing the surface area, which gives rise to the predominant crystal form in seasonal snow covers; the grains exhibit an almost spheroidal shape, except where flattened at grain bonds (Colbeck, 1982). In equilibrium, the saturation vapor pressure at the grain surfaces as a function of temperature and grain radius can be expressed by combining the Clausius-Clapeyron and Kelvin equations

$$P = P_o \exp(2\sigma_{sg} / \rho_i R T_o r) \exp (L_V / R [\frac{1}{T_o} - \frac{1}{T}]) \qquad (1)$$

where

P	saturation vapor pressure at a grain surface
P_o	reference pressure
σ_{sg}	surface free energy between the solid and gas phases
ρ_i	density of ice
R	gas constant for water vapor
T_o	reference temperature
r	radius of grain
T	temperature

This equation explains why large grains and relatively flat surfaces accumulate mass at the expense of small grains and pointed surfaces at a rate increasing with temperature. The reduction in specific surface will take place in the absence of an imposed temperature gradient (Bader et al., 1939), but an imposed gradient is necessary to explain measured growth rates (Perla, 1978; Colbeck, 1980).

The metamorphism of dry snow is thus dominated by temperature gradients that give rise to vapor density gradients in the pore space (Colbeck, 1982). The magnitude of the resultant vapor flux determines to a large extent the type of crystal morphology; the geometry of the pores and the snow particles controls where erosion and deposition take place on the ice surface (Colbeck 1983). As long as the snow is dry, heat and mass transfer occur by a hand-to-hand coupling of conduction, vapor diffusion, and possibly thin-film migration (Yoshida, 1955; Giddings and LaChapelle, 1962; de Quervain, 1963). These processes act in concert to cause grains to undergo a distillation process, leaving solutes on sublimating surfaces. These solutes diffuse along concentration gradients and are also transported as the liquid-like film migrates. The individual grains sinter together where one grows into another, and where the settlement owing to overburden pressure bring them into contact. As the grains grow, the snow also becomes more dense, concentrating chemical species on less grain surface area. Figures 1 and 2 show images of the

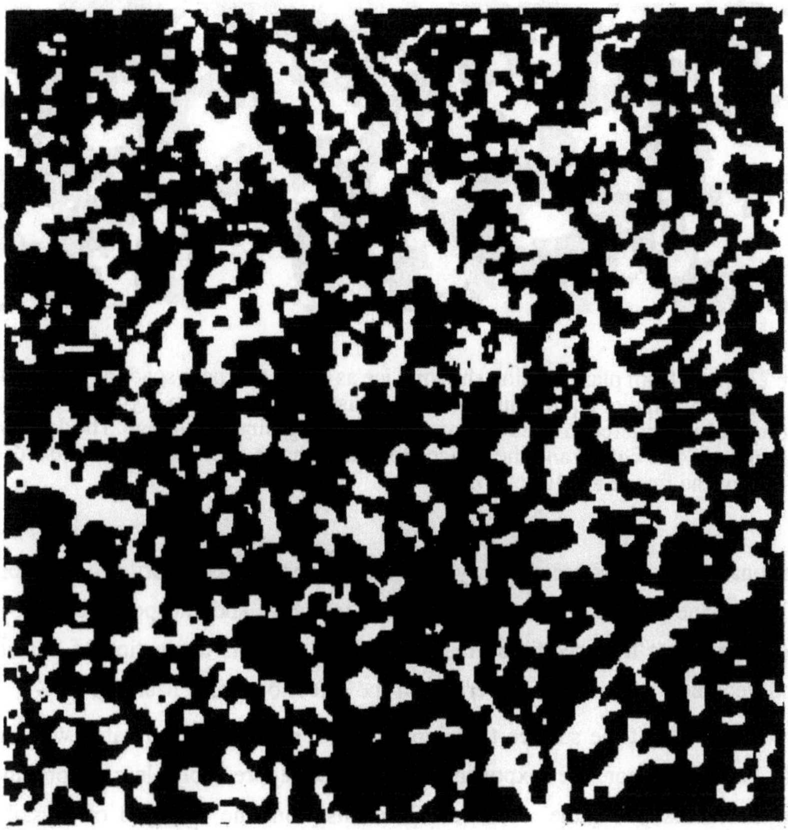

Figure 1. Image from plane section showing ice as white and pore space as black for snow in the early stages of slow growth metamorphism. Grains are small and are bonded to two or three neighbors on average. The density of this snow is 290 kg/m^3, and the surface to volume ratio is 6.1 mm^{-1}. The real dimensions of the image are 10×10 mm.

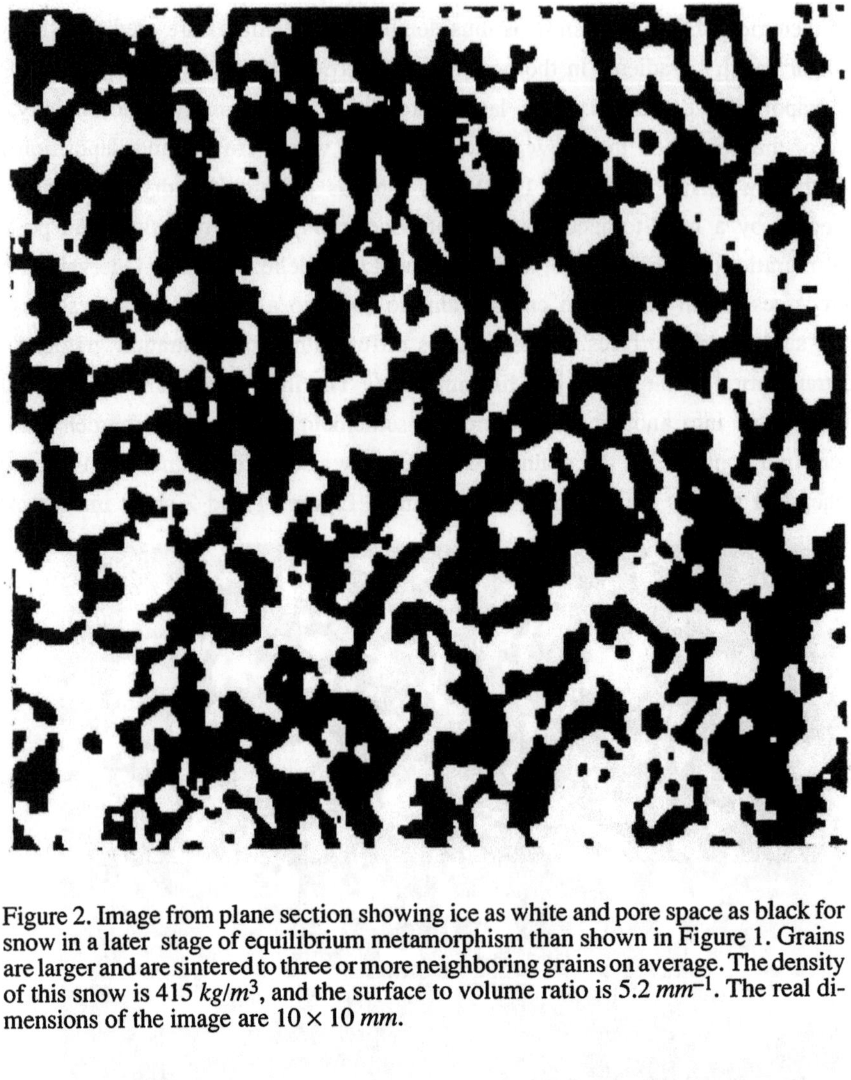

Figure 2. Image from plane section showing ice as white and pore space as black for snow in a later stage of equilibrium metamorphism than shown in Figure 1. Grains are larger and are sintered to three or more neighboring grains on average. The density of this snow is 415 kg/m^3, and the surface to volume ratio is 5.2 mm^{-1}. The real dimensions of the image are 10×10 *mm*.

two-dimensional snow structure at early and later stages of equilibrium metamorphism obtained from plane sections. The ice is shown in white and the pore space in black. One would expect that the specific surface of the ice phase is a function of density; but the size and geometry of the grains overwhelm the effect of closer packing.

Figure 3 shows that the surface-to-volume ratio is negatively correlated with density; neither the linear fit nor the exponential fit gives appreciably less error; the rms differences are almost the same. The scatter results from the variation of crystal forms departing from the equilibrium shape at different densities. This can be seen in Figure 4, which shows the depth profile of the surface-to-volume ratio in mid-winter and well into the melt season. The upper profile shows what might be expected in a snow cover with

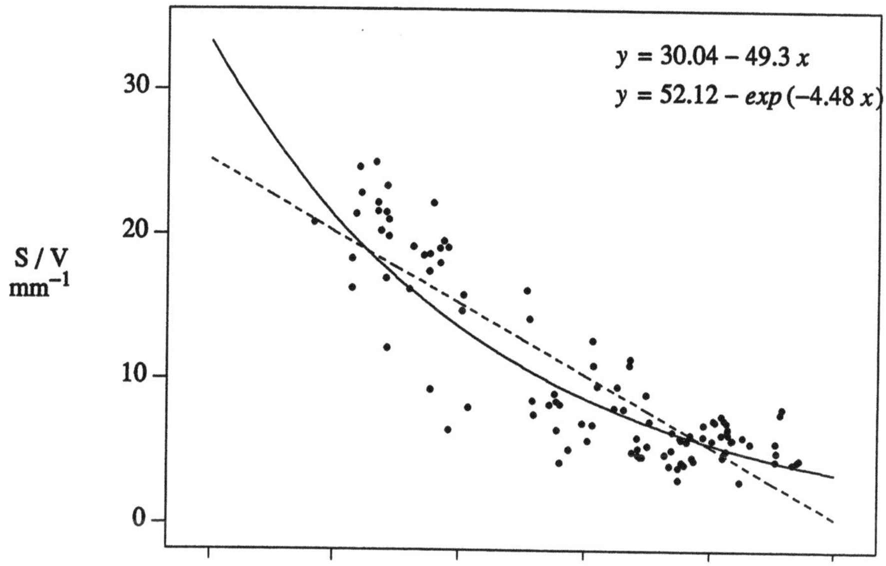

Figure 3. Specific surface (surface area to volume ratio) of the pore space plotted as a function of ice volume fraction with two lines of best fit: a linear regression and an exponential fit. Neither provides a statistically better model.

increasing density downward from the surface, and increasing grain size. A melt-freeze crust near the surface stands out sharply. The snow pack in the lower profile also had increasing density downward away from the surface, but the specific surface is almost uniform with depth, partially owing to large grain clusters formed near the surface.

The bond that form between grains represent regions of disorder whose properties are controlled by the angle of the lattices between the adjacent crystals. Low angle boundaries are those where the crystallographic orientations are nearly aligned. During the growth and sintering of spherical particles, Kuroiwa (1975) observed the crystallographic orientations between adjacent grains change to become more aligned. This movement has also been observed among grains whose crystal axes were nearly aligned in polycrystalline ice (Kuroiwa, 1969). In such a manner polycrystalline particles become single crystals, and ions incorporated into the bond regions should be excluded. On the other hand, high-angle boundaries can be considered regions where the crystallographic arrangement breaks down, and these are locations where impurities will preferentially accumulate and diffuse (Glen, 1974). The bonds regions may have a liquid-like character if the eutectic temperatures of the incorporated chemical species is exceeded (Chatterjee and Jellinek, 1971). Diffusion along concentration gradients, temperature, and type of chemical species control the composition of impurities in grain bond regions.

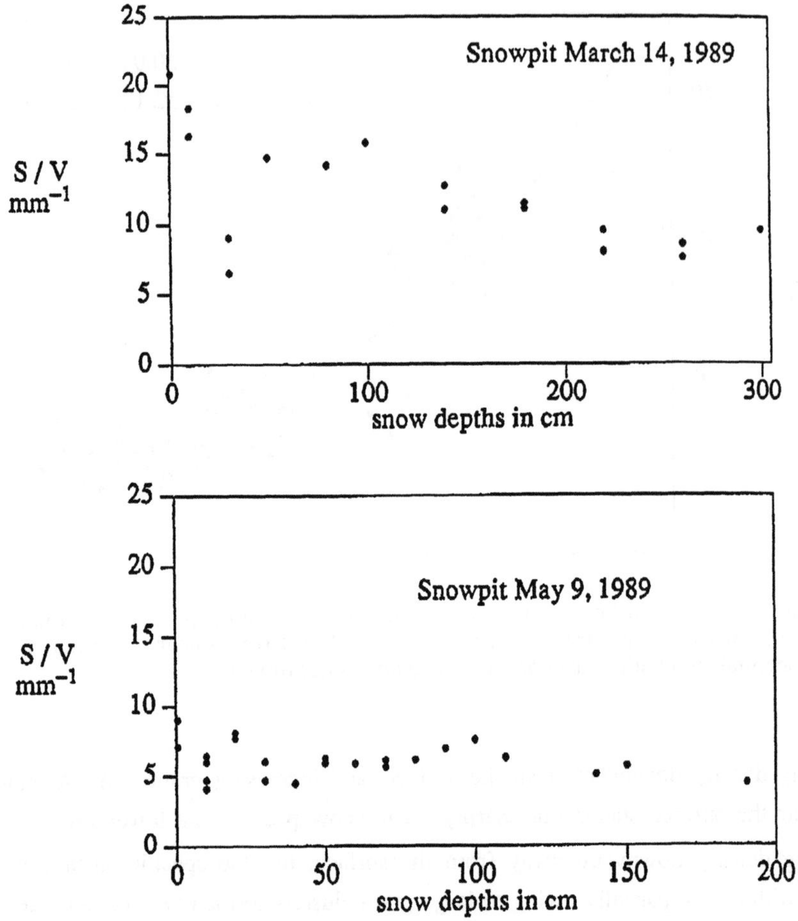

Figure 4. Depth profiles of specific surface showing the difference over time as the snow pack continues to undergo metamorphism. The two points showing low surface area at about 25 cm depth on upper graph represent a buried melt-freeze crust.

At high vapor fluxes (strong temperature gradients), rapid deposition of water molecules on the ice surface gives rise to faceted crystal shapes, whose form depends on the configuration of the ice and pore space (Armstrong, 1985; Gubler, 1985). Equation (1) cannot be used as a basis for a one-dimensional model of the formation of faceted crystals because the difference between the equilibrium vapor pressure over a rounded crystal and a flat surface is very small (Colbeck, 1983). Under the influence of strong temperature gradients the vapor flux in the pore volume can cause the local vapor densities to exceed a critical supersaturation and thereby initiate the kinetically-controlled growth of faceted crystals (Colbeck, 1983). These crystals grow to large sizes in pore spaces at the expense of smaller rounded crystals, usually near the base of the snow cover. Presumably, because growth rates are rapid, we would expect the concentration of solutes on the pore surface to be enhanced sooner than for equilibrium metamorphism, especially considering also that the bond areas of these crystals are small.

The current status of the theory of dry snow metamorphism is that models can predict with fair accuracy the increase in the average grain size, and possibly the grain size distribution. However, modeling solute release to percolating meltwaters is limited by knowledge of the change in the surface area of the grains and the number and area of bonds between grains.

2.3. Wet snow metamorphism

In the presence of liquid water, equilibrium grain growth takes place more rapidly than in dry snow, and the growth regimes can be considered in two modes; metamorphism at high liquid water contents and low liquid water contents. Grain boundaries are generally unstable when the snow pore space is saturated with liquid water; solutes present along these boundaries will be excluded into the surrounding solution. Moreover, ions diffuse away from the solid-liquid interface during freezing (e.g. Gross et al., 1975; 1987). In water-saturated snow, Colbeck (1973a) showed that grain growth (surface area decrease) depends on the inverse relation between particle radius and melting temperature

$$T_m = -\frac{2T_o\,\sigma_{sl}}{L_F\,\rho_i\,r} \tag{2}$$

where

T_m	melting temperature
T_o	reference melting temperature for flat surface
σ_{sl}	solid-liquid interfacial energy
L_F	latent heat of fusion

Colbeck (1986) later showed that current theory for grain coarsening in saturated conditions gives only fair agreement with observed particle size distributions.

In highly unsaturated snow, grains adjust their geometry, from either the dry equilibrium form with a few neighbors in contact, or from the saturated situation with few contacts, form into multigrain clusters bonded together (Colbeck, 1979a). In this case slow grain growth occurs by similar processes to those in dry equilibrium snow. There may be thin film migration within the clusters, which could contribute to growth during melt-freeze cycles. The general problem of wet snow sintering and compaction was considered by Colbeck (1979b), who found good agreement with measurements incorporating grain growth, sintering, and the effects of solutes. Once dry snow has been wetted for a period, the snow pack can experience sudden settlement, owing to intergranular glide, in response to a shift in grain geometry of multigrain clusters. Densities can increase by up to 20 percent in a day or two, depending on the overburden and the liquid water content. Figure 5 shows an image of a section through a snow specimen that has a density of 425 $Kg\ m^{-3}$, and consists of multigrain clusters. The density of the layer from which this snow was obtained was 340 $Kg\ m^{-3}$ three days prior. Other evidence for intergrain mobility can be seen in the gradual change of multigrain clusters to polycrystalline grains

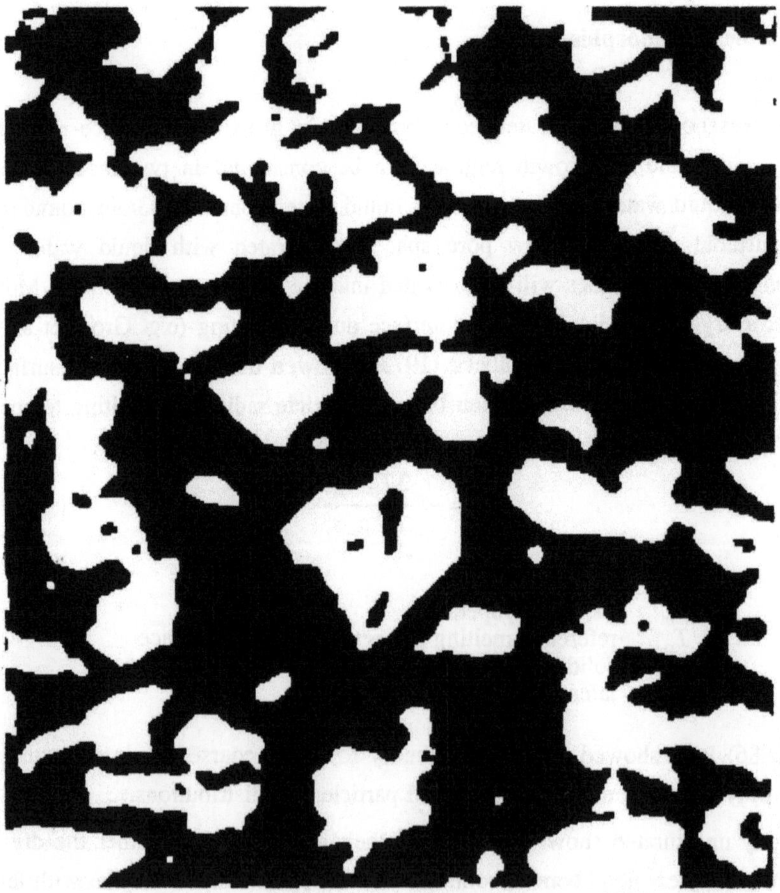

Figure 5. Image from plane section showing ice as white and pore space as black for grain clusters in late spring. Most grains in this image have four or more bonded neighbors. The density of this snow is 425 kg/m^3, and the surface to volume ratio is 3.0 mm^{-1}. The real dimensions of the image are 10×10 mm.

with fewer individual crystals (e.g. Colbeck, 1982). If a polycrystalline grain is being heated by radiation, the grain boundaries absorb the radiation preferentially and act as sources of melt, particularly if impurities are present (Glen, 1974). Colbeck (1979a) showed that the liquid would be constrained to concentrate in the veins where three or more grains meet, raising the possibility of a continuous network in grain clusters providing for the flushing of ionic impurities. The gradual adjustment of lattice alignments, in addition to grain growth and equilibrium constraints on the location of liquid water, is probably responsible for the enhanced ionic pulse observed from snows after several melt-freeze cycles (cf. Colbeck, 1981). The consequence is that the greater number of melt-freeze cycles, the greater the ionic pulse.

Additional discussion of wet snow metamorphism will be presented in the section on water flow because it occurs concurrently and controls the hydraulic properties.

2.4. Effects of layers

Virtually all seasonal and polar snow packs have layers, or bands of distinct physical properties. In dry snow the main effects of stratigraphic or textural discontinuities are to influence metamorphism by affecting temperature gradients and gas movement. For example, a dense layer can reduce the temperature gradient within its boundaries owing to the greater thermal conductivity (S.C. Colbeck, *personal communication).* In addition, an ice lens can act as a barrier to vapor transfer, sometimes causing supersaturations which initiate the growth of faceted crystals. Figure 6 shows a photomicrograph of a cross section through an ice lens in which the ice phase has a bright highlighted outline. The protrusions on the bottom of the ice lens are depth hoar crystals, growing in the downward direction. Figure 7 shows the situation where an ice lens blocked the drainage of some melt in the upper layer. There is a coarse mass of melt-freeze crystals sitting on top of the ice lens in the middle of the photomicrograph. The effects of stratigraphy on water flow will be discussed further in the next section.

3. Water Flow Through Snowpacks

Previous sections have surveyed the metamorphic changes in the snow and their relation to the microscale distribution of solutes, and the grain-scale differences between

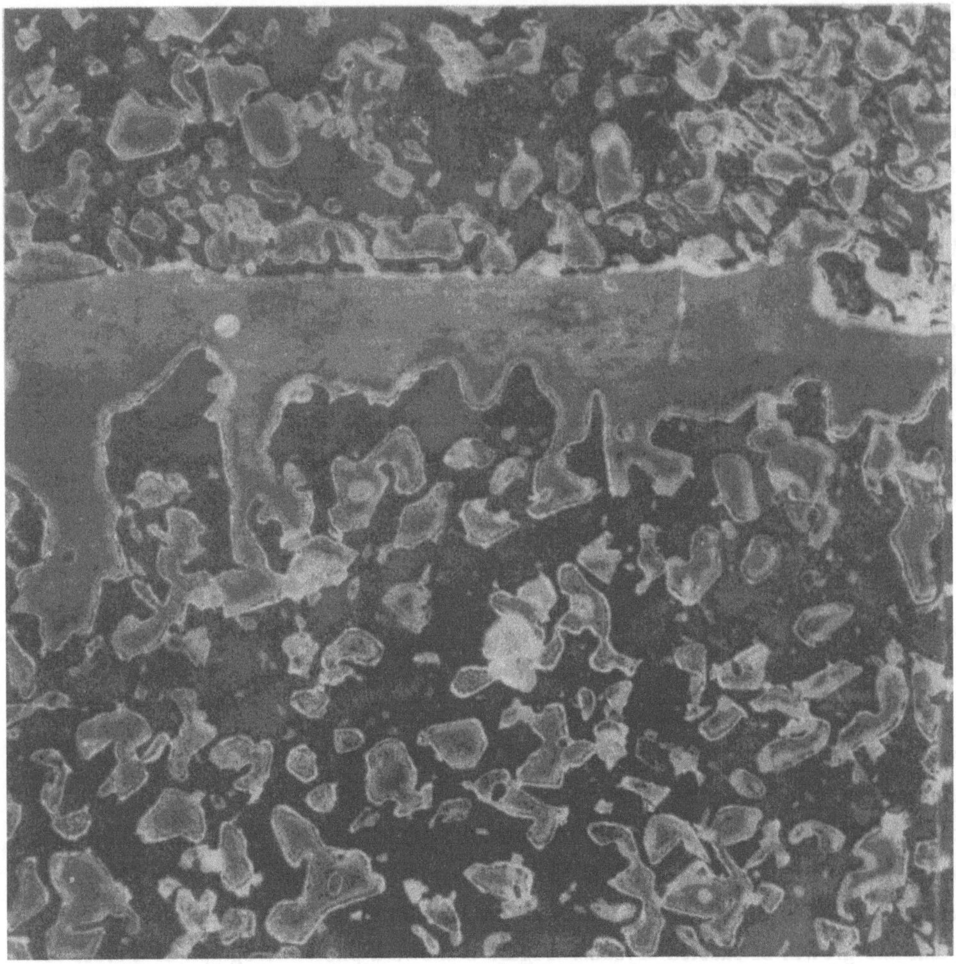

Figure 6. Photomicrograph showing cross section of ice lens with depth hoar crystals growing from the bottom. The real dimensions of the image are 10 × 10 *mm*.

dry snow and wet snow. The main effect of these processes in relation to chemical elution from snow, is to condition the pack for efficient leaching of the soluble contaminants. This section describes the flow of liquid water in snow and compares an existing model of water flow coupled with solute transport with a more complete model routing water flow through snow.

3.1. Overview

The progress of melt water removing ions from the grain surfaces and reaching the base of the pack can be quite complicated and exerts a macroscale influence on the chem-

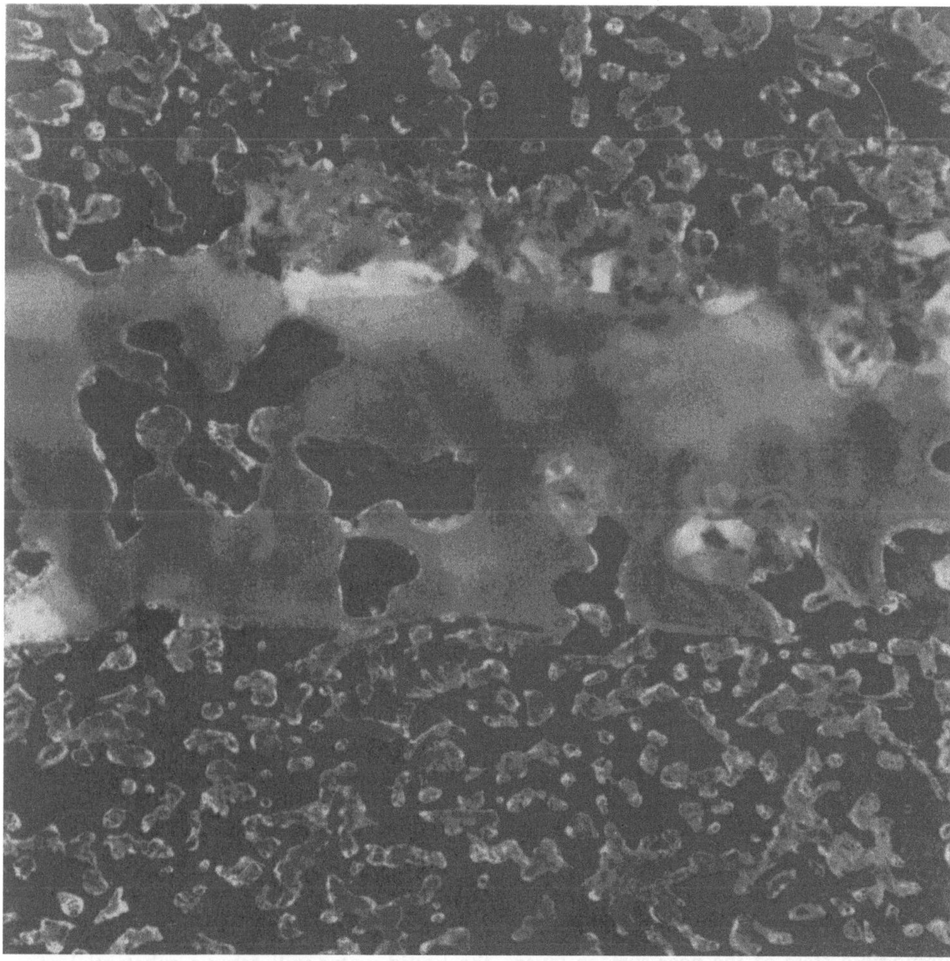

Figure 7. Photomicrograph showing cross section of ice lens with coarse grains built up on top. The real dimensions of the image are 10×10 *mm*.

ical hydrograph. A common observation comparing the chemical with the melt hydrographs of seasonal snow shows that much of the chemical load in the pack is eluted early in the melt period (see review by DeWalle, 1987). The earlier discussion on snow metamorphism describes how solutes become located on the outsides of the snow crystals, and this is why so much solute is removed with the first passage of liquid water. Figure 8 shows the results of continuously melting homogeneous snow to which tracers had been added in a laboratory experiment (Bales et al., 1989). The melt rates were similar to those found in field conditions, and the mass fraction of species eluted includes all tracers added to the snow.

The situation in the field is more complicated and involves heterogeneous snow and diurnal melt-freeze cycles. Early in the melt season, ice lenses form near the surface when melt can no longer penetrate cold layers, or when nighttime freezing causes capil-

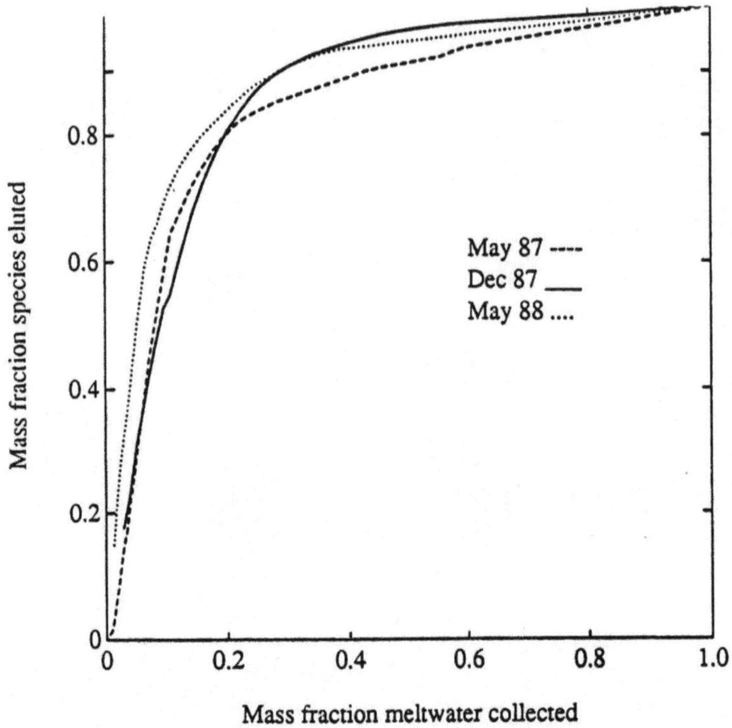

Figure 8. Results of three experiments eluting solutes from homogeneous snow at a constant melt rate in the laboratory. This shows that approximately 80 percent of the solutes are eluted in the first 20 percent of melt. From Bales et al. (1989).

lary tension toward the surface, slowing the melt wave and freezing it. Depending on the metamorphic state of the near-surface layers, the liquid water will carry a substantial portion of the chemical load because of uniform wetting and because absorption of solar radiation melts the bonds, allowing species located there to become as mobile as those on the pore surface. This cycling can cause localized zone of high solute content when an ice lens forms.

The varying fluxes of meltwater over diurnal and longer-term cycles also control the concentrations of melt discharge. Figure 9 shows the daily snowmelt hydrographs measured by lysimeters at a site on Mammoth Mountain, California, showing strong concentration peaks at the onset of melt periods, indicated here by conductivity. This figure shows that after periods when the snowcover partially refreezes a new concentration peak is observed. Once the melt wave has reached the base of the pack, the diurnal fluctuations of water flux can cause concentration and dilution of the chemical outflow. Figure 10 shows a closer look at Figure 9; the daily melt waves dilute the chemical outflow, which gradually increases in concentration as the melt flux subsides.

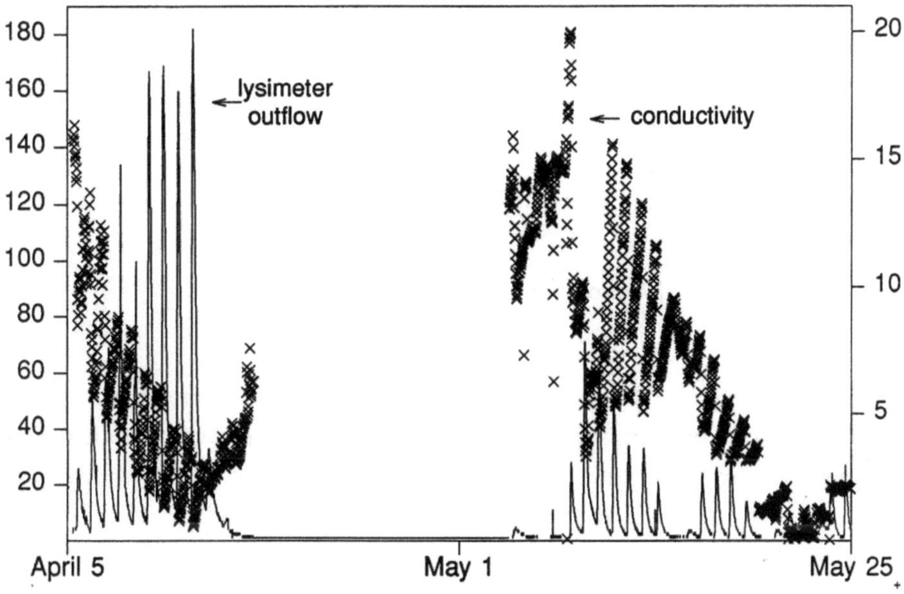

Figure 9. Lysimeter data from study plot on Mammoth Mountain, California USA, shown with measurements of meltwater conductivity during the spring, 1989. Tracer $LiCO_3$ was added to the pack as a tracer.

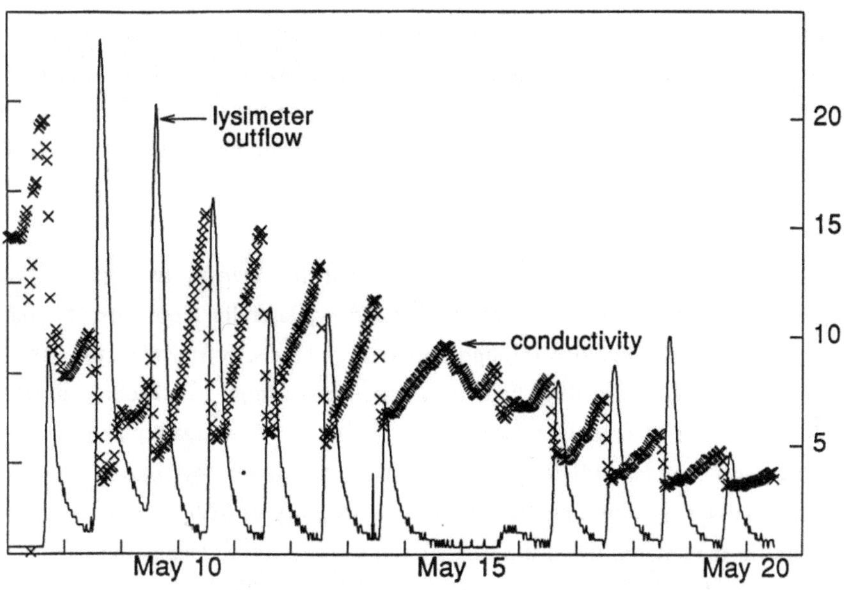

Figure 10. Lysimeter data from study plot on Mammoth Mountain, California USA, shown with measurements of meltwater conductivity during May, 1989. Higher resolution view of previous figure.

Field and laboratory evidence shows that some chemical species are removed in the melt water before others. Similar elution sequences of chemical species have been observed in the field and in the laboratory (e.g. Davies et al., 1982; Tranter et al., 1986; Brimblecombe et al., 1987; Stottlemeyer et al., 1990). Tranter et al. (1986) have suggested the anion elution order

$$SO_4^{2-} > NO_3^- > Cl^-$$

and Brimblecombe et al. (1987) proposed the cation elution sequence

$$K^+ > Ca^{2+} > Mg^{2+} > Na^+.$$

Davies et al. (1987) show in laboratory experiments that an elution order is not clearly evident until 80 percent or more of the total solute load has been removed. In the field it has been observed that preferential elution is more pronounced at greater depths (e.g. Tsiouris et al., 1985). Both field and laboratory observations show the effect is more pronounced with older snow. Some proposed hypotheses to explain preferential elution are: different solubilities of ionic species in ice, differential solid-state diffusion of ions incorporated into snow grains, inhomogeneous initial distribution of chemical species, and chromatographic separation during meltwater flow. Hewitt et al. (1990) found no chromatographic effects on elution through 30 cm columns of unsintered snow grains in laboratory experiments. They suggest that preferential elution is the result of differential exclusion of ions during grain growth, based on the results of Tsiouris et al. (1985). The role of grain bonds in this hypothesis is not considered, nor is the migration of grain boundaries with respect to the initial location of solutes. Finally, the results summarized above have not been repeatable in the field, mostly owing to the effects of inhomogeneous macroscale distributions of chemical species. For example, Bales et al. (1989) showed that species near the top of the snow cover will be eluded before those in the middle.

Heterogeneities in snow structure cause the melt to flow in two modes; two-dimensional flow by gravity, and three-dimensional flow, laterally along textural boundaries and in a variety of directions through preferred channels, or macropores. At high melt rates, the two-dimensional flow field in homogeneous snow can become unstable, resulting in fingers, like in heterogeneous snow. Where the flow takes place in fingers penetrating dry snow, solutes will remain immobile in the volume bypassed by the flow paths. In warm snowpacks flow fingers can carry the majority of melt away from the surface, while later in the season, when the background wetting front reaches the base of the pack, much of the flow reverts to more uniform advection (McGurk and Kattelmann, 1986). At this point most of any remaining chemical species are leached. Initially, there is a greater grain-growth rate in the flow fingers owing to the presence of liquid water;

later, once the surrounding snow is wet, the lower field capacity of the fingers slows the growth rate, while the surrounding snow texture coarsens rapidly.

To realistically describe the processes of melting snow, the effects of melt-freeze, ice layer formation and disintegration, stratigraphic discontinuities, and flow fingers must be included. As late as the 1980's attempts were still being made to characterize the snow cover as a homogeneous medium, usually with the result that errors were attributed to inhomogeneities (see for example Bengtsson, 1982; Jordan, 1983; Akan, 1984). The flow in homogeneous snow can be described with a nonlinear flux-concentration relationship described by Colbeck and Davidson (1973b). Colbeck (1973c) initially modeled the effects of layers by considering the flow stepwise and treating each layer as homogeneous. This work was extended (Colbeck, 1975) to consider permeability as a two-dimensional quantity, with a horizontal term and a vertical term to account for flow parallel and perpendicular to the snow strata. The problem of phase change at a planar wetting front advancing in cold dry snow was also considered by Colbeck (1976), but flow fingers and the formation of ice columns were not considered.

In an approach considering two sets of flow paths, Colbeck (1979c) treated the flow as both a uniform field and a set of preferential drainage channels. Marsh and Woo (1984a, 1984b) observed many of these processes and extended previous models to integrate and modify various approaches; later they applied the multiple-path flow model to new observations from the Canadian arctic and with measurements from the central Sierra Nevada, California with good results (Marsh and Woo, 1985). Capillary effects were added to a model to describe water flow in heterogeneous cold snow by Illangsekare et al. (1990), who simulate water flow in two dimensions. While their model simulated heterogeneous flow of water past density discontinuities, and ice lenses with holes, they failed to demonstrate that the model could simulate flow fingers. In addition, their calculation of phase change of water infiltrating into subfreezing snow uses a calibration constant to account for the rate of heat transfer, and they assume that water cannot penetrate until the snow is at the melting point of ice. However, observations in the Sierra Nevada, California USA, show that the cold content of a snow cover may not be satisfied before melt reaches the base (Kattelmann, 1985).

3.2. Solute transport coupled with water flow through homogeneous snow

This section starts with a review of the theory of Hibberd (1984), who coupled solute transport to water flow in homogeneous snow. Then the important features of the multipath flow model of Marsh and Woo (1984b) are described for comparison. A more detailed discussion of an extension of Hibberd's model is presented by Bales *(this volume)*.

In the model of Hibberd (1984), the initial conditions are a dry, homogeneous snowpack, isothermal at 0° C. The pack then begins to melt at a uniform rate. The volume flux of melt water per unit area Q is given by (Colbeck, 1978)

$$Q = k\,S^n \tag{3}$$

where $S = (S_w - S_i)/(1 - S_i)$ is the effective saturation of the pore space; S_w is the liquid water saturation and S_i is the irreducible saturation. The parameter k is a constant (Colbeck and Davidson, 1973b), and n is an exponent, which may be assigned an arbitrary value between 2.2 and 4.4, according to Denoth et al. (1979), who qualitatively noted a dependence of n on the size of grains and clusters. The velocity of the saturation wavefront, which moves through the pack is

$$v^* = \frac{S^{n-1}\,k}{\theta\,(1 - S_i)} \tag{4}$$

where θ is the porosity and S_i is the irreducible liquid water content, in the range 0.03-0.08 (Denoth et al., 1979). Using conservation of mass, the percolation velocity u^*

$$u^* = \frac{\left[\dfrac{S^n}{S + \beta}\right] k}{\theta\,(1 - S_i)} \tag{5}$$

where $\beta = \dfrac{S_i}{1 - S_i}$.

The solutes are either mobile or immobile in this formulation; Bales *(this volume)* extends this to three phases. Using two phases, a material mass balance can be written

$$\frac{\partial C_i}{\partial t^*} + (S + \beta)\left[\frac{\partial C_m}{\partial t^*} + u^*\,\frac{\partial C_m}{\partial z^*}\right] = (S + \beta)\,D^*\,\frac{\partial^2 C_m}{\partial z^{*2}} \tag{6}$$

where C_m and C_i are the mass concentrations of solute in the mobile and immobile phases, respectively, expressed per mass of that phase, and t^* is time. D^* is the dispersion coefficient and z^* is the depth measured from the surface of the snowpack. The mass balance of the solute must be expressed across the saturated wavefront simultaneously with that for the liquid

$$v^*\left[S + C_a - \beta C_{m0}\right] = S^n\,C_a - (S + \beta)\,D^*\,\frac{\partial C_a}{\partial z^*} \tag{7}$$

where C_{m0} is the initial solute concentration at any point in the pack. That is, it is the initial concentration at any point upon arrival of the saturated wavefront. Recall that initially the pack is dry. C_a is the average concentration. What has been left out is the relationship between C_i and C_m. Potential formulations are discussed by Bales *(this volume)* in another chapter. This model considers advection and dispersion only in defining chemical transport. Furthermore, the water flow the model describes can only be found in a laboratory environment; although it is helpful to identify appropriate chemical terms for fractionation and preferential elution.

Next a more realistic model of melt water flow in a natural snowpack will be described as an illustration of the difficulties one might face carrying out combined experiments to examine both the theory for water flow in heterogeneous snow and the coupled solute transport equations.

The model of Marsh and Woo (1984) follows a similar development as described above, extended to include multiple flow paths, the formation of ice lenses, and the conduction of heat to and from the ice lenses. In this case

$$\frac{dv^*}{dt^*} = u^* / [(\theta S) + (T \rho_s) / ((L_F / c_i) \rho_w)] \tag{8}$$

where

T	snow temperature
ρ_s	snow density
L_F	latent heat of fusion
c_i	specific heat of ice
ρ_w	density of water

It is assumed that on a daily basis, all of the gravitational water will reach the wetting front

$$\frac{dv^*}{dt^*} = u^*_d / [(\theta S_i) + (T \rho_s) / ((L_F / c_i) \rho_w)] \tag{9}$$

where u^*_d is the daily mean surface melt rate. To this point the model must assume that the melt front advances uniformly; further development will account for separate flow paths.

The key to routing the flow in separate paths is that the cold content in the snow that each flow regime must satisfy is related to the fraction of the unit cross-sectional area occupied by that mode of flow. The flux for finger flow and for the background wetting front can be described by

$$u^*_f = u^* (f_{df} / A_f) \tag{10}$$

and

$$u^*_b = u^* \, [(1 - f_{df})/(1 - A_f)] \tag{11}$$

where

u^*_f	finger flow flux
A_f	proportion of total area covered by fingers
f_{df}	proportion of total flow in fingers
u^*_b	flux at background front

The finger flux and background flux may be substituted into equation (9) to give two expressions for the fluxes. The terms A_f and f_{df} are obtained from measurements. A heat budgeting algorithm is used in the model to calculate the latent heat and conduction terms during freezing, melting, and heat flow, again based on the fractional areas occupied by each process. This model could be extended to many flow paths, provided that the areal geometry of the routes can be estimated. In fact, one of the problems with this flow model is that the terms A_f and f_{df} are supplied by measurement, not by the theory. It may be possible to adapt techniques described by Parlange et al. (1989) and Glass et al. (1989) to determine the geometry and extension of flow fingers. At any rate, the model can be used to calculate the difference in the depth of penetration of the flow routes, which combined with A_f can be used to calculate the volume of dry snow not leached by melt water. Once appropriate chemical terms can be formulated for water flow in snow, they may be extended to a parameterized three-dimensional surface, such as cylinders and plates for example.

4. Summary

The initial melt water percolating through the snow may be enriched with solutes in concentrations much higher than the bulk of the pack. This is because snow metamorphism prepares the solutes in a snow cover for efficient and rapid removal by flowing water. Some chemical species start out on the outsides of snow crystals where they were absorbed as gases or deposited in rime. Snow metamorphism results in the translocation of most incorporated chemical species to the outsides of the snow grains where they reside on the pore surface or in the bonds between adjacent grains. This process occurs mostly because of the migration of grain boundaries and the consumption of small grains. Except for the formation of depth hoar, the area of the pore surface usually decreases and the bond surface increases over time. Solutes located in the bond regions can be considered relatively immobile, while solutes on the pore surface, or ice-air interface, can be considered highly mobile with respect to uptake by flowing water. The chemical species with lower solubilities and larger ionic diameters are more mobile in the bond regions, and this can give rise to an elution sequence. Larger grain sizes and lower pore area usually allow gravity forces to dominate the flow of melt in a seasonal snowpack, and this combined with the concentration of solutes on the pore surface gives rise to the ionic pulse. While models of snow metamorphism can predict grain size changes, they need to be extended to allow accurate estimation of the pore surface area and the grain bond area.

At the surface of a snow cover the ice grains may be covered with a layer of liquid water and the grain bonds will preferentially melt as solar radiation is absorbed. This process mobilizes all solutes not incorporated inside the snow grains. The chemical species in the region near the surface are thus leached uniformly while melting. Melt-freeze cycles give rise to a more pronounced ionic pulse because of rapid recrystalization in the presence of liquid water. Deeper in the pack melt water responds to layers by flowing laterally along textural discontinuities and vertically in flow fingers, redistributing solutes heterogeneously. As melting progresses and the background wetting front reaches the base of the snow cover, water flow in snow becomes more homogeneous and the melt water is enriched with those species present in the basal layer and those species at the end of the elution sequence. To understand the elution sequence simple models of water flow through snow coupled with solute transport are required to examine the interplay between mobile and immobile phases of chemical species. However, they are inadequate for application in the field because of the multiple flow paths melt takes in percolating through a snow cover. Models incorporating terms describing many flow routes have not been coupled to solute transport because of the lack of suitable chemical models, and the complex geometry of the wetting fronts.

The main problems in modeling the chemical hydrographs of watersheds owing to snow melt involve process-level understanding of the solute fluxes through a snowpack and the timing and magnitude of the water and solute fluxes as functions of the energy inputs to a basin. As shown in this chapter, we are close to quantitatively linking metamorphism, water flow, and solute transport. The areal variation of surface melt provides variable source areas and melt water flow follows the patterns of surface energy exchange. For open areas with similar slope, exposure, and elevation, melt production rates may not vary significantly at scales of tens of to hundreds of meters. However, on the scale of a watershed, or even a small catchment, some areas of snow cover may be melting actively while others are not. This concept of variable source areas is not new, but the effects of spatially distributed snow melt sources on the chemical hydrograph for a watershed are only recently being considered.

5. References

Akan AO (1984) Simulation of runoff from snow-covered hillslopes. Wat Res Res 20:707-713

Armstrong RL (1985) Metamorphism in a subfreezing, seasonal snow cover, PhD. Thesis. University of Colorado Boulder CO USA

Bader H, R Haefeli, E Bucher, J Neher, O Eckel, C Tahms (1939) Der Schnee und seine Metamorphism, Beitr Geol Schweiz Geotech Ser, Hydrol 3

Bales RC (1990) Modeling in-pack chemical transformations. In: this volume Springer-Verlag

Bales RC, RE Davis, DA Stanley (1989) Ionic elution through shallow, homogeneous snow. Wat Res Res 25:1869-1877

Bengtsson L (1982) Percolation of meltwater through a snowpack. C Reg Sci Tech 6:73-81

Brimblecombe P, SL Clegg, TD Davies, D Shooter, M Tranter (1987) Observation of preferential loss of major ions from melting snow and laboratory ice. Wat Res 10:1279-1286

Chatterjee AK, HHG Jellinek (1971) Calculation of grainboundary thickness in polycrystalline ice of low salinity. J Glaciol 10:293-297

Colbeck SC, G Davidson (1973b) Water flow through homogeneous snow, IAHS Pub 106:242-257

Colbeck SC (1973a) Theory of Metamorphism of wet snow. Res Rep 313 USA CRREL Hanover NH USA

Colbeck SC (1973c) Effects of stratigraphic layers on water flow through snow. Res Rep 311 USA CRREL Hanover NH USA

Colbeck SC (1975) Water flow through a layered snowpack. Wat Res Res 11:261-266

Colbeck SC (1976) An analysis of water flow in dry snow. Wat Res Res 12 523-527

Colbeck SC (1978a) The compression of wet snow. CRREL Rep 78-10 USA CRREL Hanover NH USA

Colbeck SC (1978b) The physical aspects of water flow through snow, In: Advances in Hydroscience, 11:165-206

Colbeck SC (1979a) Grain clusters in wet snow. J Coll Int Sci 72:371-384

Colbeck SC (1979b) Sintering and compaction of snow containing liquid water. Phil Mag A 39:13-32

Colbeck SC (1979c) Water flow through heterogeneous snow. C Reg Sci Tech 3:37-45

Colbeck SC (1980) Thermodynamics of snow metamorphism due to variations in curvature. J Glaciol 26:291-301

Colbeck SC (1981) A simulation of the enrichment of atmospheric pollutants in snow cover runoff. Wat Res Res 17:1383-1388

Colbeck SC (1982) An overview of seasonal snow metamorphism. R Geo Spa Phys 20:45-61

Colbeck SC (1983) Theory of metamorphism of dry snow. J Geo Res 88:C9:5475-5482

Colbeck SC (1986) Statistics of coarsening in water-saturated snow. Ac Met 34 347-352

Davies TD, CE Vincent, P Brimblecombe (1982) Preferential elution of strong acids from a Norwegian ice cap. Nat 300:161-163

Davies TD, P Brimblecombe, M Tranter, S Tsiouris, CE Vincent, IL Blackwood (1987) The removal of soluble ions from melting snowpacks. In: Seasonal Snowcovers: Physics, Chemistry, Hydrology NATO ASI Series C 211:337-392

A Denoth, W Seidenbusch, M Blumthaler, P Kirchlechner, W Ambach, SC Colbeck (1979) Study of water drainage from columns of snow. CRREL Rep 79-1 USA CRREL Hanover NH USA

DeWalle DR (1987) Review of snowpack chemistry studies. In: Seasonal Snowcovers: Physics, Chemistry, Hydrology NATO ASI Series C 211:255-268

T Dunne, AG Price, SC Colbeck (1976) The generation of runoff from subarctic snowpacks. Wat Res Res 12:677-685

Giddings JC, E LaChapelle (1962) The formation rate on depth hoar J Geo Res 67:2377-2383

Parlange J-Y, TS Steenhuis (1989) Wetting front instability 1. Theoretical discussion and dimensional analysis. Wat Res Res 25:1187-1194

Glass RJ, TS Steenhuis, J-Y Parlange (1989) Wetting front instability 2. Experimental determination of relationships between system parameters and two-dimensional unstable flow field behavior in initially dry porous media. Wat Res Res 25:1195-1207

Glen JW (1974) The Physics of Ice. C Reg Sci Eng Monograph II-C2A USA CRREL Hanover NH USA

Gross GW, A Gutjahr, K Caylor (1987) Recent experimental work on solute redistribution at the ice/water interface. Implications for electrical properties and interface processes. J Phys 48:C1-527-C1-533

Gubler H (1985) Model for dry snow metamorphism by interparticle vapor flux. J Geo Res 90:D5:8081-8092

Hewitt AD, JH Cragin, SC Colbeck (1990) Does snow have ion chromatographic properties? Proc ESC 46:165-171

Hibberd S (1984) A model for pollutant concentrations during snow-melt. J Glaciol 30:58-65

Hoekstra P, TE Osterkamp, WF Weeks (1965) Migration of liquid inclusions in single ice crystals. Res Rep 183 USA CRREL Hanover NH USA

Illangasekare TH, RJ Walter Jr., M Meier, WT Pfeffer (1990) Modeling of meltwater infiltration in subfreezing snow. Wat Res Res 25:1001-1012

Itagaki K (1967) Particle migration on ice surfaces. Proc Physics of Snow and Ice: International Conference on Low Temperature Science 1:1:233-246

Jordan P (1983) Meltwater movement in a deep snowpack. Wat Res Res 19:979-985

Kattelmann R (1985) Macropores in snowpacks of Sierra Nevada. Ann Glaciol 6:47-49

Kumai M (1951) Electron-microscope study of snow crystal nuclei. J Met 8:151-156

Kumai M (1957) Electron-microscope study of center nuclei of snow crystals: III. J Met Soc Jap 75:49-55

Kumai M (1961) Snow crystals and the identification of the nuclei in the northern United States of America. J Met 18:139-150

Kumai M (1976) Identification of nuclei and concentrations of chemical species in snow crystal sampled at the south pole. J Atmos Sci 33:833-841

Kumai M (1985) Acidity of snow and its reduction by alkaline aerosols. Ann Glaciol 6:92-94

Kumai M (1987) Chemical properties of snow in the northeastern United States. J Phys 48:C1-625-C1630

Kuroiwa D (1962) A study of ice sintering. Res Rep 86 USA CRREL Hanover NH USA

Kuroiwa D (1969) Surface topography of etched ice crystals observed by a scanning electron microscope. J Glaciol 54:475-483

Kuroiwa D (1974) Metamorphism of snow and ice sintering observed by time lapse cine-microphotography. IAHS Publication 114:82-88

Marsh P, M-K Woo (1984a) Wetting front advance and freezing of meltwater within a snow cover 1. Observations in the Canadian Arctic. Wat Res Res 20:1853-1864

Marsh P, M-K Woo (1984b) Wetting front advance and freezing of meltwater within a snow cover 2. A simulation model. Wat Res Res 20:1865-1874

Marsh P, M-K Woo (1985) Meltwater movement in natural heterogeneous snow covers. Wat Res Res 21:1710-1716

McGurk BJ, R Kattelmann (1986) Water flow rates, porosity, and permeability in the Central Sierra Nevada. Cold Reg Hydrol Symp 359-366

Nakaya U (1956) Properties of single crystals of ice, revealed by internal melting. Res Pap 13 USA SIPRE

Nenow D (1984) Surface premelting. Prog Cry Gro Char 9:185-225

Perla R (1978) Temperature-gradient and equi-temperature snow. In: Deuxieme Recontre Internationale sur la Neige et les Avalanches. Grenoble, France 43-48

Quervain de MR (1963) On the metamorphism of snow. In: Ice and Snow MIT Press Cambridge Mass USA

Takagi S (1989) Approximate thermodynamics of the liquid-like layer on an ice sphere based on an interpretation of the wetting parameter. J Coll Int Sci 137:446-455

Tsiouris S, CE Vincent, TD Davies, P Brimblecombe (1985) The elution of ions through field and laboratory snowpacks. Ann Glaciol 7:196-201

MODELING IN-PACK CHEMICAL TRANSFORMATIONS

Roger C. Bales
Department of Hydrology and Water Resources
University of Arizona
Tucson, AZ 85721

1. INTRODUCTION

Processes of chemical changes in snow at the earth's surface are important to understanding both atmosphere-snowpack interactions and the effect of snowmelt on the biogeochemistry of snow-covered areas. Models of in-pack chemical transformations have application to processes in seasonal snowpacks, temperate-region glaciers and polar snow/ice. For example, seasonal snowpacks accumulate "wet" and "dry" deposited material for several months each year, then release much of the chemical load during the first portion of snowmelt. Good physically based models of these processes are important tools for assessing the combined effects of acidic deposition and global change on seasonally snow-covered regions of the earth. Also, good models of gas and solute redistribution in snow/ice over a variety of time scales (years to centuries) are needed to help interpret both temperate-region and polar ice cores.

Chemical species are incorporated into snow during formation in clouds, by atmospheric scavenging and riming as snow falls through the atmosphere, and by dry deposition to snow on the ground. Gas uptake by snow occurs primarily by dry deposition, due to the higher temperatures and the much longer exposure times of snowpacks compared to snow crystals in the atmosphere. For all but the most soluble or reactive gases, gaseous deposition is determined almost entirely by interactions below the snowpack surface rather than by atmospheric resistances. Scavenging of particulates by falling snow seems especially efficient [*Chan and Chung, 1986*]; deposition of particles to snow on the ground should depend on conditions both at and above the snowpack surface.

The chemical species deposited in snowpacks are redistributed by metamorphism, sublimation, and freeze/thaw cycling. Recrystallization of snow crystals is thought to leave most of the solutes on the ice-air interface, particularly in old firn. Sublimation changes concentrations of species, owing to mass loss of the ice. During snowmelt, the first fractions of the meltwater are observed to have significantly higher concentrations of dissolved species than later fractions, and the relative concentrations of different species also vary with time. This deviation from the average snowpack concentration, or *ionic pulse*, must involve both grain-surface and bulk-meltwater chemical interactions, and must be distinguished from variations resulting from non-uniform distributions. Discontinuities such as ice lenses, fractures and differing snow layers also influence meltwater movement, but should affect all species the same.

NATO ASI Series, Vol. G 28
Seasonal Snowpacks
Edited by T. D. Davies et al.

Four processes that can contribute to gas and solute composition changes with time in snow are; i) partitioning between ice, melt and gas phases; ii) chemical reactions involving oxidation and reduction, such as slow decomposition of H_2O_2 over long periods; iii) particulate dissolution/precipitation; and iv) separation by gravity, only important on small spatial scales. For all but the most reactive species, partitioning into ice is a minor effect, while there can be strong preferential changes in gas-species concentrations in wet snow because of different solubilities in water.

In order to adequately model these processes, one must describe both snowpack-scale (i.e. transport) and snow-grain-scale (both transport and reaction) processes. Both will influence the rate of chemical changes. At the microscale, different chemical models are needed depending on the snowpack temperature range. At and near the melting point, the surface region of snow grains has a liquid or liquid-like phase, and aqueous-phase chemistry should dominate. At lower temperatures, reactions occur at the ice-air interface. In general, it is the chemical properties of the ice-air *interfacial region*, rather than the *bulk-snow* properties, that exert a dominant influence on processes of chemical change in snowpacks.

2. Nature of Ice–Air Interfacial Region

Near 0°C the surface of snow grains involves gas-liquid and liquid-solid interfaces. Most chemical reactions involve the aqueous phase (Figure 1a). At temperatures below -30 to -40°C, the ice-air interface, without significant liquid water, is more important (Figure 1b).

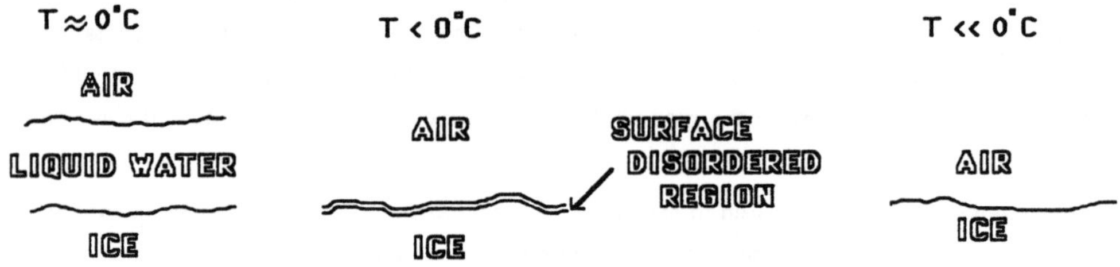

Figure 1. Simplified conceptual picture of ice-air interfacial region.

The water-to-air volume ratio (r_{wa}) of snow is the most important factor affecting chemical reactions. Water-to-air ratio is a function of density, porosity, liquid-water content, and specific surface area. Snow at 0°C can have a wide range of r_{wa} values; below 0°C, liquid water can be present at boundaries or in veins and fillets between snow grains [*Colbeck, 1979*], due in part to freezing-point depression from the presence of solutes. For dilute solutions in water, the change in freezing temperature (°C) is given by:

$$\Delta T_f = 1.86\, m$$

where m is the solute molality [*Adamson, 1979*]. Further, surface curvature may result in an equilibrium temperature between ice and liquid water that is lower than 0°C [*Colbeck, 1980*].

Due to curvature, essentially all of the liquid water in stable grain clusters (at temperatures both at and below 0°C) should be in the veins or fillets [*Colbeck, 1979*].

Considerable evidence also supports the presence of an interfacial region having liquid-like properties, or liquid film, on one or more faces of an ice grain, but its thickness and bulk properties are not well defined [*Nenow, 1984*]. This film below 0°C should be distinguished from snow at high liquid-water contents in which the veins are not sufficient to contain the water, and the entire surface will have a thick liquid film.

Investigations of the liquid-like or disordered nature of the ice-air interface have focused on determining physical properties and an equivalent physical thickness. To have properties of a bulk phase, the thickness, or amount of water present, should be sufficiently large that interfacial effects (e.g. preferential orientations) are small [*Adamson, 1979*]. A quantitative theory of the surface structure of ice was proposed by *Fletcher* [*1968*], based on thermodynamics of interfaces. His theories predict a thickness on the order of nanometers for the interfacial region, at temperatures down to about -20°C, with sharp increases near 0°C.

A liquid-like layer should be observed on crystal surfaces as temperatures approach the melting point; above 0°K crystals begin losing their perfectly faceted shape, and lose essentially all faceting at the roughening temperature. The roughening temperature for ice is thought to be not too far below 0°C, based on models on the ice surface [*e.g., Rottman and Wortis*, 1984]. It is hypothesized that a discontinuity in chemical behavior will occur at the roughening temperature, and that above that temperature the grain surface (vs. fillets and three-grain boundaries) will have a greater relative contribution to the surface-chemical behavior of the grain. Ellipsometric studies of the ice surface suggest that the ice-liquid-water interface changes from smooth to rough at a temperature of -4 to -2°C, depending on facet [*Furukawa et al., 1987*].

Golecki and Jaccard [*1978*], using proton channeling, found an intrinsically disordered surface region on the basal plane of ice that increases rapidly with temperature above -40°C. They noted that partial order is retained up to at least -1.8°C, however. Results suggest an equivalent thickness for a surface *amorphous* layer of ~100 nm at -1°C, dropping to near zero below -40°C. Various other authors have proposed liquid-like interfacial regions, with thicknesses as high as 100 nm at -1°C and 30 nm at -10°C [*Gubler, 1982*]. Values as low as 2.5-14 nm at -1°C have also been suggested [*Gilpin, 1979;1980*].

Ionic strengths in the surface liquid film may be quite high owing to the exclusion of charged species from the solid ice. For example, at a liquid-water mass fraction of 0.001 and a bulk (melted) snow ionic strength of 0.001 M, the liquid-vein water will have an equilibrium ionic strength approaching 1.0 M. If the temperature of the snow is lowered sufficiently, liquid-water content decreases and salts should precipitate out. Concentration of sulfuric acid at triple-grain boundaries, observed in polar ice [*Mulvaney et al., 1988*] results in locally high ionic strengths. In snow the phenomenon is a result of melt-freeze action; but there was no evidence for melt-freeze action in the ice-core sample. Concentration of species at ice grain boundaries is thought to be a general phenomenon, and will influence studies such as interpretation of conductivity measurements of ice cores [*Schwander et al., 1983; Legrand et al., 1987*], and interpretation of gas [*Schwander, 1989*] and ion [*Davidson et al., 1989*] concentrations in ice cores.

There is also indirect chemical evidence for an aqueous phase at the ice-air interface, though thorough, systematic investigations are yet to be done. Adsorption of CO_2 onto ice in the range of -35 to -10°C has been observed to increase with temperature [*Ocampo and Klinger, 1982*], possibly due to trapping of CO_2 molecules at the surface by mobile surface water molecules (i.e. a surface liquid-like layer). This is consistent with the observed increase in surface roughening for crystals grown above about -10°C [*Colbeck, 1985*]. At -196°C, the surface area for CO_2 adsorption is smaller than that for N_2, but CO_2 adsorption at -40°C gives a surface area equal to that for N_2 at -196°C [*Ocampo and Klinger, 1983*]. This could be associated with the greater ability of CO_2 molecules to hydrate.

Adsorption of n-alkanes onto ice at temperatures in the range of -35 to -20°C exhibit behavior similar to that for adsorption on liquid water [*Orem and Abramson, 1969*]; at lower temperatures (-96 to -35°C), adsorption energies were low, consistent with simple physical adsorption. Several observations of adsorption/absorption onto ice and snow have been made by Russian scientists. From their results it was suggested that dissolution of formic, acetic, and chloroacetic acids in a liquid-like layer occurs at temperatures from -10 to -2°C [*Nechaev et al., 1981*]. Dissolution was less than in bulk water, and it was suggested that the liquid-like layer had a structure that was intermediate between bulk water and ice. Preliminary experiments on SO_2 adsorption on ice spheres by *Sommerfeld and Lamb* [*1986*] showed an increase in adsorption in going from -11 to -3°C (consistent with gaseous absorption by liquid), but no change in going from - 30 to -11°C. This result suggests that there is little change in surface roughening below -11°C, and that the roughening temperature lies between -3 and -11°C. *Valdez et al.* [*1988*] observed that SO_2 uptake onto growing ice surfaces at -15°C was in amounts equivalent to Henry's law equilibrium between the gas phase and deposited H_2O. This is consistent with uptake into a 10-100 nm liquid-like layer at the growing interface, with SO_2 release inhibited by diffusion in the layer. At equilibrium, SO_2 should be largely excluded from the ice.

In cases where there is *liquid* water in the ice-air interfacial region, the electrical properties of the solid-liquid interface may exert an important influence on chemical reactions. There is evidence that the solid-liquid interface on ice is negatively charged at the acidic pH's expected in natural snow. For example, *Nechaev and Ivanov* [*1974*] report a pH of zero charge of 6.5 for snow at -4.0°C. The electrical double layer should contain an excess of positive versus negative counterions; that is, it should exclude anions. Models for the electrical double layer on oxide surfaces are well developed [*e.g. Stumm and Morgan, 1981*], and should apply to ice as well. The same potentiometric and electrophoretic methods should apply to develop data for modeling. However, key surface parameters such as dielectric constant [*James & Parks, 1982*] and viscosity [*Clifford, 1975*] differ in thin films versus bulk water.

3. GASEOUS UPTAKE BY SNOW

Two important applications for modeling the uptake and reaction of gases at ice/snow surfaces are: i) the dry deposition of acidic species to snowpacks and ii) the uptake and redistribution of volatile species in snow/ice. Modeling could also be applied to fluxes of gases into snow from the underlying soil. Acid deposition is of particular importance in seasonally snow-covered regions that accumulate chemicals for several months during the winter and spring, then melt during a short period of time in the spring and summer. Deposition and redistribution of volatile species is quite important in interpreting polar and temperate-region ice cores. The models that follow the existence of a liquid-like surface on ice at temperatures near 0°C, based on the evidence described above.

Gaseous deposition and uptake of highly soluble gases such as HNO_3, HCl and H_2O_2 should be limited by atmospheric transport, and there should be little diffusion of the species into the snowpack. Essentially all of the uptake should occur in the surface layers. For less-soluble, reactive gases such as SO_2, O_2, O_3, organic acids, aldehydes and other organic gases, uptake will depend on both transport into the pack and uptake at the surface of snow/ice grains. For these gases, aqueous-phase reactions should dominate over gas-phase reactions. Oxides of nitrogen present additional modeling difficulties, as gas-phase reactions are important.

3.1. GENERAL MODEL FOR GASEOUS TANSPORT IN SNOW

One-dimensional transport of a non-decaying chemical species in porous media with a mobile air phase and immobile water and ice phases can be described by:

$$(1-r_{wa})\theta\frac{\partial C_a}{\partial t}+r_{wa}\,\theta\frac{\partial C_w}{\partial t}+(1-\theta)\frac{\partial C_I}{\partial t}=(1-r_{wa})\theta D\frac{\partial^2 C_a}{\partial x^2}-(1-r_{wa})\theta u_a\frac{\partial C_a}{\partial x} \tag{1}$$

where C_a, C_w and C_I are the solute concentrations in the air (mol $m^{-3}{}_a$), water (mol $m^{-3}{}_w$) and ice (mol $m^{-3}{}_I$), respectively; θ is the porosity (volume fraction air + water); D is the dispersion coefficient ($m^2\,s^{-1}$); and u_a is the average linear velocity of air ($m\,s^{-1}$). Note that D includes the combined effects of mechanical dispersion and molecular diffusion:

$$D=\alpha_d\,u_a^b+\tau D_m \tag{2}$$

where α_d and b are constants, D_m is the molecular diffusion coefficient of the species of interest and τ is a tortousity factor. Mass transfer between air and (liquid) water is described by:

$$\frac{\partial C_w}{\partial t}=\alpha_w\,(\frac{C_a}{H}-C_w) \tag{3}$$

and transport between water and the solid ice is described by:

$$\frac{\partial C_I}{\partial t}=\alpha_I\,(C_w K_I-C_I) \tag{4}$$

where α_w and α_I are mass-transfer coefficients (s^{-1}); H is a Henry's coefficient ($m^3{}_w m^{-3}{}_a$); and K_I is an ice-water partition coefficient ($m^3{}_w m^{-3}{}_I$). At equilibrium:

$$C_a = H\,C_w \tag{5}$$

and:

$$C_I = K_I\,C_w \tag{6}$$

In the absence of liquid water at the surface, equation (5) should be replaced by an adsorption isotherm. There are analytical solutions for equation (1) with one rate-limiting step, i.e. either (3) and (6) or (4) and (5), under certain boundary conditions. Several numerical solutions have also been developed.

Figure 2a illustrates the time scales for diffusive penetration of a moderately soluble gas into *dry* snow, using equations (1) and (5), with $u_a = 0$ and $K_I = 0$. Note that after one to ten years, penetration is several meters. An H value of 1.45 $m^3{}_w m^{-3}{}_a$ was used for the calculations, which is representative of gases such as CO_2 (acidic conditions) and O_3. For *moderately soluble* species and over *long time scales*, the r_{wa} is of limited importance. (Figure 2b). Note that after 100 yr there is little difference in the diffusive penetration for 100-fold differences in r_{wa}.

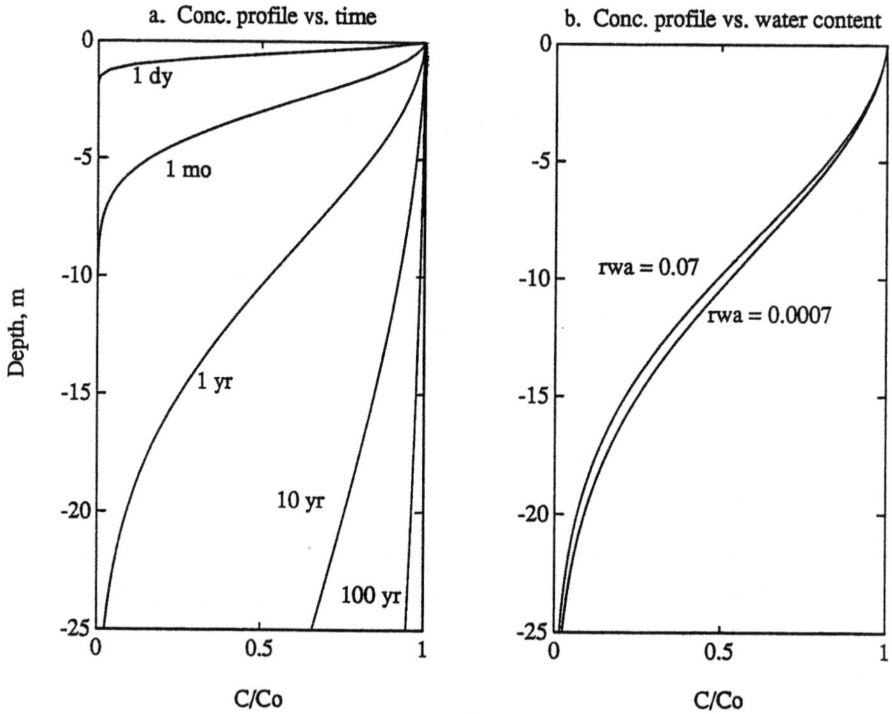

Figure 2. Diffusive transport of a sorbing gas into ice/snow. $D_m\,\tau = 1.4\times10^{-5}\,m^2\,s^{-1}$. a) penetration for $r_{aw} = 0.0007$, for times of 1 dy, 1 mo, 1 yr, 10 yr and 100 yr; and b) penetration after 100 yr, for $r_{aw} = 0.07$ and 0.0007. $H = 1.45\ m^3{}_a m^{-3}{}_w$.

3.2. Dry Deposition to Snow

Surprisingly little work has been done on the dry deposition of gases to snowpacks, with most being done with SO_2. Recent reviews have compared several reported measurements of dry deposition to snow [*Davidson, 1989*] and surfaces in general [*Voldner, 1986*], and discussed the importance of atmospheric versus surface resistances. The following example of SO_2 uptake is based on the model of *Bales et al.* [*1987*].

The simplest traditional formulation of dry deposition when applied to a snow surface is illustrated in Figure 3a. The flux of gas between some reference level and an absorbing interface (mass sink) is constant with vertical distance:

$$F = (D_m + K)\frac{dC_a}{dz} \tag{7}$$

where D_m and K are molecular and eddy diffusivity, respectively, and C_a is gas concentration. Between z_r and z_d, the resistance to gas transfer is determined primarily by eddy transport, and is termed the aerodynamic resistance, r_a. Between z_0 and the plane of final gas absorption, there is a diffusional resistance and a true surface-reaction resistance, the relative contributions of which are often not separated. This combined resistance is here referred to as the surface resistance, r_s. The total resistance to transfer is therefore:

$$r = r_a + r_s \tag{8}$$

By definition, deposition velocity is related to flux and reference-level concentration:

$$v_d = (r_a + r_s)^{-1} = \frac{F}{C_a(z_r) - C_a(z_d)} \tag{9}$$

Under most conditions, deposition of SO_2 to snow is dominated by the surface resistance [*Valdez et al.*, 1987].

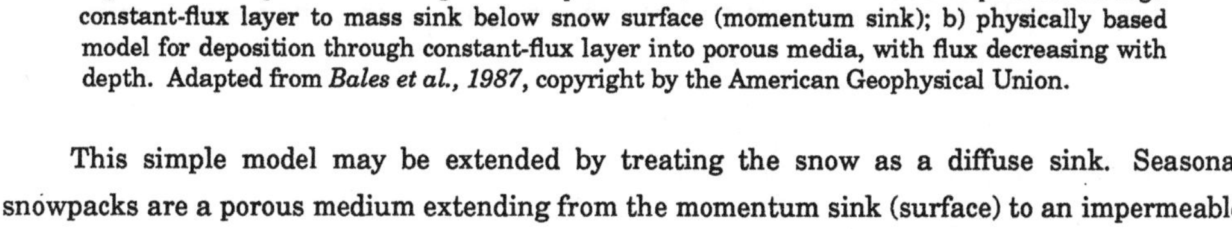

Figure 3. Conceptual model of gaseous deposition: a) traditional model for deposition through constant-flux layer to mass sink below snow surface (momentum sink); b) physically based model for deposition through constant-flux layer into porous media, with flux decreasing with depth. Adapted from *Bales et al., 1987*, copyright by the American Geophysical Union.

This simple model may be extended by treating the snow as a diffuse sink. Seasonal snowpacks are a porous medium extending from the momentum sink (surface) to an impermeable boundary (ground), as shown on Figure 3b. In shallow snow, the lower boundary will have a major effect on uptake. The constant-flux layer extends from the reference level to the snowpack surface only, beneath which flux decreases with depth.

The uptake of SO_2 within a snowpack can be described by a model incorporating diffusion, air-water partitioning and aqueous-phase reaction. This is, therefore, a *wet-snow* model that requires the existence of either bulk water or liquid-like surface layers on the snow grains. Partitioning to the snow depends on the liquid-water-to-air volume ratio (r_{wa}), given by:

$$r_{wa} = \frac{\rho_b X_m}{1 - \frac{\rho_b}{\rho_I}(1 - X_m) - \rho_b X_m} \tag{10}$$

where ρ_β and ρ_I are the bulk-snow and ice specific gravities, respectively, and X_m is liquid-water mass fraction. In very cold snow, with liquid water present as only a thin film on snow grains, X_m (and hence r_{wa}) will depend on specific surface area. The snowpack is taken to be isothermal and homogeneous.

Partitioning of SO_2 into the liquid phase is assumed to obey Henry's law:

$$C_4 = \frac{C_a}{\alpha_0 H} \tag{11}$$

where C_4 is the total aqueous concentration of S(IV) (mol m^{-3}) and α_0 is the first ionization fraction of SO_2(aq) (see Table 1). At any point within the snow, mass conservation of total sulfur requires that the change in gaseous SO_2 concentration plus the change in aqueous S(IV) and S(VI) concentrations with respect to time must be equal to the vertical diffusive transport of SO_2 gas. Equation (1) reduces to:

$$(1 - r_{wa})\theta \frac{\partial C_a}{\partial t} + r_{wa}\,\theta \left[\frac{\partial C_4}{\partial t} + \frac{\partial C_6}{\partial t} \right] = \theta D \frac{\partial^2 C_a}{\partial z^2} \tag{12}$$

The changes in aqueous concentrations of S(IV) and S(VI) species are given by:

$$\frac{\partial C_4}{\partial t} = \frac{1}{H} \frac{\partial}{\partial t} \left[\frac{C_a}{\alpha_0} \right] \tag{13}$$

and:

$$\frac{\partial C_6}{\partial t} = k' C_4 \tag{14}$$

where k' is a (s^{-1}) pseudo-first-order rate constant for S(IV) destruction by oxidation. The value for k' depends on pH, temperature and oxidant concentration. Substituting and rearranging gives:

$$\frac{\partial C_a}{\partial t} + \frac{k' r_{wa} C_a}{H \alpha_0 + r_{wa}} = D \left[1 + \frac{r_{wa}}{H \alpha_0} \right]^{-1} \frac{\partial^2 C_a}{\partial z^2} \tag{15}$$

The accumulated concentrations of S(VI) above pre-existing background may be determined from an integration from $t = 0$ to $\bar{t}$, the time of exposure to SO_2:

$$C_6 = \frac{1}{H} \int_0^{\bar{t}} \frac{k' C}{\alpha_0} dt \tag{16}$$

Table 1. Empirical Rate Laws for S(IV) → S(VI)[a]

Oxidation by hydrogen peroxide

$$-\frac{dC_4}{dt} = \frac{k_1[H_2O_2]C_4\alpha_1}{1+K[H^+]} \qquad (1)$$

Oxidation by ozone

$$-\frac{dC_4}{dt} = (k_0\alpha_0 + k_1\alpha_1 + k_2\alpha_2)C_4[O_3] \qquad (2)$$

Oxidation by oxygen

$$-\frac{dC_4}{dt} = k_1[\text{Fe(III)}]C_4\alpha_2 \qquad (3)$$

$$-\frac{dC_4}{dt} = k_2[\text{Mn(II)}]C_4\alpha_1 \qquad (4)$$

Ionization fractions

$$\alpha_0 = \frac{[SO_2(aq)]}{C_4} = \left[1 + \frac{K_{a1}}{[H^+]} + \frac{K_{a1}K_{a2}}{[H^+]^2}\right]^{-1} \qquad (5)$$

$$\alpha_1 = \frac{[HSO_3^{-}]}{C_4} = \left[\frac{[H^+]}{K_{a1}} + 1 + \frac{K_{a2}}{[H^+]}\right]^{-1} \qquad (6)$$

$$\alpha_2 = \frac{[SO_3^{2-}]}{C_4} = \left[\frac{[H^+]^2}{K_{a1}K_{a2}} + \frac{[H^+]}{K_{a2}} + 1\right]^{-1} \qquad (7)$$

[a]From *Hoffmann and Calvert,* 1985.

Expressions for the surface deposition velocity of SO_2 may thus be obtained by integration over the depth of the pack, $\bar{z}$. They give the total mass of sulfur accumulated at any depth up to time $\bar{t}$, divided by the SO_2 concentration above the snowpack, C_{a0}. Total surface deposition velocity is the sum of the value for S(IV):

$$v_4 = \frac{X_m \rho_B}{H \bar{t} C_{a0}} \int_0^{\bar{z}} \frac{C_a}{\alpha_0} dz \qquad (17)$$

and that for S(VI):

$$v_6 = \frac{X_m \rho_B}{H \bar{t} C_{a0}} \int_0^{\bar{z}}\int_0^{\bar{t}} \frac{k' C_a}{\alpha_0} dt\, dz \qquad (18)$$

When there is no oxidation, i.e., $k' = 0$, and if changes if pH are negligible, a simple analytical solution to equation (15) may be obtained [*Bales et al., 1987*]. Results for the no-oxidation would be similar to that illustrated on Figure 2. The deposition velocity of an unreactive species partitioning into liquid water in snow is linearly dependent on the amount of liquid water present and the solubility and is inversely proportional to the exposure time. For the more general case numerical procedures must be used.

The chemical reactions involved in the uptake, acid dissociation, and oxidation of SO_2 in the liquid phase of snow are presented in Table 2. The modeling approach is similar to that used for modeling aqueous-phase reactions in fogwater and hydrometeors [*Jacob and Hoffmann, 1983; Hoffmann and Jacob, 1984*]. Reactions (R1) - (R6) will all exert some control on the pH of the liquid phase in the snow. Below an SO_2 concentration of about 0.01 ppbv, atmospheric CO_2 gives an equilibrium pH at 0°C of close to 5.6; at SO_2 = 1.0 ppb, pH = 5.0; at SO_2 = 10 ppb, pH = 4.6.

Table 2. Chemical Reactions

No.	Reaction	Constant
R1	$SO_2(g) + H_2O = SO_2(aq)$	H
R2	$SO_2(aq) = HSO_3^- + H^+$	K_{a1}
R3	$HSO_3^- = SO_3^{2-} + H^+$	K_{a2}
R4	$S(IV) \rightarrow SO_4^{2-}$	See Table 1
R5	$CO_2(g) + H_2O = H_2CO_3^*$	K_{a1}
R6	$H_2CO_3^* = HCO_3^- + H^+$	K_{a2}
R7	$H_2O = H^+ + OH^-$	K_w
R8	$H_2O_2(g) = H_2O_2(aq)$	H
R9	$O_3(g) = O_3(aq)$	H
R10	am–$Fe(OH)_3(s) = Fe^{3+} + 3\,OH^-$	K_{s0}
R11	$Fe(OH)_2^+ = Fe^{3+} + 2\,OH^-$	β_2
R12	$FeOH^+ = Fe^{3+} + OH^-$	β_1

Reactions (R1-R3) and (R5-R7) are fast relative to the oxidation of S(IV) to S(VI), and can thus be expressed by equilibrium relations. Only reaction (R4), which is known to proceed by several pathways, requires a kinetic description. Table 1 lists some of the empirical rate laws that are thought to be important in snow.

Oxygen is abundant in snowpacks, and may be the primary oxidant at higher pH if a suitable metal catalyst such as iron, manganese, or copper is present. Hydrogen peroxide levels as high as 200 - 400 ppb (6 - 12 μM) have been observed in Greenland ice cores [*Neftel et al.*, 1984]; high concentrations have also been found at various depths in profiles taken from seasonal snow covers in Switzerland [*Sigg et al., 1987*]. and the Western U.S. [*M. Conklin, U. Arizona, unpublished data*]. The distribution of H_2O_2 in snow crystals is unknown, but metamorphism should result in it reaching an equilibrium distribution in snow grains. Freeze/thaw cycles should tend to concentrate H_2O_2 at the grain surface, where it could be lost to meltwater. Ozone is ubiquitous and will diffuse into snowpacks as a result of concentration gradients; in so doing, it can react with HO_2, OH and other species to produce free radicals and other decay products, in addition to reacting with aqueous S(IV). There is ample evidence of O_3 uptake by snow [*e.g., Aldaz, 1969; Galbally and Roy, 1980*], though most data resulted from the addition of extra O_3 above a non-conditioned surface. In the following analysis, O_3 and H_2O_2 decay have been neglected.

The strong dependence of v_d on r_{wa} is illustrated on Figure 4, which shows model calculations for three different oxidants. A ten-fold change in r_{wa} results in a 3.2-fold change in v_d. The experimental v_d values and profiles of SO_2 uptake versus depth on Figure 5 illustrate this dependence on r_{wa}. The dual limitations of gaseous diffusion and surface uptake/reaction can also be seen from the profiles of aqueous S(IV) and S(VI) on Figure 5. Uptake initially occurs near the surface, with the gas penetrating further into the snowpack as surface layers become more acidic.

Model calculations show the same uptake pattern as do experiments, with concentrations declining with depth (Figure 6). Because r_{wa} was not measured in the experiments, it is an adjustable parameter in the model.

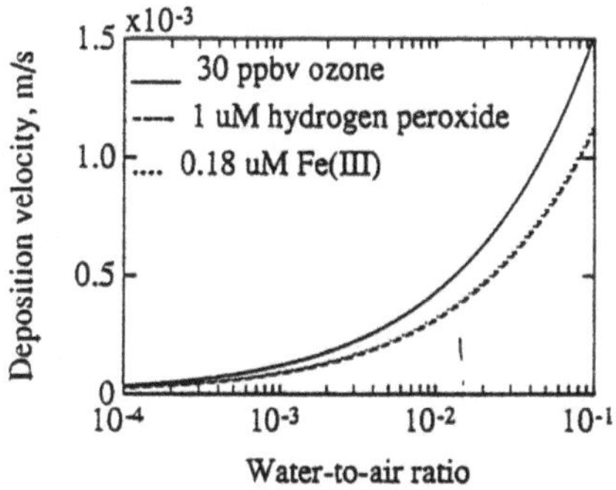

Figure 4. SO_2 deposition velocity as a function of r_{wa} for three different oxidants. SO_2 = 20 ppbv, $\bar{t}$ = 6 hr and ρ_b = 0.4. Adapted from *Bales et al., 1987*, copyright by the American Geophysical Union.

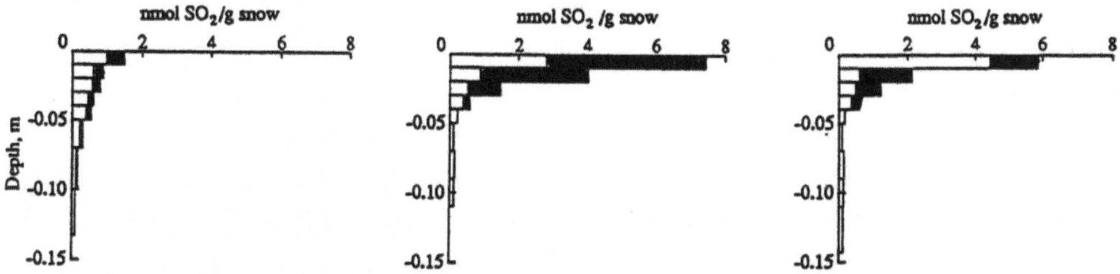

Figure 5. SO_2 concentration versus depth for three experiments [*Valdez et al., 1987*]. Adapted from *Bales et al., 1987*, copyright by the American Geophysical Union.

	a)	b)	c)
T, °C	-2.2	0.0	0.0
SO_2, ppbv	81	70	79
age, dy	112	9	10
ρ_s, $kg\,m^{-3}$	400	380	400
v_d, $10^{-4} m\,s^{-1}$	2.0	9.3	5.9

For an acidic species such as SO_2, v_d declines with time due to a drop in pH as the acid accumulates in the aqueous phase. The S(IV) → S(VI) oxidation rate also declines with pH. Figure 7 illustrates the decline in v_d over a 20-hr uptake period. For all three oxidants modeled, a ten-fold increase in time results in a 2.5–3.2-fold decline in v_d. Experimental results show this same magnitude of drop in v_d with time [*Conklin, 1990*].

Model calculations of both v_d values and uptake patterns show good agreement with experiments. A model r_{wa} value in the range of 0.01 gave good agreement with experiments at -2°C. Further, the range of v_d values observed suggests that surface resistance, which is limited by both diffusion into the snowpack and chemical reaction at the ice-water-air interface, rather than

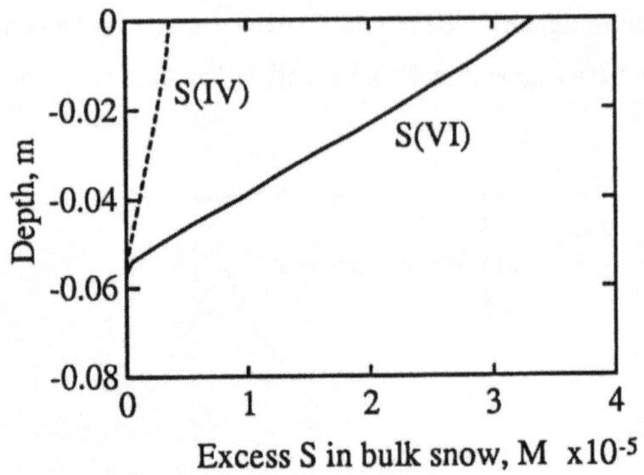

Figure 6. Profiles of S(IV) and S(VI) uptake in the presence of Fe(III) catalyst. SO_2 = 20 ppbv, r_{wa} = 0.0071, $\bar{t}$ = 6 hr and ρ_b = 0.4, Fe(III) = 0.018 μM.

atmospheric resistance is generally more important for SO_2 uptake by seasonal snowpacks. This range of v_d's and the strong dependence on r_{wa} was also observed in a field study in northern Michigan [*Cadle et al., 1985*]. From direct measurements of uptake in the snow, an average SO_2 deposition velocity of 1.5×10^{-3}m s^{-1} was reported for snow with an appreciable exposure time above -3°C, and 6×10^{-4}m s^{-1} for samples that remained below -3°C. On the other hand, SO_2 deposition velocities calculated from atmospheric measurements are generally higher than the values from snowpack measurements, and do not directly consider the snowpack properties. Reported deposition velocities are as low as 0.5–1.6×10^{-3}m s^{-1} [*Whelpdale and Shaw, 1974; Dovland and Eliassen, 1976; Barrie and Walmsley, 1978*]. *Granat and Johansson* [*1983*], using a flow-through chamber method, determined deposition from the difference between inlet and outlet concentrations and found an uptake of SO_2 corresponding to a $v_d = 1\times10^{-3}$m s^{-1}. Deposition of NO_2 and NO was negligible.

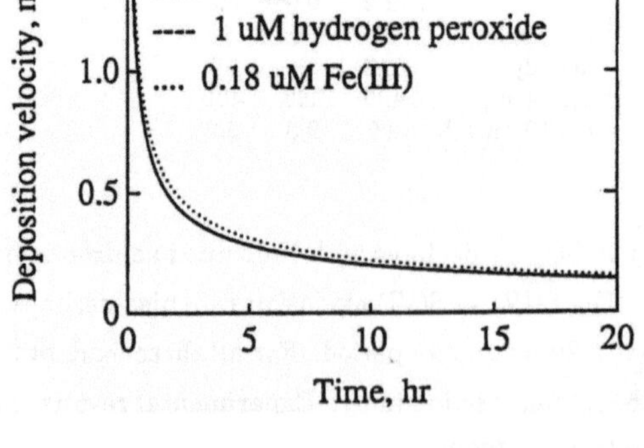

Figure 7. SO_2 deposition velocity as a function of time for three different oxidants. SO_2 = 20 ppbv, r_{wa} = 0.0071, ρ_b = 0.4. Adapted from *Bales et al., 1987*, copyright by the American Geophysical Union.

Based on the success to date with SO_2, an explicit model of this sort described could be applied to other gases. *Johansson and Granat* [*1986*] measured HNO_3 deposition to snow and found deposition velocity to be strongly temperature dependent between -18 and -2°C. *Huebert et al* [*1983*] found HNO_3 to be rapidly scavenged by falling snow. One preliminary report suggests that SO_2, NO_2 and O_3 are not strongly adsorbed by ice at -15°C, whereas HNO_3 and H_2O_2 are largely taken up [*Lee and Kinsley, 1986*]. *Colbeck and Harrison* [*1985*] measured O_3 dry deposition to grassland using micrometerological methods and found it to be inhibited by the presence of snow, but atmospheric stability also changed the measurements.

3.3. Redistribution in Ice Cores

A general model of the type described by equations (1) – (6) can also be applied to problems such as gas and solute trapping in ice cores. Snow and ice on the Earth's surface provide a record of chemical and physical conditions in the atmosphere over time scales ranging from single storms to thousands of years. Interpreting the climatic record from ice and snow requires viewing snow and glacial ice as a dynamic rather than static medium. Changes in its chemical composition result from air and energy exchanges with the atmosphere, gaseous diffusion in the pack, meltwater flux in the pack, and chemical reactions at the ice-air surface. The magnitude and importance of these processes must be understood over the time scale of interest. This is particularly true for measurements with high spatial resolution and short time scales.

The air trapped in glacial ice differs in age from the ice enclosing it and in chemical composition from the air at the times of snowfall and pore close-off. Gas exchange with the atmosphere is controlled by the physical properties of the snow and its pore space. Chemical interactions at the ice-air interface enhance retention at near-surface layers and retard diffusion of gases to deeper layers. While the process of gas mixing over depth is primarily caused by molecular diffusion, short-term atmospheric pressure changes produced by wind may force air exchange between the void volume of the snow and the atmosphere. For example, *Colbeck*, [*1988*] has shown that pressure pumping from surface winds can disrupt the thermal regime in snowpacks.

Unfortunately, our present understanding of many of these processes is insufficient to the uptake and redistribution of chemical species over long (>> hrs) time scales. Further work is needed in order to estimate both transport and chemical parameters for the temperatures of interest. In polar ice, the cold temperatures means that little or no liquid water is present, which should simplify chemical descriptions. Subtle changes over long time scales places additional burdens on modeling, however.

3.4. Ice-Water and Ice-Air Partitioning

It is generally assumed that during metamorphism, solutes are redistributed to the grain boundaries and exteriors. One parameter needed for modeling that has received some special attention is the water-ice, (and air-ice) partition coefficient. The distribution of oxidants, salts, and other species in both *wet* and *dry* snow is important for understanding the chemical behavior of snow and ice surfaces.

During rapid formation of ice crystals and snow grains, impurities should be distributed throughout the grain; slower growth or metamorphism should result in concentration of species near the surface. This is because the presence of an impurity in a crystal lattice causes a strain greater than that caused by its presence on the grain boundary, which is disordered compared to the interior of the crystal. Thus an interaction energy drives solutes in growing crystals toward the boundary. Preliminary results suggest that even freezing in liquid N_2 is *slow* and results in grain-scale concentration gradients [*Peterson, 1990*].

Fractionation of chemical species in ice and snow depends on particulate dissolution, complexation in the surface phase, and the efficiency of species exclusion from the solid ice during metamorphism. Time-lapse photographs in fact suggest that impurity pockets, or particles can migrate across the surfaces of ice spheres [*Kuroiwa, 1974*].

Relatively few measurements have been made of distribution coefficients, K_I, for the ice/water system. Measured values are generally small. The equilibrium distribtion of alkali-metal fluorides and chlorides gives K_I values on the order of $10^{-4} - 10^{-3}$ [*Gross, 1968; Gross et al., 1977*], but depend on counterion concentration. NH_4F has a K_I of 0.13 [*Gross et al., 1975*], and antifreeze glycoprotein has a reported K_I of 0.8 [*Gross et al., 1987*]. Available results of incorporation at an advancing interface demonstrate that diffusion does not play an important role in regulating the incorporation of ions into the ice; K_I bears no relation to the diffusion coefficient [*Franks, 1982*]. The reported K_I for H_2O_2 is only on the order of 0.01 [*Sigg et al, 1987*], which is somewhat surprising given the molecule's similarity to H_2O. It should also be noted that K_I can depend on freezing rate [*e.g. Smith et al., 1955; Leung and Carmichael, 1984*].

4. Solute Movement with Percolating Meltwater

Solute release from melting snow can be modeled by simple curve fitting, which is very useful when a description of meltwater chemistry is to be included into a larger watershed model. A single-term or two-term exponential function has proven to be very versitile in this regard. Alternately, use of a model is based on first principles provides greater flexibility in applying common model parameters to a wider variety of conditions. Models of this sort are particularly useful in examining the behavior of different solutes, and in comparing the effects of different melt patterns and initial conditions.

One should be able to model movement of major ions such as Na^+, K^+, Ca^{2+}, Mg^{2+}, Cl^-, NO_3^-, SO_4^{2-} independent of each other. NH_4^+ and low-molecular-weight organic acids (i.e. formic, acetic) should also be conservative in the absence of other organic matter. The natural organic matter present in snow exerts an influence over the cation composition of meltwater, and may contribute to the poor cation-anion balances observed by some surveys in forested areas. Natural organic matter in the liquid-water fraction should complex with cations, could also react with oxidants, and could thus influence chemical transport and transformations. It is also plausible that early meltwater fractions are enriched in dissolved and particulate natural organic matter, and thus affect transport of other species out of the snowpack. For trace metals such as Fe, Mn, and Cu, one needs to consider hydrolysis and precipitation. H^+ presents further problems, as it cannot be modeled independently; it must be modeled together with other ions. No effort has been made to model particulate movement in unsaturated porous media. Snowpack particulates can be important sources of dissolved ions, however.

Ionic concentrations in the first fractions of meltwater are often two to six times the average values found when 20-30 percent has melted. It is hypothesized that this occurs due to the unequal distribution of chemical species in snow, both at the *grain* and *snowpack* scales. Dry deposition, changes in atmospheric scavenging with time, freeze-thaw-drain cycles, rainfall, etc. could all explain the unequal *snowpack*-scale distribution. Ice-water partitioning, changes during metamorphism, scavenging versus in-cloud species incorporation, dry deposition, colloidal material, etc. could explain the unequal *snow-grain*-scale distributions.

4.1. Physically Based Model for Ion Release from a Snowpack

Principles of water flow through homogeneous snow [*Colbeck, 1978*], together with knowledge of the snow chemical composition, can in principle be applied to predict the composition of meltwater arriving at the base of a snowpack. This can be further coupled with an energy-balance or heat-flux model to predict the generation of meltwater within the snowpack [*e.g. Morris, 1987; Marks, 1987*]. Factors such as fractures or channels (in saturated snow) provide short circuiting of meltwater response. Ice lenses and other nonhomogeneous features are the rule rather than the exception in snowpacks, and exert a significant effect on meltwater flow [*e.g. Marsh and Woo, 1985*].

The following mathematical development is an extension of the work of *Hibberd* [*1984*], which is based on the one-dimensional theory for water percolation through a homogeneous snowpack [*Colbeck, 1978*], and the work of various soil scientists [*e.g. Smiles et al., 1978*]. The following development includes the retention of solute species in the immobile liquid or ice phases of the snowpack. *Hibberd's* work considered only advection and dispersion.

It is assumed that an initially homogeneous, dry, isothermal (0°C) snowpack melts at the surface at a uniform rate. The volume flux of meltwater (m s^{-1}) is described by:

$$Q = k\, S^{n} \tag{19}$$

where S is the effective water saturation of the pack (behind a meltwater wavefront) and k (m s^{-1}) is a constant [*Colbeck and Davidson, 1973*]. Effective saturation is:

$$S = \frac{S_w - S_i}{1 - S_i} \tag{20}$$

where S_w is the total and S_i the irreducible water saturation ($m^3{}_w\, m^{-3}{}_{a+w}$). Note that S_w is dimensionally the same as r_{wa}. The exponent n is typically assigned a value of 3.0 [*Colbeck, 1978*]. A saturation wavefront moves downward through the pack with a velocity v^* (m s^{-1}):

$$v^* = \frac{S^{n-1}\, k}{\theta(1 - S_i)} \tag{21}$$

This saturation wavefront constitutes a moving discontinuity separating the effective saturation level from the initial saturation value of zero; conservation of mass must be satisfied across the wavefront. This criterion gives the percolation velocity at the wavefront:

$$u_w = \frac{k}{\theta(1-S_i)} \frac{S^n}{S+\beta} \tag{22}$$

where $\beta = \dfrac{S_i}{1-S_i}$.

Solutes in the pack are assumed to be in one of three phases, mobile water, immobile water or ice. A material balance on solute in the pack is:

$$(1-\theta)\frac{\partial C_I}{\partial t} + f\,\theta S_w \frac{\partial C_{mw}}{\partial t} + (1-f)\theta S_w \frac{\partial C_{iw}}{\partial t} = \frac{\partial}{\partial z}\left[f\,\theta S_w D \frac{\partial C_{mw}}{\partial z}\right] - f\,\theta S_w u_w \frac{\partial C_{mw}}{\partial z} \tag{23}$$

where C_I, C_{mw} and C_{iw} are the solute concentrations in the ice ($\mathrm{mol\,m^{-3}}_I$), mobile water ($\mathrm{mol\,m^{-3}}_{iw}$) and immobile water ($\mathrm{mol\,m^{-3}}_{mw}$), respectively; f is the fraction mobile water ($\mathrm{m^3}_{mw}\mathrm{m^{-3}}_{mw+iw}$); and D is the longitudinal dispersion coefficient ($\mathrm{m^2\,s^{-1}}$).

Across the saturation wavefront, a solute mass balance must be achieved simultaneously with that for the liquid: Initial and boundary conditions must also be specified. For simple cases (i.e. analytical solutions), there appears to be little gained from using the more-involved fixed boundary condition, versus a simpler, semi-infinite formulation.

Three possible formulations for the C_{iw} term are developed: The first is linear partitioning:

$$\frac{\partial C_{iw}}{\partial t} = K_{im} \frac{\partial C_{mw}}{\partial t} \tag{24}$$

where K_{im} ($\mathrm{m^3}_{mw}\,\mathrm{m^{-3}}_{iw}$) is a partition coefficient. The second is a first-order kinetic release:

$$\frac{\partial C_{iw}}{\partial t} = \tau_1 C_{iw} \tag{25}$$

where τ_1 ($\mathrm{s^{-1}}$) is a pseudo-first-order rate coefficient (mass-transfer coefficient). The third is a kinetically limited reversible partitioning:

$$\frac{\partial C_{iw}}{\partial t} = \tau_2 (C_{mw} - C_{iw}) \tag{26}$$

where τ_2 ($\mathrm{s^{-1}}$) is a also pseudo-first-order rate coefficient (mass-transfer coefficient).

Using a linear equilibrium expression for ice-water partitioning (equation 6) and substituting into equation (23) eliminates one dependent variable. For the case of fast mobile-immobile-water partitioning (equation 24), the set reduces to a single equation. The analytical solution is straightforward for uniform initial conditions and constant velocity — i.e., after the wavefront reaches the bottom of the snowpack. Solutions are also available for solute redistribution in a moving wavefront [*Smiles and Gardiner, 1982*]. A numeral solution is needed in the general case of nonuniform initial conditions and slow transfer between mobile and immobile water.

The results of numerical calculations using a simple finite-difference code are shown on Figure 8 for the linear-partitioning case. Note the initial rise in C/C_0 at each depth, followed by a gradual drop to zero. The model mimics the behavior observed in intermediate-scale laboratory experiments (Figure 9). While the curves on Figure 8 are qualitatively consistent with observations of the flux of dissolved species in both small labaoratory reactors [*Tsiouris et al., 1985 Colbeck, 1981; Bales et al., 1989*] and in the field [*e.g. Johannessen and Henriksen, 1978; Tranter et al., 1986; Bales et al., 1989b*], there is currently an insufficient data base for validation. In particular, the K_{im}'s, D's and τ_i's are largely fitting parameters.

The movement of meltwater in a naturally layered snowcover is complicated by interaction of the wetting front with stratigraphic horizons. The infiltration of liquid water takes place in two modes: diffusion through the matrix and flow through macropores or fingers. Temporal variability between the movement of the background wetting front and the water in the flow fingers depends on the depth of the snow and the development of the macropores. The spatial variability in flow of meltwater to the base of the pack is related to ice lenses, with decreased flow beneath impermeable sections and increased flow below the more-permeable sections. Clearly, more work on modeling water flow through snow is needed. Given a model of snow melt and meltwater flux, the processes of chemical change can be described.

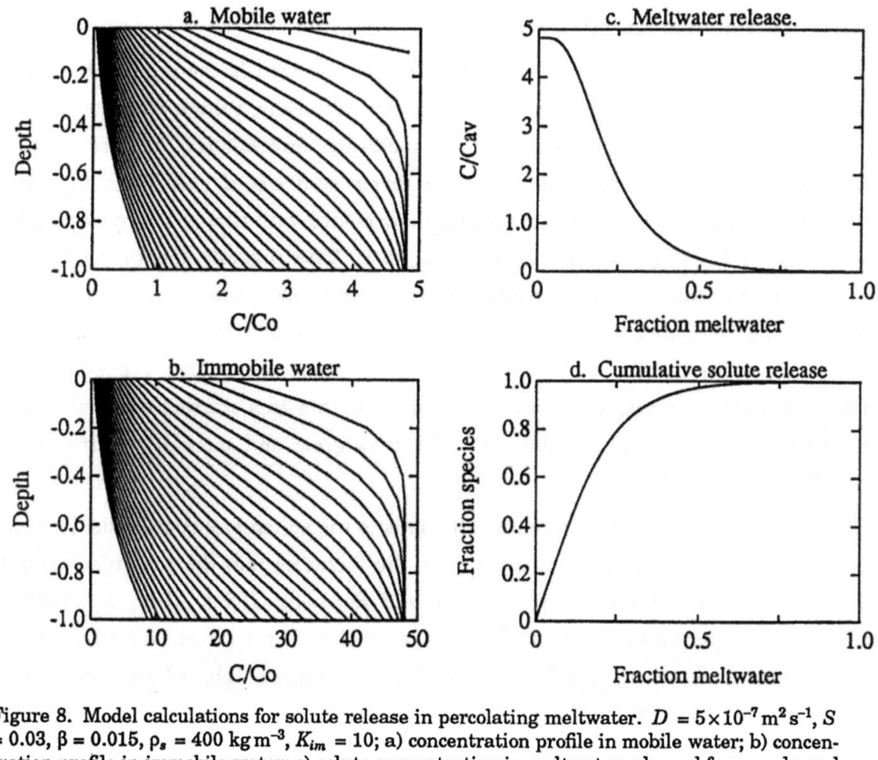

Figure 8. Model calculations for solute release in percolating meltwater. $D = 5\times10^{-7}\,\mathrm{m^2\,s^{-1}}$, $S = 0.03$, $\beta = 0.015$, $\rho_s = 400\,\mathrm{kg\,m^{-3}}$, $K_{im} = 10$; a) concentration profile in mobile water; b) concentration profile in immobile water; c) solute concentration in meltwater released from pack; and d) cumulative fraction of solute released from pack.

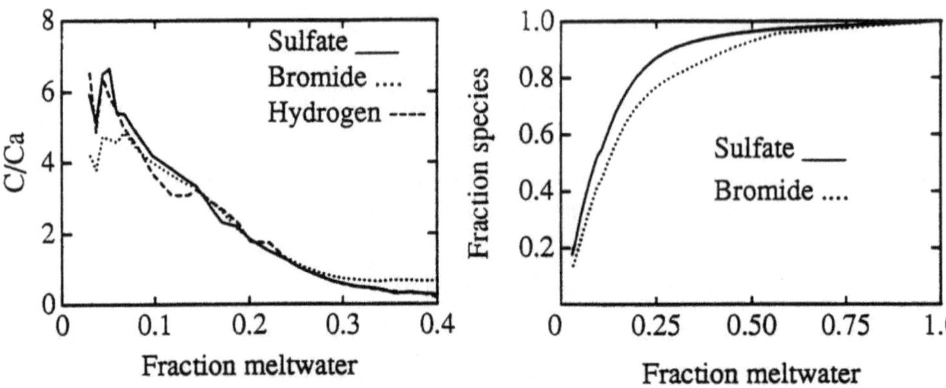

Figure 9. a) Ionic pulse for three species in snowmelt from 0.4-m pack in intermediate-scale experiment; Br^-, applied at mid-depth, lags slightly behind SO_4^{2-} and H^+, applied at the top of the pack. b) Mass fraction species collected as a function of meltwater collected in the same experiment; about 80 percent of the species applied at the top and 70–80 percent of the species applied at mid depth came out in the first 20 percent of the meltwater. Adapted from *Bales et al., 1989*, copyright by the American Geophysical Union.

4.2. Modeling Preferential Elution

A second observed, but relatively unexplained, phenomenon is the preferential elution of species in meltwater relative to bulk-snow composition. Typically, sulfate concentration is relatively higher than nitrate or chloride in the first meltwater fractions [*e.g. Tranter et al., 1986*]. Cations also exhibit this ionic pulse. For example, *Brimblecombe et al.* [*1987*] compared the ionic concentrations in the first 10 percent of natural snow melted with that of the remaining 90 percent and observed C_{10}/C_{90} ratios of approximately 2-8. Data suggest that the ratios for SO_4^{2-} were slightly higher than for NO_3^-, which were slightly higher than for Cl^-. In experiments involving melting laboratory ice at room temperature, there was little preferential elution; however initial ion concentrations in meltwater were still about three times the average [*Brimblecombe et al. 1987*]. That is, there was an ionic pulse even with a starting material of *assumed* uniform composition. Unequal distribution in the snow grain and snowpack contribute to this behavior [*Bales et al., 1989*].

One approach to modeling preferential elution is to use an ion-exclusion model to mimic the observed relative ion release. There is evidence in the literature that there is an excess of negative charge at the ice-water interface [*Nechaev and Romanov, 1981*]. Using a simple anion-exclusion model for constant u_w and an assumed f gives the pattern shown on Figure 10a. These calculations are qualitatively similar to the data on Figure 10b, which come from the intermediate-scale snowmelt experiment of Figure 9.

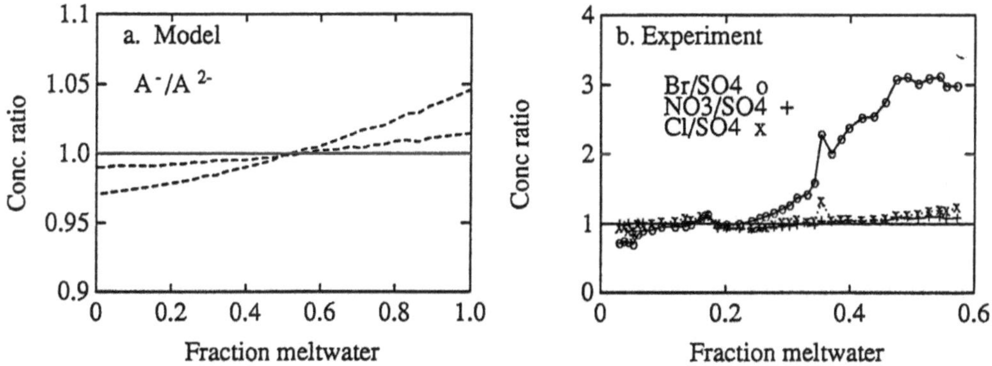

Figure 10. **a) Ionic ratios for monovalent versus divalent ion using an anion-exclusion model. b) Ionic ratios observed in intermediate-scale experiment shown on Figure 9.**

5. Summary and Research Needs for Modeling

The modeling approaches presented above seek to describe in-pack chemical transformations using statements of the physical and chemical phenomena that are based on first principles. The two examples given are for relatively simple physical systems, but illustrate the level of chemical modeling that can be applied to any physically well-defined situation.

One important application for this sort of detailed physical/chemical models is as components in integrated hydrochemical models for alpine watersheds. Models that accurately describe physical and chemical processes in seasonally snow-covered basins are essential tools for interpreting and predicting the responses of these areas to the combined effects of acidic deposition and global change. To model the ionic pulse that occurs during the first part of spring snowmelt, one needs to describe the *snowmelt* and *snowmelt-chemistry* hydrographs at different points in the watershed. The aggregation of these hydrographs from individual contributing areas then gives the whole-watershed response, which depends on melt patterns and initial conditions across the watershed.

Second, good chemical models will contribute to gaining an understanding of physical processes in snowpacks. For example, the evolution of concentration distributions in a snowpack undergoing metamorphism or melting indicates patterns of water and energy movements in the pack.

Third, physically accurate, validated models are needed as tools to aid in interpretation of snow and ice cores from both polar and temperate regions. Solute redistribution by gas and water movement occurs over time scales from days to centuries. For all but the coldest drilling sites, coupled air-flow and water-flow models are needed. One example is the redistribution and decay of species such as H_2O_2, which should show a strong annual cycle. Once incorporated into the snowpack, vertical distributions of H_2O_2 are smoothed out because of metamorphic processes

[*Sigg and Neftel, 1988*]. The concentration of H_2O_2 in surface layers is sometimes reduced as well, possibly by photolysis due to considerable exposure to the sun. Modeling this decay and redistribution will also require a description of light transmission in snow.

In general, better modeling of chemical processes in snow is limited by good *physical* models of the snowpack and by *data* for model testing. In the first example illustrated above, the chemical processes involved in SO_2 deposition to snow (e.g. reaction rates) are better understood than are the physical properties of the pack (e.g. r_{wa}). In the second example, the chemical model of solute redistribution was applied to a one-dimensional, homogeneous media; water flow in natural snow is generally not one dimensional nor is the media homogeneous.

In order to determine the influence of microscale versus macroscale properties, and to separate chemical from physical effects, it is necessary to investigate each separately, using both experiments and mathematical models. Given physically and chemically realistic models, a key step toward application is good parameter estimation. Model systems, i.e. controlled experiments, are essential for estimating chemical parameters. With the physics well defined, one can attribute observed changes to chemistry. Data from both model and natural systems is needed for model testing and validation.

Controlled experiments are also needed for key physical parameters. In the context of the models described above, further experimental work at the macroscale is needed to establish relations for intrinsic permeability, unsaturated hydraulic conductivity, dispersion and tortuosity. At the microscale there is a critical need for a better model of the ice-air interfacial region. Specifically, we need an understanding of the equivalent amount of liquid water present as a function of temperature and other properties. Ice-water partition coefficients and ice-air adsorption relations for some species are also needed.

Other areas of investigation in the earth and environmental sciences have common needs for a better understanding of chemical transformations in the surface region of ice. For example, formation of a solid solution of HCl and H_2O at 190-193°K has been reported [*Wofsy et al., 1988*], but the morphology and nature of the ice used in their experiments is not known. It was speculated that diffusion grain boundaries may have contributed significantly to the uptake [*Molina et al., 1987*]; grain boundaries may also have played a dominant role in the observed "incorporation" of HCl into the ice. A recent report in fact finds that HCl is not easily incorporated into ice crystals, but is strongly partitioned toward the grain boundaries [*Wolff et al., 1989*]. Clearly more work with both physical (laboratory) and mathematical models is needed.

6. List of Variables and Parameters

b	constant relating dispersion to velocity, dimensionless
C_a	gas-phase concentration, mol $m^{-3}{}_a$
C_I	solute concentration in ice, mol $m^{-3}{}_I$
C_{iw}	solute concentration in immobile water, mol $m^{-3}{}_{iw}$
C_{mw}	solute concentration in mobile water, mol $m^{-3}{}_{mw}$
C_w	solute concentration in water, mol m^{-3}

C_4 total S(IV) concentration in liquid water of snow, mol m^{-3}

C_6 S(VI) concentration in liquid water of snow, mol m^{-3}

D dispersion coefficient, m^2 s^{-1}

D_m gaseous diffusion coefficient, m^2 s^{-1}

f fraction mobile water, $m^3_{mw}\ m^{-3}_{w}$

F flux of gas to snow surface, mol m^{-3} s^{-1}

H Henry's coefficient, $m^3_{w} m^{-3}_{a}$

k' rate constant for S(IV) —> S(VI) conversion, s^{-1}

k intrinsic permeability, m s^{-1}

K eddy diffusivity, m^2 s^{-1}

K_I ice-water partition coefficient, $m^3_{w}\ m^{-3}_{I}$

K_{im} mobile-immobile-water partition coefficient, $m^3_{iw}\ m^{-3}_{mw}$

m solute molality, mol kg^{-1}

n exponent in unsaturated-flow equation, dimensionless

Q volume flux of meltwater, m s^{-1}

r total resistance to dry deposition, inverse of v_d, s m^{-1}

r_a atmospheric resistance to dry deposition, s m^{-1}

r_s surface resistance to dry deposition, s m^{-1}

r_{wa} water-to-air volume ratio in snow, m^3 m^{-3}

S effective water saturation $\frac{S_w - S_i}{1 - S_i}$

S_i irreducible water saturation, $m^3_{w}\ m^{-3}_{T}$

S_w saturation, $m^3_{w}\ m^{-3}_{a+w}$

T_f freezing temperature, °C

t time, s

u_a air velocity, m s^{-1}

u_w water velocity, m s^{-1}

v^* velocity of saturation wavefront in snowpack, m s^{-1}

v_d gaseous deposition velocity, m s^{-1}

v_4 deposition velocity for S(IV), m s^{-1}

v_6 deposition velocity for S(VI), m s^{-1}

x depth in snowpack, m

X_m liquid-water mass fraction in snow, $kg_w\ kg^{-1}_{T}$

z height above snowpack, m

z_d height of mass sink above snowpack, m

z_r height of reference above snowpack, m

α_0 first ionization fraction of SO_2(aq)

α_w air-water mass-transfer coefficient, m s^{-1}

α_I ice-water mass-transfer coefficient, m s^{-1}

α_d constant relating dispersion to velocity

β relative irreducible water saturation $\frac{S_i}{1 - S_i}$

ρ_b bulk-snow density, kg m^{-3}

ρ_I ice density, $kg\ m^{-3}_{I}$

ρ_s snow density, $kg_{w+I}\ m^{-3}_{T}$

θ porosity, $m^3_{a+w}\ m^{-3}_{T}$

τ tortuosity factor, dimensionless

τ_1 rate coefficient for mobile-immobile water transfer, s^{-1}

τ_1 rate coefficient for mobile-immobile water transfer, s^{-1}

7. REFERENCES

Adamson, A. W. (1979) *Physical Chemistry of Surfaces, 4th ed.*. Wiley, New York.

Aldaz, L. (1969) Flux measurements of atmospheric ozone over land and water. *J. Geophys. Res.* 74:6943-6946 .

Bales, R. C., Davis, R. E., and Stanley, D. A. (1989) Ionic elution through shallow, homogeneous snow. *Water Resour. Res.* 25:1869-1877 .

Bales, R. C., Sommerfeld, R. E., and Kebler, D. G. (1990) Ionic tracer movement through a Wyoming snowpack. *Atmos. Environ.* 11, In press .

Bales, R. C., Valdez, M. P., and Dawson, G. A. (1987) Gaseous deposition to snow: II. Physical-chemical model for SO_2 deposition. *J. Geophys. Res.* 92:9789-9799 .

Barrie, L. A. and Walmsley, J. L. (1978) A study of sulfur dioxide deposition velocities to snow in northern Canada. *Atmos. Environ.* 12:2321-2332 .

Brimblecombe, P., Clegg, S. L., Davies, T. D., Shooter, D., and Tranter, M. (1987) Observations of the preferential loss of major ions from melting snow and laboratory ice. *Water Res.* 21:1279-1286 .

Cadle, S. H., Dasch, J. M., and Mulawa, P. A. (1985) Atmospheric concentrations and the deposition velocity to snow of nitric acid, sulfur dioxide and various particulate species. *Atmos. Environ.* 19:1819-1827 .

Chan, W. H. and Chung, D. H. S. (1986) Regional-scale precipitation scavenging of SO_2, SO_4^{2-}, NO_2 and HNO_3. *Atmos. Environ.* 20:1397-1402 .

Clifford, J. (1975) Properties of water in capillaries and thin films. In F. Franks*Water, A Comprehensive Treatise*, vol. 5, New York, pp. 75-132.

Colbeck, I. and Harrison, R. M. (1985) Dry deposition of ozone: Some measurements of deposition velocity and of verticle profiles to 100 metres. *Atmos. Environ.* 19:1807-1818 .

Colbeck, S. C. (1978) The physical aspects of water flow through snow. In V. T. Chow*Adv. Hydrosci.*, vol. 11, pp. 165-206.

Colbeck, S. C. (1979) Grain clusters in wet snow. *J. Colloid Interface Sci.* 72:371-384 .

Colbeck, S. C. (1979) Water flow through heterogeneous snow. *Cold Reg. Sci. Technol.* 1:37-45 .

Colbeck, S. C. (1980) Thermodynamics of snow metamorphism due to variation in curvature. *J. Glaciol.* 26:291-301 .

Colbeck, S. C. (1981) A simulation of the enrichment of atmospheric pollutants in snow cover runoff. *Water Resour. Res.* 17:1383-1388 .

Colbeck, S. C. (1985) Temperature dependence of the equilibrium form of ice. *J. Cryst. Growth* 72:726-732 .

Colbeck, S. C., *Air Movement in Snow due to Windpumping*, 1990. in press

Colbeck, S. C. and Davidson, G., 1973. Water flow through homogeneous snow, *The Role of Snow and Ice in Hydrology. IAHS Publication 106*, pp. 242-257.

Committee on Glaciology, Polar Research Board (1983) *Snow and Ice Research: An Assessment.* National Academy Press, Washington, DC.

Conklin, M. H. (1990) Dry deposition of SO_2 to Snow: Effect of pH and Duration of Exposure. *in preparation.*

Davidson, C. I. (1989) Mechanisms of wet and dry deposition of atmospheric contaminants to snow surfaces. In H. Oeschger and C. C. Langway, Jr.*The Environmental Record in Glaciers and Ice Sheets*, John Wiley & Sons, New York, pp. 29-51.

Davidson, C. I., Harrington, J. R., Stephenson, M. J., Small, M. J., Boscoe, F. P., and Gandley, R. E. (1989) Seasonal variations in sulfate, nitrate and chloride in the Greenland ice sheet: Relation to atmospheric concentrations. *Atmos. Environ.* 23:2483-2493 .

Dovland, H. and Eliassen, A. (1976) Dry deposition on a snow surface. *Atmos. Environ.* 10:783-785 .

Fletcher, N. H. (1968) Surface structure of water and ice, II. A revised model. *Philos. Mag.* 18:156:1287-1300 .

Franks, F. (1982) The properties of aqueous solutions at subzero temperatures. In F. Franks*Water, A Comprehensive Treatise*, vol. 7, Plenum, New York, pp. 215-338.

Furukawa, Y., Yamamoto, M., and Kuroda, T. (1987) Ellipsometric study of the transition layer on the surface of an ice crystal. *J. Cryst. Growth* 82:665-677 .

Galbally, I. E. and Roy, C. R. (1980) Destruction of ozone at the earth's surface. *Quarterly Journal of the Royal Meteorological Society* 106:599-620 .

Golecki, I. and Jaccard, C. (1978) Intrinsic surface disorder in ice near the melting point. *J. Phys., C: Solid State Phys.* 11:4229-4237 .

Granat, L. and Johansson, C. (1983) Dry deposition of SO_2 and NO_2 in winter. *Atmos. Environ.* 17:191-192 .

Gross, G. W. (1968) Some effects of trace inorganics on the ice/water system. In *Trace Inorganics in Water*, Adv. Chem. Series, vol. 73, Am. Chem. Soc., Washington, D. C., pp. 27-97.

Gross, G. W. (1977) Concentration dependent redistribution at the ice-water phase boundary -- III. Spontaneous convection, chloride solutions. *J. Chem. Phys.* 67:5264-5274 .

Gross, G. W., Gutjahr, A., and Caylor, K. (1987) Recent experimental work on solute redistribution at the ice/water interface. Implications for electrical properties and interface processes. *J. Physique* 48:527-533 .

Gross, G. W., Wu, C., Bryant, L., and McKee, C. (1975) Concentration dependent solute redistribution at the ice/water phase boundary. II. Experimental investigation. *J. Chem. Phys.* 62:3085-3092 .

Gubler, H. (1982) Strength of bonds between ice grains after short contact times. *J. Glaciol.* 28:457-473 .

Hibberd, S. (1984) A model for pollutant concentrations during snow-melt. *J. Glaciol.* 30:58-65 .

Hoffmann, M. R. and Calvert, J. G. (1985) In *Chemical Transformation Modules for Eulerian Acid Deposition Models, Vol. II, The Aqueous Chemistry*, pp. . U.S. E.P.A., N.C.A.R. Interagency Agreement DW 930237

Hoffmann, M. R. and Jacob, D. J., 1984. Kinetics and mechanisms of the catalytic oxidation of dissolved sulfur dioxide in aqueous solution: an application to nighttime fog water chemistry, *SO_2, NO and NO_2 Oxidation Mechanisms: Atmospheric Considerations*, pp. 101-172, Butterworth Publishers, Boston, Massachusetts.

Huebert, B. J., Fehsenfeld, A., Norton, R. B., and Albritton, D. (1983) The scavenging of nitric acid vapor by snow. In H. R. Pruppacher, R.G. Semonin and W. G. N. Slinn*Precipitation Scavenging, Dry Deposition, and Resuspension*, Elsevier, New York, pp. .

Jacob, D. J. and Hoffmann, M. R. (1983) A dynamic model for the production of H^+, NO_3^-, and SO_4^{2-} in urban fog. *J. Geophys. Res.* 88C:6611-6621 .

James, R. O. and Parks, G. A. (1982) Characterization of aqueous colloids by their electrical double layer and intrinsic surface-chemical properties. *Surf. Colloid Sci.* 12:119-216 .

Johannessen, M. and Henriksen, A. (1978) Chemistry of snow meltwater; changes in concentration during melting. *Water Resour. Res.* 14:615-619 .

Johansson, C. and Granat, L. (1986) An experimental study of the dry deposition of gaseous nitric acid to snow. *Atmos. Environ.* 20:1165-1170 .

Kattelmann, R. C. (1985) Macropores in snowpacks of Sierra Nevada. *Ann. Glaciol.* 6.

Kuroiwa, D., 1974. Metamorphism of snow and ice sintering observed by time lapse cine-photomicrography, *IAHS Publication 114*, pp. 82-88.

Lee, Y.-N. and Kinsley, T. (1986) Adsorption of nitric acid, hydrogen peroxide, and sulfur dioxide on ice. *EOS* 67, 898 .

Legrand, M., Petit, J.-R., and Korotkevich, Y. S. (1987) D.C. conductivity of Antarctic ice in relation to its chemistry. *J. Physique* 3:605-611 .

Leung, W. K. S. and Carmichael, G. R. (1984) Solute redistribution during normal freezing. *Water, Air, and Soil Pollution* 21:141-150 .

Marks, D., 1988. Climate, Energy Exchange and Snowmelt in Emerald LAke Watershed, Sierra Nevada, Ph.D. Dissertation, 158 pp., University of California, Santa Barbara, CA.

Marsh, P. and Woo, M.-k. (1984) Wetting front advance and freezing of meltwater within a snow cover, 1, Observations in the Canadian Arctic. *Water Resour. Res.* 20:1853-1864 .

Marsh, P. and Woo, M.-k. (1984) Wetting front advance and freezing of meltwater within a snow cover, 2, A simulation model. *Water Resour. Res.* 20:1865-1874 .

Marsh, P. and Woo, M.-k. (1985) Meltwater movement in natural heterogeneous snow covers. *Water Resour. Res.* 21:1710-1716 .

Molina, M. J., Tso, T.-L., Molina, L. T., and Wang, F. C.-Y. (1987) Antarctic stratospheric chemistry of chlorine nitrate, hydrogen chloride, and ice: Release of active chlorine. *Sci.* 238:1253-1257 .

Morris, E. M. (1987) Modeling of water flow through snowpacks. In H. G. Jones and W. J. Orville-Thomas*Seasonal Snowcovers: Physics, Chemistry, Hydrology*, Reidel, Amsterdam, pp. 179-208.

Mulvaney, R., Wolff, E. W., and Oates, K. (1988) Sulphuric acid at grain boundaries in ice. *Nature* 331:247-249 .

Nechaev, E. A., Fedoseeva, V. I., and Fedoseev, N. F. (1981) Surface properties of dispersed ice (snow). *Zh. Fiz. Khim. (Russ.)* 55:1822-1826 .

Neftel, A., Jacob, P., and Klockow, D. (1984) Measurements of hydrogen peroxide in polar ice samples. *Nature* 311:43-45 .

Neftel, A., Jacob, P., and Klockow, D. (1986) Long-term record of H_2O_2 in polar ice cores. *Tellus* 38B.

Nenow, D. (1984) Surface premelting. *Prog. Cryst. Growth Charact.* 9:185-225 .

Ocampo, J. and Klinger, J. (1982) Adsorption of N_2 and CO_2 on ice. *J. Colloid Interface Sci.* 86:377-383 .

Oeschger, H. (1985) The contribution of ice core studies to the understanding of environmental processes. In C. C. Langway, Jr., H. Oeschger and W. Dansgaard*Greenland Ice Core: Geophysics, Geochemistry, and the Environment*, American Geophysical Union, Washington, D.C., pp. 9-17.

Orem, M. W. and Adamson, A. W. (1969) Physical adsorption of vapor on ice: II. N-alkanes. *J. Colloid Interface Sci.* 31:278-286 .

Peterson, C., 1990. Ion Flux through a Shallow Snowpack: Effects of Initial and Melt Conditions, M.S. Thesis, 90 pp., Dept. Hydrol. Water Resour., University of Arizona, Tucson, AZ.

Rottman, C. and Wortis, M., Brown, R. A., Kelzer, J., Stelger, U., and Yeh, Y. (1983) Enhanced light scattering at the ice-water interface during freezing. *J. Phys. Chem.* 87:4135-4138 .

Schwander, J. (1989) The transformation of snow to ice and the occlusion of gases. In H. Oeschger and C. C. Langway, Jr.*The Environmental Record in Glaciers and Ice Sheets*, John Wiley & Sons, New York, pp. 53-67.

Schwander, J., Neftel, A., Oeschger, H., and Stauffer, B. (1983) Measurement of direct current conductivity on ice samples for climatological applications. *J. Phys. Chem.* 87:4157-4160 .

Sigg, A. and Neftel, A. (1988) Seasonal variations in hydrogen peroxide in polar ice cores. *Ann. Glaciol.* 10:157-162 .

Sigg, A., Neftel, A., and Zurcher, F. (1987) Chemical transformations in a snow cover at Weissfluchjoch, Switzerland, situated at 2500 m.a.s.l.. In H. G. Jones and W. J. Orville-Thomas*Seasonal Snowcovers: Physics, Chemistry, Hydrology*, NATO Advanced Studies Institute, Reidel, Amsterdam, pp. .

Smiles, D. E. and Gardiner, B. N. (1982) Hydrodynamic dispersion during unsteady, unsaturated water flow in a clay coil. *J. Soil Sci. Soc. Am.* 46:9-14 .

Smiles, D. E., Philip, J. R., Knight, J. H., and Elrick, D. E. (1978) Hydrodynamic dispersion during absorption of water by soil. *J. Soil Sci. Soc. Am.* 42:229-234 .

Smith, V. G., Tiller, W. A., and Rutter, J. W. (1955) A mathematical analysis of solute redistribution during solidification. *Can. J. Phys.* 33:723-745 .

Sommerfeld, R. A. and Lamb, D. (1986) Preliminary measurements of SO_2 adsorbed on ice. *Geophys. Res. Let.* 13:349-351 .

Stauffer, B., Neftel, A., Oeschger, H, and Schwander, J., Stauffer, B., J.Schwander, and Oeschger, H. (1985) Enclosure of air during metamorphosis of dry firn to ice. *Ann. Glaciol.* 6:108-112 , American Geophysical Union, Washington, DC.

Stumm, W. and Morgan, J. J. (1981)*Aquatic Chemistry*. Wiley, New York.

Tranter, M., Brimblecombe, P., Davies, T. D., Vincent, C. E., Abrahams, P. W., and Blackwood, I. (1986) The composition of snowfall, snowpack, and meltwater in the Scottish Highlands – evidence for preferential elution. *Atmos. Environ.* 20:517-525 .

Tsiouris, S., Vincent, C. E., Davies, T. D., and Brimblecomb, P. (1985) The elution of ions through field and laboratory snowpacks. *Ann. Glaciol.* 7:196-201 .

Valdez, M. P., Bales, R. C., Stanley, D. A., and Dawson, G. A. (1987) Gaseous deposition to snow: I. Experimental study of SO_2 and NO_2 deposition. *J. Geophys. Res.* 92:9779-9789 .

Valdez, M. P., Dawson, G. A., and Bales, R. C. (1989) Sulfur dioxide incorporation into ice depositing from the vapor. *J. Geophys. Res.* 94:1095-1103 .

Voldner, E. C., Barrie, L. A., and Sinois, A. (1986) A literature review of dry deposition of oxides of sulphur and nitrogen with emphasis on long-range transport modeling in North America. *Atmos. Environ.* 20:2101-2123 .

Whelpdale, D. M. and Shaw, R. W. (1974) Sulfur dioxide removal by turbulent transfer over grass, snow and water surfaces. *Tellus* 26:196-204 .

Wofsy, S. C., Molina, M. J., Salawitch, R. J., Fox, L. E., and McElroy, M. B. (1988) Interactions between HCl, NO_x, and H_2O ice in the antartic stratosphere: implications for ozone. *J. Geophys. Res.* 93:2442-2450 .

Wolff, E. W., Mulvaney, R., and Oates, K. (1989) Diffusion and location of hydrochloric acid in ice: Implications for polar stratospheric clouds and ozone depletion. *Geophys. Res. Let.* 16:487-490 .

8. ACKNOWLEDGEMENTS

Much of the research described in this paper was carried out with financial support from the U.S.D.A Forest Service, Rocky Mountain Forest and Range Experiment Station. Helpful comments and suggestions from Robert Davis and Vijai Gupta are also acknowledged.

CHEMICAL CHANGE IN SNOWPACKS

Peter Brimblecombe and David Shooter[1]
School of Environmental Sciences
University of East Anglia
Norwich NR4 7TJ UK

The title of the NATO workshop held at Maratea Italy in July of 1990 "The Processes of Chemical Change in Snowpacks" may convey a misleading impression of the proceedings. Indeed some might argue that the workshop tended by and large to neglect changes of a chemical nature. The aim of this short discussion paper is to illustrate a framework on which chemical changes in snowpacks might be discussed. The paper draws on the discussions that took place at the workshop and aims to complement the much more detailed paper by Roger Bales (1991) which deals with the way chemical and compositional changes in snowpacks are modelled.

Many of the papers given at Maratea concerned themselves with spatial distribution and transformations of components in snowpacks. These were represented by the chemical differentiation found in windblown snows and the fractionation that occurs during firnification and snow melt. A purist might consider these to be not so much chemical changes as compositional ones.

Chemical transformations, characterised by reactions (see Satchell, 1977 for ways of defining reactions), are certainly important in snowpacks, but are often rather neglected. Two different kinds of chemical change can be distinguished in the material published on snowpacks. There are reactions that take place in open systems. These would include the deposition on SO_2 on snow, or the loss of HNO_3 acid from the pack. Contrasting processes that take place in a closed system might include photochemical, electrochemical and thermal reactions. Here the postulated destruction of H_2O_2 in snow, the redox reactions during freezing precipitation and dissolution might be given as examples. We can think of such closed systems as exhibiting in-pack chemical change.

In the first of the models presented by Bales (1991), the gaseous deposition of SO_2 to snow is considered under isothermal conditions with a homogeneous-water-containing snowpack and a variable water to air volume ratio. Chemical change is modelled, model with the S(IV) to S(VI) oxidation being included through the aqueous phase reactions of SO_2. Oxidants include ozone and hydrogen peroxide. In his second model of ion release from a snowpack, Bales (1991) is primarily concerned with the changing composition of meltwater as it progresses through a snowpack.

---

1. Permanent address: Chemistry Department, University of Auckland, Private Bag, Auckland NEW ZEALAND

NATO ASI Series, Vol. G 28
Seasonal Snowpacks
Edited by T. D. Davies et al.

The role of the liquid phase

It is possible to make a number of preliminary observations about the chemical reactions within snowpacks.

(1) The reactions are likely to take place in association with the liquid phase at ice grain boundaries rather than within the ice grains. Indeed both models presented by Bales assume a stationary liquid or liquid-like phase surrounding snow grains provides a medium for chemical reaction. The supposition that reactions occur in the liquid phases is reasonable because the diffusion rate of reacting solutes in ice would be many orders of magnitude slower than in the liquid phase. However monomolecular processes where diffusion would not be critical (eg photochemical degradations) can certainly be envisaged.

(2) If we accept the importance of a liquid phase in contact with ice for as a medium for chemical reactions in closed snowpacks, then we need to recognize that this system is restricted by Gibbs phase rule:

$$F = C + 2 - P$$

where F is the number of degrees of freedom, C is the number of components and P the number of phases (Denbigh, 1971). The fact that we have a solid phase (the "ice") in equilibrium with the liquid phase reduces the number of degrees of freedom the system possesses by one compared with simple solution reactions.

(3) However there are further reductions in the degree of freedom, because temperature and pressure are usually not freely chosen variables within snow packs. This means that simple systems with no degrees of freedom are easy to imagine. One system to be considered would be that of a solute and water (i.e. $C = 2$) with two phases present ($P = 2$). This would give two degrees of freedom, but as both T and P cannot usually be freely chosen the two degrees of freedom vanish. This means that the concentration of solute is fixed.

These thermodynamic restrictions lead to a very important, but also very obvious result. The liquid in contact with the ice must be a solution rather than pure water, otherwise temperature could not vary, and would have to be 0 ^{o}C. An objection to this view could argue that the radius of curvature affects the freezing point, but this is small compared with that brought about by the presence of solutes. Similarly the surface effects that lead to the "quasi liquid film" described by Fletcher (1968), may well be swamped by the effects of dissolved solutes.

Solute concentration and temperature

Bales (1991) identifies the variable X_m the liquid-water mass fraction. This important variable, together with the bulk concentration of available solute in the snow ($\bar{m}_s$) can be related to temperature (e.g. see Robinson and Stokes Eqn. 8.7):

$$T' = vRT_0^2W\bar{m}_s/1000L_0X_m = 1.860v\bar{m}_s/X_m$$

where T' is the freezing point depression, v the stoichiometric number, R the gas constant, T_0 freezing point of water, W the molecular weight of water and L_0 the latent heat of fusion at T_0. The term $\bar{m}_s/X_m$ is the molal concentration of solute in the liquid phase.

However this relationships is only accurate when the solute concentration in the liquid layer low. As freezing point depression is less than 2 °C for molal solute, concentrations have to be high if the liquid state is to be maintained at substantially sub-zero temperatures. This is in line with arguments for the presence of brines at the surface of ice grains, and indeed the observations of Mulvaney et al (1988) have suggested high concentrations of sulphuric acid to be present at grain boundaries of polar ices.

At high concentrations solution behaviour is non ideal so freezing point depression has to be related to activity rather than concentration. The relationship thus becomes complex and non-linear. Furthermore it is dependent on the specific ions present in the solute and the large freezing depressions require the variation of the latent heat of fusion (L) with temperature to be considered. This leads to polynomial equations of the form:

$$-\log(a_W) = 0.004207T' + 2.1\times10^{-6}(T')^2 + \ldots$$

where a_W is the activity of water in the system.

Equilibrium processes in the "liquid-layer"

The presence of a solute and non-ideal behaviour clearly does more than simply change the freezing point depression (as important as this variable is). The alteration of other thermodynamic variables, most notably activity, means that the properties such as solubility will be affected. Let us consider an idealized example such as the dissolution of nitric acid in a liquid layer that contains sodium chloride. Trace amounts of HNO_3 dissolve according to the equation:

$$HNO_{3(g)} = H^+ + NO_3^-$$

described by the equilbrium equation

$$K_H = aH^+.aNO_3^-/pHNO_{3(g)} = f_\pm^2 cH^+.cNO_3^-/pHNO_{3(g)}.$$

where a and c refer to the activity and concentration of the dissolved species, $f_\pm$ is the mean activity coefficient and $pHNO_{3(g)}$ the pressure of nitric acid.

The mean nitric acid activity coefficients at the freezing point of sodium chloride solutions were determined using the method of Pitzer (i.e. Harvie et al, 1984). Low temperature values of K_H for nitric acid were calculated from the equation of Brimblecombe and Clegg (1990). Assuming nitric acid to behave ideally (i.e. $f_\pm = 1$) the relative change in solubility is proportional to the square root of K_H. The solubility (normalised to that at 0^oC) is illustrated by line (b) in Fig. 1 which shows a regular increase in solubility with decreasing temperature. This is equivalent to saying that HNO_3 solubility decreases with temperature. Accounting for activity coefficients in the calculations leads to enhanced solubility at small freezing point depressions (where $f_\pm < 1$) and lowered solubility at large freezing point depressions (where $f_\pm > 1$).

A concentration scale does not give the whole picture because despite increases in solute concentration with decreasing temperature, the volume of liquid water decreases (i.e. the liquid-water mass fraction, X_m introduced by Bales). Hence the total amount of nitric acid that dissolves in the liquid present in the snowpack also decreases. The normalised amount of nitric acid present in the liquid layer in snow as a function of temperature is illustrated in Fig. 2 for cases where (a) nitric acid solubility is constant with temperature, (b) nitric acid solubility is adjusted for K_H variation with temperature and (c) the effect of activity coefficients are also included. We see that temperature effects on equilibrium and the activity coefficients have an effect on the amount of HNO_3 that can dissolve in the liquid layer, but the biggest effect is due to changes in the amount of liquid present (X_m).

Kinetics in the "liquid-layer"

The presence of salts also affects the observed rate constants for reactions, but these are not well understood except at low ionic strengths or for particular reaction systems (e.g. Laidler, 1965). At low ionic strengths the rates of reaction between oppositely charged species decreases as a function of ionic strength while the rates of reaction between like charged species increases as a function of ionic strength. Such a salt effect can be relatively large as shown in Fig. 3. The relative amount of product produced from a reaction which has a rate constant independent of temperature (a) is compared to a reaction

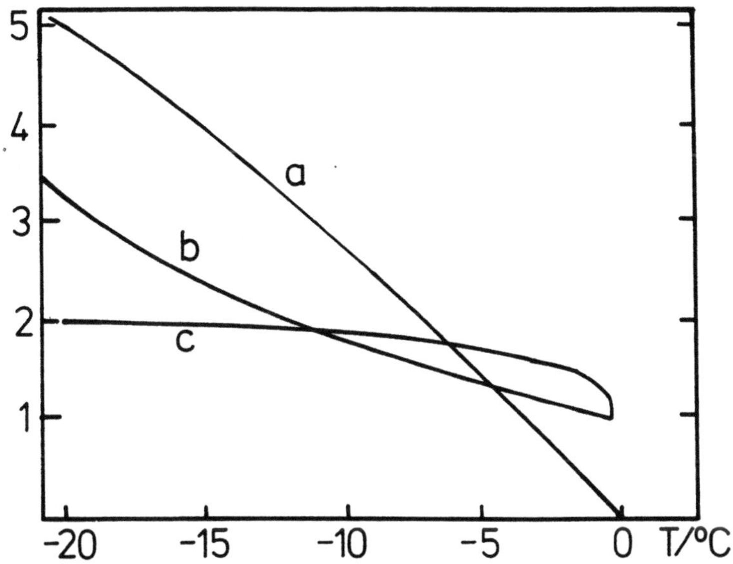

Fig. 1 Calculated equilibrium concentrations of nitric acid in sodium chloride brines associated with the liquid layer on ice as a function of temperature: (a) the molal concentration of NaCl as a function of temperature (b) concentrations nitric acid normalised to 0°C, assuming ideal behaviour (c) concentrations nitric acid normalised to 0°C, after determing the activity coefficient in low temperature sodium chloride solutions.

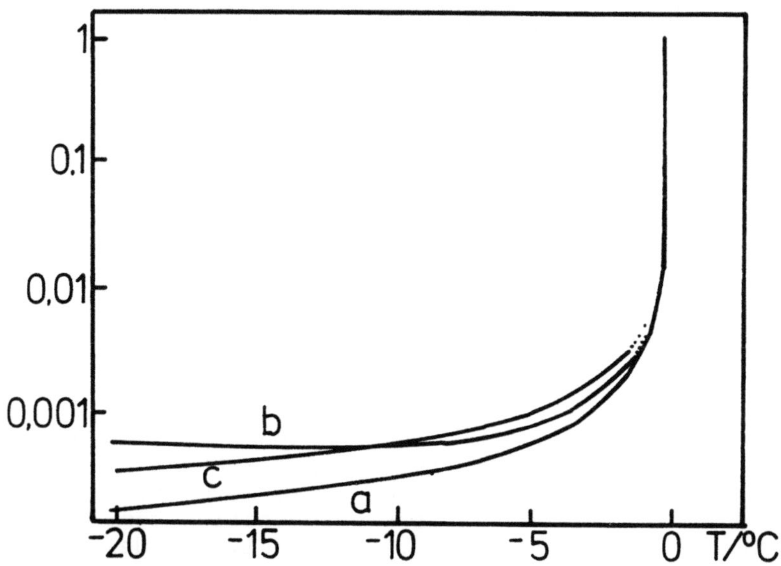

Fig. 2 Calculated equilibrium amounts of nitric acid in sodium chloride brines associated with the liquid layer on ice as a function of temperature. The ice was assumed to contain 0.001 mol/kg sodium chloride. The relative amount of nitric acid in ice (a) assuming nitric acid solubility is constant with temperature, (b) nitric acid solubility varies with temperature (after Brimblecombe and Clegg, 1990) and (c) both solubility and activity coefficient vary.

(b) that halves its rate for a 10°C fall in temperature and one (c) with a rate constant (for a reaction between like charged univalent species) that increases with ionic strength (Laidler, 1965).

Clearly salt effects can be significant. In the case of some important reactions, such as the aqueous oxidation of S(IV), there are specific effects which cause dramatic increases in the reaction rate in brines (Clarke and Radojevic, 1983).

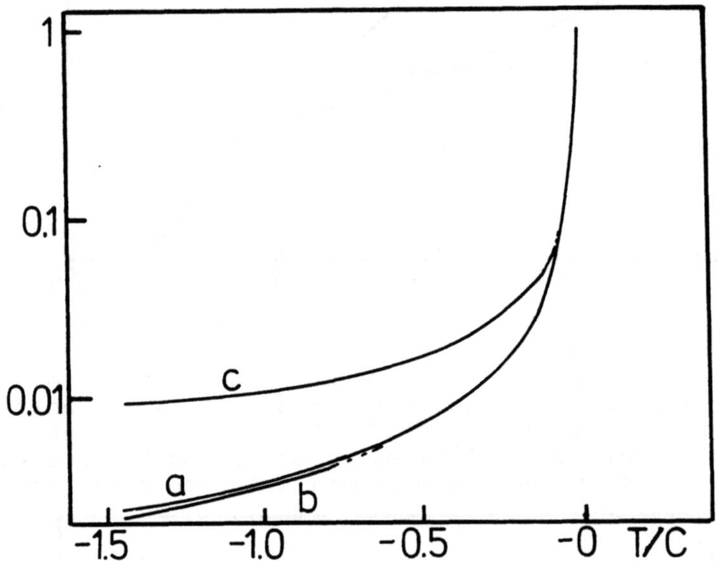

Fig. 3 Calculated amounts of product in a hypothetical reaction in the saline liquid associated with ice where the rate constant is (a) independent of temperature (b) halves its rate for a 10°C fall in temperature and (c) increases with ionic strength (for a reaction between like charged univalent species).

Ion release from snowpacks

While fractionation of ions into snowpack meltwater has been successfully modelled by Bales (1991), preferential elution is not so easily represented. Bales uses an ion-exclusion model, but differences observed between various anion-cation combinations are not always easy to account for. Previously a model based on the differences in solubility of the ions in ice was used to distinguish between the behaviour of ionic species, (Brimblecombe et al. 1988). Both models have a common theme in that ionic solubility in ice grains will depend on charge, but size and concentration effects can also be influential.

Conclusion

Reactions in snow packs are likely to occur in concentrated solutions at the grain boundaries. Concentrations may be high enough to require equilibrium and rate calculations to account for departure from ideality. As has been mentioned by Bales (1991) some of the basic coefficients used in models need to be determined by independent means. Measurements are currently so difficult that they are often taken from model interpretation of the results, thus departures from ideality might easily be overlooked.

References

Bales (1991) Modeling in-pack chemical transformations, this volume

Brimblecombe P, Clegg SL (1990) Equilibrium partial pressures of strong acids over concentrated solutions- III. The temperature variation of HNO_3 solubility. Atmos. Environ,. 24A:1945-1955

Brimblecombe P, Clegg SL, Davies TD, Shooter D, Tranter M (1988) The loss of halide and sulphate ions from melting ice. Water Res. 22A:693-700

Clarke AG, Radojevic M (1983) Chloride ion effects on the aqueous oxidation of SO_2. Atmos. Environ,. 17:617-624

Denbigh K (1971) The Principles of Chemical Equilibrium. Cambridge University Press, Cambridge

Fletcher NH (1968) Surface structure of water and ice II. A revised model. Philos. Mag. 18:1287-1300

Harvie CE, Moller N, Weare JH (1984) The prediction of solubilities in natural waters: the Na-K-Mg-Ca-H-Cl-SO_4-OH-HCO_3-CO_3-CO_2-H_2O system to high ionic strengths at 25°C. Geochim. Cosmochim. Acta 48:723-751

Laidler KJ (1965) Chemical Kinetics. McGraw-Hill, London.

Mulvaney R, Wolff EW, Oates K (1988) Sulphuric acid at grain boundaries in ice. Nature 331:247-249

Robinson RA, Stokes RH (1959) Electrolyte Solutions, Butterworths, Oxford

Satchell DPN (1977) The classification of chemical reactions. Naturwissenschaften 64:113-121.

Snow chemistry and biological activity: a particular perspective on nutrient cycling

H.G. Jones
INRS-Eau
Université du Québec
2700, rue Einstein
Ste-Foy, Québec
G1V 4C7

Abstract

The state-of-science on the effects of biological activity on snow cover is discussed. Except for some research that concerns the relationship of snow chemistry and snow algae the subject has received very little attention from snow chemists. This absence of studies relating biota directly to snow chemistry has led us to approach the subject by a theoretical treatise of nutrient dynamics in snow. The biogenic origins of nutrients in snowfall are described. The subsequent changes that could be expected in the concentrations of major nutrients (N, S, P, Ca, Mg, K) in forest snow cover have been estimated by comparing snow in open areas with that in litter-laden forest snowpacks and snow meltwaters. The influence that winter-active vertebrates and invertebrates could have on snow-nutrient fluxes have been calculated from data on the population densities of selected species and on the nutrient characteristics of excrement. Although the estimated biologically-induced fluxes are very variable due to the natural fluctuations of the snow-litter-animal ecosystems it is shown that the nutrient which is the most liable to biological activity in snow is nitrogen. N-depletion due to forest litter may attain 20-40 eq ha-1 during the spring melt period while algal demand can be 1-3 eq ha^{-1} day^{-1}. Animal fluxes are associated mostly with mammalian herbivores and insects; deer, for example, can supply up to 20 eq N ha^{-1} day^{-1} to the snow cover and collembola may also excrete the same amount under conditions related to swarming on snow.

Introduction

There is very little knowledge on the changes in snow chemistry due to the interactions between plants, animals and seasonal snow cover. In contrast, there are

NATO ASI Series, Vol. G 28
Seasonal Snowpacks
Edited by T. D. Davies et al.

many references to studies on the physical influence of snow on the distribution and population dynamics of plants and animals. Formazov (1946), for example, published a very comprehensive review on the importance of snow cover in the ecology of animals and birds. In this work he classifies living things as either chionophiles (well adapted to snow cover), chioneuphores (partially adapted to snow cover), or chionophobes (having great difficulty to function in snow covered environments). A priori, this type of work would seem to be of secondary interest to the subject of snow chemistry. However, as McKay and Adams (1981) have pointed out one must take into account the effects of snow cover on living things to determine the effects of living things on snow cover. Neither of them can be considered as separate entities in the environmental system of which both the snow and the living things are a part.

Snow cover may be considered as an extension of the atmosphere. In many regions snow meltwater is the major hydrologic reservoir for the sustenance of terrestrial and fresh water biota. Snow cover also can act physically as a habitat or biologically as a reservoir of nutrients. It thus represents an environment which both flora and fauna may use exclusively as a milieu for nutrition and reproduction (e.g.the intranivean life cycle of snow algae; Hoham, 1980), as a protective mantle for forage (e.g. the subnivean activity of small mammals; Everdun and Fuller, 1972), or as a surface for transport and food cover (e.g. the supranivean activity of larger mammals, Kelsall, 1969). In the absence of any well defined studies on the influence of biological activity on snow chemistry, the knowledge that is available on the physical interactions between snow cover and living things can give some insight into the possible repercussions on the chemical composition of snow.

The influence of plants and animals on the chemistry of snow may be direct or indirect. An example of the former is the excretion of metabolic products on, or in, snow by chionophiles and chioneuphores (DelGiudice et al, 1989). Indirect effects include the effects of trees on the distribution of snow (Jones, 1987) and the subsequent influence of the depth of the distributed snow cover on meltwater composition. In addition to the spatial aspects of the interactions between snow and living things, the extent of the biological activity is very much dependent on the temporal existence of snow covers. The life cycles of many plants and animals will be modified by the duration and depth of the snow. Populations of small mammals, for example, will fluctuate with interannual variations in snow depth (Pruitt, 1975). Even within the lifetime of a snowpack, the presence of psychrophilic vegetative populations and of certain insect and animal communities (Merriam et al, 1983) may have very little effect on the snowpack during certain periods e.g the cold accumulation phase. At other times, the activity of many organisms is stimulated when temperatures rise and free water flows in the pack. Under these conditions the chemistry of snow and meltwaters may be modified by nutrient uptake, the secretion of metabolites and by the death and decomposition of the individuals themselves.

In the last resort, the greatest biological influence by far on snow chemistry is Man himself. From the direct and volontary additions of chemicals to snow (Transportation Research Board, 1988) to the indirect effects of gaseous and particulate deposition from industrial emissions on snow (Cadle and Dasch, 1988) the impact of human activities on snow chemistry is complex. Many of the effects of Man on the chemical composition of snow in the atmosphere and on the ground are dealt with in Davies et al 1990.

In this paper we review the state-of-the-science on the influence that natural communities exert on the chemistry of seasonal snow covers. The ecological relationships between plants, animals, microbes and snow cover will be described in turn (trees, invertebrates, mammals, and microorganisms). The repercussions of the organisms on the chemistry of snow and snow meltwaters will then be given. For lack of precise quantitative studies in the literature much of the following discussion is by necessity qualitative; in some cases, however, chemical interaction between snow and living things has been quantified by combining data from different sources.

Snow cover chemistry and biological activity: definitions, chemical species and the analytical approach

The definition of what constitutes the chemical composition of snowpacks is subjective. Snow cover is physically a heterogeneous system closely related to the atmospheric and terrestrial environments. Snow will thus be subjected to inputs of animals and plants during its life on the ground and both the sampling procedures and chemical analysis of the system will be dictated by the objectives of the research in question. To study the evolution of the chemistry of snowpacks in a boreal forest environment, for example, Jones (1987) considered canopy fallout as an integral part of the forest-snow system. On the other hand, Barry and Price (1987) in a similar study on the chemistry of snow and snow meltwaters in an aspen forest considered the litterfall to be an extraneous component of the snow cover. Both approaches can be justified by the extent to which the biophysical environment will affect the snow. In the case of the boreal forest the litterfall occurs throughout the winter period while deciduous litter falls prior to the formation of the snow cover (Esseen, 1985). In such studies the methodology is by necessity site specific. Each research worker has, in effect, to estimate the degree to which biological activity may mask any of the changes in snow chemistry that could be attributed to geochemical factors only.

If studies do, or have to, take into account the impact of biological activity on the chemistry of the snow cover then the nature of the chemical components has to be qualified. The vast majority of chemical studies on snow and snow meltwaters deal with soluble inorganic species (Davies et al, 1987), soluble organics (Twickler et al, 1986) or particulate micropollutants (Gregor and Gummer, 1989). In this paper we propose to limit our study of the chemical components to the soluble inorganic and organic species that are produced or transformed in the snow cover. Solubility is operationally defined as particulates $< 0.45\ \mu m$; filters of this pore size remove all living organisms. This means, for example, that the secretion of metabolic products by the large invertebrate populations of collembola found in and on snowpacks during spring (Aitchison, 1979) would be defined as biologically-induced chemical change. Even though the subsequent amount of detritus left on the snow cover by the massive mortality of the same populations may be very significant (approx. 0.5-1 g m^{-2},) the residual particulate content is not considered by definition to be an integral part of the chemical composition of the pack. In a similar vein only the species leached from forest litter are included as chemical components of snow, the particulate material ($> 0.45\ \mu m$) being classed as extraneous matter. Soluble gases released from the soil by microbiological activity into the snowpack (e.g CO_2, Solomon and Cerling, 1987; N_2O, Rhodes et al 1986) are also considered to be part of the snow chemical makeup.

There are very few references in the published literature on the origin and evolution of biogenic chemical species in snow cover. This dearth of knowledge is not only due to the scant treatment that the subject has generally received from snow scientists but also to the complex nature of biological activity both in and on snow. Some studies have described the assimilation of simple nutrient species (e.g. NO_3) from snow and snow meltwaters by snow algae (Hoham et al, 1989). However, the large number and intricate structure of excreted metabolic products are major obstacles in the development of an overall picture of the chemical relationships between plants, animals, microbes and snow cover. Nevertheless, all these chemical species are part of the biogeochemical cycles of nutrients and can be lumped into models of nutrient cycles as components of total elemental pools and fluxes. Figure 1, for example, shows some quantitative aspects and chemical species of the nutrient pools and fluxes associated with the biota in the cycle of nitrogen in a boreal forest. It can be seen that the total pools and fluxes of nitrogen within the system are considerably greater than the input from the atmosphere. The snowcover itself

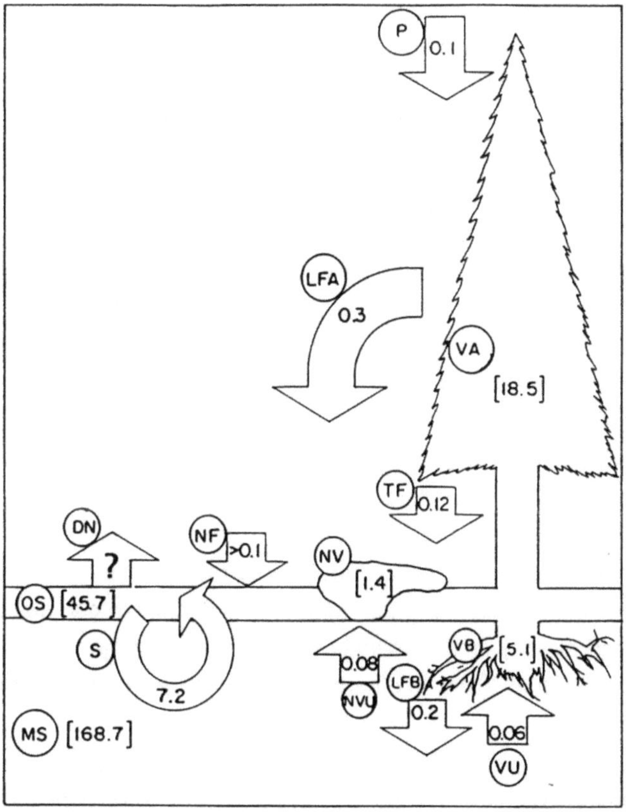

Figure 1. Nutrient pools [keq ha^{-1}] and nutrient fluxes (keq ha^{-1} $year^{-1}$) in a boreal forest system.

Pools: VA, above ground vascular tissue; VB, below ground vascular tissue; NV, non-vascular tissue; OS, Organic soil; MS, Mineral soil.

Fluxes: P, precipitation; LFA, above ground litterfall, LFB, below ground litter; TF, throughfall; VU, vascular uptake; NVU, non-vascular uptake; NF, nitrogen fixation; S, soil mineralization - assimulation, DN, denitrification.

Recalculated from data reported in Van Cleve and Alexander, 1981.

would not quantitatively influence the overall pools and annual fluxes of nitrogen. However, it may play an important role in the nutrient cycle by retaining nutrient inputs from animals and plants. The nutrients would accumulate during the winter and be released in spring. It should also be noted that the inorganic species of nutrients found in snow (e.g. NH_4, NO_3) are readily bio-available and can be used directly by plants and microbiological communities in spring.

To establish a consistent theme for this present paper we propose to study the effect of biological activity on snow as a function of the dynamics and fluxes of the major nutrients (Ca, Mg, K, C, N, S, P) in and through the snowcover system. In this manner we wish to develop a simple model of the interaction between snow and living things similar to figure 1. We believe the approach to be valid even if we have not been able to find any specific studies on the subject. However, there are many papers in which information on microbiological, plant and animal communities in winter and certain aspects of nutrient cycling in, and on, snow is reported. Our approach is to combine the available data so that the relative importance of the different communities in determining the nutrient fluxes in snow can be calculated - albeit not measured directly. In this manner we are able, in some cases, to quantify concentration changes and fluxes; in others, only conclusions of a descriptive nature can be made.

The following sections describe the global influence of biota on fresh snowfall and the degree to which the separate communities subsequently control nutrient cycles (N, S, P, K, Ca, Mg) in the snow cover and at the snow-atmosphere and snow-ground interfaces.

Snowfall and global biological activity

The chemical composition of the Earth's atmosphere is biotically controlled (Lovelock, 1979). Atmosphere-ocean and atmosphere-soil/vegetation fluxes driven by photosynthesis and the metabolic emissions of both primary and secondary producers sustain the global nutrient cycle (Mooney et al, 1987). In addition, in-situ transformations of the atmospheric species occur by non-biotic oxidation and other photochemical reactions. The return of the emitted compounds and their derivatives from the atmosphere to the surface is the result of dry deposition, occult deposition and precipitation. Dry deposition includes both the physical processes of adsorption, dissolution, and fallout, and the biotic processes of absorption, assimilation and respiration of gaseous species. Occult deposition is the result of the retention of fog and cloud water containing dissolved and adsorbed matter by vegetation.

Precipitation, rain and snow, will capture the atmospheric species during freefall (Grennfelt, 1987).

Fresh snow thus contains a host of compounds which globally reflect the current biological activity of the planet. This is the basis for determining the past climates of the earth and the influence of Man on the evolution of the Earth's atmosphere by the analysis of ice-cores from perennial ice masses. It is assumed that as the fresh snow which falls on these ice masses does not melt, the initial chemical composition of snow is not altered over time (e.g methane content, Craig et al, 1988). This, however, may be not true for some chemical species (e.g. HNO_3, Neubauer and Heumann, 1988).

The analyses of perennial ice-cores shows that the spread of agrarian societies modified the natural cycle of nutrients by changing the relative concentrations of biogenic compounds released into the atmosphere (Thompson et al, 1988). The development of the industrial revolution later led to a rapid rise in the concentrations of some atmospheric constituents (e.g. SO_2 in the S cycle) and also introduced classes of compounds that were not previously a part of the natural cycle (chlorofluorohydrocarbons, Rasmussen and Khalil, 1986). In the case of the N cycle, the emissions from industrial sources (NH_4, NO_x, N_2, N_2O, HNO_3, etc.) can mask the true contribution of biogenic sources to the overall composition of the atmosphere and precipitation. It is, however, sometimes possible to estimate directly anthropogenic sources of some compounds by stable-isotope analyses of solubles in precipitation (e.g the C cycle, $^{13}C/^{12}C$, Craig et al, 1988). In the absence of any direct measurement of the chemical signatures due to biogenic and/or industrial sources, the relative contributions have to be estimated from calculated global budgets (Bowden, 1986). Such calculations show that for many classes of compounds in the C cycle the natural sources are still the dominant control for atmospheric and precipitation chemistry. This is particulary true in the case of the organic components of the troposphere. For example, for the atmospheric gaseous compounds of the C cycle, 80-90% of methane and non-methane hydrocarbons originate from biogenic processes. (Warneck, 1987).

Organic material of both natural and anthropogenic sources is also either adsorbed onto particulates such as soot or condensed as aerosols. The scavenging of gaseous and particulate organics and aerosols from the atmosphere by snow crystals is well known (Gill and Greadel, 1983). In general, snow is a more efficient scavenging agent for aerosols than rain due to the higher surface area to weight ratio of the ice crystals relative to raindrops. Snow crystals, however, can even display higher scavenging efficiency than raindrops for volatile organics (Czuczwa et al, 1988). The reason for this is not clear but is thought to be due to the scavenging of water droplets which then freeze (Pruppacher and Klett,1981) or to the absorption of organic films on the crystal surfaces (Gill and Greadel, 1983).

Snowfall chemistry: biogenic organics and nutrient pools

The number of organic compounds that can be isolated from snow is considerable (Lunge et al, 1977) and it becomes difficult to distinguish those of natural sources from anthropogenic emissions. Some classes of compounds such as the polychlorinated biphenyls and polyaromatic hydrocarbons can be clearly identified with industrial activities and combustion of fossil fuels (Thomas, 1986). Other classes, however, behave like inorganic species i.e. they are common to both natural and industrial sources e.g the n-alkanes of the C cycle. In addition, the existence of rapid reaction mechanisms in the atmosphere (Finlayson-Pitts and Pitts, 1986) also complicates any attempts to determine the influence of natural emissions of organics on snowfall composition. In general, however, the presence of n-alkanes with an odd-to-even preference in the lengths of the carbon chain and fatty-acids with an analogous even-to-odd preference indicates that the organics originate with biogenic activity. Meyers and Hites (1982) used these criteria to show that in non-urban areas straightchain fatty acids (C_{12}, C_{14}, C_{16}, C_{20}, C_{24}, C_{26}) and hydrocarbons (C_{27}, C_{29}) from biological sources dominate (85%) the solvant-extractable organic content of winter precipitation. The dominance of palmitic acid (n- $C_{15}H_{31}COOH$) is a prominant feature of the biogenic signature of snowfalls (Fig. 2). They attributed the presence of this compound and its even carbon-number analogues in snow to microbial processes and degraded plant waxes. As the winter season progressed the concentrations diminished; in spring the concentrations again rose as biological activity increased.

Lunge et al (1977) examined the wet deposition of organic micropollutants in Norway. Although they were specifically interested in the long-range transport of pollutant aerosols, an analysis of their data shows that the major fatty-acid components (C_{12}, C_{14}, C_{16}) were partly from natural sources. The concentrations of these classes of compounds in snowfall are very low (10-100 $\mu g\ L^{-1}$). They are partially soluble and at these concentration levels would be leached during the melt to appear in the dissolved organic carbon (DOC) content of meltwaters. A calculated annual accumulation of these organics in the snowcover in a temperate zone, assuming a snowpack maximum water equivalent of 300 mm, shows that the meltwater would discharge between 3 and 30 mg m^{-2} of fatty acids and n-alkanes to the soil in spring. In the soil system these compounds are rapidly degraded. The amount discharged by snow, however, is negligible compared to the total amounts

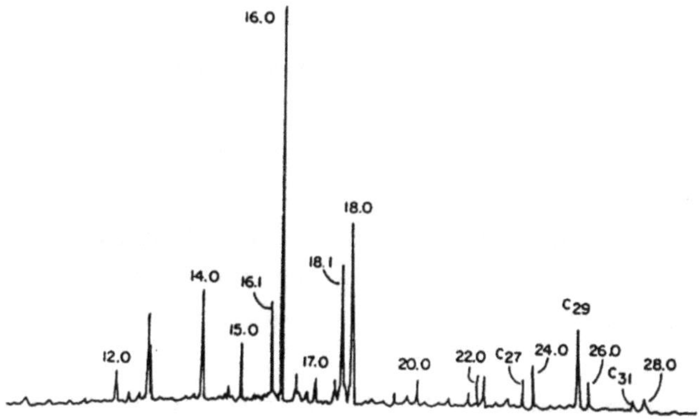

Figure 2. Biogenic hydrocarbons in snow in a semi-rural area. Analyses of n-Alkanes and methylated fatty acids by GC. The fatty acids are identified by carbon chain length and number of double bonds. n-Alkanes are labeled by C-subscripts. Palmitic acid (derived methyl ester) is the largest peak at 16.0. Taken from Meyers and Hites (1982).

of labile organics within the soil and litter horizons (Van Cleve and Alexander, 1981). As a carbon source for heterotrophic production the atmospheric organics are thus of minimal importance. However, as they may have other functions in nutrient cycling, we do not presently know if snow does play an important role as a reservoir of biogenic organics in systems with well developed soils. It is also not known what the fate of these biogenic compounds is once deposited in the pack. We have not, in effect, been able to locate any reference to work that involves the study of the long-term evolution of alkanes and fatty-acids in snow cover.

In addition to the above organic compounds, snowfall includes soluble tannic and humic matter from suspended soil particules and degraded plant residues. Low molecular-weight carboxylic acids (formic, acetic and oxalic) may also be present in the dissolved organic carbon (DOC) fraction from snow. These acids could be formed by the in-situ oxidation of a host of organics from both natural and industrial emissions. However, they are known particulary to be the result of the oxidative degradation of biogenic cyclic olefins and their occurrence in the precipitation of remote areas in the southern hemisphere lends credence to this type of source (Keene et al, 1983). Methane sulphonic acid is another organic acid of biogenic origin which is found in precipitation. It is formed by the oxidation of dimethyl sulphide from the ocean (Saltzman et al, 1983) and wetland areas (Cooper et al, 1987). It has been reported in snow samples by Fernandes (1984). The DOC levels in snowfall vary widely but are generally in the range of 1-2 mg L^{-1}. Even in remote areas DOC may show relatively high values. Twickler et al (1986) showed that the dissolved organic carbon in firn from the Greenland icecap varied from 30-110 $\mu g\ L^{-1}$. No speciation of the organic content was attempted but the deposition rate was found to exhibit a seasonal deposition pattern. Athough the reasons for the seasonality of the organic content in snow could not be definitely established it was thought to be consistent with the long range transport of biogenic material. Table 1 shows the concentrations of selected compounds and of the total organic matter thought to be derived from natural sources in snowfall recorded at different regions of the northern hemisphere.

The atmospheric pathways of the C-, N-, and S-cycles have been considerably modified by anthropogenic input of CO_2, NO_x, and SO_2. Barrie (1990) has discussed the natural and anthropogenic contributions of these N- and S-compounds and their atmospheric derivatives to snowfall while Jeffries (1990) has published a comprehensive review on the concentrations of major ions in premelt snowpacks from both remote and industrial regions. Table 2 presents an estimation of the total pools of nutrients of atmospheric origin that could be expected to be found in snow covers in different regions at the time of maximum snow accumulation. The total ion loads vary considerably according to the remoteness or the industrialization of the regions.

Table 1: Concentrations of "biogenic" organics observed in snowfall

Site	Organics	Conc.(mg L^{-1})	Reference
Greenland (Polar snow)	Total dissolved	0.11 (0.03-0.32)	Twickler et al., 1986
United States (rural)	Solvant extracted (Fatty acids; n-alkanes)[1]	0.030 (0.012-0.055)	Meyers and Hites, 1982.
Western Canada (mountain forest)	Total dissolved	3.3 (2.0-6.5)	Wallis et al., 1981
Eastern Canada (boreal forest open clearing	Total dissolved	0.46	Jones and Tranter, 1989.

[1]) Origin of major fraction assumed to be natural from remoteness of sites.

Table 2: Nutrient pools (eq ha^{-1}) in seasonal snowcovers[1]

Reference	Site	N	S	P	K	Ca	Mg
Belikova et al., 1984	Russia; Siberian forest (mean)	--	81	--	--	--	--
Davies et al., 1987	Scotland; mountain snow	9[2]	19	--	--	--	8
Barry and Price, 1987	Canada; eastern deciduous forest	42[2]	19	--	--	--	--
Jones, 1987	Canada; eastern coniferous forest	100	70	neg.[3]	5	10	4
Berg et al., 1989	California; western mountain	23	11	--	2	10	3
Suzuki, 1982	Japan; urban area			--	14	72	94

1 Pior to the snow melt period. Recalculated from data published in the original reference, or recalculated from reported concentration measurements and typical values for the SWE of a snow cover in mid-winter for the appropriate region. Values rounded.

2 $N-NO_3$ only.

3 neg. = negligible

Snowfall and living organisms

Accompanying the degraded and oxidized organic matter of biological origin in snow are viable organisms. Bacteria and other particulate organic matter can act as ice-nucleation agents (Watanabe et al, 1988). Although the major sources of ice nucleation in the atmosphere are inorganics such as fine soil particulates (clays) and sea salts (Murakami and Kikuchi, 1982) organic matter is often found at the centre of snow crystals. No studies on any further biological activity of ice-nucleation organisms once deposited in snow cover have been reported. Subsequent occurrence of microbial activity within the pack, however, may occur as a large part of the organisms in the atmosphere are probably present as spores (Meszaros, 1977).

Snow and vegetation

Snow is not the major factor in the delimitation of vegetative biomes of the Earth. It is temperature which determines the extent and productivity of the ecosystem associations (Billings, 1980). In North America, for example, the delimitation between the non-forested tundra and boreal forest in the upper latitudes (i.e. Boreal forest treeline) has been attributed to the strong gradients in air temperature across the climatic front created by Arctic and Pacific air massses in the west-east flow regime (Larsen, 1972). This results in relatively narrow ecotones (taiga) between the two more extensive biomes.

In contrast to air temperature, snow is an extremely important factor for species association and growth of vegetation on far smaller scales than those of biomes (Payette et al, 1973). This is particulary true in the tundra where the absence of trees causes snow distribution to be almost exclusively the result of wind action. The physical interaction between snow and the low-lying flora of the tundra is of little importance and the topography of the snow cover is moulded by the shape of the land. Extreme environments for plant propagation and growth are no-snow sites and those with too much snow (Billings, 1972). Both sites are characterized by very poor vegetative cover, few species and slow growth. The snow-free sites are subjected to intense cold due to the absence of protection against thermal losses and dessication; in sites with too much snow, the pack persists too long and the growing season becomes too short. The vegetative communities are thus very dependent on the spatial distribution of snow-cover without, in turn, having any significant influence on the distribution itself. Plants grow at ground level and very little plant debris is produced at the snow surfaces by wind-induced ice-crystal erosion. The snow cover is thus relatively free of biological material that may influence its chemical composition.

In the more southerly tundra boreal-forest ecotone the variations in snow depth contribute to the competition between vascular and non-vascular (lichen) plant growth (Petzold and Mulhern, 1988). Shallow snow depths (< 30 cm) favours the non-vascular species as they can better resist greater exposure to dessication and abrasion by wind.The survival of pockets of trees, e.g. black spruce and birch thickets, depends on the presence of well insulating blankets of deep snow. A mutual relationship is established whereby trees, in turn, serve to trap the snow and trigger drifting. The average snow cover depth is the dominant factor in determining the height of the most productive plant tissue. Foliage below the snow level remains lush and healthy. If, however, the trees manage to survive and exceed the average snow depth their exposed areas become subjected to severe wind-induced abrasion by ice crystals (Savile, 1972); the upper parts of the trees show extensive scar tissue, bud and branch damage, and needle mortality (Petzold and Mulhern, 1988). The amount of biological debris deposited on the snow is thus greater than in open tundra; during snowmelt this material will be leached and soluble nutrients will penetrate the upper soil and litter horizons (see "Litterfall and Throughfall" below).

In contrast to the tundra and the forest-tundra ecotone the forest system consists of tall free-standing vegetation which considerably exceeds snow cover depths, and the trees play a key role in the spatial and temporal distribution of the snow cover. Litter is also an important factor in snow chemistry (Barry and Price, 1987; Jones, 1987). Forest transition to temperate grasslands, scrub, and agricultural crop lands is accompanied by a diminishing influence of vegetation on snow cover. These ecosystems occupy an intermediate position where the vegetation can exert some influence on the chemical composition and on the distribution and retention of blowing snow (Steppuhn, 1981).

Forests and snow chemistry: distribution and interception of snow and nutrient cycling

As snow falls to the ground it can be subjected to two main redistribution phenomena at the level of the forest canopy. The first is the aerodynamic turbulence due to wind-canopy interaction (Jeffrey, 1968); the second is the direct interception and retention of snow by the canopy (Meiman, 1968). The snow that is held on the tree is referred to as "quali", the Kobuk-Valley Eskimo word to describe this type of snow (Pruitt, 1970). Turbulence causes changes in wind speed and in the capacity of the flow of air to maintain the snow crystals in suspension. Snow deposition will then be channelled into certain favoured configurations which are well-recognizable features of forest snowpacks. Quali may evaporate in dry weather thus reducing the amount of deposited snow around tree crowns; alternatively it may reach the ground either by wind removal or sliding off the branches when the snow melts.

The chemical composition of quali may also be changed by contact with the foliage of coniferous trees and even in contact with the trunk and branches of deciduous trees. Schuepp et al (1987) found that by stratifying the quali they could demonstrate the cation leaching at the foliage-snow interface. Potassium, in particular, was found in high concentrations in the underlying layers (up to 51 μeq L^{-1} of K for snow in contact with foliage compared to 2-4 μeq L^{-1} of K in the upper layers). In a comparative study of quali on natural and artificial fir the authors were able to demonstrate that the results were representative of a real exchange between the plants and the intercepted snow and were not dependent on any prior dry deposition. Dry deposited material will, of course, also be leached from foliage by melted snow (see "Litterfall and Throughfall"). There was no significant change in other nutrients e.g. NO_3, and SO_4.

Snow cover in forests will thus be distributed unevenly and snow depth will vary as a function of the type of forest and the density of the tree canopy. The trees will subsequently influence the chemical composition of the snow by litterfall and throughfall. The physical distribution of snow will modify the spatial characteristics of the original atmospheric pool of nutrients while the subsequent chemical inputs will modify the nutrient fluxes. The immediate implication of the physical influence of trees on snow depth will be the variable composition of meltwaters. This is due to the dependence of meltwater chemistry on the extent of leaching that occurs when meltwaters move from the surface to the snow-soil interface (Foster, 1978). The spatial redistribution of the nutrient pool from the snowpack to the forest floor during melt can be simulated by taking into account the depth pattern of redistributed snow and the bulk chemical concentration of the cover prior to melt (Jones et al, 1986).

Forests and snow chemistry: Trees and snow depth patterns

Depth patterns of forest snow covers have been studied extensively in forest hydrology. Snow depth and snow-water-equivalent (SWE) vary underneath the forest canopy (Woo and Steer, 1986), between stands of different wood species (Patch, 1981) and also between the forest proper and associated clearings (Golding and Swanson, 1986). From the data published by Meiman (1968) on snow distribution in forests, it can be estimated that, in general, open areas (defined as a function of tree height, H) harbour 26% more snow at maximum accumulation than the mean value for the forest floor. Golding and Swanson found that in a fir/pine forest, accumulation was greater in openings whether they were large (8-13 ha, increased accumulation, 20%) or small (0.25 to 6-H, 13-45% increase). The snow depth profile was relatively symmetrical across the diameter of small openings (1-H). In larger openings (> 3-H), however, the lack of shade and the reemission of longwave radiation by trees causes local melting and results in lower depths in the northern quadrants of clearings. The difference is smaller for deciduous forests than those with coniferous trees due to the absence of foliage during the snow season (Swanson and Stevenson, 1971).

Within the forest proper where canopy cover can vary from 100% (completely closed canopy at the defined angle of view, e.g. 30°, Patch, 1981) to 0% (open) snow depth can show even greater variability than between the mean values for clearings and the forest floor (Jones, 1987; Havas and Kubin, 1983). Havas and Kubin, for example, found that the mean snow depth under tree crowns of a an old spruce forest was 50% of that of the mean snow depth between the trees. Table 3 cites some selected examples of the variability of snow cover in different forest types as recorded in the literature.

Table 3: Influence of forest cover on snow cover distribution patterns (SWE, mm)

Reference	Forest type	Snow in clearings		Snow under forest canopy	
		Mean	Detailed distribution	Mean	Detailed distribution
Golding and Swanson, 1986	Western coniferous		112, 1H[1] clearing 119, 2H clearing 110, 4H clearing	84 79 72	
Jones, 1987	Eastern coniferous	185			75, 100% canopy closure 140, 50% canopy closure
Patch, 1981	Eastern mixed	279			229, 100% softwood canopy 254, 50% softwood canopy
Woo and Steer, 1986	Eastern coniferous	210[2]			120, at base of tree stem 180, 1m from tree stem
Zakrisson, 1987	Open coniferous		300, 5m from forest edge 275, 25m from forest edge		225, 5m from clearing edge 250, 25m from clearing edge
Berndt, 1965	Western coniferous	217		155	
Meiman, 1970	Average of 25 forest types	378[3]		300[4]	

1) H = average tree height
2) Calculated from depth data of Woo and Steer using a density of 0.3
3) Calculated from 4) and mean percent increase (26%) of SWE in clearing over that under the forest canopy.
4) Typical value for SWE in eastern forest (1m depth, density 0.3).

As the variability of forest floor snow depth is high, extensive sampling is expensive, and to reduce costs, models to simulate snow depth have been developed (e.g. regression analysis, Patch, 1981; Dickison and Daugharty, 1982; Jones, 1987: Monte Carlo technique, Woo and Steer, 1986). The model of Woo and Steer (1986) uses the probabilities of tree distribution and tree radii to simulate the depth pattern in northern spruce forests. Patch (1981) regressed data of forest cover parameters (number of stems, NS; percent stems hardwood, %SH, basal area of plot, BA; percent basal area hardwood, %BAH; overhead canopy closure, CC; and southerly canopy closure, SC) on snow depth (SD) and SWE. The variables the most useful for predicting SD and SWE were CC and %BAH. Jones (1987) attempted to correlate SD and SWE to 16 measured parametres on both softwoods and hardwoods of a small boreal forest basin in Quebec, Canada. The most significant correlations found for both SD and SWE were those associated with the basal area of softwoods (BAS) and the mean distance of softwood stems from the snow sampling point within a radius of 3m (MDS). These results are analogous to those of Dickison and Daugharty (1982) who found significant correlations between data relating the proportion of hardwoods (%BAH) to SD and SWE for predominantly hardwood forests in New Brunswick, Canada. Jones (1987) found that the correlations between measured forest parameters were relatively constant for both the accumulation and melt phases of the forest floor snowpack; the regression expression (Figure 3) was thus used to simulate SD and SWE at different study sites throughout the snow season. This contrasts with the results of Dickison and Daugharty (1982) who found that the variations of the correlation coefficients on different dates did not allow them to use the procedure to meaningfully simulate SD and SWE trends within the forest for both accumulation and melt periods. The elimination of aspect and elevation as parameters in the study of the very small watershed studied by Jones (1987) may explain the consistency of the correlations obtained by this author compared to those of Dickison and Daugharty (1982); their study site varied in altitude from 195-478 m.a.s.l.

Although these models simulate the influence of trees on forest snow cover they are empirical; they can give only an overall picture of snow distribution at any one time. More complete models on the evolution of forest snow cover would have to take into account time-dependent processes such as evaporation and localised melt which are also strongly influenced by trees (Golding and Swanson, 1986).

Snow Water Equivalent (SWE) is a parameter of models to simulate the chemical composition of snow meltwaters (Goodison et al, 1986; Stein et al; 1986). Once the SWE distribution pattern on the forest floor has been simulated by the above tree-

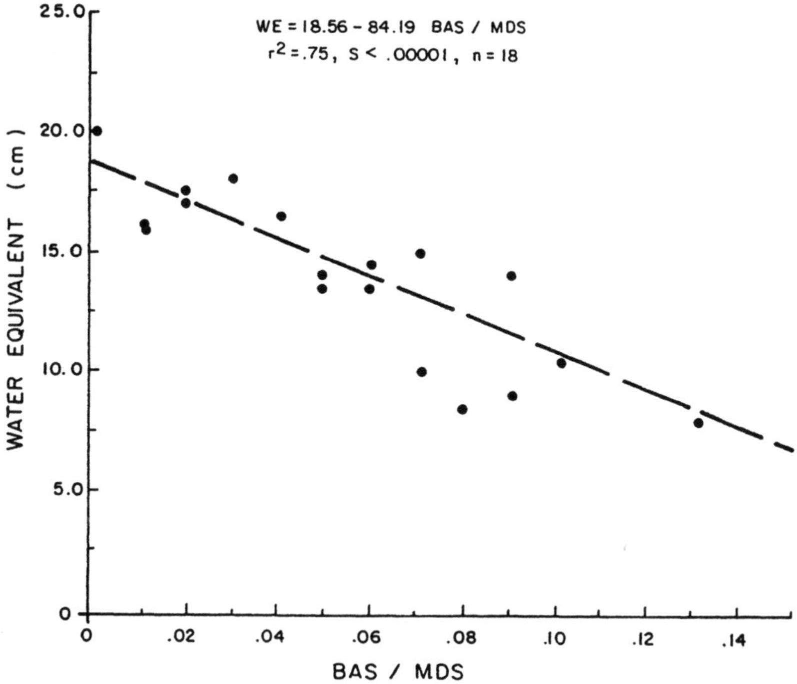

Figure 3. Influence of the forest canopy on the distribution of snow in a boreal forest: regression of snow water equivalent values on data for the basal area of softwoods (BAS) and mean distance of softwood stems from the sampling point MDS. Taken from Jones (1987).

input models the snow chemistry models can then use the output to simulate the chemical composition of the meltwaters. Figure 4 shows the tree cover, SWE of the forest floor snowpack, total nutrient pool of S in snow, and chemical composition (S-SO_4) of the first meltwaters discharged to the soil during the spring melt. It can be seen that the nutrient pools under coniferous trees can vary by 250% while the nutrient flux due to the discharge of the first 20 mm of melt at the forest floor varies by 187%. It should be noted that this is a hypothetical system and is taken only to show the physical influence of the trees on the nutrient pool of the snowpack and the chemistry of the meltwaters.

Figure 5 shows the distribution of the N-total nutrient pool in a sub-arctic boreal-forest/muskeg ecosystem at Schefferville, Québec; the pool distribution is calculated from the reported snow depth pattern (Adams and Findlay, 1966) and the avarage concentration of N-NO_3 and N-NH_4 in winter precipitation for the region (Grimard, 1985). In this system the wind plays a far greater role in snow distribution than in the temperate coniferous forests. The wind has scoured much of the snow from the open areas (muskeg) and it has become entrapped by the pockets of trees which line both sides of the muskeg. Nutrient pools are thus higher under the forest than in the open. In figures 4 and 5 the calculated nutrient pools are based only on the chemistry of snowfall. Snow chemistry in forests, however, will also be related to litterfall and to throughfall.

Forests and snow chemistry: litterfall and throughfall

Litterfall occurs through the year but is particulary heavy during autumn in deciduous forests and autumn and winter in coniferous forests (Essenn, 1987). The decomposition of litter is a complex process which gradually transforms the organic tissue into available nutrients and refractory humic matter. Litter which falls onto the forest floor rapidly becomes colonized by a succession of invertebrate and microbiological (bacteria and fungi) populations (Waring and Schlesinger, 1985). Even under snow cover the relatively mild temperatures and high humidity allow the degradation of the litter to proceed throughout the winter season (Taylor and Jones, 1990). Litter which falls onto snow, and remains in the pack will, however, be relatively unaffected by animal or microbiological activity during the cold accumulation period. Invertebrate populations are low and microbiological activity is reduced by the cold temperatures and lack of liquid water. With the onset of melt, liquid water will wet the in-pack litter and soluble organics and other chemical species will be leached from the vegetative debris (Berg and Staaf, 1981). Fresh litter can lose up to 20% of dry weight by leaching, with deciduous litter containing more

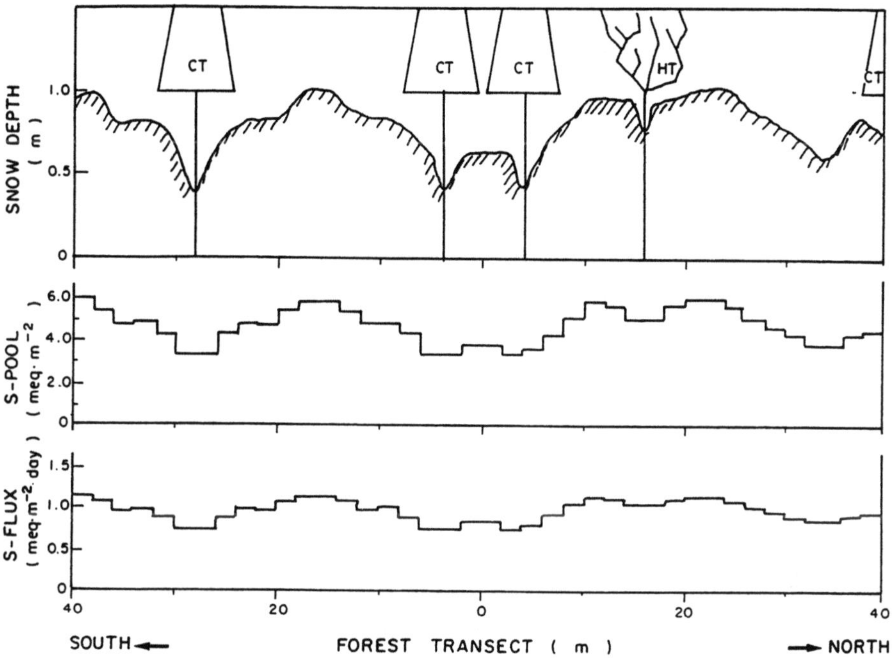

Figure 4. Snow depth, m, and the chemical pool of S-SO_4 (meq m^{-2}) and meltwater fluxes (meq $m^{-2}day^{-1}$) across a transect of a typical mixed forest.
CT, coniferous tree; HT' hardwood tree.
Simulation of pack depth is based on Woo and Steer (1986), Patch (1981), Meiman (1986), and Jones (1987).
Snow S-Pools are calculated from depth, density of 0,3, and a concentration of 20 µeq L^{-1}
Meltwaters fluxes are calculated by the model SNOQUAL1, Stein et al. (1986).

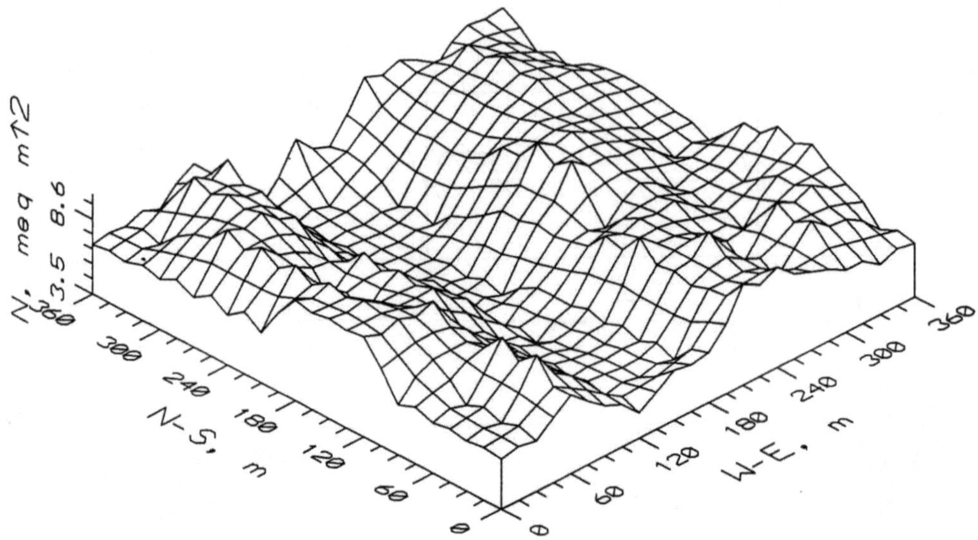

Figure 5. Distribution of N-Pool in a boreal-forest/muskeg system, Schefferville, Québec, in mid-winter.

N-pool calculated from SWE measurements (Adams and Findlay, 1966) and from mean winter values for precipitation concentrations of $N\text{-}NO_3$ and $N\text{-}NH_4$ (Grimard, 1985).
Note the low values of the N-Pool along the open muskeg (parallel to N-S axis at 140 m east) where wind removes the snow while the forest pool is high due to entrapping of snowfall and wind-blown snow

soluble components than coniferous litter (Taylor and Parkinson, 1988). Meltwater will also stimulate heterotrophic and autotrophic activity on freshly deposited organic matter and remove the soluble metabolic products. Autotrophic activity associated with the in-pack litter will not be due to the true snow algae (Hoham, 1980) but to the epiphytic algal populations and mosses which are attached to the litter when it is rejected or dislodged from the trees. Wetted litter, however, can stimulate or inhibit snow-algal activity (Hoham, 1976). During the melt, the temperature of meltwaters are constant and the chemical composition of the melt will depend on the amount of litter wetting that takes place. This, in turn, will influence the degree of biological activity and physical leaching.

Throughfall refers to the precipitation that falls through the forest canopy during precipitation events. Leaching of organic and inorganic meltwaters by diffusion from wetted tissue (Tukey, 1966, 1970), ion exchange with foliage, and/or leaching of dry deposited aerosols may take place (Dasch, 1987). Rain has far more contact with foliage than snow and the chemical composition of rainfall is affected to a greater degree than that of snow. The influence of tree canopies on the chemistry of the snow cover is, therefore, not only limited to litterfall but also to rain-on-snow events. In winter, foliage is absent from deciduous trees while the foliage of coniferous trees is dormant. There is, however, still active exchange between the coniferous foliage and rain (Cronan and Reiners, 1983) and epiphytic lichens and mosses may also influence throughfall by N-uptake (Lang et al 1976). K and Ca increases in throughfall are mostly due to exchange with exposed tissue while increases in SO_4 originate from the leaching of dry deposited material. Glatzel et al (1986) have emphasized the influence of tree foliage and branches on snow cover chemistry by, what they term, a "secondary deposition effect". Rime from fog droplets and hoarfrost which accumulate on branches are concentrated in pollutants and also provide an increased surface of ice for the dry deposition of gases and aerosols. The rime and frost then slough off onto the surface of the snow and contribute far more to the chemical load than a normal snowfall of the same SWE. This process, which is similar to the interaction between quali and trees (Schuepp et al, 1987), can account for a four or five fold increase in N and S fluxes to snow cover (N_{max}, 10 eq ha^{-1} day^{-1}; S_{max}, 9 eq ha^{-1} day^{-1}; Glatzel et al., 1986) in locations where riming is favoured (e.g. exposed ridges) to those where it is not.

Litter (foliage, small twigs, and attached vegetation) contains the major nutrients (N, S, P, Ca, K, Mg) in quantities which can vary from 0.1% to 2% of dry weight (Table 4A). As tree foliage is a very efficient captor of dry deposition, litterfall will also carry adsorbed aerosols and gaseous species of these nutrients onto the snowpack. The extent to which litter can affect the nutrient pools of snow has been determined by of litter leachate. Stottlemeyer (1987) attributed the increase of K in snow cover over the quantity deposited by bulk precipitation in a deciduous forest to bud scales and other organic debris from the trees. At the time of maximum accumulation, the K pool in the snow pack was approximately 6 eq ha^{-1} which represents a three fold increase over that (2 eq ha-1) due to the bulk precipitation (mostly snow). In a coniferous forest Fahey (1979) found that snow collected beneath the forest canopy was richer in nutrients (K, Ca, Mg) compared with snow from the open areas; the denser the stand the greater the enrichment of the snow. K was again the ion which showed the greatest enrichment factor, the concentration in snow being up to 16 times that of snow in the open (Table 4B). In the Fahey's study, however, the results

Table 4 (A, B): Typical nutrient concentrations (%, dry weight) in fresh litter (A) and concentration increases (mg L^{-1}) or decreases of soluble nutrients in snow and snow meltwaters containing litterfall (B)

Reference	Type of litter	A. Total nutrient concentrations									
		$N\text{-}NH_4$	$N\text{-}NO_3$	N_T	$S\text{-}SO_4$	S_T	$P\text{-}PO_4$	P	K	Ca	Mg
Chase and Young, 1978	Red Pine Foliage	--	--	0.93	--	--	--	0.11	0.78	0.24	0.08
	Branches			0.22				0.03	0.16	0.26	0.04
Lang *et al.*, 1982	Balsam fir: Needles	--	--	2.35	--	--	--	0.13	0.46	0.21	0.03
	Twigs			2.13	--	--	--	0.20	0.81	0.33	0.12
Courchesne and Hendershot, 1988	Mixed forest: all tissue, maple leaves dominant	--	--	0.84	--	--	--	0.02	0.12	0.11	0.01
Taylor and Jones, 1990	Balsam fir: Needles			1.77	--	0.11	--	--	--	--	--
	Twigs	--	--	0.50	--	0.04	--	--	--	--	--
	Birch: Leaves	--	--	1.41	--	0.08	--	--	--	--	--
	Lichens	--	--	0.91	--	0.09	--	--	--	--	--

Reference	Snow-litter conditions	B. Increases or decreases ( ) of nutrient concentrations									
		$N\text{-}NH_4$	$N\text{-}NO_3$	N_T	$S\text{-}SO_4$	S_T	$P\text{-}PO_4$	P	K	Ca	Mg
Fahey, 1979	Field; coniferous forest			0.0					0.38	0.47	0.07
Moloney *et al.*, 1983	Laboratory; leaching of coniferous litter						0.2		1.00	0.20	
Barry and Price, 1987[1]	Field; snow lysimeter, melt of 1985		(1.22)		(0.40)						
Stottlemeyer, 1987	Field; snowcores								0.10		
Jones, 1987	Field; snow lysimeters, melt of 1985	(0.03)	0.10		0.26		0.03		0.12	0.14	0.03
Courchesne and Hendershot, 1988	Field; snow lysimeters, melt of 1986		(1.80)		1.66				0.86	0.26	0.19
Jones and Taylor, 1990	Laboratory; leaching of balsam fir litter	(0.12)	(0.27)						0.52	0.12	
Mean		(0.75 ± 0.006)	(0.80 ± 0.86)		0.51 ± 1.05		0.12 ± 0.12		0.50 ± 0.37	0.24 ± 0.14	0.009 ± 0.008

N_T, total nitrogen; S_T, total sulphur; P_T, total phosphorus
1) Recalculated from published mass-balance data

suggested that much of the enrichment may be due to snow-canopy interaction (Schuepp et al, 1987) during the snowfall rather than to subsequent litterfall to the snow cover. Jones and Sochanska (1985) calculated mass balances for ionic species in the snowpack of a coniferous forest from daily snow core and precipitation measurements. They found that over the whole melt period the snow pack gained large amounts of ionic species that they attributed to the leaching of litterfall. They found that the overall percent increase of soluble ions in the pack over the amounts in precipitation (snow and throughfall) was greater for the nutrients associated with the intra-system terrestrial cycle than those deposited from the atmosphere, i.e. PO_4(750%) > Mn(680%) > K(619%) > Mg(187%) > NH_4(129%) > Ca(50%) > NO_3(30%) > SO_4(25%). The cations may have been mobilized directly by dissolution of salts and low molecular weight organics or been the result of ionic exchange with the hydrogen ion on weak Bronsted bases (Cronan and Reiner, 1983). The results, however, of Jones and Sochanska (1985) on the daily mass balance calculations were very variable, both losses and gains of many ions being recorded. The heterogeneity of the snow cover leads to large error deviations around the mean values (Jones, 1987) and no significant conclusions on the daily fluctuations of the ionic mass balances could be made at that time. In addition, the overall mass balances for the melt period will vary from year to year and, in some cases, may change sign (gains to losses) in different years (Barry and Price, 1987).

Leaching experiments in the field (Barry and Price, 1987; Jones and Deblois, 1987; Courchesne and Hendershot, 1988) or in the laboratory (Molony et al, 1983; Jones and Deblois, 1987) with snow lysimeters give results which can be more closely related to the interaction between snow meltwaters and tree litter than comparative studies on snow covers (Table 4B). Barry and Price (1987) carried out a comprehensive study on the concentrations and mass balances of solutes from plastic-lined (snow) and unlined (snow and soil) lysimeters in a deciduous forest. The authors reported that excess ion loads (Snowpack+precipitation-runoff for PO_4, K, Mn) were discharged from the pack in the lined lysimeters. The excesses were due to the leaching of leaves and organic debris which accumulated in the snow lysimeters before and during the formation of the seasonal snow pack. When the authors took special precautions to eliminate leaves from the lined lysimeters prior to snow cover formation, they observed decreases in the excesses of these ions. Barry and Price (ibid) also reported concomitant losses of H and NO_3 in the snow meltwaters whether the lysimeters were leaf-free or not. In contrast, SO_4 showed an alternance in its behaviour, a gain being registered in the first year (1983) when the lysimeter contained leaves and a loss in the second (1985, Table 5). Losses of ion loads were still attributed to interaction with organic matter as non-leafy litterfall interactive processes were proposed. In a similar study to that of Barry and Price, Courchesne and Hendershot (1988) allowed fresh litterfall in a maple-beech-birch forest to accumulate in snow lysimeters and compared the chemistry of the meltwaters with that of snow lysimeters with no litterfall. They demonstrated that the influence of freshly fallen litter on snow led to an increase in all nutrients except N (Table 5). H concentrations were also decreased by contact with meltwater. Courchesne and Hendershot proposed that microbiological assimilation of NO_3 and denitrification were the most probable mechanisms for the removal of N from the meltwaters. It should be noted that the study of Courchesne and Hendershot is not directly related to in-pack changes of snow and meltwater chemistry. In general, most of the litter in deciduous forests is deposited on the ground before the seasonal snow cover is laid down and litter degradation by soil organisms and invertebrates can mask any changes due to in-pack meltwater-litterfall interaction alone. The results of Barry and Price (1987), however, do show that in-pack changes in the chemistry of the meltwaters can take place in deciduous forests; but it is far less important to the chemical composition of the runoff than the interaction between meltwater and the litter horizon of the soil.

Table 5: Nutrient pools (eq ha^{-1}) due to precipitation (plus throughfall) in a forest snow cover and the overall losses or gains in the pools (eq ha^{-1}) due to the presence of litterfall

Reference	Snow cover, original atmospheric nutrient pool							
	$N\text{-}NO_3$	$N\text{-}NH_4$	N_T	$S\text{-}SO_4$	$P\text{-}PO_4$	K	Ca	Mg
Jones, 1987[1]	80	30	110	70	neg.	5	10	4
Reference	Snow cover plus litterfall: losses ( ) or gains in soluble nutrient pools over original pool in snow							
	$N\text{-}NO_3$	$N\text{-}NH_4$	N_T	$S\text{-}SO_4$	$P\text{-}PO_4$	K	Ca	Mg
Barry and Price, 1987[2]	(55)			(22)				
Courchesne & Hendershot, 1988[3]	(100)	--	--	117	--	76	45	55
Jones, 1987[4]	8	(10)	(2)	29	1	17	37	12
Stottlemeyer, 1987[5]						6		

neg. = negligible
1) Typical values for boreal forest snowpack mid-winter prior to snowmelt
2) Estimated from snow lysimeters, snowmelt of 1985
3) Estimated from snow lysimeters and lysimeters with litterfall, snowmelt of 1986; meltwaters in this system contained high concentrations of all nutrients even prior to meltwater-litter contact.
4) Estimated from snow lysimeters, snowmelt of 1985
5) Estimated from snow cores, open and wooded areas

Jones and Deblois (1987) studied the changes in the ionic loads of meltwaters in a boreal forest where the dominant stands were dense associations of balsam fir and white birch. Litterfall occurs throughout the winter. Measurements of the chemical composition of snow cover, meltwaters from snow lysimeters, and wet-only precipitation collectors were used to calculate ionic mass balances (Table 5) in a similar manner to that of Barry and Price (1987). Jones and Deblois also found that overall losses and gains could be reversed depending on the year (e.g. NO_3, loss, spring 1984, 6.1 eq ha^{-1}; gain, spring 1985, 8.3 eq ha^{-1}). They proposed that losses or gains were the net result of the physical leaching of litter and associated dry deposited material, and microbiological assimilation. Under certain conditions microbiological assimilation would exceed physical leaching to give net losses and vice-versa to give net gains. By repeating the study on daily mass balances on snow cover (see above, Jones and Sochanska, 1985) with the more precise lysimeter data, Jones and Deblois (1987) were able to show that when the pack was saturated with free water during rain-on-snow events, there was a tendency for the rate of in-pack assimilation of N, (i.e. NH_4 and NO_3), to dominate any leaching or production of the ions from the litter. On days with no rain microbiological activity decreased relatively faster than the leaching rate (including dry deposition) and gains of nutrients were recorded. K showed similar behaviour to N; other nutrients, P(PO_4), S(SO_4), and Ca, always showed gains while the results for Mg were ambivalent. From the amount of litterfall captured over winter at the boreal forest site, the composition of the litter, and the excess ion loads in the meltwaters, Jones and Deblois (ibid) estimated that 59% of the Mg content, 51% of the K content, and 41% of the Ca content of litter were released by leaching and microbiological activity. The results were, however, not statistically significant and Jones and Deblois (ibid) and Jones and Tranter (1989) had to resort to laboratory experiments to verify the observations in situ. By means of a snowmelt simulator they conclusively demonstrated that N, (NH_4 and NO_3), in-pack depletion can occur due to meltwater-litter interaction. In-pack production of K and Ca by leaching occurred consistently but, in contrast to the field experiments, no K depletion was ever found. This suggests that microbiological activity on meltwater-wetted litter in the laboratory is not the same as in natural snowpacks. It is known, for example, that in addition to the microbial communities attached to litter, the microbiological populations of natural snow cover consists of autonomous true snow-algae (Hoham, 1987) and snow-fungi (Stein and Amundsen, 1967). Rapid population growth of these communities in melting snow may be responsible for the cation depletion (K, Ca) that has been occasionaly recorded in natural snow covers (Hoham et al., 1989).

Both Moloney et al (1983) and Taylor and Jones (1990) have studied the respiration rates of wetted coniferous litter in the laboratory as a measure of microbiological activity. The results are very similar; they can be extrapolated to show that the microbial in-pack repiratory rates of litter during the melt period is approximately 1.5 mg CO_2 g^{-1} day^{-1}. This represents approximately a tenth of the respiratory activity that is measured in fresh litter in summer (Taylor and Jones, 1990). The potential of microbial activity in litter to take up nutrients has been shown in a more direct manner by Taylor and Jones (ibid). They measured significant increases in the N and S contents of certain components of a boreal forest litterfall between the end of the snow cover accumulation period and the end of the springmelt.

The leaching and microbiological degradation of meltwater-wetted litter also releases soluble organic matter. Stottlemeyer and Toczydlowski (1989) and Courchesne and Hendershot (1988) attributed the anionic deficits in snow meltwaters from deciduous forests to organics containing carboxyl groupings. Jones (1987) estimated that the mean increase of soluble organics in snow lysimeter meltwaters in a boreal forest was 3 kg ha^{-1}; the mean equivalent weight of the material was 140 g M^{-1} which is similar to the eqivalent weight of soluble humic matter (100 g M^{-1}) in natural waters (Oliver et al, 1983). Jones and Tranter (1989) subsequently found that meltwater leaching of balsam fir litter in a laboratory experiment increased the soluble organic content of snow by 185%; extrapolation of the results to the boreal forest site yielded an equivalent excess discharge of organics of 1.08 kg ha^{-1}..

All of the above studies conclude that litterfall changes the nutrient pool of snow cover; the greatest changes take place during the spring melt period. Tables 4A,4B and 5 resume the results of the above studies and permit a comparison of different litter and forest types by transposing the data on original nutrient pools and fluxes into the same units. It can be seen that in all cases the concentrations of the cationic nutrients and P-PO_4 are increased in snow meltwaters. For the cations the results are relatively consistent even though the experimental conditions and litter types were different. In general, N-NH_4 and N-NO_3 concentrations are reduced presumably by microbiological assimilation and possibly denitrification. Denitrification, however, would only occur if the wetted litter would develop anaerobic sites within the organic matrix; this can happen in the case of soil particles (Knowles, 1978). Snow meltwaters are well oxygenated and the existence of anaerobic conditions in litter is unlikely although no specific experiments have been made to resolve the issue one way or the other. S-SO_4 shows substantial increases in concentration in litter-wetted meltwaters, most probably due to the leaching of dry-deposited material, although Barry and Price (1987) do report a reduction in concentration. If wetted litter can become anaerobic then sulphate reduction could accompany denitrification. Some direct assimilation of S can also take place. Jones and Tranter (1989) in a series of laboratory experiments did observe both increases and decreases in the concentration of SO_4 of meltwaters in contact with fresh balsam fir litter. There was, however, no recognisable pattern to the fluctuations with the experimental conditions. Gamache (1990) and Hoham et al. (1989) have also observed SO_4 reduction in snow packs when snow algae were present; in this case S-SO_4 depletion would be due to algal assimilation. Finally table 5 on nutrient pools shows that litter contributes more to the soluble cationic load of snow cover than that deposited from the atmosphere. It contributes less S-SO_4 than the snowfall and is effective in reducing the N pool.

Snow and animals

To attempt to determine the influence of animals on the chemical composition of seasonal snow covers one must take into account the distribution of plants and the spatial and temporal variability of animal populations in winter and spring. Plants and animal populations are very closely associated in winter due to the increased energy expenditure needs by animals at that time of the year when the availability of food is reduced by snow cover.

Formozov (1946) has written a very comprehensive monograph on the relationship between snow cover and the ecology of mammals and birds. The physical characteristics of snow cover in different ecosystems plays a major role in the distribution of mammalian and avian species and, in particular, is the dominant factor in the rate of over-winter mortality. Foxes, for example, which can move fairly easily over open tundra snow cover conditioned by the wind will perish in forest fluffy snow cover. The soft nature of the forest snow proves a serious obstacle to the movement of mammals. Small mammals such as shrews and voles cannot effectively move through snow more than 3-6 cm deep. Deer have great difficulty in running in snow over over 20 cm deep but moose can progress with little effort in snow depths of up to 50 cm. Deep snow will again favour the small mammals; they use subnivean air spaces for foraging (Coulianos and Johnels, 1962) at the snow cover-soil interface protected from low air temperatures by the thermal insulation of the cover and from attacks by predators. On the other hand deeper snow cover reduces the foraging efficiency of the larger mammals. The deep cover increases the energy expenditure of herbivores (Kelsall, 1969) to uncover ground vegetation such as lichens; the animals have to resort to alternative sources of nutrition i.e. tree browse, bark and epiphytic lichens. Predators of the large herbivors are very dependent on snow depth and snow-surface hardness. In certain fluffy snow conditions moose and deer may easily outrun wolves. The former find themselves at a distinct disadvantage, however, when deep snow is covered with thick icy crusts (Verme, 1968). In these conditions which usually prevail near the end of winter, the large and middle-sized herbivores break through the crust, their limbs become lacerated and they fall to predators who attack rapidly on the hard icy surface.

Mammals and snow chemistry: nutrient cycling

The ecology of animals and birds is complex and very little is known about the role that the populations play in the cycling of nutrients in divers ecosystems during winter. All animals, insectivores, herbivores and carnivores, however, will directly effect the chemistry of the snow cover by both urine and solid excrement. In addition, the very active microtene rodents (e.g. voles, Kalliomaki et al 1983, 1984) of the subnivean environment can indirectly lead to changes in gas concentrations (e. g. CO_2, N_2O) in the lower layers of the snow pack. The origin of the CO_2 is not so much due to the respiration of the animals themselves but rather to microbiological activity in the soil (Bashenina, 1956). Under certain conditions of soil temperatures and snowpack structure high concentrations of CO_2 will be produced in the subnivean airspace (Pruitt, 1970). The small burrowing mammals, however, also produce vertical shafts to the surface; the tunnel networks are thus well ventilated and prevent the build-up of toxic levels of the gas (Adams, 1981).

In so far as the direct effect of the animals on the snow cover is concerned, the greater part of nutrient cycling in forests during the winter is most probably due to middle-sized herbivores i.e the forest populations of deer. Although the active populations of burrowing animals may be very high at certain periods their excrement is mostly restricted to the litter layer at the soil-snow interface where they search for insects or strip the basal tissue of vegetation (Coulianos and Johnels, 1962). In the open areas of tundra and steppe snowfields the animal populations and the food supply are so sparse that nutrient fluxes onto snow must be relatively small; these assumptions should be taken with caution as there is, in effect, no literature which deals with the specific study of nutrient fluxes in snow due to animal populations. In addition the heterogeneous nature of snow cover may give rise to a great diversity of species and population size over small areas (Formozov, 1946). The results of studies on nutrient fluxes may be site-specific and not bear any direct relationship with any real regional nutrient cycling controlled by the presence of snow.

It is possible, however, to gain some insight into the relative pools and fluxes of nutrients in snow-covered areas by combining data on the population density of animals in winter and the chemical composition of excrement. The estimation of population density of some mammals (e.g. voles and shrews) is, however, almost impossible under snow cover and even well designed trapping techniques offer at most a crude estimation of the numerical importance of the population (Fuller et al, 1969). On the other hand the displacements and fluctuations of deer populations can be measured in winter (Fowler and Dealy, 1987). These authors attempted to develop a predictive model for the population density of mule deer (Odocoileus hemionus) in the Keating winter range of northeastern Oregon. On a regional basis deer populations preferred to remain at an optimum elevation; in addition the daily movement of deer was influenced by the wind chill factor (Oliver, 1973). However, the number and movement of the deer into and out of cells (variable areas of the range delimited by certain site and associated vegetative parameters) were found to be dependent on the depth of the snow cover. As snow depth increases animal occupancy decreases. Figure 6 shows the avarage population density of mule deer during the winters of 1976-1979 in the Keating range. The isograms show the deer to be concentrated in two areas where the density of the population can attain over 200 individuals per acre (approx. 500 deer ha-1). The regional density, however, is much lower i.e. 1 animal ha^{-1}. The favoured habitat of the deer were cells where snow depth varied from 6 to 25 cm.

The chemical composition of deer excretion is also known. The majority of soluble nutrients are found in urine rather than faeces. Urinary nitrogen is mostly in the form of N-urea while faecal nitrogen is present as insoluble crude protein (8 - 26%, Margantoni and Hudson, 1989). In a recent series of articles DelGiudice et al (1987a,b,c, 1988, 1989) have reported the chemical analyses of deer bladder urine and urine collected from snow. Urinalysis is a relatively new technique for the identification and study of nutritional indices in mammals, particulary wolf and deer. The chemical characteristics of urine will change with the low digestibility of winter diets (Robbins et al, 1974) and also with the concomitant physiological state of the animals. During early undernutrition in winter deer will conserve urea nitrogen, sodium, and potassium by renal tubular reabsorption. The concentration of these components of urine will therefore decrease. As starvation progresses, however, the animals will catabolise endegenous protein, including bone collagen, and urea nitrogen, sodium, potassium, and calcium concentrations increase dramatically (DelGiudice et al 1987a). Phosphorus changes are less marked (DelGiudice et al, 1989). Table 6 shows the concentration of urea nitrogen, sodium, and potassium in white-tailed deer (<u>Odocoileus</u> <u>virginianus</u>) bladder urine and urine collected from snow in Minnesota. Deer will excrete approximately 25-30 ml of urine per kg of animal weight per day (Patenaude, 1990). From the data on the population density of deer, the average amount of urine excreted daily, the mean composition of urine,

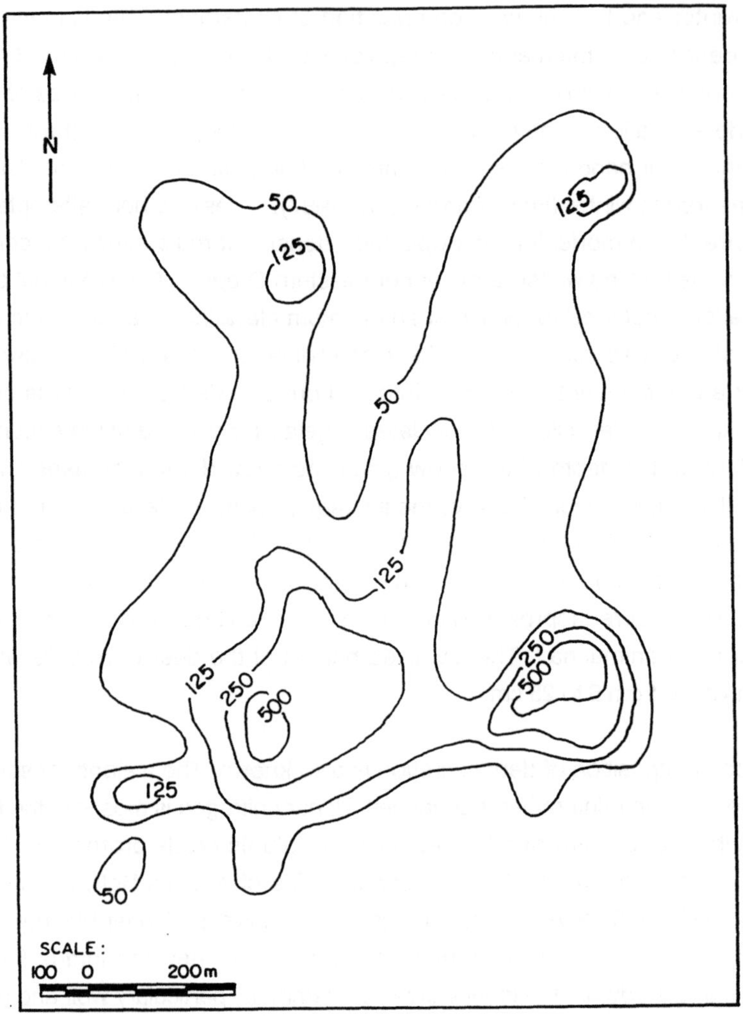

Figure 6. Population density (animals ha^{-1}) of mule deer <u>Odocoileus hemionus</u> on the Keating Winter Range, northeastern Oregon, U.S.A.

Distribution recalculated from data of Fowler and Dealy (1987) and animal mean weight, the rate of nutrient transfer from deer to snow cover can be calculated. Table 7 and Figure 7 show that the amount of N-urea that is deposited as deer urine can significantly increase the total N-pool (NH_4 and NO_3) derived from atmospheric deposition. The results show that recycling of N from vegetation to the forest floor reservoir by these animals can be a major pathway even in the winter season. Inorganic N, however, can be rapidly incorporated by vegetation while N-urea has to be first metabolised by microorganisms to N-inorganic before it becomes available for plant growth. The rate of N-urea transformation in snow is not known. In a series of controlled experiments on the stability of N-Urea in snow, DelGiudice et al (1988) reported that N-urea is stable in cold snow. When snow begins to melt N-urea concentrations decrease; from the consistency of the ratio of N-urea to another urinary metabolite (creatinine) the authors attributed the diminishing concentrations to meltwater dilution. In-pack microbiological degradation of N-urea cannot, however, be entirely ruled out.

Table 6: Chemical composition (meq L^{-1}) of deer bladder urine, of urine collected from snow 24, 48, 72 and 120 hours after deposition, and snow[1]

Element[2]	Bladder urine	Urine in snow				Snow
		24 hr	48 hr	72 hr	120 hr	
N	351	73.6	68.3	68	21.4	0.01[3]
P	5	--	--	--	--	--
K	78	16.6	15.3	15.8	4.6	0.001
Ca	150	--	--	--	--	--

1) Original data taken from DelGuidice <u>et al.</u>, 1988, 1989.
2) Recalculated from DelGuidice <u>et al.</u>, 1988, 1989.
3) No data by DelGuidice, calculated mean value for mid-western snowcover from Stottlemeyer, 1987.

Many forest systems are not suitable habitat for deer and they will be replaced by other herbivore species. In, for example, low-lying boreal forest moose is the dominant large herbivore. This species is also a member of the cervidae but, in contrast to deer, covers its territory at much lower population densities. The occupation of good moose habitat in the southern Canadian boreal forest can vary from approximately 0.001 to 0.005 animals per hectare (Crete, 1990). Moose, however, produce relatively higher quantities of excrement per animal as they weigh more than deer of comparable age and sex; Table 7 shows the contribution of moose to the pool of nutrients in a typical boreal snow cover.

Table 7: Typical nutrient pools (eq ha^{-1}) and fluxes (eq ha^{-1}day^{-1}) in a forest snow cover on a mid-winter day due to herbivore activity

	Pool				Fluxes[5]			
	N	P	K	Ca	N	P	K	Ca
Snowcover[1]	54	neg.	2.7	5.4				
Deer[2]					14	0.7	3.2	
Hare[3]					0.18	9×10^{-4}	0.04	0.08
Moose[4]					0.01	6×10^{-5}	0.003	0.006

neg. = negligible
1) Hypothetical snowcover (150 mm i.e. maximal-depth tolerance, 50 cm, for browsing deer)
2) Population density 20 animals ha^{-1}; Fowler and Dealy, 1987
3) Population density 10 animals ha^{-1}; Alain, 1990
4) Population density 0.005 animals ha^{-1}; Crête, 1990
5) Fluxes are calculated from data on deer urine (also used for hare and moose, DelGuidice et al., 1988, 1989.

In addition to deer and moose, other mammals may occur in large enough numbers (e.g. hares) to contribute appreciably to nutrient cycling. As previously mentioned, faeces are also a means of waste product metabolism and and are a part of the herbivore pathway for nutrient cycling in many ecosystems. Walker (1985) has shown, for example, how vegetative patterns in the treeless high tundra can be correlated with the occurrence of different mammalian faeces. The availability of total faecal nitrogen (crude protein) to plants, however, is less than that of urinary nitrogen due to differing microbial degradation rates. Faecal nitrogen, therefore, is still a potential source of nitrogen for plant production but over a longer time frame than urinary nitrogen. Pruitt (1970) has, in effect, pointed out the role of hares in localised nutrient cycling in shrub (alder and birch communities) during winter by the accumulation of faecal pellets around the shrubs utilised the most heavily. Although urine is the main element in soluble nutrient waste, the faeces may also contain some smaller amounts of nutrient leachable-species. There is, however, no data on the leachates of faeces in snow but they may be, under certain circumstances, a source of soluble chemical species in the snowpack (Jones, 1987). This author concluded, for example, that the inclusion in snow cores of hare faeces which are deposited in profusion in boreal wooded areas in spring, was responsible for the relatively high concentrations of nutrients in some snow samples. Concentration ranges in the snow, particulary of phosphates, were thus very large (Jones and Sochanska, 1985) and Jones (1987) resorted to subjective snow coring (i.e. avoiding the collection of animal faeces) to obtain a better picture of the evolution of nutrients deposited by atmospheric and vegetative fallout in wooded areas. The method, however, is flawed to some extent as urine-wetted snow is essentially

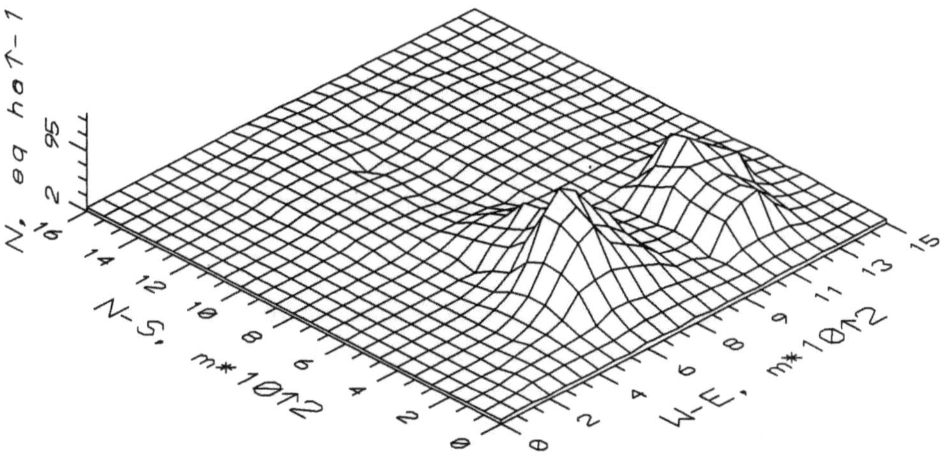

Figure 7. Typical N-pool and N-fluxes in a part of the Keating, Winter Range, northeastern Oregon for a mid-winter's day (winters 1976-1979) due to deer.

N-pool is the background value of snow cover, 8 eq ha^{-1}

N-fluxes are superimposed on the pool and vary up to 175 eq ha^{-1} day^{-1}.

Fluxes calculated from a typical population density distribution (Fowler and Dealy, 1987) and the N-urea content of deer urine (DelGuidice et al., 1987 a b c).

Note 1. the two high flux regions here deer browse close to the Powder River which runs along the W-E axis. 2. The low background value of the N-pool in snow due to low depths and low nitrogen content of the cover.

impossible to distinguish from urine-free snow in litter-laden boreal-forest snow covers. The study site used by Jones was a dense balsam fir stand; this is typical of very large areas of the boreal forest in which deer are not common and most herbivore cycling of nutrients is due to small mammals (eg. shrews, voles, hare) and birds. By using a similar method to that for the estimation of nitrogen cycling by deer the contribution of small mammals to the fluxes of N to snow in boreal forest systems has been determined (Table 7).

Table 7 permits a comparison of the effects of different herbivores on snow nutrient pools. The table shows that deer can contribute substantially to the nutrient pool in certain areas; at a density of 20 animals ha-1 the animals could double the N- pool of a typical snowpack in three days. On the other hand, moose would only increase the N-pool of the pack by 0.06% in the same period. The figures are, however, not directly comparable. Deer are gregarious species and move over territory in herds. Population densities can vary from 0 to an extreme high density of 500 animals ha-1 (Fowler and Dealy, 1987). Moose have a more homogeneous distribution but of low density. In a large region of 10^5 km^2 of the boreal forest the figures shown for moose in table 7 would be a reasonable estimate of the true effect of the animals on nutrient cycles over a winter season. The figures for deer, however, would be only valid for small areas of 5-10 ha and at certain times only. Over a very large area the values for deer could be one or two orders of magnitude less. Table 7, nevertheless, does give an idea of the extent of animal-driven nutrient cycling in forest ecosystems.

In open snowfields the density of animals and birds is generally less than that of the forest ecosystems. In certain situations, however, some populations may be appreciable; e.g. polar snowfields close to the ocean frequented by penquins. Christie (1987) proposed that the elevated nitrogen concentrations that he measured in snow overlying two moss-dominated communities in the South Orkney Islands were due to the birds, either by direct excrement, wind blown dust from the colony, or ammonia volatilization. When bird activity on the snow was high the nitrogen concentrations rose from 40 $\mu g\ L^{-1}$ to approximately 100 $\mu g\ L^{-1}$. The penquin populations thus contributed 60% of the direct input of N to the moss beds during snowmelt. The major contribution was from ammonia. The ammonia is the result of the rapid decomposition of urine in the colony, the volatilization of the gas, and gaseous dry deposition and wet deposition to the snowpack. NH_4/NO_3+NO_2 ratios rose from 2.9 in accumulated winter snow to 9.9 in the summer pack when the colony was active. Total average input of inorganic N to the two moss beds was 184 mg m^{-2} $year^{-1}$ (131 eq ha^{-1}) of which 64 mg^{-2} $year^{-1}$ (45.7 eq ha^{-1}, 35%) was due to snow (including penquin contribution of 50 mg m^{-2} $year^{-1}$, 35.7 eq ha^{-1}) and 120 mg

m^{-2} $year^{-1}$ (85.7 eq ha^{-1}, 65%) due to nitrogen-fixing epiphytic cyanobacteria. On the other hand, birds may effectively remove potential nutrients from open snowfields. Hendricks (1987), showed in a study of water pipits at high altitudes in Wyoming that the population of arthropods available to the birds on a snowfield was greater than that in adjacent alpine meadows. Although absolute population densities are different, the arthropods are easy to locate against the snow background and foraging efficiency of the birds is high. The author, however, stated that, in spite of the abundance of available prey on the snow, the birds generally avoided the snowfield. Edwards and Banko (1976), on the other hand, observed that bird activity on Alaskan snow patches was highest when arthropod populations peaked and total nutrient transfer from snow cover to non-snow areas was maximised under these conditions. The pool of particulate nutrients (total living and dead organisms) was approximately C, 17.8 eq ha^{-1}; N, 3.1 eq ha^{-1}; P, 0.08 eq ha^{-1}; Ca, 0.003 eq ha^{-1}, and Mg, 005 eq ha^{-1} (Edwards and Banko, 1976). According to our definition of biologically-induced chemical changes in snow, however, we cannot establish if this arthropod-bird relationship actually decreases or increases the quantity of soluble nutrients in snow. Arthropod population density on snow is relatively low, 3-4 animals m^{-2}, (Hendricks, 1987; avarage observations per day over 6 days). The quantity of matabolic products is negligible compared to the pool in the snow and the excreta of the birds probably more than compensates for arthropod removal.

Snow and invertebrates

The severe effects of low temperatures on invertebrates are reflected by the restricted flora and fauna in cold regions. Downes (1965) in his treatise on the adaptations of insects in the Arctic states that on the southern tundra the number of insect species is about 5% of that of a corresponding temperate area, and in the high arctic, little over 1%. In the Antarctic with a rigid cold polar climate very few species of insects are found. In temperate zones with well defined winter seasons and snow cover the majority of invertebrate populations with inter-annual life cycles will survive the cold by overwintering in specialized microhabitats, developing cold-hardiness, becoming dormant, or using behavioral techniques such as cocoon spinning (Danks, 1978). With the onset of the snow cover the activity of invertebrates will thus decrease drastically in early winter. It will recover, albeit at very low levels, when the insulating effect of the snow cover permits a stable hiemal threshold and in spring when the snowpack depth decreases (e.g. Aranea; Aitchison, 1978).
In spite of the very low overall populations of invertebrates in winter, many species have been observed to remain active beneath or on the snow surface (e.g. Collembola, summer population density, 80,000 m^{-2}; Hagvar, 1984: winter population density, 1000 m^{-2}, Aitchison, 1979). Those species which forage or

need, at some time of their life cycle, to reside on the snow surface i.e. true snow insects, were first noted by Brauer (1871) while Frey (1913) distinguished true snow insects from those that accidentally find themselves deposited on the snow. An example of the latter type of situation is the arthropod fallout on alpine snowfields by ballooning spiders (Crawford and Edwards, 1986); the aeolian-transported spiders from lower altitudes which are unfortunate enough to be trapped on snow will become too chilled to move, go into a torpid state and become easy prey for birds or true snow arthropods. Flying insects in alpine regions will avoid snowfields; Ion and Kereshaw (1989) found that the ratio of catches of flying-insect traps on lichen-heath and adjacent snowfields in the Selwyn/MacKenzie mountains of Canada was 160:1. The true snow invertebrates are those which leave the subnivean environment for the snow surface when conditions are either favourable (for forage) or absolutely necessary (for copulation). Most of these species are inhabitants of the temperate coniferous forest. If conditions become too severe at the surface, large scale mortality may occur (e.g. collembola) or retreat to the subnivean milieu made. Hagar found that winter-active spider, Bolphantes index, makes its webs in crevices in the snow which is honeycombed with numerous passages down to the soil surface. As these particular animals cannot burrow in snow they have to rely on the openings to descend to the relative warmth of the subnivean air spaces when temperature or wind conditions are unfavourable.

Invertebrates and snow chemistry: nutrient cycling

To evaluate the effect of true snow insects on the chemical composition of snow we have to combine data on population density and metabolic activity per animal. There are publications which allow the estimation of of some surface-snow invertebrate populations but data on the metabolic activity of these populations does

not exist. It is, however, possible to calculate rates of release of some metabolic products to obtain a first approximation of invertebrate-snow chemical fluxes.

As most subnivean invertebrates are never found on the snow surface the surface populations are extremely low (Table 8). From the observations of Aitchison (1979a,b,c,d) on subnivean soil samples and pit-trap catches one can estimate the subnivean populations of some invertebrate populations in southern Canada. Invertebrate population density varied from 50 m^{-2} (diptera) to 2,500 m^{-2}, (acari). In contrast, Hagvar has estimated the surface populations of some true snow invertebrate species in coniferous forests of Scandanavia to be from 0.01 m^{-2}, highest value for Chionea araneoides (Hagvar, 1971) to 1 m^{-2}, highest value recorded for Bolyphantes index (Hagvar, 1973) and Diamesa permacer (Hagvar and

Ostbye, 1973). Surface-snow/subnivean population ratios are thus very low ($2.5x10^3$:1) and we would expect the metabolic activity on the surface snow to be correspondingly low. In addition, the surface populations are very dependent on temperature and wind conditions. Figure 8 shows the population density of the tipulid Chionea araneoides on surface snow under different conditions of temperature and wind. This organism is typical of true snow invertebrates in that they show maximal activity in a supercooled state (Somme and Ostbye, 1969). The animals will leave the snow surface when the temperature goes below 6°C; if the animals cannot find more favourable conditions they will develop chill-coma and become torpid. Although normal surface-snow populations of invertebrates are low some species will invade the surface of the snowpack in very large numbers during the spring. Of particular importance is the very high population densities of collembola which can cause the snow cover to become black (Isotoma saltans), red (Onychiurus alborufescens) or yellow (Onychiurus cocklei) in spring (Chapman, 1954). The animals have to attain the snow surface to mate and the energy expenditure is high. In contrast to the higher animals, invertebrates can excrete the majority of wasted N-containing metabolites as non-urea compounds e.g. NH_4 (Persson, 1980, quoted by Rosswall, 1981).

Table 8: Typical nutrient pools (eq ha^{-1}) and fluxes (eq $ha^{-1}day^{-1}$) in a forest snow cover during a warm spring day due to invertebrate activity

	Pool		Fluxes[4]	
	N	P	N	P
Snow[1]	54	neg.		
Collembola[2]			40	4.8
Other invertebrates[3]			1.2	0.14

neg. = negligible
1) Hypothetical snowcover (150 mm, SWE)
2) Collembola populations, maximal activity, mating period, 2 x 10^7 individuals ha^{-1} (2 x 10^3 m^{-2}); estimated from Aitchison, 1979
3) Araneida and Tipulidae populations, estimated from Hagvar 1973 and 1971 respectively.
4) Estimated using respiratory-excretion relationships in Handbook of environmental data and ecological parameters, 1979.

As the animals respire, NH_4 is released at the snow surface and the quantity of ammonia can be calculated. Assuming the population of collembola to be 2,500 m^{-2}, the mean dry weight per animal to be 0.06 mg (Custer et al,1986) and the mean respiratory rate per animal to be 0.55 μl O_2 hr^{-1} (HEDEP, 1979) one can use a series of equations which relate repiratory rates and animal weight to the excretion rates of different forms of nutrients (HEDEP, 1979). Table 8 shows the calculated NH_4, total N, and total P that would be released at the snow surface by a collembola swarm. The amount of the excretory products incorporated into the snow would depend on

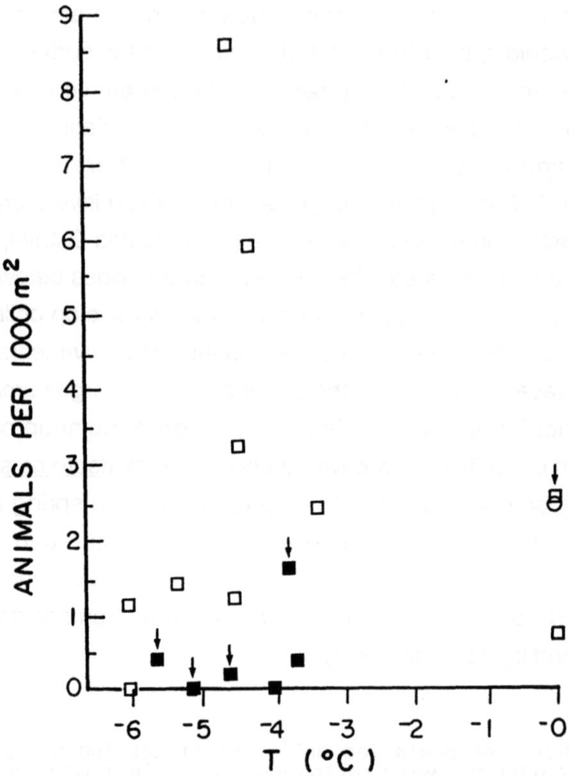

Figure 8. Population density of the tipulid <u>Chioea araneoides</u> Dalm on surface snow according to conditions of temperature and wind. Population densities on windy days are the solid symbols, arrows indicate falling snow. Taken from Hagvar (1971).

many factors. The fact, however, that collembola invade the snow cover during melt when the snow is wetted would favour the dissolution of the metabolites into the meltwaters. From table 8 it can be seen that the high activity of collembola populations could lead to extremely high nitrogen and phosphorus fluxes (N; 40 eq ha^{-1} day^{-1}: P; 4.8 eq ha^{-1} day^{-1}) on the snow pack surface. Collembola populations are, however, very patchy on snow cover and the true population density over an area of a hectare may be one or two orders of magnitude smaller than the figure used for the calculation of the N fluxes. Using a similar method to calculate the influence of other invertebrates on snow (e.g. Aranea, 1 animal m^{-2}; weight, 2.0 mg, respiratory rate, 3 μl O_2 hr^{-1}) leads to a small flux of approximately 1.2 eq ha^{-1} of N, and 0.14 eq ha^{-1} of P per day.

Microbiological activity and snow chemistry

Microbiological activity may take place on or under the snow cover. The activity under the pack is the respiration of soil organisms and associated processes such as nitrification and denitrification. Respiration and denitrification give rise to CO_2 and to N_2O+N_2 respectively which diffuse up through the snow under the influence of high concentration gradients in the upper soil horizons. If the snow cover is relatively deep (> 50 cm) the lower part of the snowpack will experience high concentrations of these gases due to the absence of any convection currents and the limited rate of diffusion in the small interstitial air spaces. Solomon and Cerling (1987) measured the CO_2 profiles in soil and snow cover in a montane region of Utah. They found that the lower 0-20 cm of the pack showed consistent concentrations of > = 4000 ppmv when the snow depth was over 1.5 m (January-April). During the melt period the concentration gradient in the pack was 25 ppmv CO_2 cm-1. Meltwater percolating down through the pack contains progressively greater amounts of dissolved CO_2 as the equilibrium conditions change. The pH of the waters will diminish and increased weathering of soil and calcerous particulates previously deposited on the snow by wind (DeWalle et al, 1983) will occur. As soon as the pack disappears soil surface concentrations of CO_2 return to normal atmospheric values. In the case of the montane soil the rate of CO_2 release from the soil was less under the snow than when the soil was snow free. In other systems, however, CO_2 production may be higher underneath the snow pack than in snow-free periods (Schinner, 1981).

Rhodes et al (1986) studied NO_3 concentrations in a snow covered forested watershed in the Sierra Nevada and concluded that denitrification occurred throughout the winter. N_2O concentrations in the snow pack were not measured; however, from the results of a summer field experiment at the same site, and the temperature differences between the soil surface in summer and winter, it can be estimated that the rate of N release from the ground to the snow pack would be of the order of 0.3 eq ha-1 day-1. Reynolds (1983) from a comparative study of new snow and old snow has suggested that NH_4 is another gaseous N-species that could be released from soil to snow cover. In all of the above examples the in-pack chemical changes are due to the low diffusion rates of the gaseous emissions through the pack and the physical dissolution of the gases by meltwaters; in the case of ammonia surface capture by ice may also occur.

True in-pack chemical changes due to microbiological activity is mostly due to the productivity of snow algae, snow fungi, and other microbial flora and fauna. Snow algae (Hoham, 1971) have been the subject of more attention than the fungal (Stein and Amundsen, 1967) and bacterial (Visser, 1973) communities and the ecology of snow algae is consequently far better understood than that of their heterotrophic counterparts. Hoham (1987) has reviewed the ecology, distribution and composition of snow algae and associated psychrophilic microbiological organisms (fungi, protozoa and rotifers); these decomposers and primary consumers may depend to a large extent on the autotrophic production of the algal populations.

Snow algae

Snow algae have been reported in all continents except Africa (Hoham, 1971, 1980, 1987; Marchant, 1982; Gerrath and Nicholls, 1974; Kol, 1968; Kobayashi, 1967). Kol (1968) has described 110 species of snow algae; the species belong to many phyla, the Chlorophyta (green algae), the Euglenophyta, Chrysophyta, and Pyrrophyta (Dinoflagellates), the Cyanophyta (blue-green algae), and the Cryptophyta. The majority of the snow algae, however, belong to the Chlorophyta and are responsible for the coloration of snowfields particulary in springtime and, in the case of alpine fields, in the summer (Hoham, 1971; Kol, 1968; Fukushima, 1963). Chlamydomonas nivalis (Bau.) Wille, for example, is a green alga that is often found in high-altitude snowfields above the treeline in summer. It gives a characteristic red coloration to the snow at the end of its growth period due to the production of a secondary pigment (l'astaxanthine, Viala 1966); secondary pigment formation may be stimulated by nitrogen depletion in the snow (Czygan, 1970; Bidigare, personal communication). Green coloration of snow is also due to the Chlorophyta (genus Chloromonas); in general these algae are found in coniferous forest snowpacks and are shade-loving species (Hoham et al, 1979, 1983; Hoham and Mullet, 1977; Hoham, 1975). Red snow in deciduous stands due to the dinoflagellate Gymnodinium pascheri has been reported (Gerrath and Nicholls, 1974) while yellow and orange coloration of snow caused by green algae occurs in partially shaded coniferous stands (Hoham, 1980). Both Pollock (1970) and Kol (1975a,b) have also related the various colorations of mountain snow cover (Pollock, North America; Kol, High Tatra Mountains of Hungary) to different microbiological communities.

The life cycles of snow algae are very complex and difficult to elucidate due to the transient nature of the snow habitat and the uncertainties in the identification of the organisms. Hoham and coworkers (Hoham, 1974a, 1974b, 1975, 1980; Hoham and Mullet, 1977; Hoham et al, 1979, 1983) have emphasised the need for the correct identification of the different stages in the life cycle of snow algae. They have thus devoted much time to the reassignment of incorrectly classified stages to the correct species. Figure 9 shows the life cycle of a typical snow alga (Hoham, 1987). When air temperatures rise meltwater is produced and nutrients are mobilised; light penetrates the snow and stimulates the germination of resting zygospores at the soil/snow interface in seasonal snow covers or at the old-snow/new-snow interface in permanent snowfields. Zoospores are released which swim through the interstitial meltwater towards the surface; the depth of snow and water content is important for zoospore migration. The zoospores divide at a certain depth below the snow surface to form daughter cells which are asexual (zoospores) or sexual (gametes). Colored bands of algae develop in which populations may reach 500,000 cells mL^{-1} (Hoham and Blinn, 1979) or 1,000,000 cells mL^{-1} (Hoham et al., 1989). Sexual reproduction gives rise to new zygotes which return to the soil or permanent snow surface where they lie dormant until the cycle recommences with the following spring or summer melt. The time span of the vegetative phase may be very short, a few days to a few weeks. However, there is not enough knowledge on the nutrient evolution of snow and algal populations to clearly identify the factors, nutrient or otherwise (Hoham, 1985), which control population density and the different stages in the life cycle.

Snow algae and snow chemistry

For many years workers in the field of snow algae have carried out some chemical analyses of snow as a measure of the quality of the algal habitat. Conductivity and pH are the most common parameters and have been used in some cases to classify the snow as it relates to different algal communities (Kol, 1968). Hoham and Mohn (1985) have shown that the chemistry of the snow is a factor in snow-algal ecology by demonstrating that different species do have different optimum pH requirements for growth. Algae will develop when the physical and chemical conditions are right but in the process of colonizing the habitat will themselves change the chemical environment. Algal blooms will, for example, increase the pH of the snow by

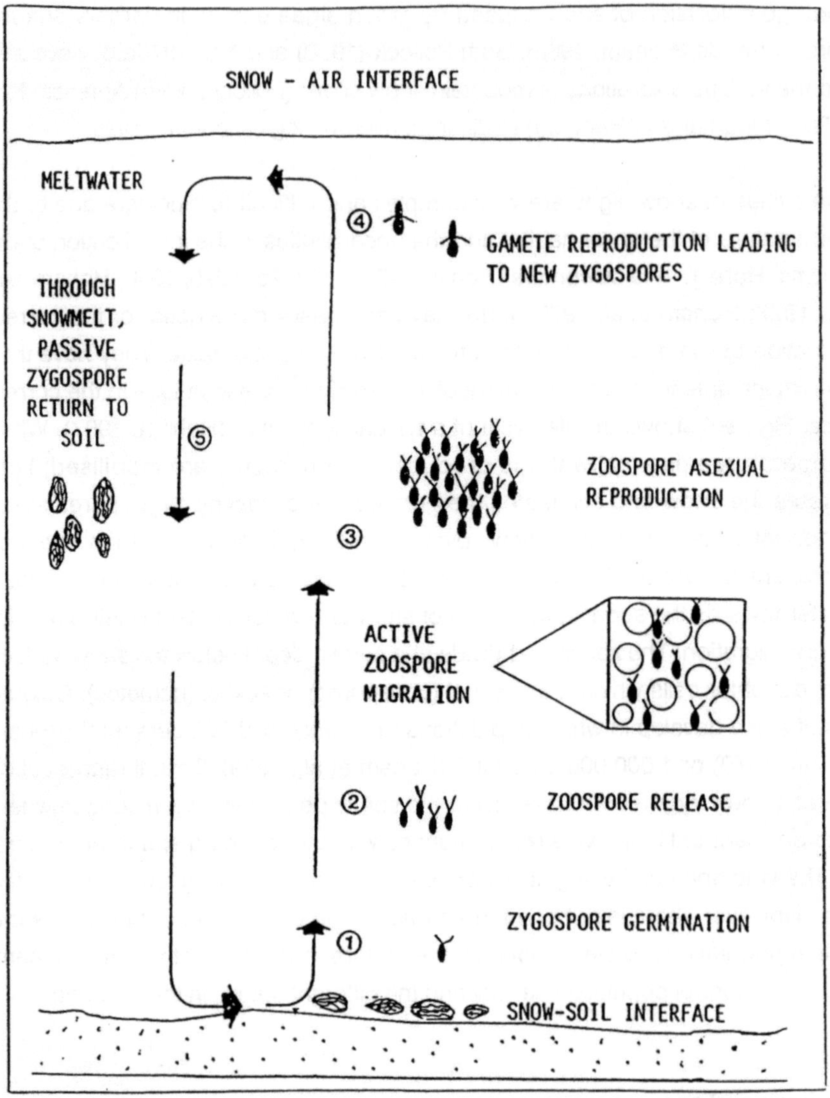

Figure 9. Life cycle of snow algal flagellates (modified after Gamache, 1990).

assimilating CO_2 during photosynthesis (Hoham, 1987). Newton (1982) found that Chlymadomonas nivalis tends to colonize snow areas with high conductivity but it could not be determined in this case if the increase in ionic species was due to algal metabolic products or a result of physical processes in the algal-rich areas.

In recent years, the relationships between the nutrient contents of snow and snow algal populations are beginning to draw attention (Gerrath and Nicholls, 1974; Hoham, 1976; Fjerdingstad et al, 1978; Hoham, 1987; Hoham, 1989; Hoham et al, 1989; Gamache, 1990). Attempts are being made to directly relate the biological activity to nutrient dynamics; it is from these studies that most of the knowledge on snow ecology and nutrient dynamics is being obtained. Gerrath and Nicholls (1974) in a study on a bloom of Gymnodinium pascheri in Ontario, Canada, analysed the snow for nutrients. N concentrations were low (2.6 μeq L^{-1}) but Ca and P levels were high (P, 0.4; Ca, 1000 μeq L^{-1}) compared to typical snowpacks for the region. The authors concluded that nutrient levels in the snow were high enough to support the algal population. The chemistry of the snow, however, suggests that the snow and ice from the ditch where the bloom was occurring was contaminated with surface water. Fjerdingstad et al (1978) studied the concentration of 29 different heavy metals and other ionic species by Chlamydomonas nivalis in the snow of eastern-Greenland and Spitzbergen. They found that the concentration factors (ppm-algae/ppm snow) for the major nutrients could vary appreciably (S, 317-840; Ca, 190-3800; K, 1880-4381; P, < 500). No measurements for N were reported. Hoham (1976) proposed that dust and coniferous litter covering snowbanks could be potentially important sources of nutrition for snow algae. In a series of laboratory experiments he followed the growth of two snow algae in a medium containing different concentrations of litter extract. He demonstrated that Chloromonas pichinchae (Lagerh.) Wille gave relatively large positive growth responses to leachates of coniferous needles, bark, and pollen. On the other hand, Raphidonema nivale Lagerh. responded with a low positive response, neutral or negative response according to the medium/leachate ratio. Inhibition of growth is probably due to tannins and phenol. Hoham (ibid) also found that C. pinchinchae grew best in meltwaters from snow taken from under the forest canopy. These meltwaters contained more P (0.56 μeq L^{-1}) than snow from open areas (0.06 μeq L^{-1}). Although light is also a factor in the relative distribution of algal species (Hoham, 1975), the above results do correspond to observations in the field where C. pinchinchae is much more abundant than R. nivale in snow under coniferous trees. These results also tie in with those of Moloney et al (1983) and Jones (1987) who reported increases in $P\text{-}PO_4$ when litter is wetted (Table 4B).

Litter can thus supply some nutrients to the snow and algae will incorporate them. Both algae and litter (Table 5), however, will deplete the snow in N and only the atmospheric pool and additional fluxes of N from invertebrates and mammals can supply the nutrient demand. The demand for N, in effect, can be an appreciable part of the amount available in the snow. Figure 10 shows the frequency distribution of N (NH_4 + NO_3) concentrations in the snowpack of a boreal forest site for areas with snow algae and those without (Gamache, 1990). It can be seen that the presence of algae considerably depletes the N pool in the snow. These results are very similar to Hoham et al 1989 (Table 9). The table shows the depletion of nutrients in snow in two forested areas (Lac Laflamme, Quebec, [LL, QUE.]; Whiteface Mountain, NY, [WF, NY]) due to snow algae; N is the nutrient the most in demand (depletion of 15.8 μeq L^{-1}, LL; 3.0 μeq L^{-1}, WF) while S depletion is also appreciable(4.9 μeq L^{-1}, LL; 2.0 ueq L-1, WF). Gamache (1990) distinguished the real biological depletion of the nutrients, from any preferential elution of nutrients that could have occurred by physical leaching in those areas with algae, by calculating the concentration ratios of nutrients to Cl -a conservative ion. The depletion of nutrients is real and related to the algal populations.

Table 9: Differences in the mean chemical composition (nutrient concentrations, μeq L^{-1}) of snow with algae (SA) and snow without algae (SN) in two forested regions of eastern North America[1]

Site	Snow	n^2	Nutrient concentrations meq L^{-1}							
			$N\text{-}NO_3$	$N\text{-}NH_4$	N_T	$S\text{-}SO_4$	$P\text{-}PO_4$	K	Ca	Mg
Lac Laflamme, Québec (eastern boreal forest)	SN	23	13.4	6	19.4	10.8	0	0.7	3.5	1.6
	SA	23	2.3	1.5	3.8	5.9	0	1.9	1.9	1.5
	Δ^3		(11.1)	(4.5)	(15.6)	(4.9)	0	1.2	(1.6)	(0.1)
Whiteface Mountain, NY (subalpine forest)	SN	3	9.5	6.5	16.0	11.4	0	58.2	11	2.5
	SA	3	7.7	5.3	13.0	9.4	0	31.3	7	3.3
	Δ		(1.8)	(1.2)	(3.0)	(2.0)	0	(26.9)	(4)	0.8

1 Recalculated from data by Hoham et al., 1989; Hoham, 1987; Germain, 1990.
2 Number of samples
3 Δ = SN-SB, losses ( )

Are the measured depletions of N consistent with the knowledge on in-pack algal production? What kind of nutrient decreases should we expect from photosynthesis in snow? Mosser et al (1977) measured the productivity of <u>Chlamydomonas</u> <u>nivalis</u> in melted snow from the Beartooth Mountains of Montana with ^{14}C. The estimated

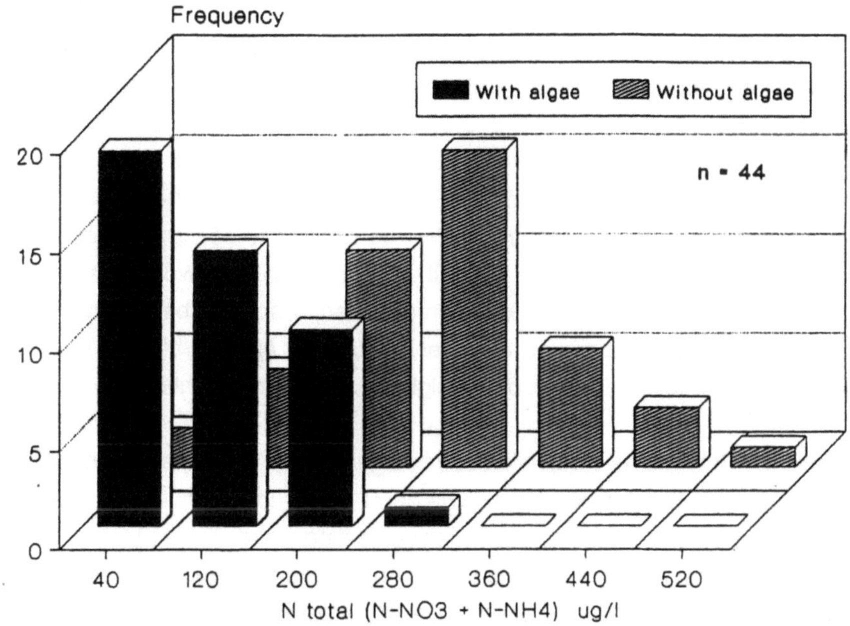

Figure 10. The distribution frequency of N total ($N\text{-}NO_3$ and $N\text{-}NH_4$) concentrations ($\mu g\ L^{-1}$) in a boreal forest snowpack during the spring melt. Snow classed as algal snow is snow with populations > 4 000 cells ml^{-1} (Modified after Gamache, 1990).

rates of production were 0.05-0.97 (mean, 0.31) μg C mm^{-3} hr^{-1}. If the population of active cells is 10^5 ml^{-1} and the cell diameter is 10 μm then the amount of C incorporated (for 8 hours of sunlight) is approximately 1 mg L^{-1} day^{-1}. According to the Redfield Ratio a fixation rate of this amount would require a concomitant assimilation of N equal to approximately 125 μg L^{-1} day^{-1} which is 16 μeq L^{-1} day^{-1}. This is of the same order of the concentrations that are found in the pack. If we further assume that the pack is 100 mm SWE and the algae occupy a horizontal band equivalent to 15 mm SWE then the daily flux of N depletion would result in a depletion of 2.4 eq ha^{-1} day^{-1}. Algal depletion of N can, therefore, be an important factor in the nutrient cycle of the snow cover, particulary as algal blooms usually occur late in the life of the pack when pack height and nutrient contents can be low.

A similar calculation for sulphur depletion due to algal activity leads to an assimilation of 2.0 μeq S L^{-1} day^{-1}; this is also of the same order of the concentration depletions that are actually recorded in the snow samples containing algae (Table 9). The input of S-SO_4, however, from leached debris and dry deposition will generally mask the assimilation of S by algae and overall changes in the pack will be more attenuated with regard to those of N.

Snow chemistry and biological activity: a perspective

We would like to finish this discussion with an example that synthesizes the above sections and allows an overall perspective of nutrient cycling in snow cover. We now have enough data to simulate the nutrient dynamics of a snow-biological ecosystem. As our example we have taken the phenology of a 1 km long transect in a mixed open-closed boreal forest in a temperate zone during the spring period. Figure 11 shows the evolution of the atmospheric N-total pool during the cold period prior to the start of melt. The pool builds up with successive snowfalls until melt begins on day 30. The pool is greater in the open forest area (fig. 11, op) of the transect than under the densely closed coniferous canopy (cl).

During the first melt period (6 days) the only biological activity of note is the increased depletion of the N-pool from days 47 to 53 at the 500 m point where a large amount of litter had previously been deposited at the open-closed transition zone. In the intermediate cold period that follows the depth of the snow allowed deer to move into the area at point 250 m on day 55 to browse for a period of 2 days. The deposited excrement contributes to the pack and is also leached in the following melt which starts on day 60. With the onset of this last massive melt and warm weather, collembola swarm for 1-2 days in an area at point 700 m; the N-pool rises

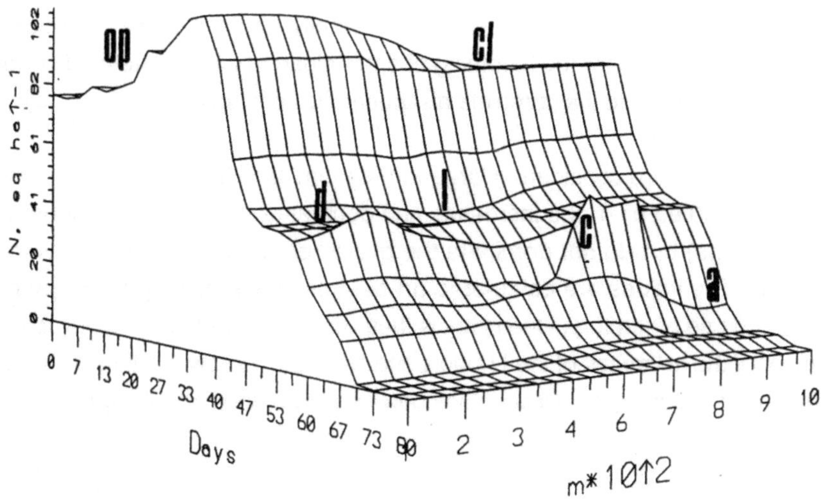

Figure 11. Phenology of a boreal forest snowpack. The evolution of the N-pool (eq ha^{-1}) in a transect of 1 km over 80 days prior to complete disappearance of the snowpack.

<u>op</u>, open forest; <u>cl</u>, closed coniferous canopy. <u>l</u>, the depletion of the N-pool in the melting snowpack due to litter at edge of closed canopy at 500 m; <u>d</u>, the influence of deer, population 20 animals ha^{-1}, residence time on transect at 250 m = 2 days; <u>c</u>, the influence of invertebrates, collembolla, 2 days of mating activity at 700 m, population density, 2 x $10^3 m^{-2}$. <u>a</u>, algal activity at end of snowmelt season under closed canopy at 900 m.

Background snow cover atmospheric N-pool calculated for SWE ≃ 300 mm, N = 30 µeq L^{-1}.

Biological increase or decrease into N-pool calculated from 1) Fowler and Dealy (1987) DelGuidice et al. (1987) for deer; 2) Jones (1987), Barry and Price (1987), and Courchesne and Hendershot (1988) for litter; 3) Aitchison (1979a), Custer et al. (1986), and HEDEP (1979) for collembola, and 4) Hoham et al. (1989), Hoham (1987) and Gamache (1990) for snow algae.

but is rapidly discharged by the meltwaters. Finally on days 63-64 the pack has diminished enough to allow the blooming of snow algae which deplete the N-pool by 2-5 eq a few days prior to the complete disappearance of snow.

In this example we are considering large forest areas (km ^{2}) with a fairly substantial N-pool of 60-100 eq ha-1; the increases and decreases in the pool are substantial (-10 to 40 eq ha) but do not dominate the overall contribution of the snow to the forest floor. The values are calculated on best estimates for typical population densities and use of territory. Chemistry is thus very dependent on ecology and any refinement of the estimates of the N-pools and N-fluxes in snow is ultimately a better knowledge of the ecological relationships between snow and living things.

Snow chemistry and biological activity: conclusion on the state-of-science

It has been shown that plants, animals and microbiological activity can influence the nutrient pools of snow cover by depositing the nutrients on, or assimilating the nutrients from the snow. The influence is far greater in forested areas than on open snowfields. The influence of plants is of an extensive nature, that of animals is more intensive due to mobility. We have discussed those types and species of plants and animals for which there is enough data on the ecology and nutrient composition to evaluate the effect of their activity on seasonal snow covers. The effects can range from negligible to "overwhelming" depending on the resident nutrient pools in the snow cover and the fluxes due to population density.

The state-of-science in this field of study is at a level quasi nonexistent. The knowledge on the ecology of snow and living things far outweighs our knowledge on snow chemistry. We know , for example, much more on the physiology and growth of snow molds which parasitize graminaceous plants under snow cover (Newsted et al, 1985) than we do on the chemical relationships between snow and these important psychrophilic fungi. We do not have any precise recommendations as to what the relative priority of future subjects to be studied should be. The biogeochemistry of snow has not developed to the point where we have enough of a picture to propose research that would fill in any knowledge gaps. We can only offer a general recommendation that a research worker in the field of snow chemistry should carefully evaluate the biological aspects in his, or her, particular area of interest. One should thus emphasize methodology and an interdisciplinary approach to one's problem. If snow scientists do not take a holistic approach to their research studies they may misinterpret the processes for the chemical changes in the very material they are interested in.

This paper opened the discussion on snow and biological activity with a reference to the reflection by McKay and Adams (1981) on snow and living things. It seems even more appropriate to finish the paper with it as well. The cycle is complete.

References

Adams WP and Findlay BF (1966) Snow measurement in the vicinity of Knob Lake, central Labrador-Ungava, winter 1964-1965. In: Proceedings of the Eastern Snow Conference, 23rd Annual meeting, Hartford, Connecticut, Feb. 10 and 11, pp 26-40.

Aitchison CW (1979a) Winter-active subnivean invertebrates in southern Canada. I. Collembola. Pedobiologia 19:113-120.

Aitchison CW (1979b) Winter-active subnivean invertebrates in southern Canada. II. Coleoptera. Pedobiologia 19:121-128.

Aitchison CW (1979c) Winter-active subnivean invertebrates in southern Canada. III. Acari. Pedobiologia 19:153-160.

Aitchison CW (1979d) Winter-active subnivean invertebrates in southern Canada. IV. Diptera and Hymenoptera. Pedobiologia 19:176-182.

Aitchison CW (1978) Spiders active under snow in southern Canada. Symp Zool Soc Lond 42:139-148.

Alain M (1990) personal communication, Minister des loisirs, chasse et peches, Govt. Quebec, Canada.

Barry PJ and Price AG (1987) Shortterm changes in the fluxes of water and of dissolved solutes during snowmelt. In: Seasonal Snowcovers: Physics, Chemistry, Hydrology. Editors, Jones, H. G. and Orville-Thomas, W. J. NATO-ASI series C: Mathematical and physical sciences 211: 501-530. D. Rheidel Publishing Company, Dordrecht.

Belikova TV, Vasilenko VN, Nazarov IM, Pegoev AN and Fridman ShD (1981) Characteristics of background sulfate pollution of the snow cover on the territory of the USSR. Meteorologiya i Gidrologiya, 9:47-55.

Berg N, Marks D, Dawson D, McGurk B, Melack J, Setaro F and Bergman J (1989) Evaluation of methods for measurement of snowfall and collection of snow for chemical analysis. California Air Resources Board, Final Report, July 1989.

Berg B and Staaf H (1981) Leaching, accumulation and release of nitrogen in decomposing forest litter. In: Terrestrial Nitrogen Cycles. Clark, F.E. and Rosswall, T., Eds. Ecol. Bull. (Stockholm) 33:163-178.

Berndt HW (1965) Snow accumulation and disappearance in lodgepole pine clearcut blocks in Wyoming. J Forestry 63:88-91.

Billings WD (1980) Plants in high places. Natural History 90(10):83-88.

Billings WD (1972) Arctic and Alpine vegetation: plant adaptations to cold summer climates. In: Arctic and Alpine Environments, Editors, Ives, J. D. and Barry, R. D., Methuen and Co. Ltd, London. pp 403-444.

Bourgeois JC, Koerner RM and Alt BT (1985) Airborne pollen: a unique air mass tracer, its influx to the Canadian High Arctic. Annals of Glaciology 7:109-116.

Bowden WB (1986) Gaseous nitrogen emissions from undisturbed terrestrial ecosystems: An assessment of their impacts on local and global nitrogen budgets. Biogeochemistry 2:249-279.

Brauer,F (1871) Insecten-leben im winter. Vereine zur Verbreitung naturwissenschaftlitcher Kenntnisse im Wein. Marz 22, 1871.

Cadle SH and Dasch JM (1988) Wintertime concentrations and sinks of atmospheric particulate carbon at a rural location in northern Michigan. Atmospheric Environ. 223:1373-1381.

Chapman JA (1954) Observations on snow insects in western Montana. Can Ent 84:357-363.

Chase AI and Young HE (1978) Pulping, biomass, and nutrient studies of woody shrub and shrub sizes of tree species. Bulletin 749, Life Sciences and Agriculture Experiment Station, University of Maine at Orono. 36 pages including tables and figures.

Christie P (1987) Nitrogen in two contrasting Antarctic bryophyte communities. Journal of Ecology 75:73-93.

Cooper WJ, Coope DJ, Saltzman ES, DeMello WZ, Savoie DL, Zika RG and Prospero JM (1987). Emissions of biogenic sulfur compounds from several wetland soils in Florida. Atmospheric Environ. 21:1491-1495.

Coulianos CC and Johnels AG (1962) Note on the subnivean environment of small mammals. Arkiv for zoologi 15:363-370.

Courchesne F and Hendershot WH (1988) Cycle annuel des elements nutritifs dans un bassin-versant forestier: contribution de la litiere fraiche. Can J For Res 18:930-936.

Craig H, Chou CC, Welhan JA, Stevens CM and Engelkemeir A (1988). The isotopic composition of methane in Polar ice cores. Science 242:1535-1539.

Crawford RL and Edwards JS (1986) Ballooning spiders as a component of arthropod fallout on snowfields of Mount Rainier, Washington, USA. Arctic and Alpine Research 18:429-437.

Crete M (1990) personal communication, Minister des loisirs, chasse et peches, Govt. Quebec, Canada.

Cronan CS and Reiners WA (1983) Canopy processing of acidic precipitation by coniferous and hardwood forests in New England. Oecologia 59:216-223.

Custer TW, Osborn RG, Pitelka FA and Gessaman JA (1986) Energy budget and prey requirements of breeding lapland longspurs near Barrow, Alaska, U.S.A. Arctic and Alpine research, 18:415-427.

Czuczwa J, Leuenberger C and Giger W (1988) Seasonal and temporal changes of organic compounds in rain and snow. Atmospheric Environ 22:907-916.

Czygan FC (1970) Blutregen und Blutschnee: Stickstoffmangelzellen von Haemotococcus pluvialis und Chlamydomonas nivalis. Arch Mikrobiol 74:69-76.

Danks HV (1978) Modes of seasonal adaptation in the insects. I. Winter survival. Can Ent 110:1167-1205.

Dasch JM (1987) Measurement of dry deposition to surfaces in deciduous and pine canopies. Environmental Pollution 44:261-277.

Davies TD, Tranter M and Jones HG (1991) The chemical composition of seasonal snowcover: a state-of-the-science review. A NATO Advanced Research Workshop, Maratea, Italy, July 25-31, 1990. Published by Rheidel Inc.

Davies TD, Brimblecombe P, Tranter M, Tsiouris S, Vincent CE, Abrahams P. and Blackwood IL (1987) The removal of soluble ions from melting snowpacks. In: Seasonal Snowcovers: Physics, Chemistry, Hydrology. Editors, Jones H.G. and Orville-Thomas, W. J. NATO-ASI series C: Mathematical and physical sciences 211:337-392. D. Rheidel Publishing Company, Dordrecht.

DelGiudice GD, Mech LD and Seal US (1989) Physiological assessement of deer populations by analysis of urine in snow. J. Wildl. Manage. 53(2):284-291.

DelGiudice GD, Mech LD and Seal US (1988) Chemical analysis of deer bladder urine and urine collected from snow. Wildl Soc Bull 16:324-326.

DelGiudice GD, Mech LD, Seal US and Karns PD (1987a) Winter fasting and refeeding effects on urine characteristics in white-tailed deer. J Wildl Manage 51:860-864.

DelGiudice GD, Seal US and Mech LD (1987b) Effects of feeding and fasting on wolf blood and urine characteristics. J Wildl Manage 51:1-10.

Dickison RBB and Daugherty DA (1982) A comparison of forest cover measures as estimators of snow cover and ablation. In: Proceedings of the Canadian Hydrology Symposium '82, Hydrological processes in forested areas. Fredricton, N.B., June 14 & 15, pp 113-129.

Downes JA (1965) Adaptations of insects in the Arctic. Ann Rev Ent 10:257-274.

Edwards JS and Banko PC (1976) Arthropod fallout and nutrient transport: a quantitative study of Alaskan snow patches. Arctic and Alpine Research, 8:237-245.

Esseen Per-Anders (1985) Litter fall of epiphytic macrolichens in two old Picea abies forests in Seden. Can J Bot 63:980- 987.

Everdun LN and Fuller WA (1972) Light alteration caused by snow and its importance to subnivean rodents. Can J Zool 50:1023-1032.

Fahey TJ (1979) changes in nutrient content of snow water during outflow from Rocky Mountain coniferous forest. Oikos 32:422-428.

Fernandes AR (1984) Organic acids in snow and rain in Great Britain. M.Sc thesis, University of East Anglia, Norwich, U.K.

Fjerdingstad E, Vanggaard L, Kemp K and Fjerdingstad E (1978) Trace elements of red snow from Spitzbergen with a comparison with red snow from east Greenland (Hudson land) Arch Hydrobiol 84:120-134.

Formazov AN (1946) Snow cover as an integral factor of the environment and its importance in the ecology of mammmals and birds. Moscow Soc. Naturalists, Materials for Fauna and Flora U.S.S.R. Zool. Section, New Ser. 5:1-152. English translation: Prychodko, W. and Pruitt, W. O. (1963). Occasional Paper 1, Boreal Inst. Univ. Alberta, Edmonton.

Foster PM (1978) The modelling of pollutant concentrations during snow-melt. Laboratory note, RD/L/N 46/78. Central Electricity Research Laboratories, Leatherhead, England. 14 pages plus figures.

Fowler WB and Dealy JE (1987) Behaviour of mule deer on the Keating Winter Range, Res. Pap. PNW-RP-373. Portland Oregon: U.S. Departement of Agriculture, Forest Service, Pacific Northwest Research Station; 1987. 25pp.

Frey R (1913) Beitrage zur Kenntnis der Arthropoden-fauna im winter. Med Soc Fauna Flora Fennica 39: 106-121.

Fukushima H (1963) Studies on cryophytes in Japan. J Yokahama Municipal University, Ser. C, Nat Sci 43:1-146.

Fuller WA, Stebbins LL and Dyke GR (1969) Overwintering of small mammals near Great Slave Lake, northern Canada. Arctic 22: 34-55.

Gamache S (1990) M.Sc. thesis, Institut national de la recherche scientifique, Ste Foy, Quebec, Canada, in press.

Germain L (1990) M.Sc. thesis, Institut national de la recherche scientifique, Ste Foy, Quebec, Canada, in press.

Gerrath JF and Nicholls KH (1974) A red snow in Ontario caused by the dinoflagellate Gymnodinium pascheri. Can J Bot 52:683-685.

Gill PS, Graedel TE and Weschler CJ (1983) Organic films on atmospheric aerosol particles, fog droplets, cloud droplets, raindrops, and snowflakes. Rev Geophys Space Phys 21:903-920.

Glatzel G, Kazda, M and Makart G (1986) Winter deposition rates of atmospheric trace constituents in forests: assessement of total input. In: Atmospheric Pollutants in Forest Areas. Georgrr, H.W., ed., D. Reidel Publishing Company. pp. 101-108.

Golding DL and Swanson RH (1986) Snow distribution patterns in clearings and adjacent forest. Water Resourc Res 22:1931-1940.

Goodison BE, Louie PYT and Metcalfe JR (1986) Snowmelt acidic shock study in south Central Ontario. Water Air and Soil pollution, 31:131-138.

Gregor JD and Gummer WD (1989) Evidence of atmospheric transport and deposition of organochlorine pesticides and polychlorinated biphenyls in Canadian Arctic snow. Envir Sci Technol 23:561-565.

Grennfelt, P. (1987). Deposition processes for acidifying compounds. Environ Tech Letters 8:515-527.

Grimard Y (1985) Sommaire des donnees de qualite des eaux de precipitation 1984. Report PA-19, Govt. Quebec, Ministere de l'environnement, Quebec, Canada.

Hagvar S (1984) Effects of liming and artificial acid rain on Collembola and Protura in coniferous forest. Pedobiologia 27:341-354.

Hagvar S (1973) Ecological studies on a winter-active spider Bolyphantes index (Thorell) (Araneida, Linyphidiidae) Norsk. ent. Tidsskr. 20:309-314.

Hagvar S (1971) Field observations on the ecology of a snow insect, Chionea araneoides Dalm. (Dipt. Tipulidae) Norsk. ent. Tidsskr. 18:33-37.

Havas P and Kubin E (1983) Structure, growth and organic matter content in the vegetation cover of an old spruce forest in northern Finland. Ann Bot Fennici 20:115-149.

HEDEP (Handbook of environmental data and ecological parameters) (1979) Editor in chief, Jorgensen, S.E.; Editorial Board, Friis, M.B., Henriksen, J., Jorgensen, L.A., Jorgensen, S.E. and Mejer, H.F. International Society for Ecological modelling. ISBN 87 87257 16 5.

Hendricks P (1987) Habitat use by nesting water pipits (Anthus Spinoletta): a test of the snowfield hypothesis. Arctic and Alpine research, 19:313-320.

Hoham RW (1989) Snow microorganisms and their interaction with the environment. In: Proceedings of the Western Snow Conference, Fort Collins, Colorado, April 1988, pp. 31-35.

Hoham RW, Yatsko C, Germain L. and Jones HG (1989) Recent discoveries of snow algae in upstate New York and Québec Province and preliminary reports on related snow chemistry. In: Proceedings of the Eastern Snow Conference, 46th Annual meeting, Quebec City, Quebec, June 8 & 9, pp 196-200.

Hoham RW (1987) Snow algae from high-elevation temperate latitudes and semi-permanent snow: their interaction with the environment. In: Proceedings of the Eastern Snow Conference, 44th Annual meeting, Fredericton, N.B. June 3 & 4, pp 73-79.

Hoham RW and Mohn WW (1985) The optimum pH of four strains of acidophilic snow algae in the genus Chloromonas (Chlorophyta) and the possible effects of acid precipitation. J Phycol 21:603-609.

Hoham RW, Mullet JE and Roemer SC (1983) The life history and ecology of the snow alga Chloromonas polyptera comb. nov. (Chlorophyta, Volvocales). Can J Bot 61:2416-2429.

Hoham RW (1980) Unicellular Chlorophytes - Snow algae. In: Phytoflagellates. Editor Cox, E. Elsevier North Holland, Inc. pp 61-84.

Hoham RW and Blinn DW (1979) Distribution of Cryophilic algae in an arid region of the American Southwest. Phycologia 18:133-145.

Hoham RW, Roemer SC and Mullet JE (1979) The life history and ecology of the snow alga Chloromonas brevispina comb. nov. (Chlorophyta, Volvocales) Phycologia 18:55-70.

Hoham RW and Mullet JE (1977) The life history and ecology of the snow alga Chloromonas cryophilia sp. nov. (Chlorophyta, Volvocales). Phycologia 16:53-68.

Hoham RW (1976) The effect of coniferous litter and different snow meltwaters upon the growth of two species of snow algae in axenic culture. Arctic and Alpine Research 8:377-386.

Hoham (1975) The life history and ecology of the snow algae Chloromonas pichinchae (Chlorophyta, Volvocales). Phycologia 14:213-226.

Hoham RW (1974a) New findings in the life history of the snow alga Chlainomonas rubra (Stein et Brooke) comb. nov. (Chlorophyta, Volvocales). Syesis 7:239-247.

Hoham RW (1974b) Chlainomonas kolii (Hardy et Curl) comb. nov. (Chlorophyta, Volvocales), a revision of the snow alga, Trachelomonas kolii, Hardy et Curl, (Euglenophyta, euglenales). J Phycol 10:392-396.

Hoham RW (1971) Laboratory and field studies of snow algae of the Pacific Northwest. Ph.D. thesis, University of Washington.

Ion PG and Kershaw GP (1989) The selection of snowpatches as relief habitat by woodland caribou (Rangifer tarandus caribou), MacMillan Pass, Selwyn/MacKenzie Mountains, N.W.T., Canada. Arctic and Alpine Research, 21:203-211.

Jeffrey WW (1968) Snow hydrology in the forest environment. In: Snow hydrology, proceedings of a workshop seminar, University of New Brunswick, February 28 & 29 Canadian National Committee for the International Hydrological Decade, Queen's Printer, Ottawa, Cat. # M26-103 pp.1-19.

Jeffries DS (1990) Snowpack storage of pollutants, release during melting, and impact on receiving waters. In: Acid Precipitation, Norton, S.A. ed. Springer-verlag, New York, NY, 4: in press.

Jones HG and Tranter M (1989) Interactions between meltwater and organic-rich particulate material in boreal forest snowpacks: Evidence for both physico-chemical and microbiological influences. In: Proceedings of the 6th International Symposium on water-rock interaction, Malvern, U.K., August 3-8, 1989; Miles, D.E., ed. A.A. Balkama, Rotterdam; PP.349-352.

Jones HG (1987) Chemical dynamics of snow cover and snowmelt in a Boreal forest. In: Seasonal Snowcovers: Physics, Chemistry, Hydrology. Editors, Jones, H. G. and Orville-Thomas, W. J. NATO- ASI series C: Mathematical and physical sciences 211: 531-574. D. Rheidel Publishing Company, Dordrecht.

Jones HG and DeBlois C (1987) Chemical dynamics of N- containing ionic species in a boreal forest snowcover during the spring melt period. Hydrological processes 1:271-282.

Jones HG, Sochanska W, Stein J, Roberge J. and Plamandon A.P. (1986) Snowmelt in a boreal forest site: an integrated model of meltwater quality (Snoqual1) Water, Air and Soil Pollution, 31:431-439.

Jones HG and Sochanska W (1985) The chemical characteristics of snow cover in a northern boreal forest during the spring run-off Annals of Glaciology, 7:167-174.

Kalliomaki M, Courtin GM and Clulow FV (1984) Thermal index and thermal conductivity of snow and their relationship to the winter survival of the meadow vole, Microtus pennsylvanicus. In: Proceedings of the Eastern Snow Conference, 41th annual meeting, Washington, D.C., June 7 and 8, 1984; pp 153-164.

Kalliomaki M, Courtin GM and Clulow FV (1983) The role of snow in the winter survival of the meadow vole, Microtus pennsylvanicus. In: Proceedings of the Eastern Snow Conference, 40th annual meeting, Toronto, Ontario, June 2 and 3, 1983; pp 180-182.

Kawecka B and Drake BG (1978) Biology and ecology of snow algae. 1. The sexual reproduction of Chlymadomonas nivalis (Bauer) Wille (Chlorophyta, Volvocales) Acta Hydrobiol 20:111-116.

Keene WC, Galloway JN and Holden JD (Jr) (1983) Measurement of weak organic acidity in precipitation from remote areas of the world. J Geophys Res 88:5122-5130.

Kelsall JP (1969) Structural adaptations of moose and deer for snow. J Mammal 5:302-310.

Knowles R (1978) Common intermediates of nitrification and denitrification and the metabolism of nitrous oxide. In: Microbiology-1978, Washington, D.C., American Society for Microbiology, 367-371.

Kobayashi Y (1967) Coloured snow with Chlamydomonas nivalis in the Alaskan arctic and Spitzbergen. Bull Nat Sci Mus 10:207-210.

Kol E (1975a) Cryobiological researches in the High tatra I. Acta Bot Acad Sci Hungariacas 21:61-75.

Lang GE, Reiners WA and Shellito GA (1982) Tissue chemistry of Abies balsamea and Betula papyrifera var. cordifolia from subalpine forests of the northeastern United States. Can J For Res 12:311-318.

Lang GE, Reiners WA and Heier RK (1976) Potential alteration of precipitation chemistry by epiphytic lichens. Oecologia 25:229-241.

Larsen JA (1972) Ecology of the northern continental forest border. In: Arctic and Alpine Environments, Editors, Ives, J. D. and Barry, R. D., Methuen and Co. Ltd, London. pp 342-370.

Lunde G, Gether J, Gjos N. and Lande MB (1977) Organic micropollutants in precipitation in Norway. Atmospheric Environ. 11:1007-1014.

Marchant HJ (1982) Snow algae from the Australian Snowy Mountains. Phycologia 21:178-184.

McKay GA and Adams WP (1981) Snow and living things. In: Handbook of Snow; Principles, Processes, management and Use. Editors, Gray, D. M. and Male, D. H. Pergamon Press Toronto pp 3-31
Merriam G, Wegner J and Caldwell D (1983) Invertebrate activity under snow in deciduous woods. Holarctic Ecol 6:89-94.
Meszaros A (1977) On the size distribution of atmospheric aerosol particles of different composition. Atmospheric Environ 11:1075-1081.
Meyers PA and Hites RA (1982) Extractable organic compounds in midwest rain and snow. Atmospheric Environ 16:2169-2175.
Mieman JR (1968) Snow accumulation related to elevation, aspect and forest canopy. In: Snow hydrology, proceedings of a workshop seminar, University of New Brunswick, February 28 & 29 Canadian National Committee for the International Hydrological Decade, Queen's Printer, Ottawa, Cat. # M26-103, pp 35-47.
Moloney KA, Stratton LJ and Klein RM (1983) Effects of simulated acidic, metal-containing precipitation on coniferous litter decomposition. Can J Bot 61:3337-3342.
Mooney HA, Vitousek PM and Watson PA (1987) Exchange of materials between terrestrial ecosystems and the atmosphere. Science 238:926-932.
Morgantini LE and Hudson RJ (1989) Nutritional significance of Wapiti (Cervus Elaphus) migrations to alpine ranges in western Alberta, Canada. Arctic and Alpine Research, 21:288-295.
Mosser JL, Mosser AG and Brock TD (1977) Photosynthesis in the snow: The alga Chlamydomonas nivalis (Chlorophycae). J Phycol 13:22-27.
Murakami M and Kikuchi K (1982) Some considerations on the center nuclei of snow crystals. National. Inst. Polar. Res. Japan, Special Issue No. 24, Proceedings of the 4th symposium on polar meteorology and glaciology, 1982, pp 175-183.
Neubauer J and Heumann KG (1988) Nitrate trace determinations in snow and firn core samples of ice shelves at the Weddel Sea, Antarctica. Atmospheric Environ 22:537-545.
Newsted WJ, Huner NPA, Insell JP, Griffith M. and van Huystee (1985) The effects of temperature on the growth and polypeptide composition of several snow mold species. Can J Bot 63:2311-2318.
Newton APW (1982) Red-coloured snow algae in Svalbard-some environmental factors determining the distribution of Chlamydomonas nivalis (Chlorophyta volvocales). Polar Biol 1:167-172.
Oliver BG, Thurman EM and Malcolm RL (1983) The contribution of humic substances to the acidity of coloured natural waters. Geochim Cosmochim Acta 47:2031-2035.
Oliver JE (1973) Climate and Man's environment. New York, NY: John Wiley and Sons, Inc.
Patch JR (1981) Effects of forest cover on snow distribution in the Nashwaak experimental watershed project. In: Proceedings of the Eastern Snow Conference, 38th Annual meeting, Syracuse, N.Y. June 4 & 5, pp 76-87.
Patenaude, R. veterinary officer, Govt. of Quebec, Canada, personal communication, may, 1990.
Payette S, Filion L and Ouzilleau J (1973) Relations neige- vegetation dans la toundra forestiere du Nouveau Quebec, Baie d'Hudson. Naturaliste can., 100: 493-508.
Persson T (1980) Contribution of soil fauna to nitrogen mineralization in a coniferous forest. Possible effects of food selection by the fauna. In: Processes in the nitrogen cycle, Rosswall, T. ed., SNV PM 1213, Stockholm, Swedish Environment Protection, pp 135-141.
Petzold DE and Mulhern T (1988) Vegetation-depth relationships in the boreal forest-tundra ecotone of eastern Canada. In: Proceedings of the Eastern Snow Conference, 45th Annual meeting, Lake Placid, N.Y. June 8 & 9, pp 42-49.
Pollock R (1970) What colors the mountain snow? Sierra Club Bull, 55:18-20.
Pruitt WO (Jr) (1975) Life in the snow. Nature Canada, October-December, pp 42-49.

Pruitt WO (1970) Some ecological aspects of snow. In: Ecology of the subarctic regions, Proceedings of the Helsinki symposium, June 27 to August 3, pp 83-100.
Pruppacher HR and Klett JD (1981). Microphysics of clouds and precipitation. D. Rheidel Publishing Co. Dordrecht, Holland.
Rascher CM, Driscoll CT and Peters NE (1987) Concentration and flow of solutes from snow and forest floor during snowmelt in the West-Central Adirondack region of New York. Biogeochemistry 3:209-224.
Rasmussen RA and Khalil MAK (1986) The behavior of trace gases in the atmosphere. The Science of the Total Environment 48:169-186.
Reynolds B (1983) The chemical composition of snow at a rural upland site in mid-Wales. Atmospheric Environ 17:1849-1851.
Rhodes JJ, Skau CM and Greenlee DL (1986) The role of snowcover on diurnal nitrate concentration patterns in streamflow from a forested watershed in the Sierra Nevada, Nevada, USA. In: Proceedings: Cold regions hydrology symposium, American Water Resources Association, Fairbanks, Alaska, July, 1986. pp 157-166.
Robbins CT, Prior RL, Moen AN and Visek WJ (1974) Nitrogen metabolism of white-tailed deer. J Anim Sci 38:871-876.
Rosswall T (1981) The biogeochemical nitrogen cycle. In: Some perspectives of the major biogeochemical cycles. Likes, G.E., ed. SCOPE; 17. John Wiley & Sons, New York. pp 25-49.
Saltzman ES, Savoie DL, Zika RG and Prospero JM (1983) Methane Sulfonic acid in the marine atmosphere. J Geophys Res 88:10897-10902.
Savile DBO (1972) Arctic adaptations in plants. Monograph No. 6, Research branch, Canada Department of Agriculture. ISBN 0-662-01699-8.
Schinner F (1983) Litter decomposition, CO_2 release and enzyme activities in a and on a windswept ridge in an alpine environment. Oecologia 5: 288-291.
Schuepp PH, Barthakur NN, Schemenauer RS and McGerrigle DN (1987) In: 18th Conference, Agricultural and Forest Meteorology, and 8th Conference, Biometeorology and Aerobiology, American Meteorological Society, W. Lafayette, Indiana, Sept. 15- 18, 1987. pp 253-255.
Solomon DK and Cerling TE (1987) The annual carbon dioxide cycle in a montane soil: Observations, modeling, and implications for weathering. Water Resour Res 23:2257-2265.
Somme L and Ostbye E (1969) Cold-hardiness in some winter active insects. Norsk ent Tidsskr 16:45-48.
Stein J, Jones HG, Roberge J and Sochanska W (1986) The prediction of both runoff quality and quantity by the use of an integrated snowmelt model. In: Modelling snowmelt-induced processes. Ed. E.M. Morris. IHAS publication 155:347-358.
Stein J and Amundsen CC (1967) Studies on snow algae and snow fungi from the Front Range of Colorado. Can J Bot 45:2033-2045.
Steppuhn H (1981) Snow and agriculture. In: Handbook of Snow; Principles, Processes, management and Use. Editors, Gray, D. M. and Male, D. H. Pergamon Press Toronto pp 60-126.
Stottlemeyer R and Toczydlowski D (1990) Pattern of solute movement from snow into an upper Michigan stream. Can J Fish Aquat Sci 47:290-300.
Stottlemeyer R (1987) Snowpack ion accumulation and loss in abasin draining to Lake Superior. Can J Fish Aquat Sci 44:1812-1819.
Suzuki K (1982) Chemical changes of snow cover by melting. Jap J Limnology 43:102-112.
Taylor BR and Jones HG (1990) Litter decomposition under snow cover in a balsam fir forest. Can J Bot 68:112-120.
Taylor BR and Parkinson D (1988) Does repeated freezing and thawing accelerate decay of leaf litter. Soil Biol Biochem 20:657-665.
Thomas W (1986) Accumulation of organic and inorganic trace substances in the surface snow cover. In: Modelling snowmelt- induced processes. Ed. E.M. Morris. IAHS Publication 155:359-371.

Thompson LG, Davis ME, Mosley-Thompson E and Liu K-b (1988) Pre-Incan agricultural activity recorded in dust layers in two tropical ice cores. Nature 336:763-768.
Transportation Research Board (1988). Deicing chemicals and snow control. Report TRB/TRR-1157. Washington, DC. 62pp.
Tukey HB (Jr) (1966) Leaching of metabolites from above-ground plant parts and its implications. Bull Torrey Botanical Club 93:385-401.
Tukey HB (Jr) (1970) The leaching of substances from plants. Ann. Rev. Plt. Physiol 21:305-324.
Twickler MS, Spencer MJ, Lyons WB and Mayewski PA (1986) Measurement of organic carbon in Polar snow samples. Nature 320:156-158.
Van Cleve K. and Alexander V. (1981) Nitrogen cycling in tundra and boreal ecosystems. In: Terrestrial Nitrogen cycles, Clark, F.E. and Rosswall, T. (eds.) Ecol Bull 33:375-404.
Verme LJ (1968) An index of winter severity for northern deer. J Wildl Manage 32:566-574.
Viala G (1966) L'astaxanthine chez le Chlymadomonas nivalis Wille. Compt. Rend. Hebd. Seances Acad Sci 263:1383-1386.
Vincent WF (1988) Microbial ecosystems of Antarctica. Cambridge University Press, Cambridge, U.K.
Visser SA (1973) The microflora of a snow depository in the City of Quebec. Environ Letters 4:267-272.
Walker DA (1985) Vegetation and environmental gradients of the Prudhoe Bay region, Alaska. Cold Regions Research and Engineering laboratory (CRREL), U.S. Army, CRREL Report 85-14.
Waring RH and Schlesinger WH (1985) Forest Ecosystems. Concepts and management. Academic press Inc.Orlando, Florida. 340 pages.
Warneck P (1987) The atmospheric aerosol. In: Chemistry of the natural atmosphere. Warneck, P. International Geophysics series Vol. 41. Academic Press Inc London. pp 278-373.
Watanabe M, Watabe S and Arai S (1988) Interaction of an antinucleating chemical and an ice nucleation active bacterium: A case study with an n-octylbenzyldimethylammonium salt and Erwinia-ananas. Agric Biol Chem 52:1869-1872.
Woo M-K and Steer P (1986) Monte Carlo simulation of snow depth in a forest. Water Resourc Res 22:864-868.
Zakrisson K-A (1987) Snow distribution on forest clearcuts in the Swedish mountains. Nordic Hydrology 18:185-192.

DISCUSSION PAPER ON THE REPORT BY H.G. JONES ENTITLED "SNOW CHEMISTRY AND BIOLOGICAL ACTIVITY: A PARTICULAR PERSPECTIVE ON NUTRIENT CYCLING"

P.J. Barry
Chalk River Laboratories
Chalk River, Ontario
K0J 1J0
CANDA

Prof. Jones has provided a thoughtful and thought provoking review of this most difficult of topics in snow chemistry. It is, and will be for years to come, the definitive monograph. Little is left to the discussant but to cross a "t" here and there or highlight a feature or two he thinks to be important.

The review is comprehensive. It starts up in the air which is the source of all primary constituents of snow of which the organic components, if not of anthropogenic origin, are formed as a result of biological processes. A middle section deals with the interactions between the vegetation and the inorganic constituents of snow; the fate of the organic fractions after the snow has deposited seems not to have been investigated. The final section deals with changes in the snowpack caused by animals. These take the form of inputs of organic compounds derived from animal excreta; the lot of a snowflake is not always a happy one as it shares something of the lot of humans.

The investigations reported in each of the three sections have clearly been designed to fulfill quite different purposes. There is therefore little to no connection between them. Prof. Jones, however, attempts to convert the essentially static information of the third section into a dynamic flux model similar to that described in the middle section for the inorganic constituents. The attempt is an ingenious start that is worth developing. The bulk of the work done into snow chemistry belongs to the middle section and it is this that provides the meat of the review and stimulates the major part of this discussion. Driven by concerns about the effects of pollutants from the atmosphere on the life of rivers and lakes, the emphasis of this work is on fluxes through the biosphere and the chemical transformations that the several entities undergo.

Nevertheless even this extensive literature has its limitations. Most evident of these is the strong bias for the work to be done in the forest and in the coniferous forest in particular. There is scant mention of deciduous forests and none at all of grasslands and other surfaces.

NATO ASI Series, Vol. G 28
Seasonal Snowpacks
Edited by T. D. Davies et al.

In Table 3 for example, all the items listed are for the coniferous forest. One is said to refer to an eastern mixed forest but the data belongs to a clearing and a softwood canopy.

The different forest types must be clearly distinguished. Three canopy properties seem to be important:

- the types of surface exposed, leaves in the one case, dormant twigs in the other,
- the density of the cover provided by the canopy, and
- the thickness of the canopy relative to its height above the forest floor.

The mature deciduous forest has a thin canopy and a large open space between it and the floor. The canopy of some coniferous species reaches almost down to the floor leaving little free space. Clearly the age of the forest as well as its speciation is important. The nature of the forest floor and the composition of the litter layer and organic soil horizon are also influenced by the form of the primary canopy. These variables then exert profound effects on the chemical transformations that are possible in the snow as it lies on the ground as well as when it melts and runs away. The distinction between type then is fundamental and pervasive.

Barry and Robertson (1989) deployed up to 25, 9 and 7 bulk collectors (area 1400 cm^2) respectively in an mature aspen stand, a small clearing in the forest and, for a limited period, in a coniferous stand.

Starting on 1987 December 22, and going through to 1988 March 3, the collectors were emptied 10 times mostly following a heavy snowfall. The collections for December 22 and 26 were of wet snow and freezing rain respectively. The rest were of more or less dry snow. The collectors in the coniferous stand were only operated to January 19. The ratios of the volumes (SWE) and volume weighted mean concentrations of selected ions collected in the open to those in the forest are shown in Table 1 (aspen forest) and Table 2 (coniferous forest). With more or less dry snow the deciduous forest receives about 15% less SWE than does the open while the loss in the coniferous forest is about half. The weighted mean concentrations of the 3 ions are not much altered in the deciduous forest relative to those in the open. The concentrations are, however, much enhanced when the snow is wet or rain falls. In the coniferous forest the effects are greater and highly variable. The number of samples collected in the coniferous forest was too small to draw stronger conclusions than this.

Table 1

Ratios of the Volumes and the Volume Weighted Mean Concentrations of Common Ions at the Open Site to those in the Aspen Forest

Date	SWE	Nitrate	Sulphate	Hydrogen
Dec 22	1.16	0.62	0.71	0.84
Dec 26	1.35	0.39	0.44	0.57
Mean of 8 from Dec 26 to Mar 3	1.14	0.97	0.95	1.02
cv	0.05	0.03	0.07	0.03

SWE = snow water equivalent

cv = coefficient of variation = $\frac{\text{standard deviation}}{\text{mean}}$

Table 2

Ratios of the Volumes and Volume Weighted Mean Concentrations of Common Ions at the Open Site to those in the Coniferous Forest

Date	SWE	Nitrate	Sulphate	Hydrogen
Dec 22	3.19	0.98	0.51	1.90
Dec 26	1.81	0.13	0.04	0.65
Jan 12+19	1.96	0.95	1.91	2.86

Table 3

Spatial Variability of the Volumes of Snow and of the Concentrations of Selected Ions

	SWE	Nitrate	Sulphate	Hydrogen
OPEN SITE				
Mean CV	0.029	0.046	0.079	0.051
Standard Deviation 0.009	0.043	0.071	0.029	
CV for Dec 26	0.05	0.13	0.19	0.12
ASPEN SITE				
Mean CV	0.067	0.058	0.088	0.058
Standard Deviation 0.02	0.038	0.065	0.024	
CV for Dec 26	0.14	0.28	0.39	0.33
CONIFEROUS SITE				
Mean CV	0.43	0.14	0.35	0.24
CV for Dec 26	0.73	0.54	0.43	0.50

Table 4

Ratios of the Volume Weighted Mean Concentrations of Monovalent Cations at the Open Site to those in the Aspen Forest

Date	Sodium	Ammonium	Potassium
Dec 26	0.74	0.46	0.09
Mean of 8 from Dec 26 to Mar 3	0.85	0.85	0.71
CV	0.19	0.25	0.18

Besides affecting the absolute amounts of snow and its constituents reaching the ground the canopy also affects the spatial distribution. This effect was investigated by first calculating the coefficient of variations amongst the spatially distributed collectors for each collection period. Finally the mean of each of the CV's together with a standard deviation amongst the set were calculated. Again, the collections for Dec 22 and 26th were omitted from these calculations. The results are shown in Table 3 which also shows the CV for each site for December 26th. The spatial variability when the snow is dry is surprisingly small both in the open and in the aspen stand. The spatial variability of SWE is greater in the forest than in the open but is still small with CV less than 7%. The 3 ions appear to have roughly the same spatial variability in the open as in the aspen stand but the CV is still less than 10%. Uniformity of the deposit in the aspen stand is due to the distance between the primary canopy and the forest floor which is sufficient to allow the random oscillations of dry snowflakes as they fall to obscure the patterns of abstraction in the canopy. When the snow is wet or rain falls the trajectories of the falling droplets remain more or less straight and so preserve much of the canopy shapes when they reach the ground. By contrast spatial variability is relatively large in the coniferous stand even for dry snow.

Table 4 shows the ratios of the volume weighted mean concentrations of monovalent cations at the open site relative to those in the aspen forest. The canopy is again a source of these ions, the more so for potassium and for wet snow and rain. However, the main item of note in this table is the large variability in the ratios between events (about 20%) compared to that in Table 1 for the two anions and hydrogen ion (about 5%). The reason for this difference is unknown.

If the review by Prof. Jones has a deficiency it is that a section on experimental techniques and methodologies could have been useful. So far we have discussed the effect of the primary forest canopy on the composition of the falling snow. Below this, however, are secondary and tertiary canopies provided by the shrubs and the herbs respectively. Low growing vegetation will eventually be covered as the pack grows in thickness but the taller shrubs will provide a potential for throughfall chemical changes throughout the winter. Investigations of the chemical changes induced at this level seem not to be made and certainly there are serious difficulties to making the appropriate measurements in situ. The problems encountered would also affect studies of other surface types such as grasslands. A prominent plant in the herbal canopy of our aspen stand is fern. It is quite likely that the dead fronds would be a potent source of chemical entities to the falling and, later to the standing snow in the pack. Required perhaps are laboratory measurements of leaching from the relevant plant materials under various conditions combined with computational techniques similar to those used by Prof. Jones to estimate fluxes in the final section of this review.

Similar problems await the would-be investigator of chemical changes in the snow pack. Sampling problems arise because samples cannot be easily collected where emergent vegetation occurs and this might form a potent matrix for stimulating chemical change. In an open location, the buried vegetation will get into the sample and cause chemical changes while the snow melts prior to analysis. Several workers have sought to get over the latter problem by sampling vertical slices taken on the edge of a pit. This technique can provide samples from the upper levels more or less free of vegetation but eventually as the bottom of the pack is approached the problem of contamination of the sample reappears. The longer pieces of vegetation can be removed but numerous small fragments from the fragile dried dead leafs and fronds remain.

The final section of the review is most noteworthy because of the attempt to estimate fluxes from concentrations reported in the literature. This is especially to be applauded. It is new work and thereby departs from the rest of the review which closely follows the literature. This part recognizes and draws attention to a major deficiency of much of the literature not only that pertaining to snow chemistry but also to so much that is written about environmental pollution in general. This is the need to estimate fluxes and the essential sterility of concentrations when reported alone.

One more discussion point emerges from the review and it appears to be especially relevant to this last section. This is the very definition of what constitutes the chemical composition of snow. Jones' definition is what is present in the melt water that passes a 0.45 μm filter. There are some problems with such a definition. One certainly is that not all the workers whose results form part of the review have necessarily ensured this definition applied to their measurements. Those of Barry and Price (1987) did not for their samples went unfiltered into the Dionex or the AA. However, it is essentially a definition of the composition of snowmelt water which may very well have changed from that of the snow during the melting and storage before analysis. In fact the evidence provided by the reviewer suggests that this indeed is the time in nature when most of the chemical reactions occur and the greatest source of chemical change outstripping by far all others put together is that which occurs when the meltwater comes into contact with the litter in the forest floor.

References

Barry PJ, Robertson E (1989) Chemistry of snow pack accumulation and melt in a deciduous forest, 46th Annual Eastern Snow Conference, Quebec City

of excretory products which are not related to urea. To understand animal impacts on the nutrient cycle in snow on a global scale, more information is needed with respect to their population sizes, geographical distributions in snow regions, and potential impacts on nutrient cycling and biogeochemistry. For example, moose may have an impact in Canadian boreal forests, but would not be a factor in alpine regions of the western United States (Cascade and Sierra Nevada Mtns) or in the Adirondack Mtns, New York. Mule deer are present in western North America, but are absent in eastern North America. Many of these larger forms would be sparse in treeless areas lacking protection or food and would be replaced by other species which can survive such zones (arctic polar bears and antarctic penguins). On a global scale, hares, small rodents and certain insects may have a more predictable impact on the nutrient cycle because of their more general distributions in and on snow. However, even some insects such as collembola and stoneflies, and snow worms, Mesenchytraeus (Odell, 1949), are localized at most in distribution. More emphasis should be placed on smaller invertebrates such as protozoa and rotifers. These animals are ubiquitous in snow and have been reported in snowfields where other microbiological activity (algae, etc.) is prevalent (Hoham, 1980; Pollock, 1970). At this time, it is not that clear how much the animals mentioned above contribute to the nutrient cycle of the snowpack throughout the snow accumulation period and the snowmelt which follows.

Secondly, Jones has presented considerable information on microbiological activity of the snow algae and their interaction with snow chemistry. The life cycle of snow algal flagellates was reviewed and evidence was presented for nitrogen depletion and utilization of other nutrients for growth in localized snowfields containing these microorganisms (Hoham et al., 1989). More data is needed on a worldwide basis on the interactions of snow algae and nutrient dynamics. Many snow chemistry studies do not include ortho-phosphate levels which are difficult to detect, but ortho-phosphate is often the limiting factor for growth of freshwater algae (Sze, 1986). It is also clear that snow algae are worldwide in distribution and are known from localities where there are annual snowfalls which result in recurring snowpacks (Kol, 1968). However, these microorganisms do not normally appear in snowpacks until at least one week after the snowmelt begins. Thus their part in nutrient cycling in snow does not begin until that time.

Discussion on "SNOW CHEMISTRY AND BIOLOGICAL ACTIVITY: A PARTICULAR PERSPECTIVE ON NUTRIENT CYCLING"

Ronald W. Hoham
Department of Biology
Colgate University
Hamilton, New York 13346
U.S.A.

H. Gerry Jones, INRS-Eau, Ste-Foy, Québec, Canada, has made a major attempt to demonstrate how nutrient cycling in snow is related to chemical and biological activity. His comprehensive manuscript includes topics on snow cover chemistry and biological activity, snowfall and global biological activity, snowfall chemistry (biogenic organics and nutrient pools), snowfall and living organisms, snow and vegetation, forests and snow chemistry (distribution and interception of snow and nutrient cycling, trees and snow depth patterns, litterfall and throughfall), snow and animals (mammals and invertebrates and nutrient cycling), and microbiological activity and snow chemistry (general microbial activity, snow algal biology, and snow algae and snow chemistry).

There is the need to ask more specific questions as to the importance and significance of the above factors on a worldwide scale versus more regional or local phenomena. For example, are there significant differences in chemical and biological activity which can be measured between deciduous and coniferous forests, snowpacks of different physical properties, and alpine, subalpine, temperate and polar ecosystems? How do these differences relate to snow accumulation, snowmelt and runoff when it occurs, and what effects do these have on streams, rivers and lakes?

Peter Barry, Chalk River Nuclear Laboratories, Chalk River, Ontario, Canada, will discuss forests and snow chemistry in a separate commentary. The remainder of this discussion will focus on biological activity in snow. Firstly, the animal populations mentioned in Jones' manuscript include mammals such as deer, moose, hares, and smaller burrowing rodents, as well as birds. Emphasis has been placed on their contributions to the nutrient cycle in snow through the release of excrement and urine. In addition, invertebrates such as insects and arachnids may contribute to the nitrogen and phosphorus cycle in snow through the release

NATO ASI Series, Vol. G 28
Seasonal Snowpacks
Edited by T. D. Davies et al.

Literally nothing is known about the interactions of other microorganisms such as bacteria, fungi, lichens, protozoa and rotifers with nutrient cycling in snow. Specialists are lacking in this part of snow biology, and the sparse information on these organisms in snow was apparent at a recent NASA workshop on the "Exobiology and Future Mars Missions" (Hoham, 1989). One group of microorganisms not mentioned in Jones' manuscript is the sea-ice algae. Much is known about them (Horner, 1985), and the scientists who have worked with these algae should participate with snow physicists, chemists and biologists in future meetings in snow science.

References

Hoham, R. W. 1980. Unicellular chlorophytes - snow algae. In Cox, E. R. (Ed.), Phytoflagellates, pp. 61-84, Elsevier North Holland, Inc., New York.

Hoham, R. W. 1989. Snow as a habitat for microorganisms. In McKay, C. P. and W. L. Davis (Eds.), Exobiology and Future Mars Missions, NASA Conf. Publ. 10027, pp. 32-33, Ames Res. Center, Moffett Field, California.

Hoham, R. W., Yatsko, C., Germain, L. and Jones, H. G. 1989. Recent discoveries of snow algae in Upstate New York and Québec Province and preliminary reports on related snow chemistry. In Lewis, J. (Ed.), Proceedings of the 46th Annual Eastern Snow Conference, pp. 196-200, Québec City, Québec, Canada.

Horner, R. A. (Ed.) 1985. Sea Ice Biota. CRC Press, Boca Raton, Florida, 215 pp.

Kol, E. 1968. Kryobiologie. Biologie und Limnologie des Schnees und Eises I. Kryovegetation. In Elster, H.-J. and W. Ohle (Eds.), Die Binnengewasser, Vol. 24, E. Schweizerbart'sche Verlagsbuchhandlung, Stuttgart, 216 pp.

Odell, N. E. 1949. Ice-worms in Yukon and Alaska. Nature, 164:1098.

Pollock, R. 1970. What colors the mountain snow? Sierra Club Bull., 55:18-20.

Sze, P. 1986. A Biology of the Algae. W. C. Brown Publ., Dubuque, Iowa, 251 pp.

CONTROLS ON THE COMPOSITION OF SNOWMELT

Martyn Tranter,
Oceanography,
University,
Southampton SO9 5NH,
UK.

1. Introduction

The composition of snowmelt has attracted much attention during the eighties, largely because of the possible links between snowmelt chemistry and the episodic acidification of streamwaters and lakes during the Spring thaw (Wigington et al, in press). Episodic acidification may result in fishkill, as has been observed, for example, in Scandinavia and Canada (Wigington et al, in press), and also may give rise to other deleterious effects on aquatic ecosystems (Baker et al, in press).

Johannessen and Henriksen (1978) were the first environmental chemists to publish material on solute fractionation from ice crystals into snowmelt. However, it should be noted that there is a substantial body of physical chemistry literature concerning solute inclusion and exclusion from ice crystals during melting and freezing (e.g. Jaccard and Levi, 1961; Gross, 1968; Fletcher, 1970; Addison, 1975). Johannessen and Henriksen (1978) observed that snowmelt may contain up to 5 times the solute content of the original snow, based on field lysimeter studies and laboratory experiments with snow columns (see Figure 1). They defined the concentration factor, CF, as the ratio of the concentration of a particular species, i, in snowmelt, M, to the concentration in the parent snow, P. Hence,

$$CF_i = M_i / P_i$$

CF is dependant on the mass fraction of snowmelt considered. Figure 1 shows that initial fractions of snowmelt may have CF values of 3.5 to 5 at the onset of snowmelt, and that different

NATO ASI Series, Vol. G 28
Seasonal Snowpacks
Edited by T. D. Davies et al.

ions have different CF. Thereafter, solute scavenging by the early snowmelt fractions and the subsequent depletion of solute in the remaining snowcover gives rise to snowmelt fractions with low CF values, which approach 0.2 to 0.5 depending on the ion of interest. In this particular experiment, the first 25% of snowmelt contained higher concentrations of solute than the initial snowcover.

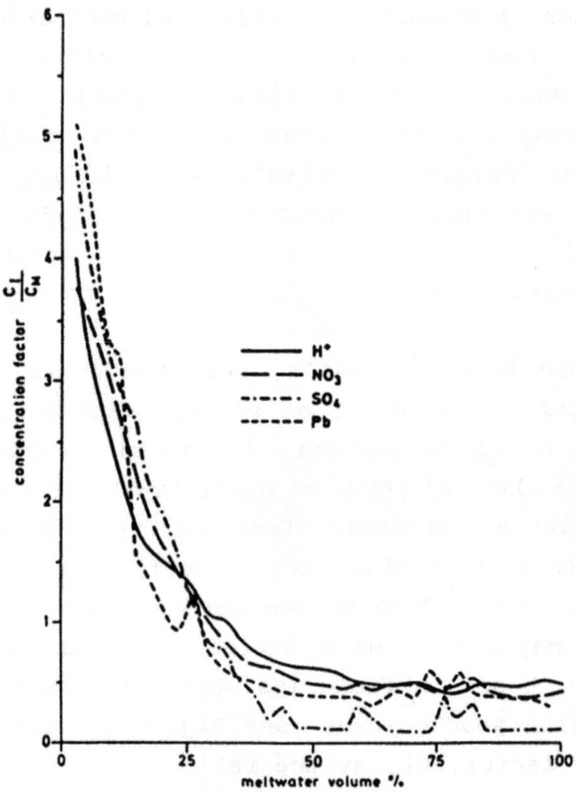

Figure 1. Fractionation of ions from snow into meltwater (from Johannessen and Henriksen, 1978).

The notion of ions fractionating from ice crystals into meltwater has been firmly established in the literature ever since, and has been observed by many workers in snow covered terrains throughout North America, Europe and Asia. However, we still appear to understand little about what actually controls the precise composition of snowmelt, other than in a qualitative manner. As an illustration of this point, various maximum values of CF are to be found in the literature. As noted above, variable snowmelt volumes may partially explain the variation in maximum CF, but variable volumes cannot explain all the variability. Snowmelt has been observed to contain concentrations of solute which are similar to those in the original snowcover (i.e. CF = 1; Berg, 1986), whereas concentrations which may be up to 20 times greater have also been observed (i.e. CF = 20; Suzuki, 1982). Even the simple observation of variable maximum CF values is largely unexplained by the existing mainstream literature. It is not surprising that, as yet, no robust model exists which is able to predict snowmelt composition given a range of snow column and meteorological conditions.

A way forward in our understanding and interpretation of the large range of initial CF values is provided by mass balance studies, which have shown that processes other than fractionation may deplete or augment the solute content of snowmelt (Jones, 1987; Jones and Deblois, 1987). These processes include the dissolution of inorganic and organic debris, and microbial reactions. Hence, the snowmelt-ice crystal (or snowcover) system may be non-conservative for a variety of dissolved species, and examination of CF values alone may give rise to misleading interpretations of the processes in operation during snowmelt.

The composition of snowmelt is also usually considered in isolation from other potential precipitation inputs to the snowpack. In many snow-covered watersheds, for example those of south-eastern Canada and the Adirondacks, the thaw may coincide with rainstorms. Hence, waters percolating through the snowcover may be mixtures of rain and snowmelt. The interactions between rain and snow have scant coverage in the mainstream literature, yet are potentially important in controlling "snowmelt" composition in these and other watersheds.

There is no systematic study of the factors which control snowmelt composition in the mainstream literature. Neither is there an exhaustive range of publications from which to draw such a study. Hence, the following is a summary of an incomplete literature, which hopefully will help to stimulate further research on snowmelt composition in the near future.

2. Location of solute in natural ice crystals

Before examining the chemical composition of snowmelt, it is necessary to understand where solute is located in ice crystals. An immediate reaction is to ask "how can solute be found in ice crystals" since ice is a solid? In this review, we operationally define "solute" in the solid phase as those ions or aerosol which can readily and rapidly dissolve on thawing, for example, chloride and fluoride, sea salt and other soluble aerosols. Other types of aerosol, for example flyash, are less readily soluble - their dissolution kinetics are relatively slow - and hence contribute solute to snowmelt over longer timescales.

Solute may be present within and on recently deposited snow crystals (Barrie, this volume). However, almost immediately after deposition, metamorphism of ice crystals commences (Davis, this volume). Vapour released from sites of high curvature and/or high temperature tends to leave behind less volatile solute on the ice crystal surface. Because solute does not travel with the vapour, and because solute is not readily accommodated into the ice lattice (Bales, this volume), solute is likely to be segregated from grains interiors to locations nearer the grain surface during metamorphism (Colbeck, 1981, 1987; Davis, this volume). Wet snow contains discrete droplets of liquid residing in triple junctions and there may also be veinlets along grain boundaries. Further, it is likely that ice crystals contain a quasi-liquid surface layer at temperatures > - 35 °C (Ushakova and Troshkina, 1974). The combination of these facts leads to the conclusion that it seems likely that much of the solute in snowcover may be found either in the quasi-liquid layer and/or in water between grain boundaries, and that progressive metamorphism of ice crystals increases the likelihood of solute being found in these locations.

Hence, transfer of water through snowcover and in combination with the melting of ice crystals usually results in solute acquisition by the meltwaters, due to the leaching of solute rich near-surface brines and conservative acquisition of solute already present in the ice crystal.

3. Solute conserving and non-conservative reactions in the snowmelt-ice system

It may seem strange at first to think of fractionation as a solute conserving reaction. However, let us focus on the conservation of solute during the melting and freezing of a snow (or ice crystal)-meltwater system. Allowing the system to experience successive melt and freeze cycles, exclusion of solute from ice crystals into solution during melting, and conversely, inclusion of solute into or onto ice crystals during freezing, does not lead to solute loss or gain by the system *per se*. Hence, fractionation may be considered as a solute conserving reaction. We define solute conserving reactions as those which do not lead to a change in the quantity of a particular solute in a snow/ice and meltwater system. These reactions or processes include fractionation, preferential elution, and the scavenging of solute by rain-on-snow. On the other hand, non-conservative reactions add or subtract particular solute from the snow/ice and meltwater system, and include reactions such as rain-on-snow, microbial activity, dry deposition, photochemical, adsorption and dissolution reactions, and the loss of rain/meltwater from the system. By considering the quantity of solute in the system, processes which alter the water equivalence of the system (e.g. sublimation) can be ignored.

In the following sections, fuller consideration of these solute conserving and non-conservative reactions is given, together with some account of the spatial and temporal distribution of the reactions throughout the snowpack. We begin with the conservative reactions.

3a) Solute conserving reactions

i) Fractionation

Concentration factors spanning a considerable range are to be found in the literature. Before considering the reasons for these variable CF values, a note of caution is needed. It is important to be aware that many snowmelt studies do not include the calculation of the mass balance of ionic species in the snow and meltwater system. Hence, there is usually some uncertainty as to whether or not non-conservative reactions influence the concentration factor. For example, if neutralisation reactions such as the weathering of carbonaceous aerosol occur within the system (see below), then the apparent concentration factor of the hydrogen ion will be small, while that of calcium and bicarbonate will be enhanced. Depending on the extent of weathering, the concentration factors may bear little resemblance to those of the fractionation reaction alone.

There is a scarcity of well-defined snowmelt studies in the literature, so we will begin by examining laboratory snowmelt experiments, since in some cases, these approximate most ideally to solute conserving systems.

Laboratory studies: Johannessen and Henriksen (1978) conducted laboratory experiments on the progressive melting of snow to demonstrate that ions fractionate from snow or ice crystals into meltwater. Maximum CF in initial meltwater fractions ranged from 3 to 5, dependant on ion (see Table 1).

Marsh and Webb (1979) investigated controls on CF using crushed ice cubes which contained various concentrations of sulphuric acid. Even crushed ice exhibits fractionation on melting. The initial maximum CF is independent of P (the concentration of solute in the original ice crystal), but increases with the depth of ice. Further, discrete bands of solute increase CF, whereas greater rates of melting produce lower CF. A criticism of the work, as is common to most other snowmelt studies, is that

Table 1. Range of initial CF values found in laboratory studies of snowmelt composition.

Author	Maximum initial CF	snow or lab ice
Johannessen and Henriksen, 1978	3 - 6	snow
Marsh and Webb, 1978	2 - 5	lab ice
Colbeck, 1981	2 - 5	snow
Brimblecombe et al, 1986	2 - 4	snow
Tsiouris et al, 1985	5 - 11	snow
Brimblecombe et al, 1988	2 - 4	lab ice
Bales et al, 1989	2 - 17	snow
Hewitt et al, 1989	2 - 6	snow
Hewitt et al, 1989	1 - 2	lab ice

replicate experiments were not performed, so that the statistical significance of the differences between experiments cannot be assessed. Further, although crushed ice exhibits similar fractionation properties to natural snowcover, the microscale physico-chemical properties of neither the crushed ice nor the natural ice crystal during melting are known with certitude. Hence, extrapolation from crushed ice experiments to natural snowcovers must be conducted with caution, even though qualitatively similar results may be obtained (cf Hewitt et al, 1989).

Foster (1978) attempted to model the experimental results of both Johannessen and Henriksen (1978) and Marsh and Webb (1978). He concluded that inhomogeneous percolation is the most important factor in controlling the magnitude of the concentration factor. Drainage through experimental ice columns (Foster, 1978; Marsh and Webb, 1979) and natural snowcover (Seip, 1980; Jones, 1985; Marsh and Woo, 1985; English et al., 1986, 1987) may occur through preferred channels, rather than uniformly through the whole matrix. This is an important observation - it appears that the opportunity for solute leaching, partly as a result of snowpack hydrology, may limit the solute content of meltwaters in laboratory ice columns (Foster, 1978; Colbeck, 1981; Bales et al, 1989), and also in natural snowcover (Tranter et al, 1988a; Bales et al, in press).

From these three basic studies, a variety of other laboratory have arisen. Colbeck (1981) added NaCl to sieved snow, and performed three snowmelt experiments, which examined the impact of diurnal melt-freeze cycles and solute banding (ie non-homogeneous distributions of solute throughout the column). Snowmelt derived from a column with NaCl added to the snow surface alone gave rise to CF of 5 in the first meltwater fractions, where CF was calculated with respect to the average snow concentration (if the CF was calculated with respect to the dilute, underlying bulk of the snow, CF of ~18 would arise). Similar effects of concentrated snowfall atop leached snowpack having a pronounced effect on meltwater composition have been observed in the field (Tranter et al, 1988b). A shallow snow column with four discrete NaCl bands gave rise to CF of approximately 2.5 in the first meltwater fractions. Adding the same amount of NaCl to the bottom of the same shallow snow column gave rise to initial CF of approximately 4. In the latter experiment, pronounced peaks in meltwater concentration occurred at the breakthrough of meltwater on each day.

Brimblecombe et al (1986) examined the partial melting of a variety of small snow samples (~500g). The partial melting of 33 samples gave rise to a meltwater fraction, 10% of the initial mass of snow, which contained concentrations of ions which spanned three orders of magnitude. Initial CF of between 2-4 were recorded, dependant on ion. It was noted that sulphate ions fractionated into the initial meltwater to a greater extent than nitrate ions. Similarly, nitrate ions usually fractionated to a greater extent than chloride ions. Hence, differential fractionation of ions occurs, and this may give rise to preferential elution of ions from snowpacks under some circumstances (Davies et al, 1987). Preferential elution is covered in greater detail below.

The concentration factor is useful for describing the magnitude of fractionation, but more detailed information on the mechanism of fractionation can be obtained by use of the fractionation factor, F, defined as

$$F_i = M_i / R_i$$

where R is the concentration of species i in the residual ice crystals, rather than in the parent ice crystals. Brimblecombe et al (1988) demonstrated that the variation in the fractionation factor of melt derived from laboratory ices of known composition can be modelled as the mixing of two components, a concentrated, surficial brine and a dilute, grain interior. They note that ice with low solute concentrations most readily approximates the two component situation, whereas ice of higher solute loading deviates from a simple two component case, which can be explained by their being a concentration gradient of solute between the surficial brine and grain interior. This may suggest that solute exclusion during metamorphism is more efficient or more rapid for ice crystals of low initial solute loading.

Tsiouris et al (1985) conducted two controlled laboratory melt experiments on homogenised, natural snow, aimed at understanding the effects of diurnal melt-freeze cycles on snowmelt composition. The control experiment, run with no diurnal melt-freeze cycle, confirmed that snowpack hydrology has a large effect on snowmelt composition. Melt collected near the centre of the base of the snow column had dissimilar discharge and composition in comparison with melt collected from the remainder of the base. Concentration factors in the first fractions ranged from 5-11, depending on collection point and ion. Further, discharge from both areas was not constant, despite attempts to keep the rate of ablation constant. The experiment examining diurnal melt-freeze cycles again showed the influence of collection point and ion on CF, and showed that as the ablation rate decreased, and discharge consequently decreased, the ionic strength of the meltwaters increased. This phenomena was also observed in the field. Hence, in agreement with other studies, the rate of ablation may also be a control on meltwater composition, and this effect may be observed in meltwater composition on time scales of a few hours.

Bales et al (1989) conducted a series of laboratory snowmelt experiments using natural snows, which had been sprayed with various dilute salt solutions or dosed by fumigation with SO_2. The object of these experiments was to observe how different mesoscale features (i.e. banding of solute) in the snow column effected the scavenging efficiency of meltwater. In smaller scale experiments using 14 cm snow columns, initial CF values ranged from 2-9, dependant on ion and experimental conditions. Melt-freeze cycles gave rise to the higher CF values. Larger scale experiments in a cold room, using 120-155 kg of snow with a depth of 340-400 cm, gave rise to initial CF of 6-17, dependant on experiment and ion. Highest CF were observed at the lowest melt rate. These experiments are unique in having the potential to compare and contrast the composition of meltwater with the solute content of the snow, both before, during and after leaching, and hence allow an evaluation of the efficiency of solute transport through the snow column. Solute is most efficiently scavenged from the top of the snow column. It seems possible that solute eluting through the snow column may partially exchange or diffuse into an "immobile" layer at the ice crystal surface. This immobile layer may be similar to the "surficial brine" of Brimblecombe et al (1987, 1988), or may be solute locked within the bonds between snow grains which form during the metamorphism of wet snow (Davis, this volume).

Hewitt et al (1989) noted that the initial CF derived from the melting of homogenised snow, between 2-6 dependant on ion, were greater by a factor of two than those derived from the melting of frozen droplets of similar composition to the homogenised snow. They conclude that the modelling of the loss of solute from snow should take into account the simultaneous physical and chemical changes which occur during metamorphism.

Table 2. Range of initial CF found in field studies of snowmelt composition.

Author	Maximum initial CF	Country
Johannessen and Henriksen, 1978	3 - 7	Norway
Skartveit and Gjessing, 1979	2 - 5	Sweden
Suzuki, 1983	20	Japan
Berry et al, 1984	5 - 7	Canada
Cadle et al, 1984	3 - 7	USA
Cadle et al, 1987	~5	USA
Berg, 1986	1	USA
Semkin and Jeffries, 1988	7 - 10	Canada
Peters and Driscoll, 1989	6 - 9	USA

In summary, Table 1 shows that in laboratory studies, initial maximum CF of 2-17 are recorded, although values in the range of 2-6 are more typical.

Field studies: Fractionation of solute into meltwaters has been observed in many field studies throughout North America, Europe and Japan. Normally, CF is simply reported, and little information on the size or control of the value is reported. CF is also calculated for different volumes of snowmelt, which does not allow easy comparison of CF values. Table 2 summarises field studies to date. Initial maximum CF vary from 1-20, but typically range from 3-7. In all these studies, it is difficult to be certain that at least some of the variability might not arise from non-conservative reactions. Further, it is often difficult to address the influence of factors such as snowpack hydrology, solute distribution and melt regime.

Bales et al (in press) examined solute movement through natural snowcover with the aid of artificial tracers added at various depths. They noted the heterogeneous nature of snowpack hydrology, and the influence of ice lenses in diverting meltwater flow. Hence meltwater catches by lysimeters may not be representative of snowmelt chemistry in the snow column immediately above. Less than 10% of tracer added to the column was collected in the lysimeter below, indicating lateral advection of solute. Tracer was observed to be concentrated above ice lenses. Initial CF of 5-10, dependant on ion, were reported, which are of similar magnitude to those of parallel laboratory studies (Bales et al, 1989).

ii) Preferential elution

Both laboratory and field observations demonstrate that under most conditions, ions do not fractionate into meltwaters in the ratio at which they are found in the parent snow or ice sample (Davies et al, 1982; Brimblecombe et al, 1985; Davies et al, 1987). Figure 2 illustrates that both sulphate and nitrate may be eluted preferentially with respect to chloride, and that dilute meltwaters are therefore proportionally enriched in chloride. Preferential elution of solute has been observed from snowpacks in North America (Semkin and Jeffries, 1988) and Europe (Johannessen and Henriksen, 1978, Babiakova and Bodis, 1986, Tranter et al, 1986, 1988a). Non-conservative reactions may promote preferential elution, while the meso-scale distribution of solute in the column or the hydrology of the snowpack can either promote or diminish the influence of preferential elution on snowmelt composition (Tranter et al, 1987; Bales et al, 1989, in press).

Experiments with artificial ices, with concentrations of ions spanning a concentration range of five orders of magnitude (1 - 10,000 μmol l^{-1} for sulphate and chloride) leave little doubt that preferential elution (hereafter, PE) of solute occurs, and that the dissimilar solubility of ions in ice crystals is one of the principal controls on PE (Brimblecombe et al, 1988) in conservative systems with homogeneous, meso-scale distributions of solute.

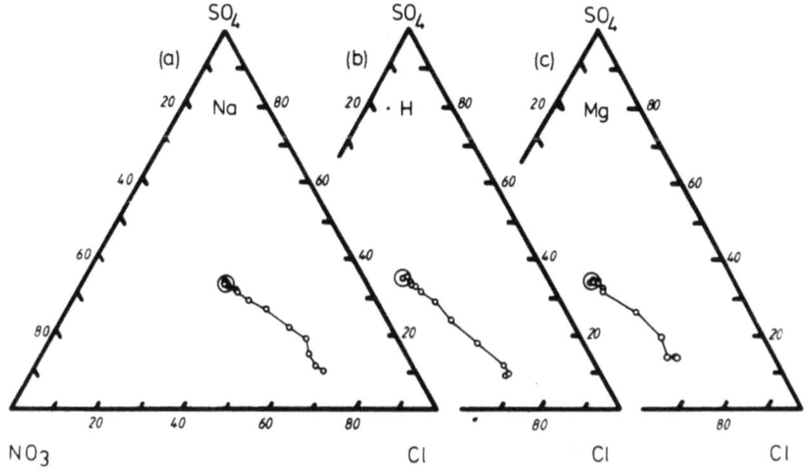

Figure 2. Preferential elution of sulphate and nitrate with respect to chloride. Anion concentrations in the original laboratory ice were each at 100 μmol l^{-1}. The counter ion, sodium, magnesium or hydrogen, has little influence on the PE of these strong acid anions. The first 13 meltwater fractions in the central circle had similar proportional anionic composition, but the later fractions become progressively proportionally enriched in chloride (from Brimblecombe et al, 1987).

Brimblecombe et al (1988) note that PE is most readily observed when the solute content of ice crystals is low, as shown in Figure 3. Sulphate is eluted more rapidly than chloride at environmental concentrations (1 - 10 μmol l^{-1}), whereas at high concentrations (> 100 μmol l^{-1}) sulphate and chloride show little difference in their tendency to be lost from melting ice. Ions that are more soluble in ice tend to be removed less readily from the artificial ice than ions which are not particularly soluble. For example, fluoride is relatively soluble in ice, and the solubility of fluoride can be enhanced by the presence of the ammonium ion. Consequently, when ammonium ions are present, the fractionation factor for fluoride is considerably reduced (Brimblecombe et al, 1988). The presence of other counter ions, such as the hydrogen ion, increases the fractionation factor to a value similar the other halide ions.

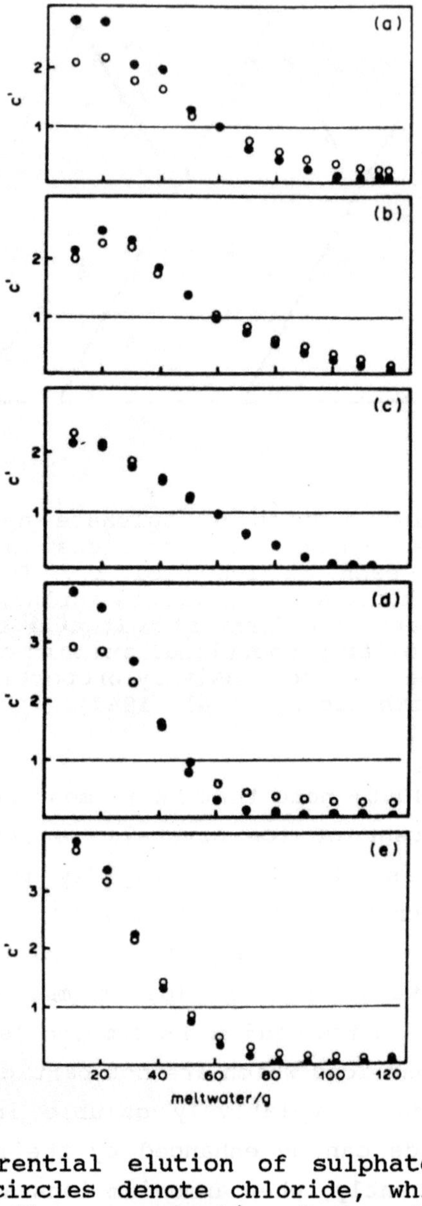

Figure 3. Preferential elution of sulphate with respect to chloride. Open circles denote chloride, while filled circles denote sulphate. The concentration factor (c') of sulphate is greater than that of chloride in initial melt fractions of experiment (a), where the concentration of both ions in the initial laboratory ice is 1.0 μmol l^{-1}. The concentration factors become similar as the initial ice becomes more concentrated, at 100 μmol l^{-1} (b) and 10,000 μmol l^{-1} (c). Experiments (a) - (c) were conducted at air temperatures of 20°C, while experiments (d), at initial ice concentrations of 1 μmol l^{-1}, and (e), at 100 μmol l^{-1}, were conducted at 4.5°C (from Brimblecombe et al, 1988).

Hence, the simplest explanation for the preferential elution of ions from melting homogeneous snows and ices is that of the mixing of two different solutions. One solution is a concentrated surficial brine and the other an ice which contains low concentrations of sodium chloride. The initial high concentrations found in meltwaters are a product of the melting or leaching of these surficial brines (Brimblecombe et al, 1988). The mixing of solute rich and dilute components has been observed in natural snowcovers (Tranter et al, 1988a), as illustrated in Figure 4. The notion of a surficial brine can be related to the existence of an "immobile layer" of solute hypothesised in other studies (Bales et al, 1989, this volume; Davis, this volume). Earlier suggestions of ice having chromatographic properties (Tranter et al, 1986) have been shown to be unfounded (Hewitt et al, 1989). Retardation of ions was not observed when aqueous solutions were applied to columns of snow and artificial ice.

A model that places most of the solute associated with the ice into an intergranular brine, but allows some compositional variation within the grain, explains many of the observations of the magnitude of PE in other near-conservative laboratory and natural snowpacks. For example, sulphate is normally eluted preferentially with respect to chloride in most systems (Tsiouris et al, 1985; Brimblecombe et al, 1985, 1986, 1987, 1988). This most likely represents the fact that sulphate is more concentrated into surficial or intergranular brines than chloride (Mulvanney et al, 1988). Further, it is the difference in the proportional ionic composition of grain interiors and surficial brine which controls the magnitude of PE. In cases where there is much solute in the surficial brine, PE is difficult to observe until much leaching has occurred.

In many snow-covered catchments, ablation rates will not be uniform (Johannessen and Henriksen, 1978). Further, snowpack hydrology will tend to cause mixing of meltwaters within the snow column (Tranter et al, 1988a), a process enhanced by the presence of ice lenses (Bales et al, in press). Differing mesoscale snowpack chemistry might also give rise to apparent PE (Bales et

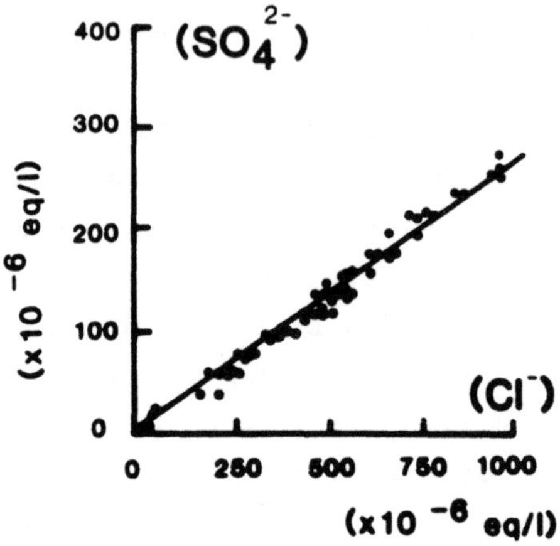

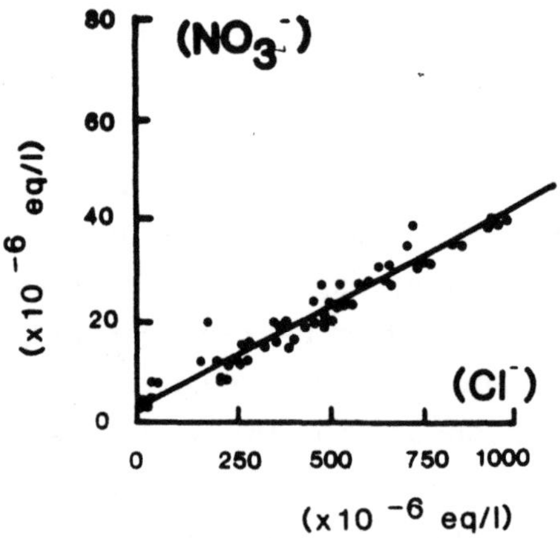

Figure 4. The influence of snowpack hydrology on meltwater chemistry. Meltwater samples were collected in a snowpack at four different locations on two consecutive days. Concentrations of strong acid anions define conservative mixing lines, i.e. the mixing of concentrated and dilute components (from Tranter et al, 1988a).

al, 1989). In particular, solute near the top of the snow column is scavenged more efficiently than solute lower down the column. Thus, in many situations, it may be that only meltwaters issuing from leached snowcover which have noticeably different proportional anion composition (Brimblecombe et al, 1988).

iii) Freezing

As yet, the effects of freezing on meltwater composition have only been poorly documented in the literature, but the following influences have been noted. Adams and Allen (1987) note that meltwaters trapped in slush layers on lake ice may become concentrated during freezing. It is likely that diurnal melt-freeze cycles may lead to some freezing of meltwaters lower in the snow column, and therefore solute in the remaining meltwater may become more concentrated. This may in part explain the higher concentration factors found in laboratory experiments which have examined such cycles (eg Colbeck, 1981). Freezing which leads to the formation of ice lenses may modify snowpack hydrology (Bales et al, in press) and therefore influence snowmelt composition indirectly.

iv) Scavenging of solute by rain on snow

There is currently some debate concerning the scavenging of solute during rain on snow events. This ultimately may be a methodological problem, involving difficulties in adequately characterising the contemporaneous changes in the composition of rainfall and rain-meltwater mixtures.

Rainfall can account for up to 50% of the solute flux from snowpacks in some catchments (Semkin and Jeffries, 1988) and observed that rainfall can cause a rapid response in water delivery to lysimeters (Semkin and Jeffries, 1986). Brown et al (1985) showed from isotopic evidence that the majority of water draining a snowpack following rainfall might be rainwater, which does not precludes the scavenging of solute by infiltrating rain. English et al (1987) make the pragmatic assumption that under

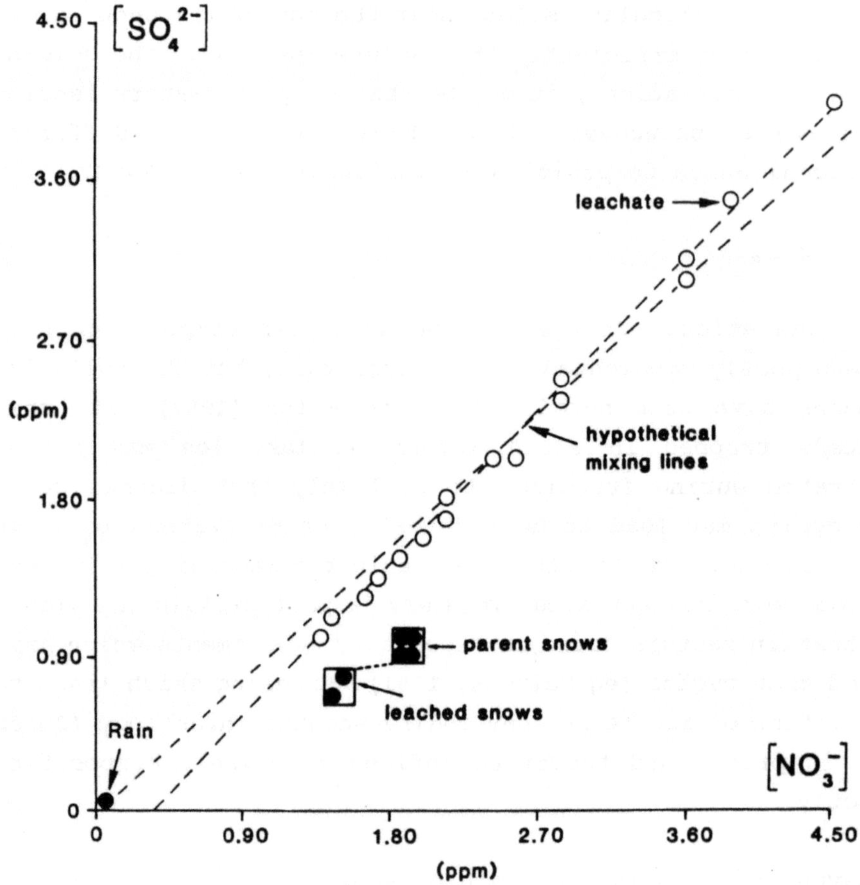

Figure 5. The leaching of solute from snow by artificial rain in laboratory experiments. Note that the leachate, which was originally deionised water, contains more solute than the parent snow (from Jones et al, 1989).

some circumstances, rainfall does not scavenge solute from snowcover. For example, scavenging of solute is likely to limited when there is heavy rainfall, which may form new preferential flowpaths (Marsh and Woo, 1984; Kattleman, 1986), and allow relatively little opportunity for solute exchange between ice crystals and rainwater (Jones et al, 1989).

However, under some circumstances, rainfall may well scavenge solute. This is most likely to occur before an efficient drainage network is established in the snowpack. As shown in Figure 5, laboratory experiments, using dilute aqueous solutions as a proxy for rainfall and homogenised snow, consistently show solute scavenging (Jones et al, 1989). Similar larger scale experiments further demonstrate solute scavenging by artificial rain on snow (Tranter et al, in prep).

3b) Non-conservative reactions

Non-conservative reactions are of importance, both to snow chemists and biogeochemists, because they may modify the pH, base cation content and nutrient status of meltwaters issuing from snowcover. In unforested Alpine and sub-Arctic snowcover, one might anticipate a relatively greater influence of inorganic processes, because sources of organic material are more limited and wind transport of inorganic particulate material may be more pronounced. By contrast, in forest snowcover there is greater likelihood of a biologically mediated influence because of the greater organic content of the snowcover (Jones, this volume).

i) Dissolution of inorganic and organic particulate materials

Delmas (1989) examined the effects of local dust on the acidity of meltwaters issuing from Alpine snowcover. As shown in Figure 5, the pH of the meltwater was higher than predicted from the composition of freshly fallen snow. Mass balance calculations suggested that Ca was being generated within the snow-meltwater system, whereas hydrogen ions were being lost. Laboratory experiments confirmed the hypothesis of local, carbonaceous dust being an effective neutralising agent for the snowmelt. The laboratory experiments suggested that the location of the dust within the snow column had some control on the composition of the snowmelt. Dust near the base of the snow column was more effective at neutralising snowmelt than similar concentrations of dust near the column surface (see Figure 6). This presumably reflects the greater rock-meltwater contact times for dust located at the base of the snow column, giving greater opportunity for neutralisation to occur.

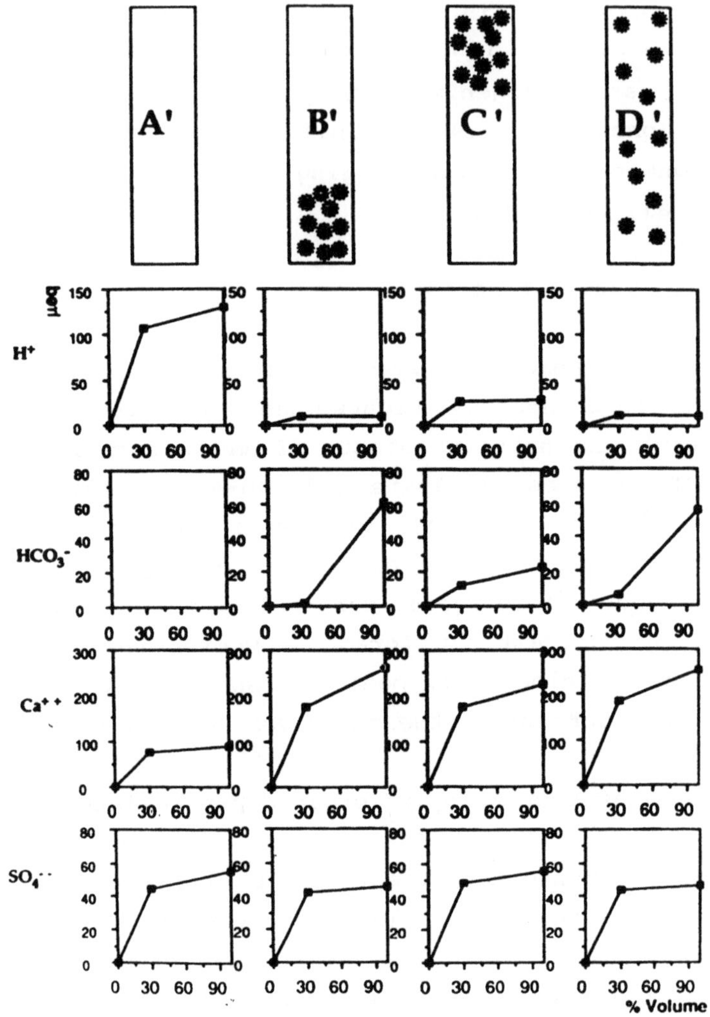

Figure 6. The influence of calcareous particles on snowmelt composition. The black stars in the top boxes (A' - D') denote the presence and distribution of particles in the snow column. In comparison with the snowmelt experiment containing no added particles (A'), it can be seen that addition of particles decreases the hydrogen ion content of meltwaters, while increasing the calcium content (from Delmas, 1989).

Similar types of reactions between weatherable aerosol and snowmelt might lead to partial neutralisation of snowmelt in the

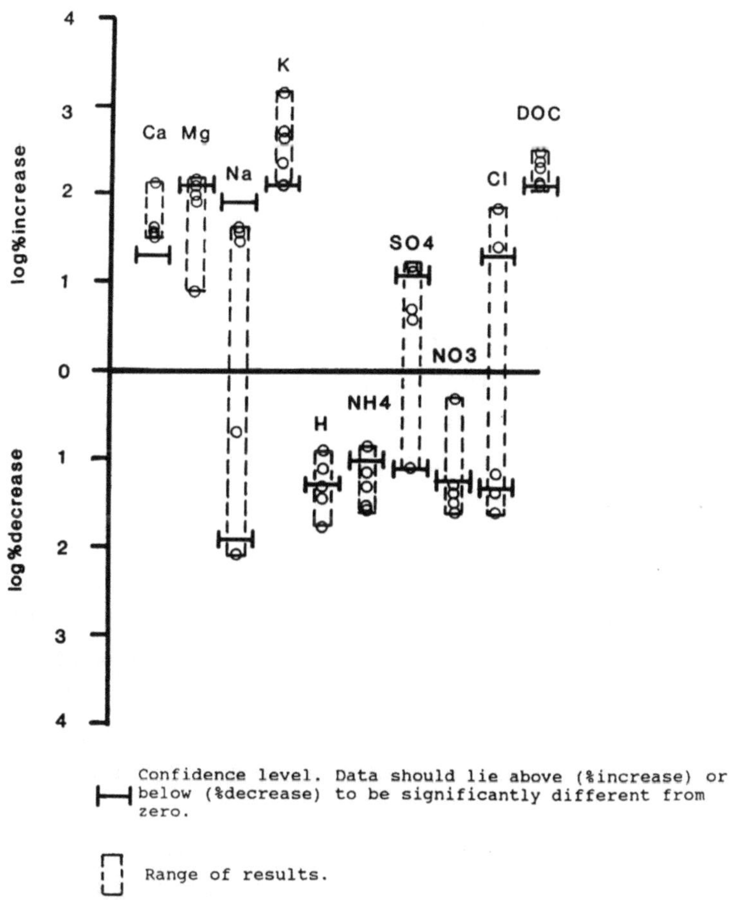

Figure 7. The influence of inorganic and organic particulate material on snowmelt composition. Snowmelt experiments show that K and DOC may be derived from the dissolution of organic debris within the snow column (from Jones and Tranter, 1989).

Sierra Nevada, USA, where strong acid anions seem to fractionate to greater extent than hydrogen ions (Berg, 1986; Williams et al, in press).

In urban locations, the pH of snowfall may be increased by the scavenging of aerosol emitted from cement works (Allan and Jonasson, 1978. Certain types of flyash may also partially neutralise acidic snow (Kumai, 1985). Calcareous soil particles may also reduce the acidity of snowmelt (Munger, 1982). Hence, a wide range of aerosol has the potential to act as neutralisation agents, and the critical factors to examine are the acidity of the fresh snowfall, the quantity and reactivity of the aerosol, together with the distribution of the aerosol within the snow column.

In some cases, it is not possible to determine whether or not inorganic or organic particulate material is the source of ions to snowmelt (Jones, 1987). However, it is likely that ions such as K and DOC (see Figure 7) are primarily derived from the dissolution of organic debris, particularly in forest snowcover (Jones, 1987; Jones and Tranter, 1989).

All the above types of dissolution reaction serve to reduce the apparent CF value for hydrogen ions and increase the apparent CF for base cations, in particular, those of Ca and K. Mass balance studies clearly show the increase in base cations and the reduction in the hydrogen ions (Jones, 1987; Delmas, 1989).

On occasions, "negative" fractionation of solute into snowmelt can occur. This has been observed in the Wasach Mountains of Utah (Messer et al, 1982), where alkaline snow gives rise to more alkaline snowmelt. Hence, there is an apparent negative fractionation of hydrogen ions. The reason for this phenomenon seems to be that alkaline soil dust and/or gypsum aerosol from the Great Salt Lake interacts with the snowmelt to give rise to a more alkaline solution.

ii) Adsorption of solute by cation exchange

Clay minerals are known to exhibit cation exchange reactions, especially the 2:1 clays (eg Drever, 1981). Clay minerals are likely to be found in the snow column, because they may be the condensation nuclei for ice crystals (Schemenauer, 1981), and may be scavenged from the atmosphere during snowfall, and dry deposited onto the snowpack surface. Bales (pers. comm.)

suggested that were there to be a sufficient loading of clay minerals in the snowpack, for example, when snowfall scavenges Saharan dust plumes, then cation exchange reactions on clays may modify the cation composition of meltwaters. This interesting idea has not been pursued to date.

iii) Microbial/biological activity

Recent work on the mass balance of nitrogen species in snowcover in forest sites, and analogous experiments in the laboratory, all suggest that microbial activity within snowpacks during periods when there is an appreciable water content in the snow cover is such that there are losses of both nitrate and ammonia from the snow-meltwater system (Jones and Deblois, 1987; Jones, this volume). Hence, in common with most near-surface aquatic systems, there may be biological mediation of meltwater chemistry within snowcover. At present, it is difficult to quantify the magnitude of microbial activity on snowmelt composition. Other possible biological effects on snowmelt composition are covered in detail by Jones (this volume).

iv) Other non-conservative processes

It is apparent that rain-on-snow will almost certainly, even if temporarily, add solute to the snow-meltwater system. The other potential non-conservative processes to effect meltwater composition are dry deposition and photo-chemical reactions. These processes have yet to receive attention in the existing literature, although Tranter et al (1988a) speculated that dry deposition might have influence on the composition of dilute snowmelt on timescales of hours.

4. Discussion

The above account of field and laboratory studies of controls on the composition of snowmelt has been largely descriptive in nature. As yet, there is no robust model of snowmelt chemistry which includes the important influences of snowpack hydrology, metamorphism and non-conservative reactions in snow-meltwater systems. First steps in developing such models are covered by Bales (this volume). However, it is encouraging to conclude that there is consensus among often disparate studies as to the major influences on snowmelt composition. These are itemised below.

Snowpack hydrology has the potential to route meltwater horizontally rather than vertically, to mix meltwaters of different compositions within the snow column, to carry meltwater in preferred channels or macropores, to limit the opportunity for solute scavenging, and hence to mitigate the high concentration factors found in some laboratory studies. Snowpack hydrology is a major control on snowmelt composition (Foster, 1978). If the ablation rate has an effect on snowpack hydrology (for example, high ablation rates might give rise to more channelised flow), then the ablation rate is also an important determinant on snowmelt composition.

The micro-scale distribution of solute within and on ice crystals is not known with certitude, yet this is of fundamental importance to snowmelt composition, since it contributes to the ease of solute scavenging by meltwaters. An associated uncertainty is how changes in ice crystal morphology, packing, and the liquid water content, which accompany metamorphism, contribute to the ease of solute scavenging (Davis, this volume).

The meso-scale distribution of solute within the snow column has an important influence on snowmelt composition, since solute scavenging is more efficient for solute found near the surface (Tranter et al, 1988b; Bales et al, 1989, 1990). This presumably results surface layers experiencing more complex melting, and from there being relatively more uniform or matrix flow of meltwater near the surface, compared to relatively more channelised or macropore flow at depth.

Preferential elution, the differential fractionation of ions into snowmelt, seems to be a fundamental property of snowmelt-ice systems, but the process may be confounded by the meso-scale distribution of solute and snowpack hydrology. It seems likely that the microscale distribution of solute throughout ice crystals differs by ion.

The conditioning of the snowpack by prior melt-freeze events seems to have a large impact on meltwater composition. These observations are consistent with the early laboratory studies of the influence of solute banding and ablation rate. Solute redistribution by diurnal-melt freeze cycles results in readily scavengable solute being relocated to the near base of the snow column, or may lead to partial freezing of meltwater at depth. Concentrations of ions in subsequent snowmelt are then likely to be elevated by the presence of the redistributed solute or by discharge of the concentrated meltwater.

Intuitively, it is anticipated that greater snow column depths should give rise to more concentrated snowmelt, since the snowmelt will have greater opportunity for solute scavenging. However, the influence of snow column depth on meltwater composition has only been reported in one laboratory study (Marsh and Webb, 1979).

Rain-on-snow is a poorly researched phenomenon, yet has the potential to mitigate the high concentration factors found in other field and laboratory studies by partially removing solute from the pack. There is some uncertainty about the extent of solute scavenging by rain in natural snowpacks, but the few laboratory studies to date clearly demonstrate solute scavenging.

Non-conservative processes, such as dissolution and biological activity may have an impact on snowmelt composition, which is dependant on the initial solute loading of the system, and the magnitude and type of the dissolution and biological activity. In relatively dilute snowpacks, dissolution of inorganic aerosol may partially neutralise the acidity of the snowmelt. In more acidic snowpacks, local aerosol of both natural and anthropogenic origin may also act as neutralising agents. Within snowpacks in forested areas, or where there is a high input of organic matter to the snowpack, biological activity has the potential to modify the nutrient status of snowmelt. It is likely that future studies of snowpack and snowmelt chemistry will consider the role of biological influences in greater detail.

5. Conclusions

Despite the disparate nature of field studies and laboratory experiments, it is reassuring to find that there is some agreement on the most important controls on snowmelt composition. These are snowpack hydrology, the micro- and meso-scale distribution of solute within the snow column, the occurrence of melt-freeze cycles, the influence of rain-on-snow, the occurrence of dissolution reactions, and the possibility of biological activity. As yet, there is no robust model of snowmelt composition which accounts for all these influences. Hence, our understanding of these controlling factors is largely qualitative at present.

6. References

Adams WP and Allan C (1987) Aspects of the chemistry of ice, notably snow, on lakes. In: Jones HG and Orville-Thomas WF (ed) Seasonal Snowcovers: Physics, Chemistry, Hydrology, NATO ASI Series C, Vol. 211, D. Reidel, Derdrecht, p393.

Addison JR (1975) Electrical properties of saline ice at 1 kHz down to -150 °C. J App Phys 46:513-522.

Allan RJ and Jonasson IR (1978) Alkaline snowfalls in Ottawa and Winnipeg, Canada. Atm Env 12:1169-1173.

Babiakova G and Bovis D (1986) Accumulation and evolution of sulphate and nitrate levels in snow. IAHS Publ. 155:271-281.

Baker JP, Bernard DP, Christensen SW, Sale MJ, Freda J, Heltcher K, Rowe L, Scanlon P, Stokes P, Suter G and Warren-Hicks W (in press) Biological effects of changes in surface water chemistry. NAPAP State of Science/Technology Report 13.

Bales RC (this volume) Modeling in-pack chemical transformations.

Bales RC, Davis RE and Stanley DA (1989) Ion elution through shallow homogeneous snow. Wat Res Resour 25:1869-1877.

Bales RC, Sommerfeld RA and Kebler DG (in press) Ionic tracer movement through a Wyoming snowpack. Atms Env.

Barrie (this volume) Processes of incorporation of chemicals into snowfall.

Barry PJ, Brown RM, Cornett RJ, Killey RWD, Price AG, and Kelly G (1984) Water chemistry during snowmelt in a northern basin. Geophysica 20:137-155.

Berg N (1986) Snow chemistry in the Central Sierra Nevada. Wat Air Soil Poll 30:1015-1021.

Brimblecombe P, Tranter M, Abrahams PW, Blackwood I, Davies TD, Vincent CE (1985). Relocation and preferential elution of acidic solute through the snowpack of a small, remote, high-altitude Scottish catchment. Annal Glac 7:141-147.

Brimblecombe P, Tranter M, Tsiouris S, Davies TD and Vincent CE (1986) The chemical evolution of snow and meltwater. In: Morris EM (ed) Modelling snowmelt-induced processes, IAHS Publication 155, p283.

Brimblecombe P, Clegg S, Davies TD, Shooter DS and Tranter M (1987) Laboratory observations of the preferential loss of major ions from melting snow. Wat Res 21:1279-1286.

Brimblecombe P, Clegg S, Davies TD, Shooter D and Tranter M (1988) The differential loss of halide and sulphate ions from melting ice. Wat Res 22:693-700.

Brown RM, Prioce AG and Workman W (1985) An isotope and energy balance study of the snow-melt process. Anal Glac 7:148

Cadle SH, Dasch JM, Grossnickle NE (1984) Northern Michigan snowpack - a study of acid stability and release. Atms Env 18:807-816.

Cadle SH, Dasch JM and Kopple RV (1987) Composition of snowmelt and runoff in northern Michigan. Env Sci Tech 21:295-299.

Colbeck SC (1987) Sow metamorphism and classification. In: Jones HG and Orville-Thomas WF (ed) Seasonal Snowcovers: Physics, Chemistry, Hydrology, NATO ASI Series C, Vol. 211, D. Reidel, Derdrecht, p1.

Colbeck SC (1981) A simulation of enrichment of atmospheric pollutants in snow cover runoff. Wat Resour Res 17:1383-1388.

Davis (this volume) Links between snowpack chemistry and snowpack physics.

Davies TD, Brimblecombe P, Tranter M, Tsiouris S, Vincent CE, Abrahams P, and Blackwood I (1987) The removal of soluble ions from melting snowpacks. In: Jones HG and Orville-Thomas WF (ed) Seasonal Snowcovers: Physics, Chemistry, Hydrology, NATO ASI Series C, Vol. 211, D. Reidel, Derdrecht, p337.

Davies TD, Vincent CE and Brimblecombe P (1982) Preferential elution of strong acids from a Norwegian ice cap. Nature, 300:161-164.

Delmas V (1989) Chimie de la neige et de la fonte printaniere au Casset (Alpes du sud). Influences des poussieres minerales. Ph. D. Thesis, University of Paris VII, UFR de Chimie, Doctorat de Chimie de la Pollution.

Drever J (1988) Geochemistry of natural waters, Prentice Hall.

English MC, Jeffries DS, Foster NW, Semkin RG, and Hazlett PW (1986) A preliminary assessment of the chemical and hydrological interaction of acidic snowmelt water with the terrestrial portion of a Canadian Shield catchment. Wat Air Soil Poll 31:27-34.

English MC, Semkin RG, Jeffries DS, Hazlett PW and Foster NW (1987) Methodology for investigation of snowmelt hydrology and chemistry within an undisturbed Canadian watershed. In: Jones HG and Orville-Thomas WF (ed) Seasonal Snowcovers: Physics, Chemistry, Hydrology, NATO ASI Series C, Vol. 211, D. Reidel, Derdrecht, p467.

Fletcher NH (1970) The chemical physics of ice, Cambridge University Press, New York.

Foster PM (1978) The modelling of pollutant concentrations during snowmelt. CEGB Report RD/L/N 46/78 Job No. VC455.

Gross GW (1968) Some effects of trace inorganics on the ice/water system. In: Gould RF (ed) Trace inorganics in water, Am Chem Soc, Advances in Chem Ser 73:27-97.

Gross GW, Wu C, Bryant L, McKee C (1975) Concentration dependent solute redistribution at the ice/water phase boundary. II Experimental investigation. J Chem Phys 62:3085-3092.

Hewitt AD, Cragin JH and Colbeck SC (1989) Does snow have ion chromatographic properties? In: Proc 46th Eastern Snow Conf, ISBN 0-920081-12-6, p165.

Jaccard PC and Levi L (1961) Segregation d'impuretes dans la glace. Z Angew Math Phys 12:70-76.

Johannessen M and Henriksen A (1978) Chemistry of snow meltwater: changes in concentration during melting. Wat Resour Res 14:615-619.

Jones HG (1985) The chemistry of snow and meltwaters within the mesostructure of a boreal forest snwocover. Annal Glac 7:167-174.

Jones HG (1987) Chemical dynamics of snowcover and snowmelt in a boreal forest. In: Jones HG and Orville-Thomas WF (ed) Seasonal Snowcovers: Physics, Chemistry, Hydrology, NATO ASI Series C, Vol. 211, D. Reidel, Derdrecht, p531.

Jones HG (this volume) Snow chemistry and biological activity: a particular perspective on nutrient cycling.

Jones HG and Deblois C (1987) Chemical dynamics of N-containing ionic species in a boreal forest snowcover during the spring melt period. Hydro Process 1:271-282.

Jones HG and Tranter M 1989. Interactions between meltwaterr and organic-rich particulate material in boreal forest snowpacks: evidence for both physico-chemical and microbilogical influences. In: Miles DL (ed) Proc. 6th Int. Sym. Water-Rock Interaction, Malvern, 1989, Balkema, p349.

Jones HG, Tranter M and Davies TD (1989) The leaching of strong acid anions from snow during rain-on-snow events: evidence for two component mixing. IAHS Publications 179:239-250.

Kattleman R (1986) Measurements of snow layer water retention. In: Cold region hydrology symposium, American Water Resources Association, Fairbanks, Alaska 1986, 377-386.

Kumai M (1985) Acidity of snow and its reduction by alkaline aerosols. Annal Glac 7:92-94.

Marsh ARW and Webb AH (1979). Physico-chemical aspects of snow-melt. CEGB Reports, RD/L/N, 60/79, 12pp.

Marsh P and Woo M-k (1984) Wetting front advance and refreezing of meltwater within a snow cover, 1. Observations in the Canadian Arctic. Wat Resourc Res 20:1853-1864.

Marsh P and Woo M-k (1985) Meltwater movement in natural heterogeneous snow covers. Wat Resourc Res 21:1710-1716.

Messer JJ, Slezak L and Liff CI (1982) Potential for acid snowmelt in the Wasatch Mountains. Utah Water Research Lab, Utah State Library, Logan, UWRL/Q-82/06.

Mulvaney R, Wolff EW and Oates K (1988). Sulphuric acid at grain boundaries in Antarctic ice. Nature 331:247-249.

Munger JW (1982) Chemistry of atmospheric precipitation in the north central United States: influence of sulfate, nitrate, ammonia and calcareous soil particulates. Atms Env 16:1633-1645.

Peters NE and Driscoll CT (1989) Temporal variations in solute concentrations of meltwater and forest floor leachate at a forested site in the Adirondacks, New York. In: Proc 46th Eastern Snow Conf, ISBN 0-920081-12-6, p45.

Schemenauer RS, Berry MO and Maxwell JB (1981) Snowfall formation. In: Gray DM and Male DH (ed) Handbook of snow, Pergamon Press, p129.

Seip HM (1980) Acid snow - snowpack chemistry and snowmelt. In: Hutchinson TC and Havas M (ed) Effects of acid precipitation on terrestrial ecosystems, Plenum Publishing Co, p 77.

Semkin RG and Jeffries DS (1985) Storage and release of major ionic contaminants from the snowpack in the Turkey Lakes Watershed. Wat Air Soil Poll 30:215-222.

Semkin RG and DS Jeffries (1988) Chemistry of atmospheric deposition, the snowpack, and snowmelt in the Turkey Lakes Watershed. Can J Fish Aquat Sci 45:38-46.

Suzuki K (1983) A chemical study of snow in Sapporo. Geog Rev Japan 56:171-184.

Skartveit A and Gjessing YT (1979) Chemical budgets and chemical quality of snow and runoff during spring snowmelt. Nord Hydrol 10:141-154.

Tranter M, Abrahams PW, Blackwood IL, Brimblecombe P and Davies TD (1988b) The impact of a single, black snowfall on streamwater chemistry in the Cairngorms, Scotland. Nature 332:826-829.

Tranter M, Abrahams P, Blackwood I, Davies TD, Brimblecombe P, Thompson IP, and Vincent CE (1987) Changes in the composition of streamwater during snowmelt. In: Jones HG and Orville-Thomas WF (ed) Seasonal Snowcovers: Physics, Chemistry, Hydrology, NATO ASI Series C, Vol. 211, D. Reidel, Derdrecht, p575.

Tranter M, Brimblecombe P, Davies TD, Vincent CE, Abrahams PW and Blackwood I (1986) The chemical composition of snowpack, snowfall and meltwater in the Scottish Highlands - evidence for preferential elution. Atms Env 20:517-525.

Tranter, M, Davies TD, Brimblecombe P, and Vincent, CE (1988a) The composition of acidic meltwater during snowmelt in the Scottish Highlands. Wat Air Soil Poll 36:75-90.

Tranter M, Tsiouris S, Davies TD and Jones HG (in prep) Laboratory leaching of solute from snowcover by rain. Submitted to Hydrological Processes.

Tsiouris S, Vincent CE, Davies TD and Brimblecombe P (1985) The elution of ions through field and laboratory snowpacks. Annal Glac 7:196-201.

Ushakova LA and Troshkina ES (1974) Role of the liquid-like layer in snow metamorphism. IASH Pub 114:62-65.

Williams MW, Melack JM and Tonnessen K (in press) Solute chemistry of snowmelt and runoff in an Alpine basin, Sierra Nevada. Wat Resour Res.

Wigington PJ, Davies TD, Tranter M and Eshleman KN (in press). Episodic acidification of surface waters due to acidic deposition. NAPAP State of Science/Technology Report 12.

CHEMICAL COMPOSITION AND FLUXES OF WET DEPOSITION AT ELEVATED SITES (700 - 3105 m a.s.l.) IN THE EASTERN ALPS (AUSTRIA)

H. Puxbaum, A. Kovar and M. Kalina
Institute for Analytical Chemistry
Technical University Vienna
Getreidemarkt 9
A-1060 Vienna
Austria

INTRODUCTION

Current research involving snow chemistry is performed at "cold" remote sites mainly to assess the historical evolution of the chemical climate (recently reviewed by Lorius et al. 1989) and the eventual impact of anthropogenic emissions in the last decades (recently reviewed by Wagenbach 1989).

Namely in or close to source regions snow chemistry at "temperate" sites has been used to investigate the impact of source regions on snow composition (e.g. Laird et al. 1986) and to assess the fluxes of nutrients and pollutants during snow melt (discussed in Tranter et al. 1987). Snow pit studies at a remote Scottish mountain range have revealed that episodically occuring long range transport of polluted air masses lead to the formation of highly polluted "black snow" layers (Tranter et al. 1988).

Only a few studies have been performed until now to assess the air/snow transfer mechanisms of the atmospheric contaminants, a field which receives increasing attention namely also in polar regions (e.g. Davidson et al. 1987). This type of research is hampered to a large extent due to the lack of sampling instrumentation for the representative collection of the various forms of suspended ice and snow particles as well as subcooled droplets from snow delivering cloud systems. Several ground based mechanistic field studies about gas/aerosol transfer to snowing clouds have been started in the last years at sites in USA (Whiteface) (Mohnen 1988), Switzerland (Rigi) (Staehelin priv. comm.) and Great Britain (Great Dun Fell) (Dore et al. 1990). Wintertime airborne cloud studies were performed in Canada (Isaac and Daum 1987).

Most of our information about the chemical composition of wet deposition in the Alpine region is derived from snow pit samples from glacier fields (e.g. Zobrist 1983, Delmas et al. 1988, Neftel et al. 1987, Psenner and Nickus 1986, Ronseaux and Delmas 1988, Vitovec 1988, Wagenbach et al. 1988).

With a few exceptions High Alpine glaciers are temperate snow fields which start to melt during late spring. Snow pit samples from such glaciers are generally collected in early spring and contain information about the snow composition of the "winter" period. To obtain data about the chemical

NATO ASI Series, Vol. G 28
Seasonal Snowpacks
Edited by T. D. Davies et al.

composition of wet deposition of the "summer" period precipitation has to be sampled preferably on an event basis with suitable collectors.

The first group reporting an estimate of annual wet fluxes of pollutants and nutrients for a High Alpine site were Psenner and Nickus (1986). They found that wet fluxes of major ionic constituents at a High Alpine glacier field were in the same order of magnitude as compared to the respective fluxes at prealpine lower sites. Considering recent views about critical loads in sensitive ecosystems (Schulze et al. 1990), the nitrogen input at High Alpine sites as estimated by Psenner and Nickus (1986) exceeds the critical load for sensitive environments significantly.

In this work we report about a two year record of the chemical composition and fluxes of wet deposition determined from samples collected at a daily basis at the mountain peak observatory "Hoher Sonnblick" (SBO) 3106 m a.s.l. The objective of this study was to accumulate data of two full seasonal cycles at a high altitude site where snowfall dominates the wet deposition process. By comparing the high altitude precipitation composition and fluxes with data from lower sites in the region basic information about the transfer processes of pollutants at high elevation sites could be derived.

EXPERIMENTAL

Precipitation Sampling in the Eastern Alps

In the northern and central part of the Eastern Alps a "wet - only" precipitation sampling network is operating collecting samples on a daily basis since 1982/83. A detailed description of this network has been given elsewhere (Puxbaum et al. 1988a). The sampling stations are generally situated in valleys or at mountain slopes at elevations from 700-1700 m a.s.l.

The location of the stations is presented in Figure 1. A short description of the sampling site characteristics is given in Table 1.

The stations are generally situated outside of the center of villages in some distance from the main motorways in the case of the valleys with dense traffic. The stations Achenkirchen, St. Koloman, Werfenweng and Innervillgraten are situated more remote from lager local sources and can be regarded as "regional" background stations. The daily control of the samples is carried out by local observers usually in the morning hours. The collection procedure follows the recommendation of the Austrian Ministry for Health and Environment (1984). The samples are stored in polyethylene bottles at 0-4°C and transported in biweekly intervals to the central laboratories in Vienna. There the samples were stored at 0-4°C again and analysed generally within 1-4 weeks after receipt.

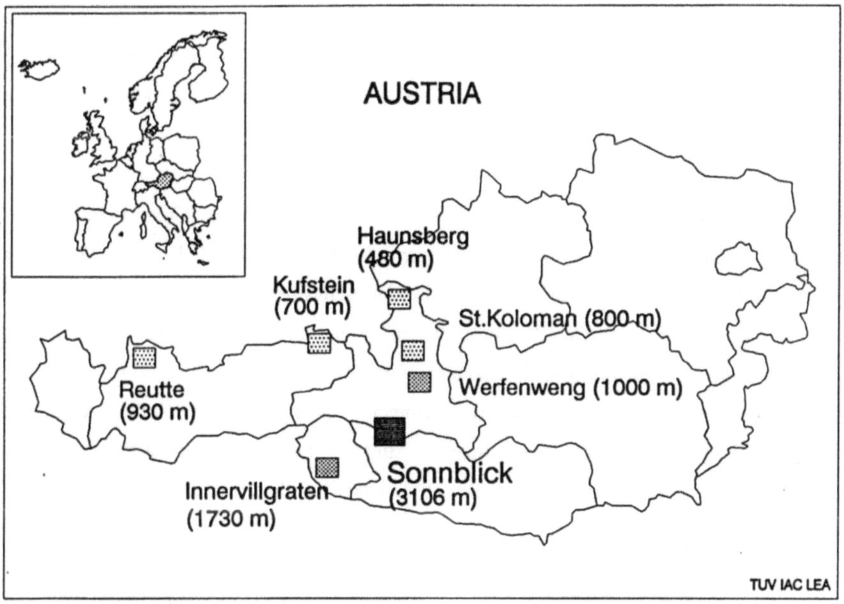

Fig. 1 : Wet-only precipitation network in the Eastern Alps (Austria).
In brackets : elevation above sea level

Tab. 1 : Sampling site characteristics -
type A : Northern Alps, stations in valleys and plains open to the north or northwest.
type B : Central Alps, stations in "Inner Alpine" regions.
SBO : Sonnblick Background Observatory

station	elevation m a.s.l.	region	characteristics
type A :			
Reutte	930	Lechtal	valley with some local industry.
Kufstein	700	Inntal	industrialized valley, heavy traffic and densely populated.
St.Koloman	900	Tennengau	mountain slope exposed to upslope rain.
Haunsberg	800	Salzachtal,	hilly terrain north of Salzburg.
type B :			
Werfenweng	900	Pongau Flachgau	narrow side valley south of the chains of the Northern Alps.
Innervillgraten	1730	Osttirol	narrow side valley south of the main chains of the Central Alps, remote from major local sources.
SBO:			
Sonnblick	3106	Hohe Tauern	summit in the Eastern Central Alps, meteorological observatory

Samples for all stations were analysed at the Institute for Analytical Chemistry at the Technical University of Vienna (IAC-TU Wien) with the exception of the station St. Koloman, which is part of the "EMEP" network. The respective samples were analysed by the group "Air Hygiene" at the Umweltbundesamt, Vienna (UBA).

In 1987 in a cooperation with the state government of Salzburg and the Central Institute of Meteorology and Geodynamiccs an additional sampling facility was installed at the Sonnblick Observatory (SBO) at 3106 m a.s.l. As the other sites the station has been equipped with a "WADOS" (wet and dry only precipitation sampler, Kroneis, Vienna). General features of the sampler are : diameter of wet deposition collection funnel - 250 mm, suitable for wintertime operation, wintertime proven lidmechanics, sensitive wet deposition (conductometry) sensor with two heating steps. The sampler was tested during a long term intercomparison study at the Meteorological Observatory in Hamburg (Winkler et al. 1989). The authors of this study conclude, that the WADOS was the most reliable instrument in the field experiment.

The determination of the daily precipitation amount is performed at all sites except at SBO by on site evaluation of the collected volume from the WADOS. For these samples we assume a catch efficiency of 90% with an uncertainity of ± 15%.

Because of the different meteorological conditions at the mountain top as compared to the valley sites the sampling efficiency of the WADOS at SBO is not comparable to the test conditions at lower sites.

Due to the lower temperatures at SBO approximately 90% of the wet deposition events are occuring as snowfall and due to the exposed situation at the mountain top many of the deposition events are accompanied with strong winds. The problems of sampling under such conditions have been reviewed recently by Vet (1987) and need not to be repeated here.

Our approach to derive "realistic" deposition data from the WADOS precipitation samples at SBO is as follows: At SBO two standard meteorological precipitation collectors are installed north and southwards of the observatory (500 cm^2 ombrometers - no wind shield). The daily precipitation amount is taken as average of the two instruments. The WADOS is used as collection vessel - the daily fluxes are calculated from the analytical data from the WADOS sample and the precipitation amount from the met ombrometers. At the nearby glacier field (Goldberggletscher) 2 "Totalizers" are installed at 2580 and 3100 m a.s.l. With these instruments monthly wet fluxes are recorded. The Totalizer water equivalent data are cross checked with the mass balance of the glacier field for the winter period and show a general agreement of the two methods. For this reason we assume that Totalizer precipitation data are representative for the regional precipitation amount. As the annual record from the Totalizer is around a factor 2 higher than the record from the ombrometers at SBO we use the monthly accumulated Totalizer precipitation amount for calculating regional wet fluxes (WADOS monthly precipitation weighed means are multiplied with the wet precipitation amounts from the Totalizer). With this procedure we estimate the uncertainity of the regional wet flux of ionic constituents to +/- 25% relative.

ANALYSIS

In all samples conductivity (25°C) and pH were determined by standard methods (Ministry for Health and Environment, 1984). Anions were determined by "suppressed mode" ion chromatography using a Dionex D10 ion chromatograph. Monovalent cations were determined at the "IAC" by "single column" ion chromatograph on a WESCAN "ICM" analyser under the conditions described by Tsitouridou and Puxbaum (1987). For the determination of the divalent cations (Ca^{2+}, Mg^{2+}) atomic absorption spectrometry was used.

For the anion and cation chromatography 250 μl sampling loops are used. Practical detection limits referring to 3 s of the signals of 6 injections of standard samples with ionic contents close to the detection limit were generally well below the concentrations in the precipitation samples. Quality control of the analytical data is performed by comparing calculated vs. measured conductivities and by the cation vs. anion balances of individual samples.

RESULTS AND DISCUSSION

Annual averages of concentrations and wet fluxes

The precipitation weighed annual mean concentrations (PWAMC's) of the main ionic constituents from the alpine network stations are listed in Tab. 2 for the years 1987/88 and 1988/89.

As it was reported for an earlier observation period (1984/85) (Puxbaum et al. 1988a) the NO_3^- and SO_4^{2-} concentrations at the stations type "A" show an increasing trend from west to east, and a decreasing trend from north to south. An unexpected result from this study was, that in the north - south transect the PWAMC's at SBO were not signigicantly lower than at the inneralpine sites at lower elevations (Werfenweng and Innervillgraten). Namely great similarities are found for the PWAMC's of most of the ionic constituents for SBO and Innervillgraten.

Tab. 2 : Yearly precipitation averaged concentrations for observation periods 87/88 and 88/89

Site Year	Prec. mm	pH	H^+	NH_4^+	Na^+	K^+	Ca^{2+} μequ/l	Mg^{2+}	Cl^-	NO_3^-	SO_4^{2-}
Reutte											
87/88	1447	4.8	17	25	8	3	31	4	11	24	28
88/89	1467	4.9	12	34	24	6	44	6	32	28	42
Kufstein											
87/88	1337	4.5	32	36	4	2	21	4	8	35	40
88/89	1337	4.4	40	41	6	2	19	4	13	39	49
Haunsberg											
87/88	1213	4.8	18	56	7	3	41	6	21	40	51
88/89	1198	4.5	28	60	8	4	47	6	19	46	64
St.Koloman											
87/88	1673	4.7	22	31	—	—	44	6	20	42	44
88/89	1437	4.4	36	26	—	—	49	8	32	31	68
Werfenweng											
87/88	1050	5.3	5	21	27	3	37	13	26	25	31
88/89	1103	4.8	18	30	15	3	48	15	13	29	48
Innervillgraten											
87/88	863	4.8	16	29	3	1	20	3	8	16	29
88/89	779	4.9	13	25	5	2	24	4	5	14	29
Sonnblick											
87/88	1375* 2872°	4.8	15	17	6	2	18	3	7	15	25
88/89	1428* 3092°	4.8	14	17	9	2	25	3	9	12	26

* Ombrometer

° Totalizer

Tab. 3 : Yearly sums of wet deposition for observation periods 87/88 and 88/89

Site Year	Prec. mm	H^+	NH_4^+	Na^+	K^+	Ca^{2+}	Mg^{2+} mequ/m²	Cl^-	NO_3^-	SO_4^{2-}
Reutte										
87/88	1447	24	36	12	4	44	6	16	35	40
88/89	1467	18	49	35	8	65	10	47	41	61
Kufstein										
87/88	1337	43	47	5	2	28	5	10	46	54
88/89	1337	53	55	8	3	26	5	17	51	65
Haunsberg										
87/88	1213	21	67	8	3	49	8	25	49	61
88/89	1198	34	72	10	4	56	8	22	56	76
St.Koloman										
87/88	1673	36	51	—	—	73	10	34	71	73
88/89	1437	52	37	—	—	70	11	46	45	98
Werfenweng										
87/88	1050	5	22	29	3	39	14	27	26	33
88/89	1103	19	33	16	4	53	17	15	32	52
Innervillgraten										
87/88	863	14	25	2	1	17	3	7	14	25
88/89	779	10	19	4	2	18	3	4	11	23
Sonnblick										
87/88	2872	42	49	16	5	52	8	21	42	72
88/89	3092	43	52	27	6	76	9	29	36	80

In Fig. 2. we compare the ranges of PWAMC's at the stations of type "A", type "B" and SBO for the two years of the experimental period. The resulting patterns are relatively similar for the mineralic components (K^+, Na^+, Mg^{2+}, Ca^{2+}) as well as for Cl^-. The patterns for H^+, NH_4^+, NO_3^- and SO_4^{2-} are somewhat different. NH_4^+ at SBO tends to be lower than at the inneralpine and the prealpine stations.

NO_3^- and SO_4^{2-} show a different behaviour at the sites under consideration. The NO_3^-/SO_4^{2-} concentration ratio exhibits a decreasing gradient from North to South with two typical clusters : Stations type A and Werfenweng have NO_3^-/SO_4^{2-} ratioes in the range of 0,8-1 while SBO and Innervillgraten have ratioes around 0.6 (Tab.4).

Tab. 4 : NO_3^-/SO_4^{2-} [equ/equ] relationships from PWAMC's of stations type "A" (arithmethic mean), Werfenweng (WER), Innervillgraten (INV) and SBO

site	1987/88	1988/89
"A"	0.87	0.62
WER	0.81	0.60
INV	0.55	0.48
SBO	0.58	0.45

We want now to continue our discussion about the gradients between the different types of station comparing the monthly averaged wet fluxes of the components under investigation (Fig. 3). Annual wet fluxes of Calcium at SBO are comparable with the highest fluxes of the component at the prealpine lower sites. Wet fluxes of free acidity, ammonium, sulfate and nitrate are comparable to the mid and lower fluxes within the range observed at the prealpine sites. For all above mentioned ions the wet fluxes at SBO are typically higher than the fluxes at the inneralpine lower sites (Tab. 3, Fig.3). A marked feature at all sites are the high Ca^{2+} - fluxes which are comparable to the respective SO_4^{2-} - fluxes. This leads to the conclusion that a large fraction of potential free acidity becomes neutralized by alkaline aerosol components. We hypothesise, that for "alkaline" events even NH_4^+ may be deliberated to the gas phase by reaction (1) which may occur in subcooled droplets or in the collected snow sample after melting.

$$CaCO_3 + (NH_4)_2SO_4 = CaSO_4 + 2\,NH_3 + CO_2 + H_2O \quad (1)$$

The wet deposited loads of total N and total S at the high alpine site are considerably high: Around 1,2 g S/m^2 (12 kg S/ha) and 1,2 g N/m^2 (12 kg N/ha) are deposited annually. Those values exceed critical loads for sensitive environments considerably (Schulze et al. 1990).

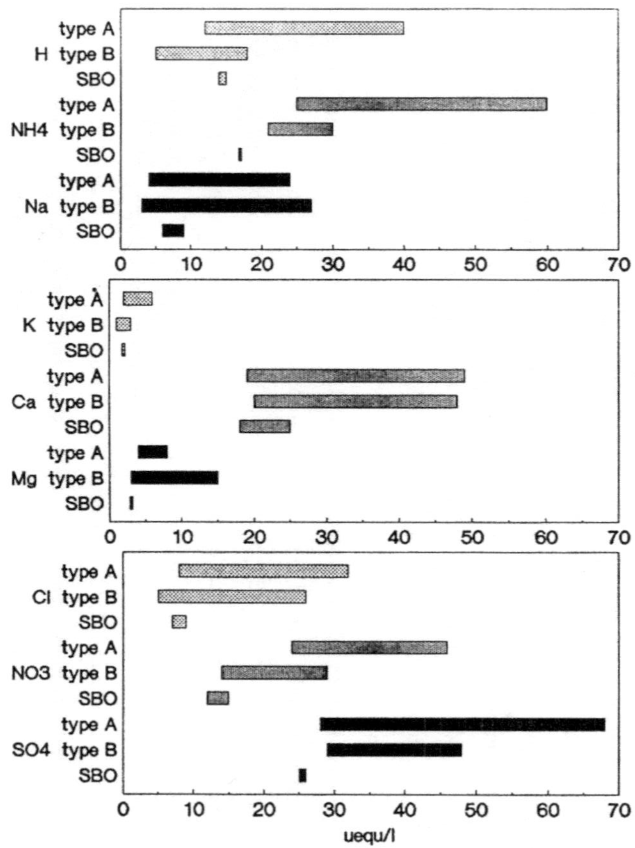

Fig. 2: Range of precipitation weighed annual mean concentrations of major ions for observation periods 87/88 and 88/89 for different types of stations

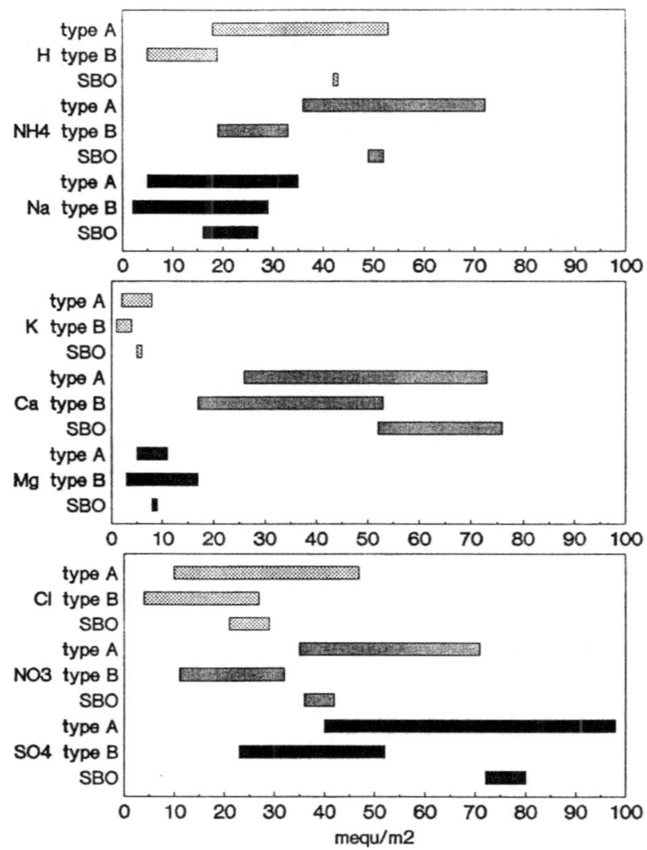

Fig. 3: Range of annual wet fluxes of major ions for observation periods 87/88 and 88/89 for different types of stations

Fowler et al. (1988) have discussed an altitudinal effect of ionic concentrations which tend to increase with increasing elevation for wet deposition from orographic clouds. Such an effect was recently reported to occur also during snowfall events - those observations were made at Great Dun Fell which extends from 150-847 m a.s.l. (Dore et al. 1990). Considering the yearly averaged data from the Alpine wet deposition network we find that higher elevation sites in Austria exhibit markedly lower concentrations of the major ions in wet deposition as compared to the lower sites. Other observations from several mountain slopes in the Austrian Alps indicate, that (yearly averaged) sulfate bulk deposition tends to stay constant along a respective slope at elevations from 500-2000 m a.s.l., while sulfate concentrations decrease markedly with increasing altitude (Smidt et al. 1990, Smidt pers. comm.). This leads us to the conclusion that in the high Alpine region other forms of precipitation activity, namely from convective events, dominate the composition of the precipitation during the course of a year.

Seasonal variations of concentrations and wet fluxes

The precipitation weighed monthly mean concentrations (PWMMC's) of most of the major ions (namely SO_4^{2-}, NO_3^- and NH_4^+) undergo large seasonal variations at all sites with low values during winter namely at SBO (Fig. 4a-c). During spring, summer and autumn PWMMC's of sulfate at SBO at frequently comparable to the respective concentrations observed at pre- and inneralpine sites. For nitrate large similarities exist with the inneralpine sites, while at prealpine sites PWMMC's are frequently a factor of 2 higher as compared to SBO. The concentration fluctuations of H^+ and Ca^{2+} appear to be quite irregular, indicating that the influence of neutralizing (alkaline mineralic) material is highly variable in the individual events (Fig. 4 d,e). To check the relationships between individual ions in the monthly averaged cases we have plotted the NO_3^-/SO_4^{2-}, NH_4^+/SO_4^{2-}, Ca^{2+}/SO_4^{2-} and $H^+/NO_3^- + SO_4^{2-}$ ratioes for the SBO data set in Fig. 5 a-d. The NO_3^-/SO_4^{2-} ratio at SBO tends to be around 0,5 during the warm season. While in winter 1987/88 no change in this relationship occured, during 88/89 a marked increase to values around 1 was found (Fig. 5a). During both years a "spring maximum" with values around 1 occured in the month Feb., March in 87/88 and in March, April in 88/89. At the other two sites the NO_3^-/SO_4^{2-} ratio shows a higher degree of variability although the Innervillgraten graph exhibits some similarities to the SBO graph. An increase of the NO_3^-/SO_4^{2-} ratio during the cold season as compared to the warm season has been reported for several US sites (discussed in Dasch 1987) and has been attributed to a different scavenging behavior of droplets vs. snow flakes with regard to SO_2 and HNO_3. Our data (Fig. 5a) indicate that there is an increase of the NO_3^-/SO_4^{2-} ratio during the cold season at all sites under investigation as compared to the warm season although this effect is not evident in every individual month. We want also include here a remark of one of the reviewers (Trevor Davies), that when comparing results from sites at different elevations, one has to consider that the snowfall season is considerably longer at the higher sites. The main reason for the scatter of the monthly average concentrations of the major ions (and thus also of the

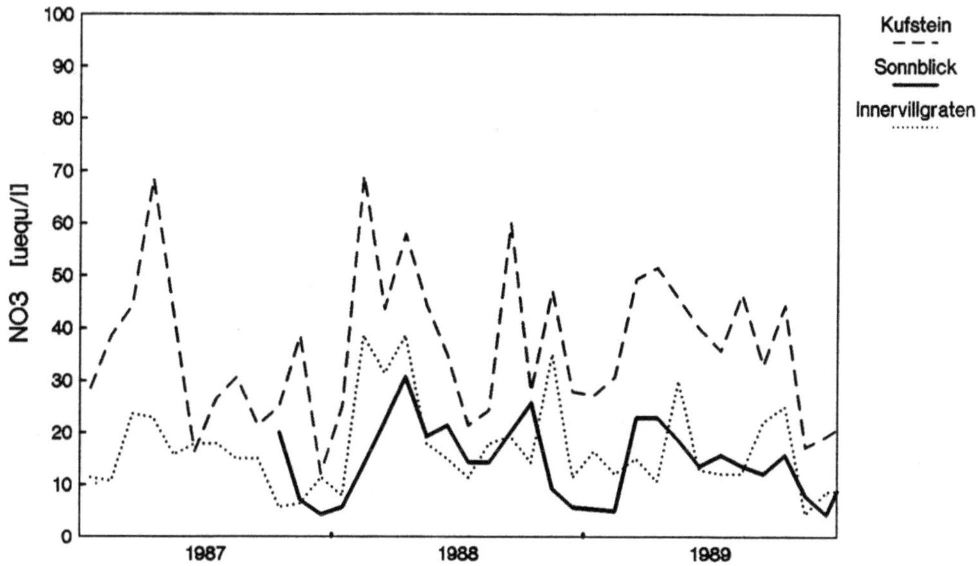

Fig. 4a: Time series of the precipitation weighed monthly mean concentrations of NO_3^-

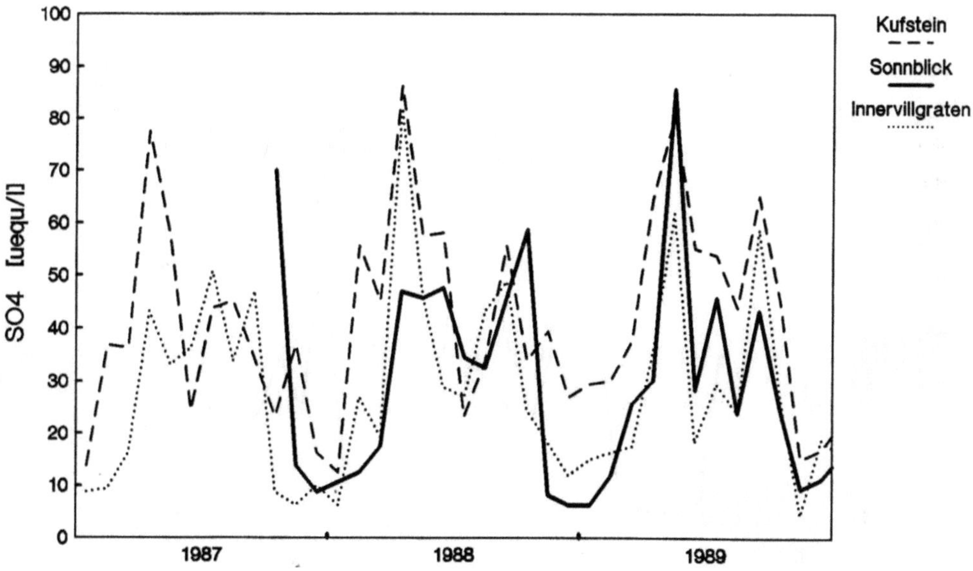

Fig. 4b: Time series of the precipitation weighed monthly mean concentrations of SO_4^{2-}

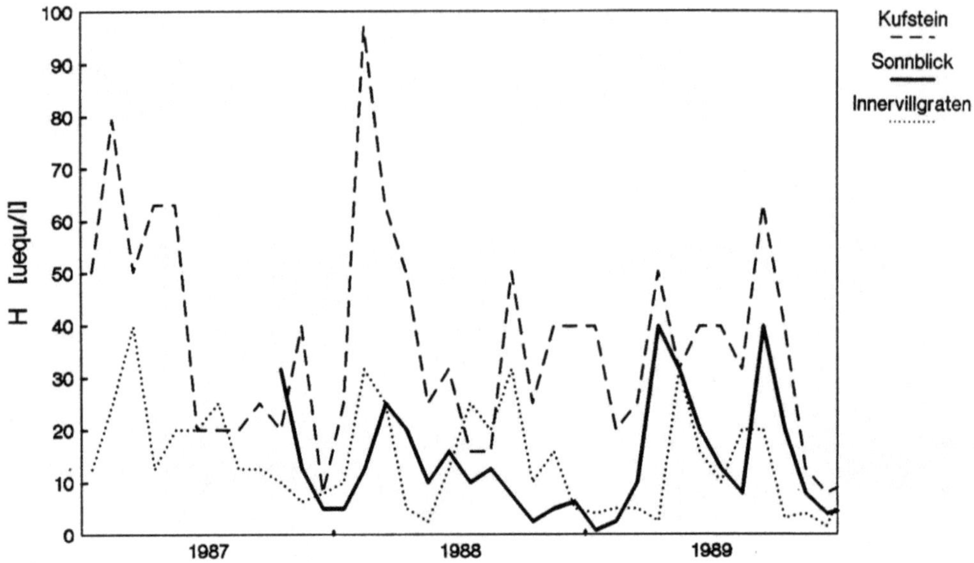

Fig. 4c: Time series of the precipitation weighed monthly mean concentrations of H^+

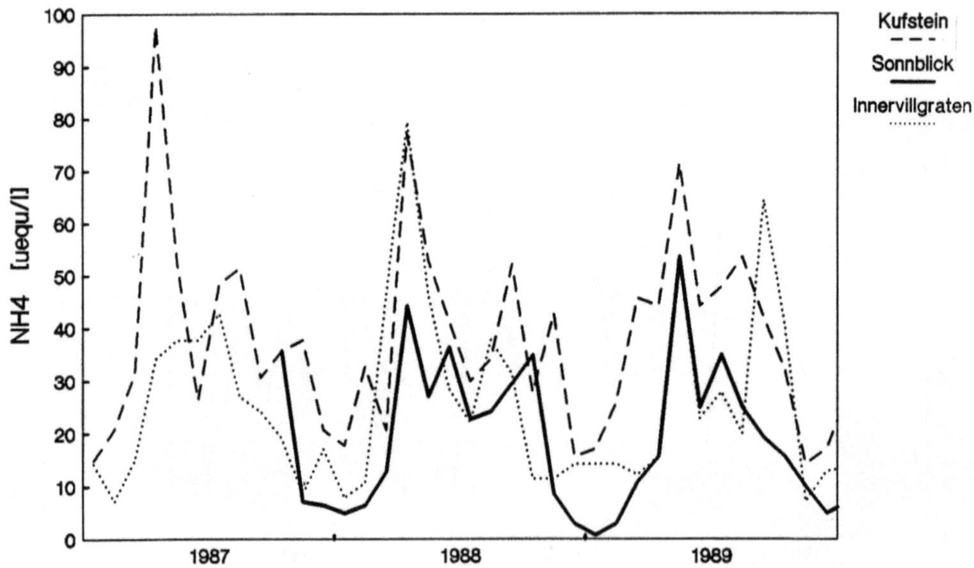

Fig. 4d: Time series of the precipitation weighed monthly mean concentrations of NH_4^+

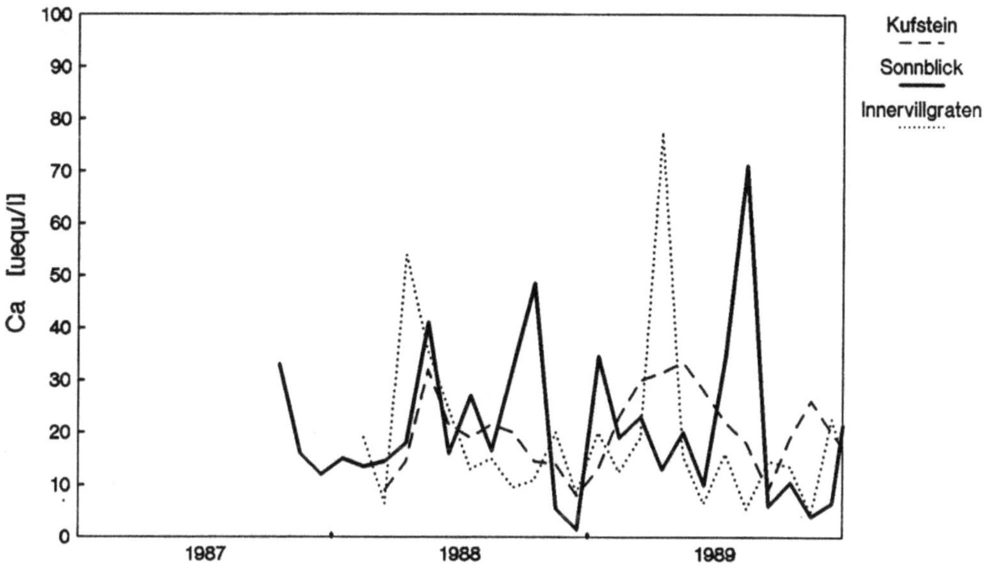

Fig. 4e: Time series of the precipitation weighed monthly mean concentrations of Ca^{2+}

concentration ratios) is the episodic nature of the precipitation events (see also Puxbaum et al. 1988a). A detailed analysis of the single events will be presented in a further paper.

A check of the NH_4^+/SO_4^{2-} relationship (Fig. 5b) indicates, that a large fraction of the SO_4^{2-} at SBO is likely to be derived from aerosol scavenging of ammonium bearing sulfate aerosols. (The minimum in Jan. 90 is derived from only two cases with extreme high Ca^{2+}). As the repeated increase of the NO_3^-/SO_4^{2-} ratio at SBO occured as a "spring maximum" and as during this period a general increase of oxidants is observed over Europe (Penkett 1988) we think that the main reason for the increase of the NO_3^-/SO_4^{2-} ratio is a higher availability of HNO_3 in spring time.

A major factor masking the possibilities of interpretations of the influence of the air chemistry on the snow chemistry are the frequent cases of high mineralic content of the precipitation samples. In the PWAMC's the sum of the soluble mineralic ion concentrations (equ) at SBO exceed the corresponding SO_4^{2-} concentrations in both years of investigation. Although absolute concentrations af Ca^{2+} are higher during the warm season at SBO, the Ca^{2+}/SO_4^{2-} relationship during the cold season tends to be >1 while in the warm season it fluctuates generally between 0,3 and 0,9 in the PWMMC 's (Fig 5c). As we can assume that the neutralisation of acidity in precipitation samples by mineralic aerosols is a "tertiary process", all types of acidity (derived from NO_3^-, SO_4^{2-} and possibly Cl^-) will be a affected by this process. If - in the complete absence of mineralic aerosols - the potential "free acidity" of the yearly

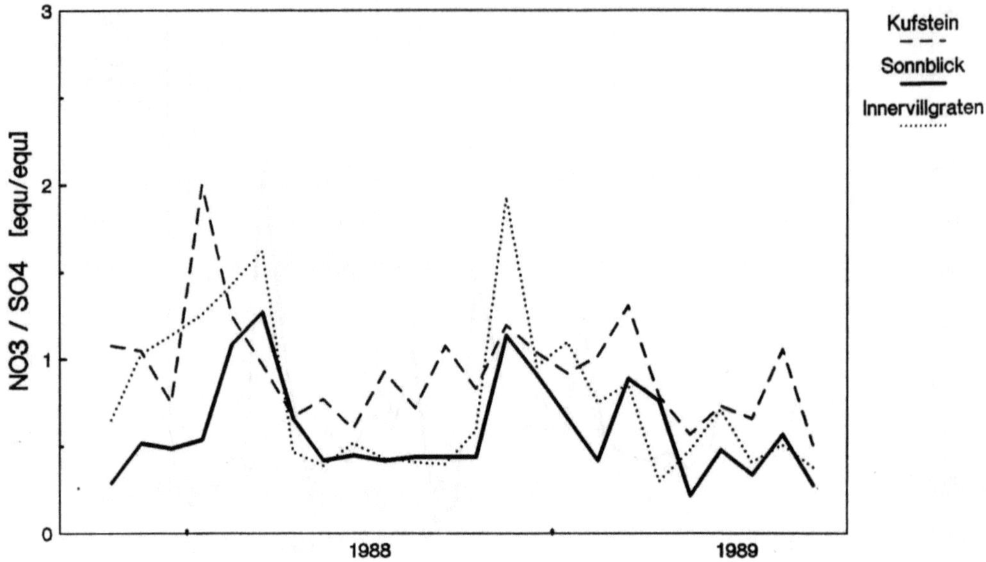

Fig. 5a: Time series of the concentration ratios NO_3^-/SO_4^{2-} for the observation period Oct. 87 to Sept. 89

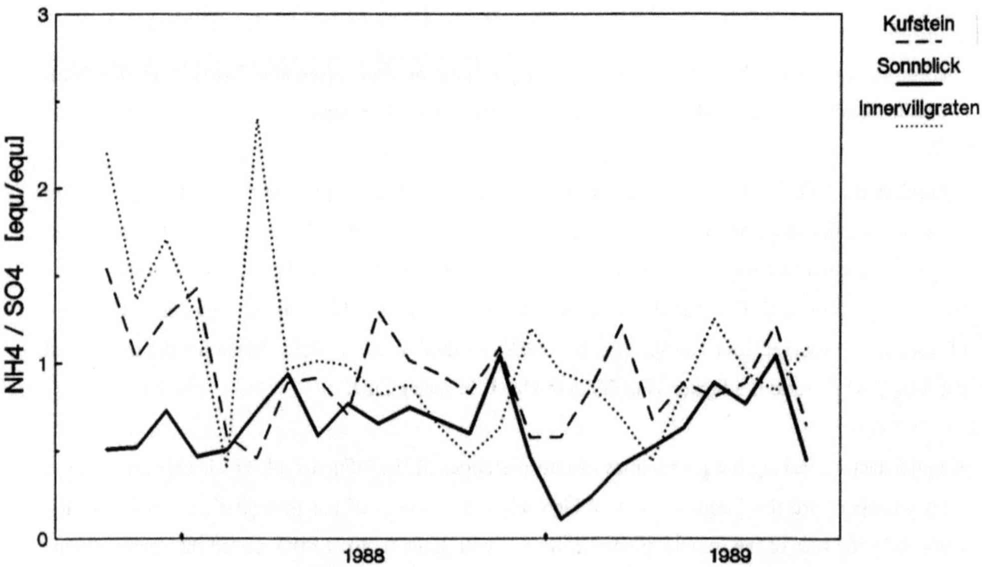

Fig. 5b: Time series of the concentration ratios NH_4^+/SO_4^{2-} for the observation period Oct. 87 to Sept. 89

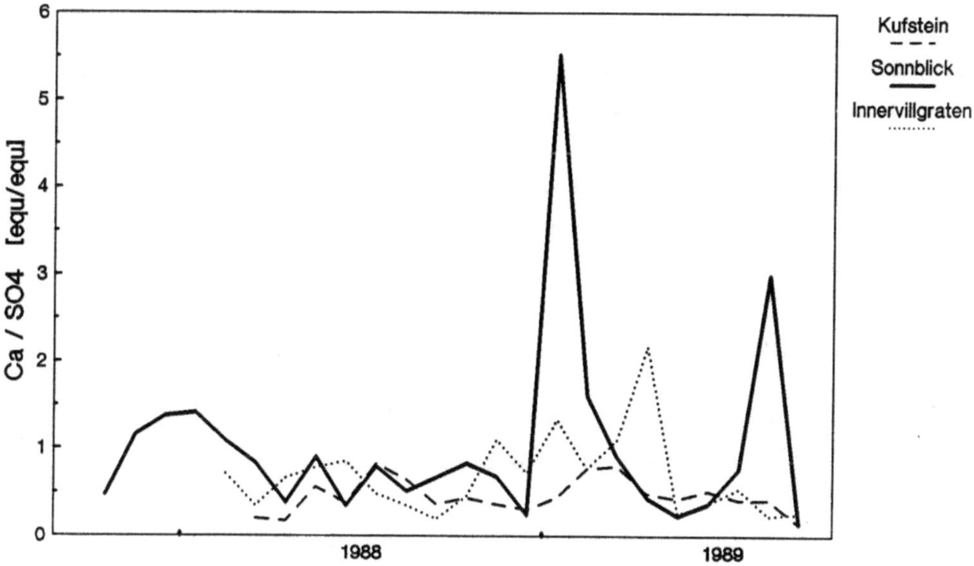

Fig. 5c: Time series of the concentration ratio Ca^{2+}/SO_4^{2-} for the observation period Oct. 87 to Sept. 89

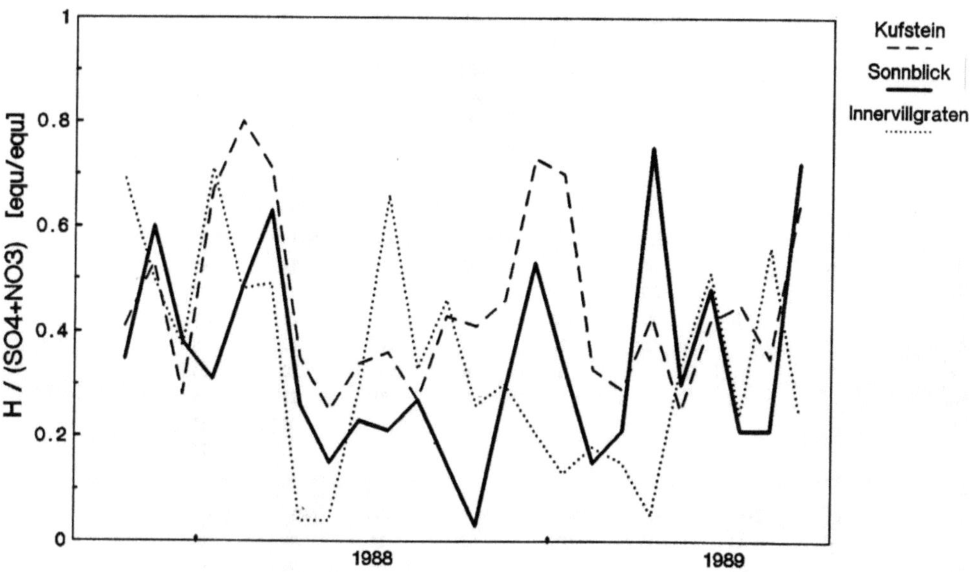

Fig. 5d: Time series of the concentrations ratio $H^{+}/(SO_4^{2-} + NO_3^{-})$ for the observation period Oct. 87 to Sept. 89

averaged precipitation would be "$[H^+]$" = $[SO_4^{2-}]$ + $[NO_3^-]$ + $[Cl^-]$ - $[NH_4^+]$, a doubling of the acidity load by wet precipitation at SBO would result. To assess the relative contribution of the measured free acidity we have plotted the $H^+/SO_4^{2-} + NO_3^-$ ratio of the PWMMC's in Fig. 5d . This relationship undergoes large fluctuations during all seasons at all sites. The relative amount of free acidity in relation to the sum of SO_4^{2-} and NO_3^- varied between 2 and 70% . From this result we conclude that at all sites under consideration no regularities concerning the preferential occurence of free acidity during certain seasons exist.

Single case statistics and quality control

For the observation period a correlation matrix for the deposition values of the individual events has been calculated for the major ions for the sites Kufstein, Innervillgraten and SBO (Tab 5).

Tab. 5 : Correlation matrix of daily event ion deposition at the stations Kufstein, Innervillgraten and Sonnblick; observation periods 1987/88, 1988/89

Kufstein

	NH_4^+	Na^+	K^+	Ca^{2+}	Mg^{2+}	Cl^-	NO_3^-	SO_4^{2-}
H^+	.64	.30	.31	.14	.23	.43	.85	.82
NH_4^+	1.	.40	.57	.53	.61	.52	.82	.87
Na^+		1.	.47	.33	.54	.79	.47	.33
K^+			1.	.54	.58	.53	.46	.48
Ca^{2+}				1.	.89	.49	.42	.47
Mg^{2+}					1.	.63	.53	.53
Cl^-						1.	.61	.47
NO_3^-							1.	.83

Innervillgraten

	NH_4^+	Na^+	K^+	Ca^{2+}	Mg^{2+}	Cl^-	NO_3^-	SO_4^{2-}
H^+	.63	.10	.06	-.04	.02	.13	.72	.63
NH_4^+	1.	.28	.39	.20	.30	.28	.77	.82
Na^+		1.	.29	.82	.78	.65	.24	.57
K^+			1.	.21	.29	.17	.19	.28
Ca^{2+}				1.	.94	.73	.20	.56
Mg^{2+}					1.	.71	.28	.64
Cl^-						1.	.28	.55
NO_3^-							1.	.76

Sonnblick

	NH_4^+	Na^+	K^+	Ca^{2+}	Mg^{2+}	Cl^-	NO_3^-	SO_4^{2-}
H^+	.48	.14	.18	.00	.05	.24	.79	.54
NH_4^+	1.	.32	.46	.36	.44	.49	.71	.92
Na^+		1.	.70	.37	.48	.40	.37	.41
K^+			1.	.34	.37	.37	.37	.44
Ca^{2+}				1.	.89	.67	.26	.42
Mg^{2+}					1.	.58	.33	.52
Cl^-						1.	.48	.52
NO_3^-							1.	.73

At the three sites SO_4^{2-} is highest correlated with NH_4^+, followed by NO_3^- and H^+. Nitrate shows the highest degree of correlation with H^+ at SBO and Kufstein, whereas at Innervillgraten similarly high correlations exist with SO_4^{2-}, NH_4^+ and H^+. Ca^{2+} at all sites is best correlated with Mg^{2+}, followed by Cl^-. Na^+ at SBO correlates highly with K^+ but only weakly with Cl, while at Kufstein and Innervillgraten Na+ correlates only weakly with K^+ but fairly well with Cl^-. There is practically no correlation between Ca^{2+} and H^+ at all sites.

Together with the information from the previous paragraph we derive following conclusions: At all sites a strong relationship exists for NH_4^+ and SO_4^{2-}. Nitrate is the important carrier of free acidity. Mineral aerosols may differ in their composition at the various sites: For SBO the Ca^{2+}/Mg^{2+} containing aerosol carries also Cl^-, while Na^+/K^+ rich cases are not related to Ca^{2+}/Mg^{2+} events. In Kufstein and Innervillgraten local sources might be the reason for the Na^+/Cl^- correlations. The generally high correlations among the ions H^+, NH_4^+, NO_3^- and SO_4^{2-} indicate that under polluted conditions those four ions generally increase. Mineralic aerosol may accompany those events however in an uncorrelated way.

Quality control has been performed by plotting calculated vs measured conductivity (Fig. 6) and ionic balances (Fig.7). For the conductivity checks r^2 are generally >0,95, the slope being relatively close to 1 (Tab 6). In the ionic balances the majority of cases lies on the 1:1 slope, however a certain number of cases tends to a relatively high cation excess, namely at the stations Innervillgraten and SBO (Fig 7). This is reflected in a weaker correlation and slopes quite below b=1 (Tab 6).

Tab. 6 : Summary of quality control (measured vs. calculated conductivity, anion vs. cation balance) observation periods 1987/88, 1988/89

site	cond. bal.		ionic bal.		
	r^2	b	r^2	b	n
Kufstein	0.97	1.1	0.94	1.0	301
Innervillgraten	0.94	1.0	0.73	0.8	191
Sonnblick	0.95	1.1	0.84	0.8	233

r^2 ... squared correlation coefficient
b ... slope of linear regression
n ... frequency

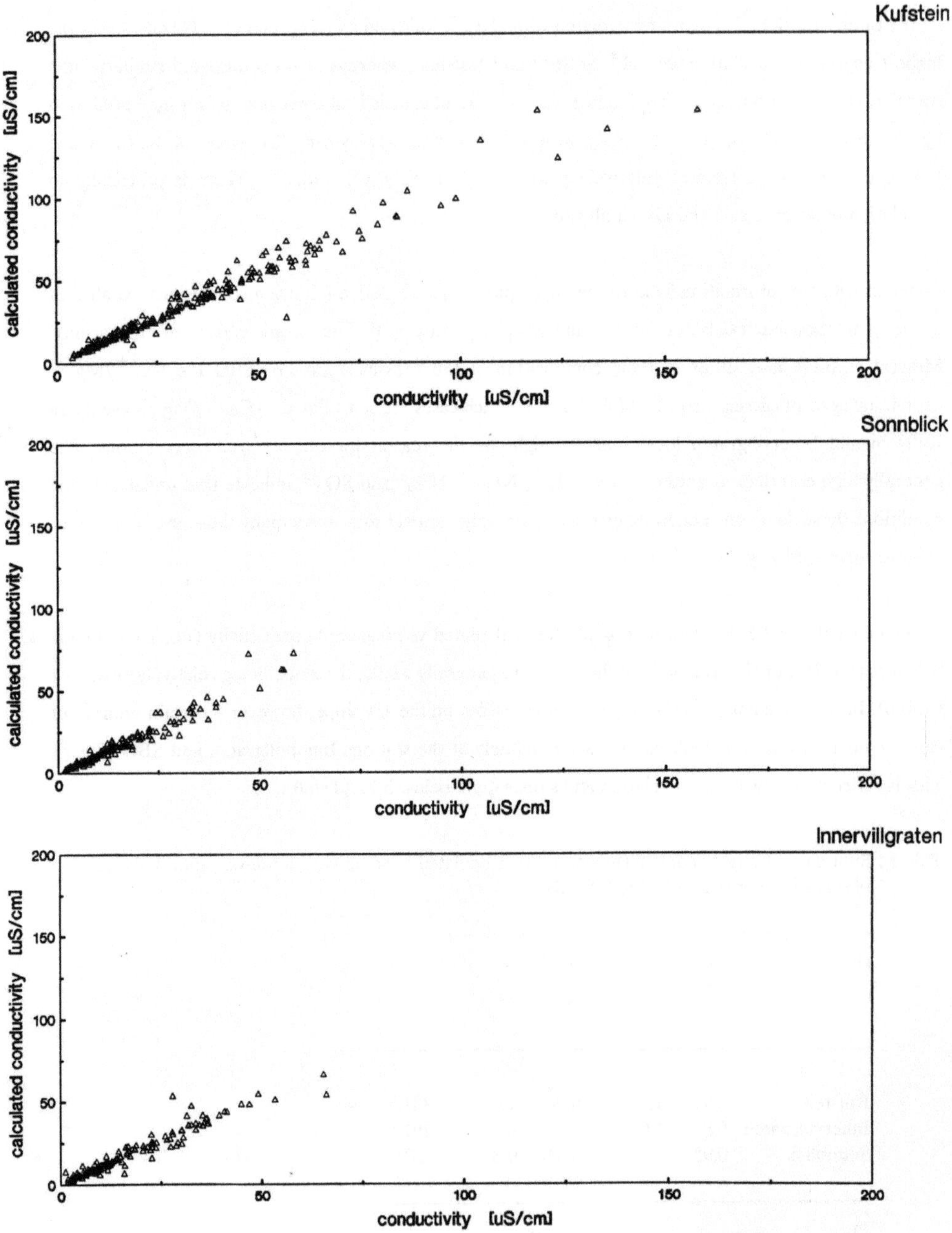

Fig. 6: Wet precipitation quality control - measured versus calculated conductivity for the observation period Oct. 87 to Sept. 89

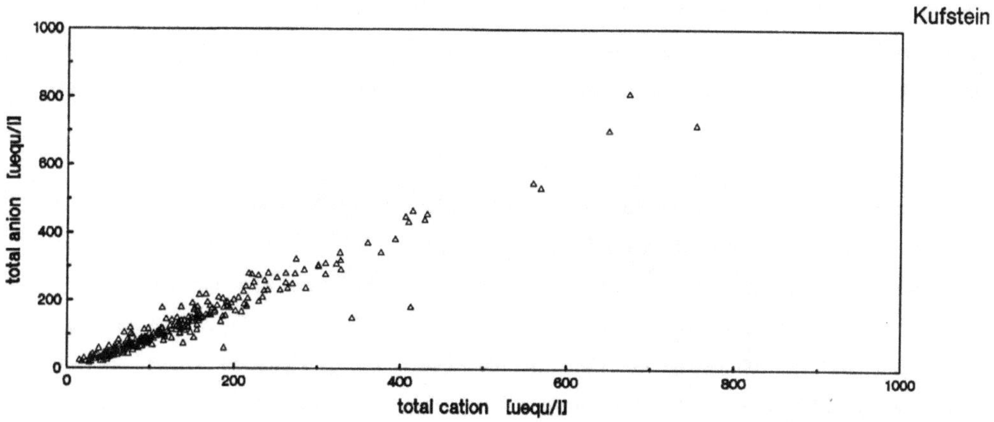

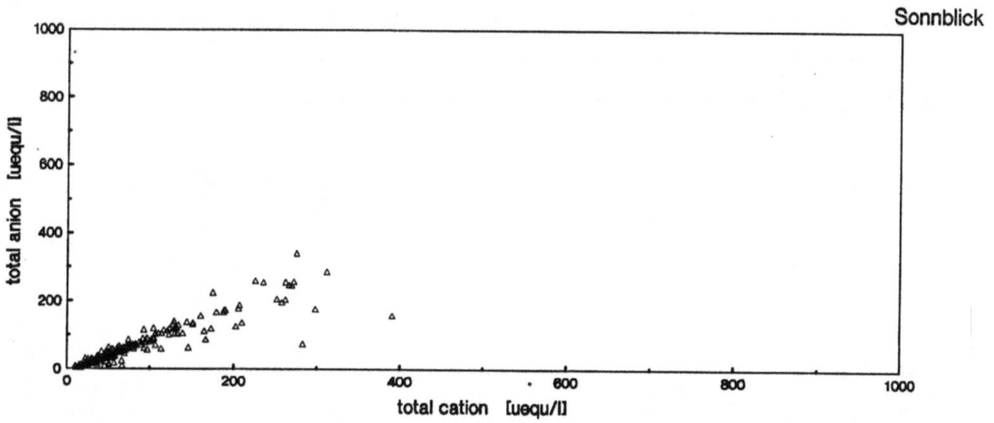

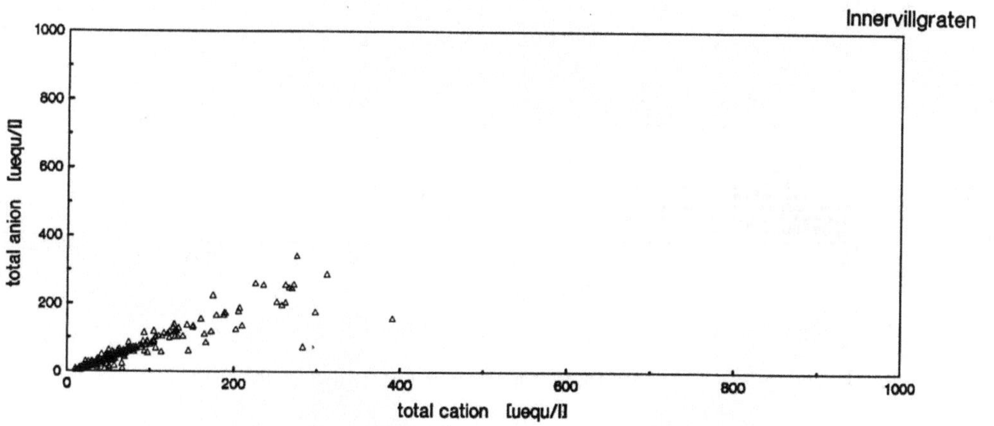

Fig. 7: Wet precipitation quality control - cation versus anion balance for the observation period Oct. 87 to Sept. 89

Our first assumption was that under conditions close to neutrality a certain fraction of the anions is present as HCO_3^-. As this ion is not analysed we miss it in the ionic balance. Relative large amounts of HCO_3^- have been identified in dust bearing layers in alpine snow in France by Maupetit and Delmas (1990).

To check our assumption, that the missing anions are related to HCO_3^- - alkalinity from mineral aerosols we have compared ion balances from "Ca^{2+}-rich" and "Ca^{2+}-poor" but "SO_4^{2-}-rich" cases which we have chosen from the Ca^{2+}/SO_4^{2-} - scattergram from SBO (Fig. 8) in the following way: The "Ca^{2+}-rich" cases are all cases with Ca^{2+} > 100 μequ/l, the "Ca^{2+}-poor - SO_4^{2-}-rich" cases are cases with SO_4^{2-} > 100 μequ/l but Ca^{2+} < 100 μequ/l. The precipitation weighed average concentrations of the investigated cases is shown in Tab. 7. The effect of the mineral aerosols becomes evident in the cation excess of 60 μequ/l, while the "acid cases" show a slight anion excess of 7 μequ/l. While the average SO_4^{2-} - concentration is relatively similar in both cases, the "Ca^{2+}-rich" case has significantly less NH_4^+ and also NO_3^- - contents. Cl^- remains surprisingly constant.

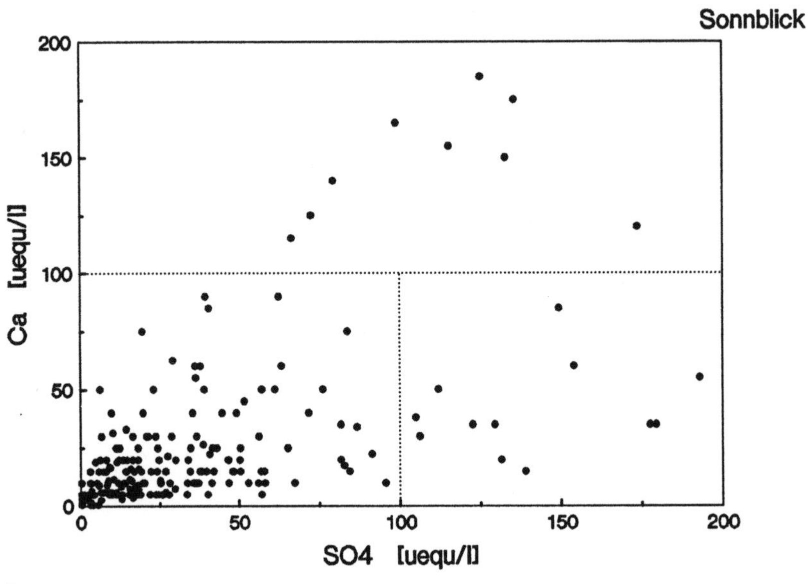

Fig. 8: Scattergramm SO_4^{2-}/Ca^{2+}, SBO, observation period Oct. 87 to Sept. 89

Tab. 7 : Case study for "Ca^{2+}-rich" and "SO_4^{2-}-rich, Ca^{2+}-poor" cases at SBO

	conditions		
	Ca^{2+} < 100 μequ/l SO_4^{2-} >100 μequ/l	Ca^{2+} > 100 μequ/l	
cases	12	9	
precipitation amount	4.7	3.3	mm
pH	4.1	5.8	
H^+	81	1.7	μequ/l
NH_4^+	102	52	μequ/l
Na^+	14	22	μequ/l
K^+	3	6	μequ/l
Ca^{2+}	36	156	μequ/l
Mg^{2+}	10	21	μequ/l
Cl^-	33	35	μequ/l
NO_3^-	75	45	μequ/l
SO_4^{2-}	145	119	μequ/l
tot. cation	246	259	μequ/l
tot. anion	253	199	μequ/l
tot. ion	499	458	μequ/l

The pH of the "Ca^{2+}-rich" case was 5,8. A check on the pH - depending solubility of CO_2 in water (considering the pseudo - Henry - Law equilibrium for 25°C) shows, that only 12 μequ/l CO_2 can be dissolved in water at pH 6,3, the formed H_2CO_3 beeing dissociated to a degree of 50%. Thus an observed concentration of 60 μequ/l $HCO3^-$ would require a pH of 7,04. It is therefore unlikely, that the observed anion deficit of 60 μequ/l in the "Ca^{2+}-rich" case originate from HCO_3^- ions. Another group of candidates to explain the anion deficit are organic acids. Formic and acetic acid are only to a very small extent soluble in cloud droplets of pH 4,1 ("acid case"), however readily soluble at pH 5,8 ("Ca^{2+}-rich" case) (Winiwarter et al. 1988). Atmospheric concentrations of 0,3 ppb each of formic and acetic acid would be sufficient to cause a content of 60 μequ/l organic acids in droplets of a cloud with 0,5 g/m^3 liquid water content at pH 5,8. Atmospheric concentrations of organic acids in rural air in Austria are around 0,9 ppb formic and 0,5 ppb acetic acid (Puxbaum et al. 1988b). Similarly high concentrations can also be present at elevations of 3 km over Central Europe (Hartmann et al. 1989, Winiwarter pers.comm.). Thus it is reasonable to assume, that unaccounted organic acids are responsible for the anion deficit observed in the "Ca^{2+}-rich" case.

Intercomparison of SBO precipitation composition with other data from elevated sites

In Tab. 8 we compare SBO PWAMC's with data from Whiteface mountain (WF), NY, USA (Mohnen 1988). We find, that during the warm season anion concentrations at SBO are comparable to the corresponding WF data, while in the cold season NO_3^- and SO_4^{2-} are typically higher at WF. NH_4^+/SO_4^{2-} ratioes at WF are below 0,5, but above 0,6 at SBO. The concentrations of mineralic constituents (RK in Tab. 8) are a factor of 5 higher at SBO during the warm season as compared to WF, during the cold season a factor of 2,6. The results indicate that the continental European air masses responsible for the incorporation of trace constituents into precipitating clouds contain a higher fraction of NH_4^+ neutralized SO_4^{2-} aerosol and a much higher fraction of soluble mineralic material as compared to the North-Eastern USA.

Tab. 8 : Intercomparison of precipitation chemistry at SBO (3106 m a.s.l. , PWAMC's for 1987/88) and Whiteface Mountain, NY, USA (1500 m a.s.l. , PWAMC's for 1987) (Data from Mohnen 1988)

site	H^+	NH_4^+	RK	SO_4^{2-} µequ/l	NO_3^-	RA	UN
cold season							
SBO	16	11	24	18	14	9	(10)
WF	34	13	9	27	27	6	---
warm season							
SBO	13	28	36	39	18	7	(13)
WF	32	15	7	36	14	5	---

WF 1987 RK ... rest of cation
SBO 1987/88 RA ... rest of anion
UN ... unidentified, presumably organic acids and HCO_3^-

CONCLUSIONS

Annual wet fluxes of major ionic constituents at a high alpine site (3106 m a.s.l.) in the eastern part of the central alps are in the same range as compared to the wet fluxes observed at lower prealpine sites situated well within the mixing layer (700 - 1000 m a.s.l.)

Precipitation weighed annual mean concentrations of major ions at the SBO site (3106 m a.s.l.) are generally comparable to the lower end of the range of the respective ion-concentrations observed at the pre- and inneralpine lower sites. (Tab.1, Fig. 2.)

Annual wet fluxes of sulfate and calcium at SBO are comparable with the highest fluxes of those components at the prealpine lower sites. Wet fluxes of free acidity, ammonium and nitrate are comparable to the mid and lower fluxes within the range observed at the prealpine sites. For all above mentioned ions the wet fluxes at SBO are typically higher than the fluxes at the inneralpine lower sites (Tab. 2, Fig. 3).

The concentrations of the major ions sulfate, nitrate and ammonium undergo large seasonal variations at all sites, with very low values during winter namely at SBO. During spring, summer and autumn precipitation weighed monthly mean concentrations (PWMMC's) of sulfate at SBO are frequently comparable to the respective concentrations observed at pre- and inneralpine sites. For nitrate large similarities exist within the inneralpine sites, while at prealpine sites PWMMC's are frequently a factor of 2 higher as compared to SBO.

Large scale mixing and transport processes appear to be the dominating factor for the observed similarities in sulfate concentrations at lower and high alpine sites during the spring to autumn period. Nitrate concentrations at SBO compare well with levels found at the inneralpine lower sites, while prealpine sites experience much higher concentrations. It appears that the different behavior of NO_3^- as compared to SO_4^{2-} is rather a spatial than an altitudinal effect.

Our results for SBO agree well with observations from snow pit samples from the wintertime accumulated snow cover at temperate glacier fields, where very low concentrations of the major ions are observed during the winter and strong increases during the beginning spring season (Psenner and Nickus 1986, Neftel et al. 1987, Delmas et al. 1988, Vitovec 1988, Gäggeler et al. 1990, Maupetit and Delmas 1990).

The chemical composition of wet precipitation at the SBO site leads to following conclusions: The relatively high NH_4^+ : SO_4^{2-} ratio (0,65 - 0,68) in the PWAMC's is an indicator, that SO_4^{2-} in precipitation is to a considerable extent at (at last 2/3) derived from aerosol scavenging of ammonium sulfate. Thus the major cause for free acidity is likely to be HNO_3 scavenging. There are no indications, that during the warm season a larger fraction of SO_4^{2-} is derived from reactive SO_2 scavenging as compared to the cold season.

Soluble mineralic aerosols play a critical role in neutralizing a large fraction of the free acidity in the wet precipitation . Without the neutralizing action of the mineralic aerosols, the flux of free acidity at SBO would be doubled.

Acknowledgements

We thank W. Winiwarter, R. Böhm, T. Davies and F. Maupetit for helpful discussions. The work was gratefully supported by the "Amt der Salzburger Landesregierung", "Zentralanstalt für Meteorologie und Geodynamik" and the "Landesforstdirektion Tirol". The data from St. Koloman were made available from the "Umweltbundesamt" (Wien). Special thanks are adressed to the helpful work of the observers at the sampling sites.

REFERENCES

Austrian Ministry for Health and Environment (1984) Richtlinie 11, "Immissionsmessung des nassen Niederschlags und des sedimentierten Staubes, Wien

Dasch JM (1987) On the Difference Between Sulfate and Nitrate in Wintertime Precipitation. Atmos Environ 21:137-141

Davidson CI, Honrath RE, Kadane JB, Tsay RS, Mayewski PA, Lyons WMB, Heidam NZ (1987) The Scavenging of Atmospheric Sulfate by Arctic Snow. Atmos Environ 21:871-882

Delmas VM, Ronseaux F, Delmas RJ (1988) Chemical Composition of the Seasonal Snowcover at a Southern French Alps Site: Some Preliminary Results. In: Unsworth MH, Fowler D (eds) Acid Deposition at High Elevation Sites. Kluwer, pp 511-516

Dore AJ, Choularton TW, Fowler D, Crossley A (1990) The Influence of Altitude on Wet Deposition by Snowfall. Int Conf on Acidic Deposition, its Nature and Impacts. Glasgow, Sept. 16-21

Fowler D, Cape JN, Leith ID, Choularton TW, Gay MJ, Jones A (1988) Wet Deposition and Altitude, The Role of Orographic Cloud. In: Unsworth MH, Fowler D (eds) Acid Deposition at High Elevation Sites, Kluwer, Dordrecht, pp 231-257

Gäggeler HW, Baltensperger U, Jost DT (1991) Deposition Study of Particulate Air Pollutants at an Alpine Snowfield. In: EUROTRAC Symposium 1990, Garmisch-Partenkirchen

Hartmann WR, Andreae MO, Helas G (1989) Measurements of Organic Acids Over Central Germany. Atmos Environ 23:1531-1533

Isaac GA, Daum PH (1987) A Winter Study of Air, Cloud and Precipitation Chemistry in Ontario, CDN. Atmos Environ 21:1587-1600

Laird LB, Taylor HE, Kennedy VC (1986) Snow Chemistry of the Cascade - Sierra Nevada Mountains. Environ Sci Technol 20:275-290

Lorius C, Raisbeck G, Jouzel J, Reynaud D (1989) Long Term Environmental Records form Antartic Ice Cores. In: Oeschger H, Langway CC (eds) The Environmental Records in Glaciers and Ice Sheets. Wiley-Interscience (Dahlem Workshop Reports), pp 343-361

Maupetit F, Delmas R (1991) Snow Chemistry of Four Elevated Glaciers in the French Alps. In: EUROTRAC Symposium 1990, Garmisch-Partenkirchen

Mohnen VA (1988) Exposure of Forests to Air Pollutants, Clouds, Precipitation and Climatic Variables - A Preliminary Assessment. Internal Report, ASRC, SUNY, Albany

Neftel A, Sigg A, Zürcher F (1987) Acid Deposition in a Snow Field at 2500 m a.s.l. in Switzerland. In: Physico-Chemical Behavior of Atmospheric Pollutants, Air Pollution Research Report 2, Reidel, Dordrecht, pp 500-510

Penkett SA (1988) Indications and Causes of Ozone Increase in the Troposphere. In: Rowland FS, Isaksen ISA (eds) The Changing Atmosphere. Wiley-Interscience (Dahlen Workshop Reports), pp 91-103

Psenner R, Nickus U (1986) Snow Chemistry of a Glacier in the Central Alps (HIntereisferner, Tyrol, Austria). Z Gletscherkunde und Glazialgeologie 22:1-18

Puxbaum H, Vitovec W, Kovar A (1988a) Chemical Composition of Wet Deposition in the Eastern Alpine Region. In: Unsworth MH, Fowler D (eds) Acid Deposition at High Elevation Sites. Kluwer, pp 419-430

Puxbaum H, Rosenberg C, Gregori M, Lanzerstorfer C, Ober E, Winiwarter W (1988b) Atmospheric Concentrations of Formic and Acetic Acid and Related Compounds in Eastern and Northern Austria. Atmos Environ 22:2841-2850

Ronseaux F, Delmas RJ (1988) Chemical Composition of Bulk Atmospheric Deposition to Snow at Col de la Brenva (Mt. Blanc Area). In: Unsworth MH, Fowler D (eds) Acid Deposition at High Elevation Sites, Kluwer, Dordrecht, pp 491-510

Schulze E-D, de Vries W, Hauhs M, Rosen K, Rasmussen L, Tamm CO, Nilsson J (1990) Critical Loads for Nitrogen Deposition on Forest Ecosystems. Water, Air and Soil Pollut 48:451-

Smidt S, Herman F, Leitner J (1990) Höhenprofil Zillertal. FBVA-Bericht 44, Forstliche Bundesversuchsanstalt Wien

Tranter M, Abrahams PW, Blackwood IL, Brimblecombe B, Davies TD (1988) The Impact of a Single Black Snowfall on Streamwater Chemistry in the Scottish Highlands. Nature 322:826-829

Tranter M, Davies TD, Abrahams PW, Blackwood I, Brimblecombe B, Vincent CE (1987) Spatial Variability in the Chemical Composition of Snowcover in a Small, Remote, Scottish Catchment. Atmos Environ 21:853-862

Tsitouridou R, Puxbaum H (1987) Application of a Portable Ion Chromatograph for Field Site Measurement of the Ionic Composition of Fog Water and Atmospheric Aerosols. Intern J Environ Anal Chem 31:11-22

Vet RA (1987) Acid Deposition Monitoring in Snowfall: State of the Science and Proposed Research. In: ICAIR Workshop on Acid Deposition Monitoring for Snowfall and Snowpack, Workshop Summary Report TR-898-24A by Life Systems Inc., Cleveland, Ohio

Vitovec W (1988) Untersuchungen zur Schneechemie des Wurtenkeesgletschers. Diploma Thesis, Technical University Vienna

Wagenbach D (1989) Environmental Records in Alpine Glaciers. In: Oeschger H, Langway CC (eds) The Environmental Records in Glaciers and Ice Sheets. Wiley-Interscience (Dahlem Workshop Reports), pp 343-361

Wagenbach D, Münnich KO, Schotterer U, Oeschger H (1988) The Anthropogenic Impact on Snow Chemistry at Colle Gnifetti, Swiss Alps. Ann Glaciol 10:183-187

Winiwarter W, Puxbaum H, Fuzzi S, Facchini MC, Orsi G, Beltz N, Enderle K-H, Jaeschke W (1988) Organic Acid Gas and Liquid Phase Measurements in Po Valley Fall-Winter Conditions in the Presence of Fog. Tellus 40B:348-357

Winkler P, Jobst S, Harder C (1989) Meteorologische Prüfung und Beurteilung von Sammelgeräten für die Nasse Deposition. BPT-Bericht 1/89, Gesellschaft für Strahlen- und Umweltforschung, München

Zobrist J (1983) Die Belastung der Schweizerischen Gewässer durch Niederschläge. VDI-Berichte 500:159-164

DISCUSSION ON "CHEMICAL COMPOSITION AND FLUXES OF WET DEPOSITION AT ELEVATED SITES (700-3105 m.a.s.l.) IN THE EASTERN ALPS (AUSTRIA)."

F. Maupetit
Laboratoire de Glaciologie et Géophysique de l'Environnement
Domaine Universitaire, BP 96
38402 Saint Martin d'Hères, France

T.D. Davies
School of Environmental Sciences
University of East Anglia
Norwich NR4 7TJ, UK

The presentation of a two-year continuous data set of the chemical composition of wet deposition at different sites in the Austrian Alps is of great interest. Glaciochemical data from snowpits and cores at alpine sites generally cover only the winter/early spring period. The full seasonal pattern of precipitation chemistry at high alpine sites is a welcome addition to knowledge.

The data confirm the behaviour of acidic species in the alpine atmosphere: measured acidity is mainly linked with nitrate while sulphuric acid is partially neutralised by ammonia, as suggested by others (Neftel et al., 1987; Delmas, 1989; Maupetit and Delmas, 1990).

This paper also clearly emphasizes that the occurrence of Saharan dust inputs to the high alpine regions is a major factor of alpine precipitation chemistry. The soluble fraction of those carbonated particles are able to neutralize partially or even completely the acidity of snowpacks (Delmas, 1989; Maupetit and Delmas, 1990). This neutralizing impact of Saharan dust particles needs an accurate titration method of acidity/alkalinity. This aspect introduces the problem of the ionic balance which is not obtained for the Innervillgraten and Sonnblick sites. Three reasons could explain this cation excess:

a) Ca excess
Ca has been measured by atomic absorption. It could be assumed that atomic absorption measurements includes a part of insoluble Ca due to dust particles leading to a cation excess. However some tests have been performed on Saharan dust snow samples and associated meltwater, concluding that Ca was always in the soluble part (Delmas, 1989).

b) Unmeasured anions: organic acids
The authors propose that organic acids are responsible for this anion deficit. Organic acids (acetic and formic) have been measured on an event basis for a three month period

NATO ASI Series, Vol. G 28
Seasonal Snowpacks
Edited by T. D. Davies et al.

(February to April 1990) on a glacier in the French Alps (Maupetit, unpublished data). These two acids were present in the samples at very low concentrations (0.5 to 1.5 $\mu eq.l^{-1}$). When dust was present, in some samples, acetic and formic acids had higher concentrations (up to 8.8 $\mu eq.l^{-1}$ for formic acid and 3.5 $\mu eq.l^{-1}$ for acetic acid for a major Saharan dust event collected in March 1990; see below). Thus it is difficult to believe that those acids should balance the 60 $\mu eq.l^{-1}$ cation excess reported by the authors.

c) Unmeasured anion; HCO_3^-
The third possibility to explain the anion deficit is the missing measurement of HCO_3^- in the dust-containing samples. A titration technique has been developed for the measurement of H^+/HCO_3^- in polar ice (Legrand et al, 1982); very high concentrations of HCO_3^- have always been measured in dust-containing samples (up to 100 $\mu eq.l^{-1}$)(Delmas, 1989; Maupetit and Delmas, 1990).

The example of a major Saharan dust event collected in March 1990 on a glacier from the French Alps is given below (units: $\mu eq.l^{-1}$):

Na^+ = 75.5	Cl^- = 64.2
NH_4^+ = 44.0	NO_3^- = 44.3
K^+ = 5.6	SO_4^{2-} = 88.0
Mg^{2+} = 17.2	HCO_3^- = 164.3
Ca^{2+} = 219.5	$HCOO^-$ = 8.8
	CH_3COO^- = 3.5
Cations = 361.8	Anions = 373.1

This sample was a Ca-rich case quite similar to those reported by the authors. It is clear that the only ion able to balance the high Ca concentrations is HCO_3^- and not the organic acids. This missing measurement of HCO_3^- should not be attributed to the pH measurement technique itself. It should be rather due to the problems of ionic stability of unfiltered liquid samples containing dust. The samples are kept in the liquid state between the sampling date and analysis, which may be performed after one and a half months. The stability of ionic acidic species is not established in the case of the presence of large amounts of dust particles. A pH measurement should be performed in the field to check this hypothesis. The samples should also be filtered just after collection and kept in the frozen state to reduce the eventual instability problems.

The interpretation of the data raises some potentially interesting questions, which might be profitably explored in future work. Because of the difference in elevation between the sampling locations, the direct comparability of the observations needs to be assessed. For example, the length of the snow season is likely to be considerably longer at higher elevation, or individual precipitation events may occur as snow at one site and as rain at another. Such differences in precipitation regime may lead to contrasts in precipitation composition at different locations.

Altitudinal differences in precipitation amount or rate may also play a role in composition differences. There is a (weak) inverse relationship between precipitation amount and ion concentration. It is likely that there is more precipitation at height in winter, because of the "orographic effect", but it is possible that this pattern is different in the summer months because of the greater frequency of convective precipitation. Another altitude-related factor which may influence relative compositions is enhanced riming at greater height.

It will be very interesting to assess the role of episodes. If a small number of episodes dominate mean monthly precipitation composition, then the interpretation of the data in climatological terms will be very difficult. In such an eventuality, the interpretation will be more appropriately conducted on a meteorological time-scale. Such meteorological-scale analysis will also be necessary to elucidate the climatological-scale influence which is manifest in the observed great interannual variability.

References

Delmas V, (1989). Chimie de la neige et de la fonte printaniere au Casset (Alpes du sud): influence des poussières minérales. Thèse de nouveau régime de l'Université Paris 7. Published by the Laboratoire de Glaciologie et de Géophysique de l'Environnement, Grenoble.

Legrand M R, Aristarain A J and Delmas R J, (1982). Acid titration of polar snow, Analytical Chemistry, 54, pp 1336-1339.

Maupetit F and Delmas R J, (1990). Glaciochemical study of four glaciers from French Alps: preliminary results, Proceedings of EUROTRAC symposium 90, P. Borrell ed., SPB Academic Pub., The Hague, pp 39-41.

Neftel A, Sigg A and Zurcher F, (1987). Acid deposition in a snow field at 2500 m.a.s.l in Switzerland, In: Physicochemical behaviour of atmospheric pollutants, Air Pollution Research Report 2, D. Reidel Publ. Comp., pp 500-510.

INFLUENCE OF URBAN AREAS ON THE CHEMISTRY OF REGIONAL SNOW COVER

Keisuke SUZUKI
Department of Geography
Tokyo Metropolitan University
Setagaya, Tokyo 158
Japan

1. INTRODUCTION

Since the Industrial Revolution, a large amount of pollutants have been released into the atmosphere, and pollutants in snow and rain are increasing. An urban area is an important source of pollutants. Some portion of pollutants is deposited near the source, but the overall deposition pattern of pollutants is affected by local as well as remote sources. Pollutants are accumulated in the snow cover during cold periods (Suzuki, 1982; Barrie and Vet, 1984), and the chemical study of snow cover can be a useful tool for determining deposition chemistry in urban areas.

Chemical studies of urban snow cover has been reported by Winchester <u>et al</u>. (1967), Oliver <u>et al</u>. (1974), Forland and Gjessing (1975), Landsberger <u>et al</u>. (1983), Landsberger and Jervis (1985). McBean and Nikleva (1986) reported the influence of Vancouver on the chemistry of mountainous snow cover.

Concerning the salting for de-icing roads, Oliver <u>et al</u>. (1974) reported the concentrations of Cl^- and Pb in urban snow. They concluded that high concentrations of Cl^- and Pb in street snow are induced by the road salting and vehicle exhausts. Landsberger and Jervis (1985) described the analytical techniques of heavy metals in urban snow. Forland and Gjessing (1975) discussed the rate of dry deposition of SO_4^{2-} and Mg^{2+} using of the surface snow cover in Bergen, Norway. They

NATO ASI Series, Vol. G 28
Seasonal Snowpacks
Edited by T. D. Davies et al.

concluded that the mean rate of dry deposition of SO_4^{2-} in the city area is about three times larger than for the stations in the surroundings, also for Mg^{2+} the maximum rate of deposition is found in the city area. McBean et al. (1986) concluded that the pH was depressed to 5 and SO_4^{2-} and NO_3^- concentrations are increased in the mountainous snow cover near Vancouver.

It is considered that the urban area affects the chemistry and physics of snow cover. This paper reports the influences of urban area on the snow cover including the interior of city.

2. URBAN AREAS COVERED WITH SEASONAL SNOW COVER

The number of cities that is located in the main snow region is very small. The relationships between the mean monthly temperature of January (below 0 °C) and the mean monthly precipitation of January in the main cities of the world and some Japanese cities are shown in Figure 1. The mean monthly precipitation of January in the continental cities is less than

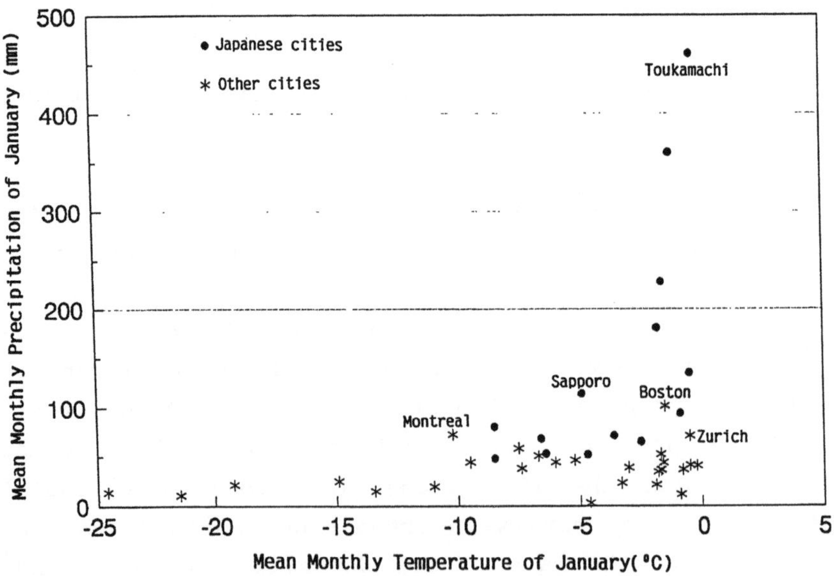

Figure 1. Relationships between the mean monthly temperature and the mean monthly precipitation of January.

100 mm. The mean monthly precipitation of January in Boston, Montreal and Zurich is 101 mm, 72 mm and 71 mm respectively, but the mean annual maximum depth of snow cover in these cities is less than 30 cm. On the other hand, the mean annual maximum depth of snow cover in Japanese cities is more than 100 cm. Especially, the mean annual maximum depth of snow cover at Toukamachi is 246 cm. Toukamachi has a population of 48,000 in 1985. About 20 million people live in the Japan Sea side area of the Japanese Islands where the mean annual maximum depth of snow cover is more than 50 cm. The Japan Sea side area of the Japanese Islands is the only region in the world which has such a high population under this type of snow conditions.

There are many Japanese cities where the mean annual maximum depth of snow cover is more than 100 cm. However, air temperature in the cities on the Japanese main island can rise over 0 °C during the snow period. So, the snow cover sometimes melts, and pollutants are discharged from snow cover. It can be concluded that chemical constituents are largely lost from the snow cover in cities on the Japanese main island. However, the retention of chemical constituents occurs in the snow cover before the spring melt period in cities in Hokkaido, northern island of Japan. Sapporo that is capital city of Hokkaido, has a population 1,543,000 in 1985. Sapporo is the largest city where the mean annual maximum depth of snow cover is more than 100 cm.

3. INFLUENCE OF URBAN AREA ON THE ACCUMULATION OF SNOW

Distributions of the depth of snow cover and the concentration of chemical constituents were studied in Sapporo. To avoid any direct influence from roads and residences, parks and baseball fields were selected as measuring and sampling sites. Twenty sites were selected (Figure 2). At each site, a trench was dug and the depth of snow cover was measured. Cores of whole snow cover were removed from the sides of the trench using a clean plastic pipe 58 mm in diameter. The snow cores were placed in a

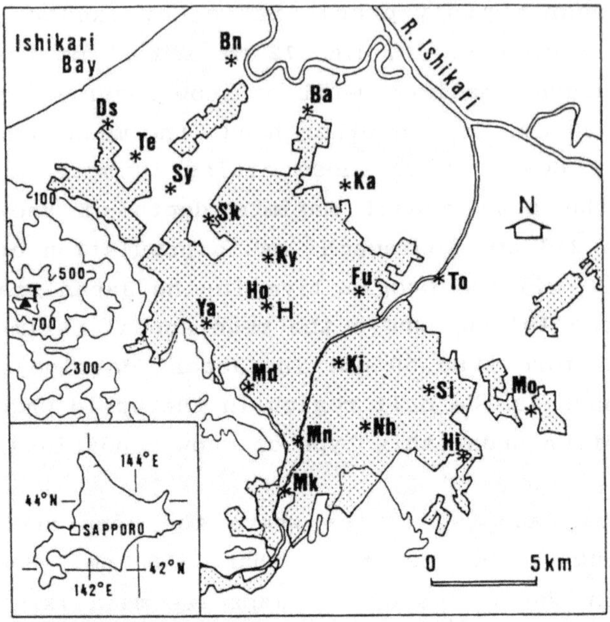

Figure 2. Location of study area. Stars indicate sampling sites. (H) and (T) show the location of Hokkaido University and Mt. Teine. Shaded parts represent the densely inhabited district in 1980. (Suzuki, 1987)

polypropylene box and allowed to melt. The meltwater weight was measured, and the water equivalent of the snow cover was obtained by dividing the weight of a core by the cross-sectional area of the plastic pipe. The density of the snow cover was obtained by dividing the water equivalent by the depth of the snow cover.

The distributions of depth and density of snow cover are shown in Figures 3 and 4. Uneven distribution of depth and density of snow cover were observed. Much of the snow deposited in the northern part of the urban area, and relatively little snow accumulated at the city center. This tendency is considered to be the result of the controlling effects of northwesterly winds and drifting snow by the buildings in the urban area. The density of the snow cover composed of hail and rimed crystals

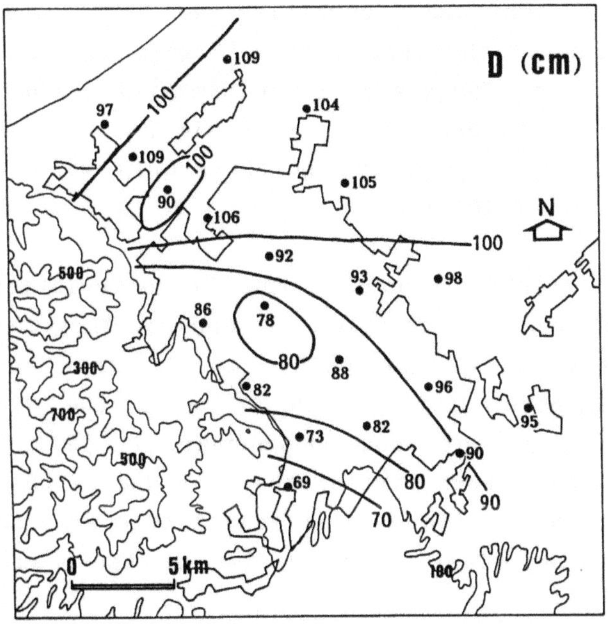

Figure 3. Distribution of depth of the snow cover.

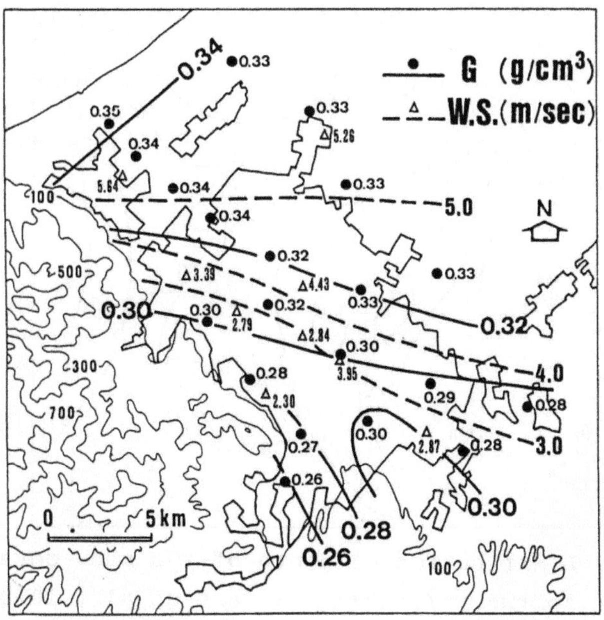

Figure 4. Distribution of density (G) of the snow cover and the mean wind speed (WS) of WSW-NW direction. (Suzuki, 1987)

is therefore larger. Snow densities near the seashore are larger than those of the inland area (Figure 4). Figure 4 also shows that the mean wind speed of the prevailing wind direction for snowfall (WSW-NW) exhibits a trend similar to snow density. This result confirms that the northwesterly wind transports hail and rimed crystals inland.

Naruse et al. (1978) reported the characteristics of the distribution of snow accumulation in Sapporo. They measured the depth of snow cover at the points in an urban and a rural district of Sapporo City. Figure 5 shows the distribution pattern of the depth of snow on February 7, 1977. Much snow deposited in the northern part of the city during the period of a northwesterly monsoon, and relatively little snow accumulated at the city center. They obtained the similar result to this study.

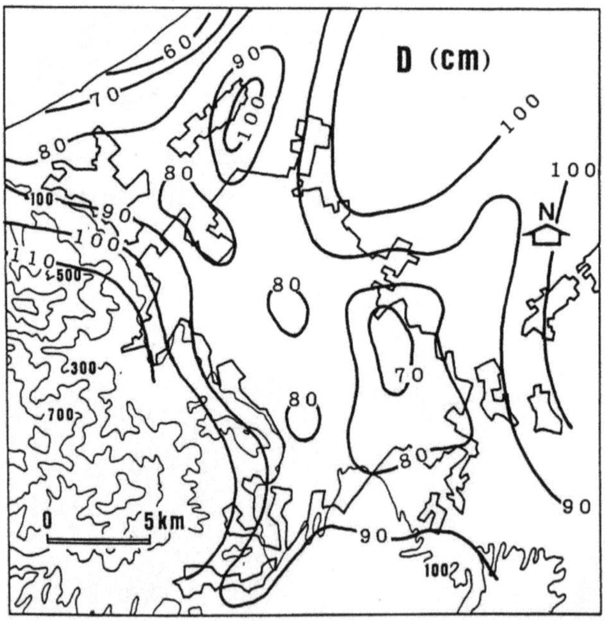

Figure 5. Distribution of depth of the snow cover on February 7, 1977. (Naruse et al., 1978)

4. SNOW COVER AS A CHEMICAL RESERVOIR

In order to confirm the conservation of chemical constituents in the snow cover during the cold period, the chemistry of the snow cover was frequently measured at Sapporo in 1981. Snow trenches were dug at the campus of Hokkaido University and cores of whole snow cover were removed from the sides of the trench using a clean plastic pipe 58 mm in diameter.

The snow cover lasted from December 13, 1980 to April 8, 1981. Three periods of heavy snowfall occurred in December and January. In February, little snow fell, and snowmelt began in the middle of March. After the snowfall at the end of January, water equivalent, conductivity, pH and concentrations of major ions in the snow cover maintained constant values until the snowmelt season. The sequence of water equivalent and concentrations of Na^+, Cl^- and SO_4^{2-} in the snow cover are shown in Figure 6. From these measurements, water equivalent and concentration of major ions were determined to be retained by the snow cover. The conservation of chemical constituents in the snow cover during cold period is also confirmed in 1979-1980 winter in Sapporo (Suzuki, 1982).

5. SOURCES OF MAJOR IONS IN THE SNOW COVER

Snow samples collected at twenty sites were allowed to melt. Immediately after melting, each meltwater sample was filtered through a 1-μm pore size membrane filter which was previously washed with 300 ml of distilled water. Conductivity and pH were measured potentiometrically at 25 °C. Na^+, K^+, Mg^{2+} and Ca^{2+} were determined by atomic-absorption spectrophotometry. Cl^- and SO_4^{2-} were determined by spectrophotometry.

The sources of the chemical constituents are indicated from the spatial variation in the concentrations of the chemical constituents in the snow cover. The relationships between concentrations of Na+ and other constituents of snow cover are

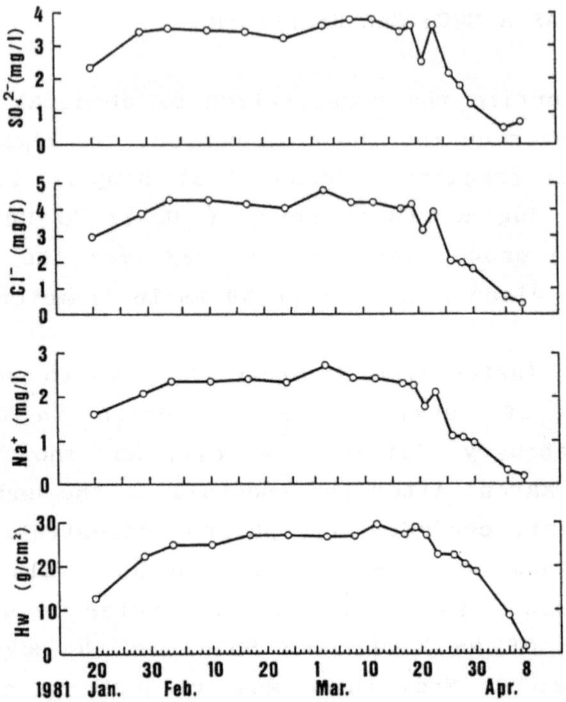

Figure 6. Sequence of water equivalent (Hw) and concentrations of Na^+, Cl^-, SO_4^{2-} in the snow cover at the campus of Hokkaido University. (Suzuki, 1987)

shown in Figure 7. The ratios of Cl^-/Na^+ at all sampling sites are nearly equal to that of sea water, and Na^+ and Cl^- are considered to come from sea water. The ratios of Mg^{2+}/Na^+ and K^+/Na^+ in the snow cover are a little greater than that of sea water. However, the concentrations of Mg^{2+} and K^+ in the snow cover bear proportional relations with Na^+ concentration. The mean enrichment coefficients of Mg^{2+} and K^+ in the snow cover are calculated to be 1.16 and 1.6, respectively. The predominant source of Mg^{2+} and K^+ in the snow cover is therefore considered to be sea water. The SO_4^{2-} concentrations at all sampling sites are larger than those of sea water. In Sapporo, only a small part of the SO_4^{2-} in the snow cover comes from sea water, and anthropogenic emission is considered to play an important role. Scatter in the ratios of Ca^{2+}/Na^+ is very large, that is to say, Ca^{2+} concentrations in the snow

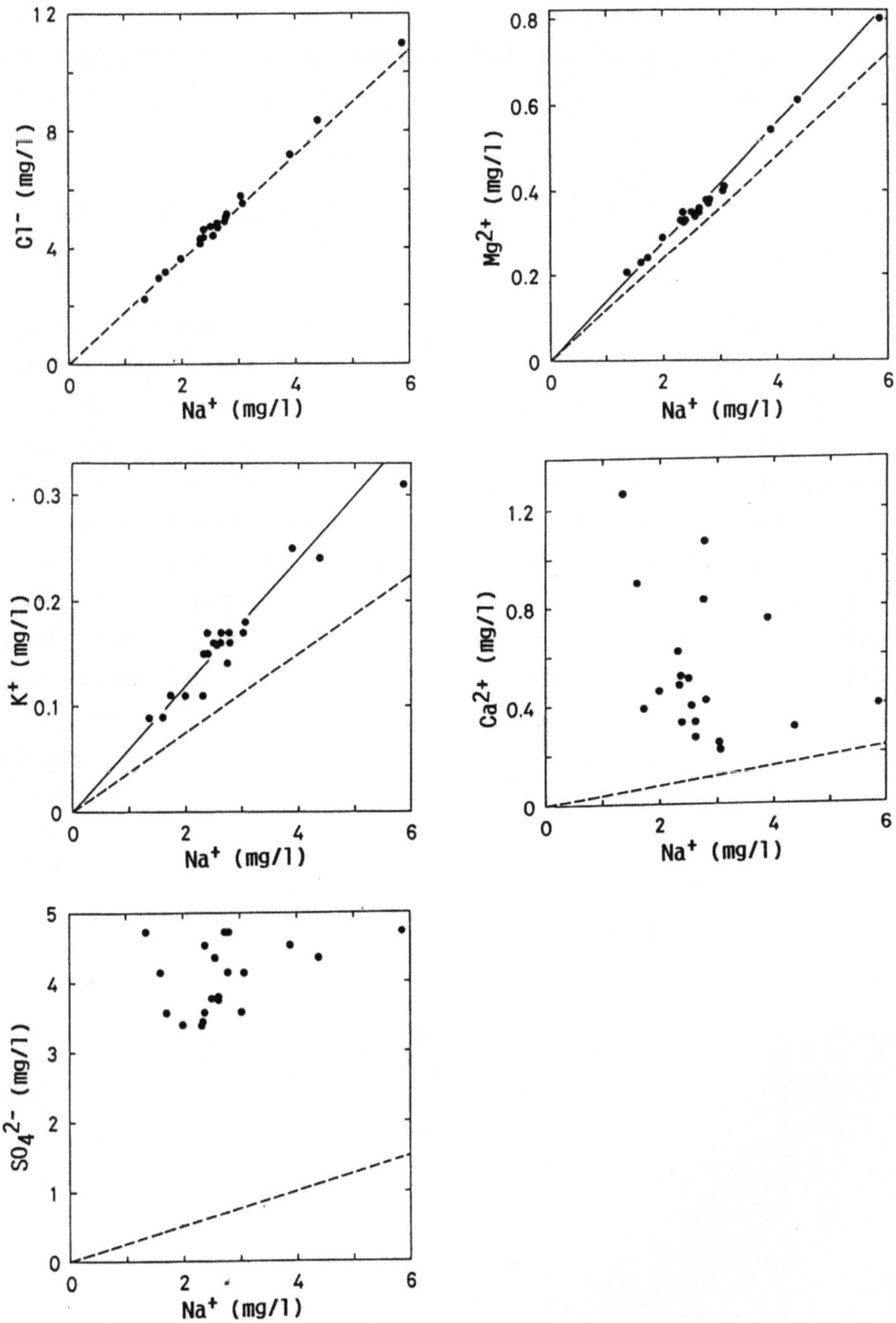

Figure 7. Relationships between concentrations of Na^+ and other constituents of snow cover. The broken lines show the ratio of each concentration to Na^+ concentration in the sea water.

cover vary with location to a great degree. This result suggests that the source of Ca^{2+} exists in the neighborhood of the study area.

6. SPATIAL DISTRIBUTION OF CONCENTRATIONS Cl^- AND SO_4^{2-} IN THE SNOW COVER

The concentrations of Na^+, Mg^{2+} and K^+ in the snow cover are proportional to Cl^- concentration. The distribution of Cl^- concentration in the snow cover is shown in Figure 8. Spatial distributions of Na^+, Mg^{2+} and K^+ concentration in the snow cover show a trend similar to that of Cl^- concentration. The Cl^- concentration near the seashore is larger than that inland, and is lowest south of the city center of Sapporo. These results correspond to the distribution of hail and rimed crystals along the seashore (Magono et al., 1965, 1966). Snow particles are almost all generated over the Sea of Japan. The fall velocities of hail and rimed crystals are larger than those of non-rimed crystals, therefore hail and rimed crystals

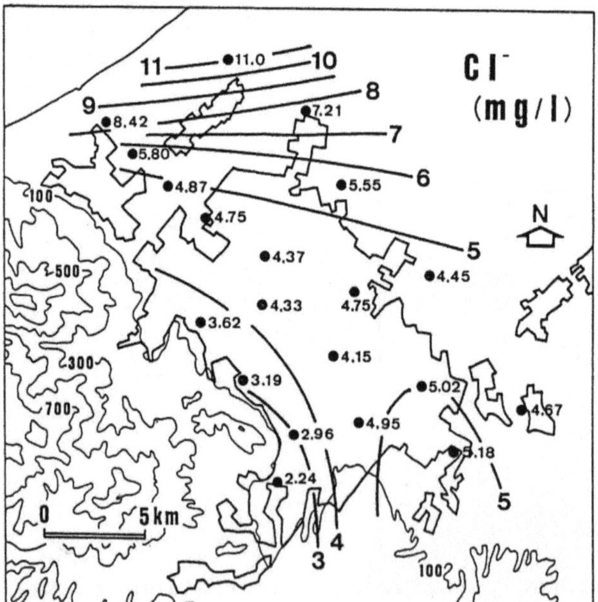

Figure 8. Distribution of Cl^- concentration in the snow cover. (Suzuki, 1987)

are observed along the seashore. Snow crystals grown by the sublimation process have extremely low concentrations of chemical constituents. Snow crystals collect cloud droplets containing sea salt particles over the Sea of Japan, and the concentration of chemical constituents in the snow increases. In order to transport hail and rimed crystals inland, sufficient wind force is required. As mentioned above, the densities near the seashore are larger than those of the inland area, and isoplethes of Cl^- concentration show a trend similar to the density of the snow cover. The wind pattern causes the patterns of Cl^- concentration and density of snow cover.

Sulfate is added to the atmosphere as anthropogenic emissions, from H_2S formed in nature by biological action, and as sea spray (Eriksson, 1960). The anthropogenic emissions are nearly equal to natural emissions (Kellog et al., 1972). The anthropogenic emissions are considered to play a more important role in Sapporo, as mentioned above (Figure 7). The concentration of nss SO_4^{2-} in the snow cover is obtained by the equation:

$$nssSO_4^{2-} = (SO_4^{2-})snow - (Na^+)snow \times (SO_4^{2-}/Na^+)sea\ water$$

The nss SO_4^{2-} comes from natural sources (excluding sea water) and from anthropogenic sources. At Mt. Teine (1024 m), situated to the west of Sapporo, the amount of SO_4^{2-} in snow released from residential and industrial areas is negligible. The mean concentration of nss SO_4^{2-} in snow collected at Mt. Teine was calculated to be 1.67 mg/l (Kimura, 1981). Height of cloud base was about 1000 m during this study, and SO_4^{2-} was scavenged by the in-cloud process. There are no anthropogenic sources of SO_4^{2-} in the windward side and neighborhood of Mt. Teine. This value (1.67 mg/l) is estimated to be a background value of nss SO_4^{2-}. The value in excess of background is estimated to originate from combustion of fossil fuels for house heating and from automobile exhausts in Sapporo.

Spatial distribution of nss SO_4^{2-} concentration in the snow cover is shown in Figure 9. The concentrations of nss SO_4^{2-} in

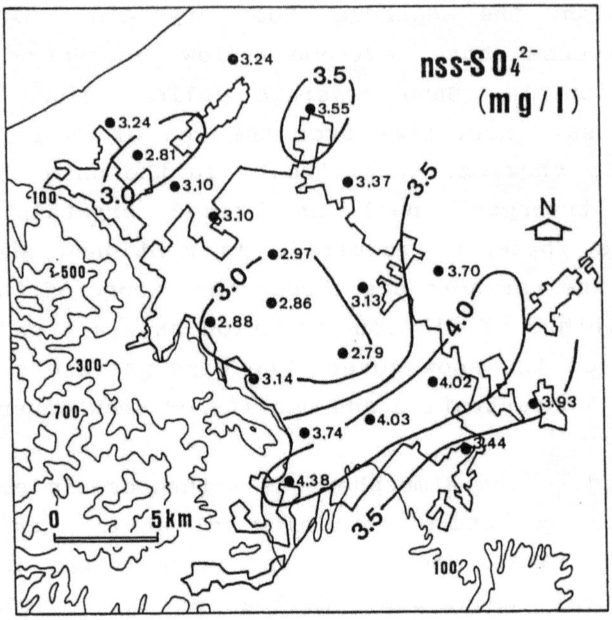

Figure. 9. Distribution of nss-SO_4^{2-} concentration in the snow cover.

the snow cover at all sampling sites are larger than the background value estimated at Mt. Teine. The nss SO_4^{2-} concentration is lowest in the city center, and the area of high concentration spreads southeast of Sapporo. The amount of sulfur emitted to the atmosphere is considered to be larger in the city center. The mean concentration of SO_2 in the air for the period from December 15, 1980 to February 15, 1981 (obtained from the Sapporo Air Pollution Observatory) is illustrated in Figure 10. The mean concentration of SO2 is highest in the city center. Point sources of SO_2, e.g. pulp plants, are not found in the city center. The major source of sulfur is due to the combustion of fossil fuel for residential heating and automobile exhausts.

The air temperature in cities is known to be higher than in suburbs, and air circulation is generated by the urban heat island effect (e.g. Bach, 1970). SO_2 released to the atmosphere

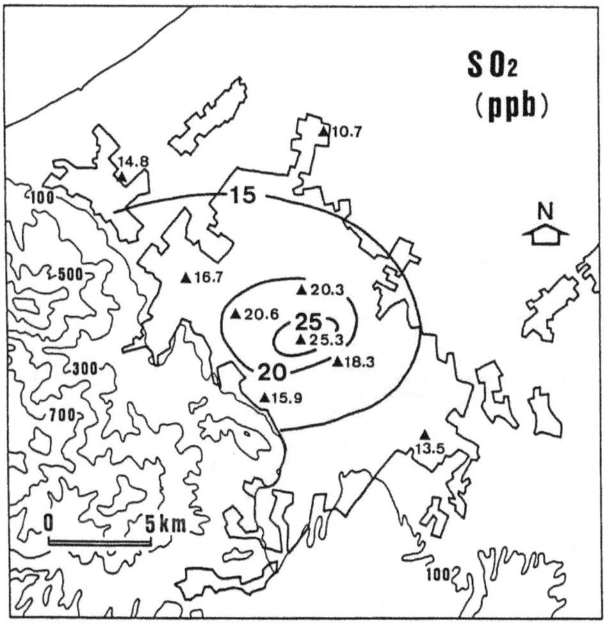

Figure 10. Mean concentration of SO_2 in the atmosphere for the period from December 15, 1980 to February 15, 1981. (Suzuki, 1987)

in the city is considered to be transported with this circulation. The heat island effect is strongly manifested in snowy cold regions such as Sapporo, in particular in calm weather conditions. Calculations of the field of horizontal divergence under calm conditions in Sapporo show that the air converges and ascends in the city center and flows in from the suburbs. If such circulation occurs frequently, the low concentration of nss SO_4^{2-} in the city center can be explained. However, high concentrations of nss SO_4^{2-} in the snow cover are only observed in the area southeast of the city center.

The wind rose and the percentage of snowfall for each wind direction at the Sapporo District Meteorological Observatory for the period from December 15, 1980 to February 15, 1981 are given in Figure 11. The prevailing wind directions are from the northwest or southeast. The existence of a high concentration band of nss SO_4^{2-} at northwest area of the city center is

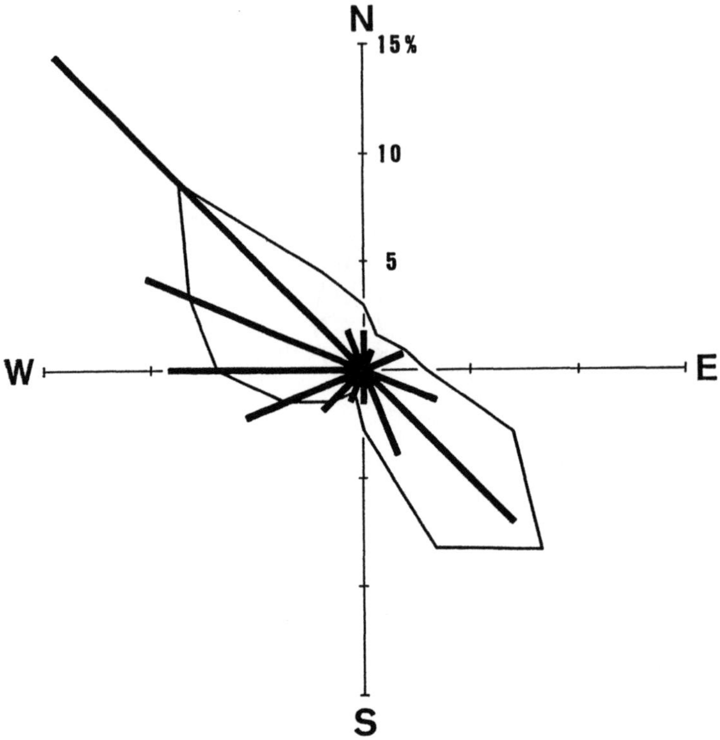

Figure 11. Wind rose (closed line) and snowfall for each wind direction (solid lines) for the period from December 15, 1980 to February 15, 1981. (Suzuki, 1987)

suggested by this result, but such a band is not observed. The distribution of nss SO_4^{2-} is influenced not only by wind conditions, but also by snow deposition patterns. Most of the SO_4^{2-} in the atmosphere is scavenged below cloud by snow. Since the snow crystals which collect SO_4^{2-} fall in northwest winds, a high concentration band of nss SO_4^{2-} in the snow cover is only observed in the area southeast of the city center.

7. CONCLUSIONS

The influences of the urban area (Sapporo City) on snow cover have been discussed. From the measurement of depth and density of snow cover, uneven distribution of depth and density of snow cover were recognized. Namely, much snow is deposited in the northern part of the urban area, and relatively little snow accumulated at the city center. This tendency is considered to be result of the controlling effects of northwesterly winds and drifting snow by the buildings and housing in the urban area.

After the heavy snowfall, water equivalent, conductivity, pH and concentrations of major ions in the snow cover maintained constant values until the snowmelt season. Chemical constituents are retained in the snow cover during the cold period. After that, the chemical study of snow cover can be a useful tool for determining the influence of urban areas on regional snow covers.

Because the SO_2 concentration in the atmosphere is highest in the city center of Sapporo, SO_2 is considered to be released to the atmosphere by combustion of fossil fuels for house heating and automobile exhausts. However, the concentration of nss SO_4^{2-} in the snow cover is lowest in the city center, and a high concentration band of nss SO_4^{2-} is located in the area southeast of the city center. This distribution is caused by an urban heat island effect which causes SO_4^{2-} to rise and by a snow deposition pattern. Northwesterly winds bring a large amount of snow, and SO_4^{2-} is scavenged below cloud by these snow particles. An influence of the urban area and a snow deposition pattern cause the distribution of nss SO_4^{2-} concentration in snow cover.

REFERENCES

Allan, R. J. and Jonasson, I. R. (1978): Alkaline snowfalls in Ottawa and Winnipeg, Canada. Atmospheric Environment, 12, 1169-1173.

Bach, W. (1970): An urban circulation model. Archiv fur Meteorologie Geophysik und Bioklimatologie, Ser. B, 18, 155-168.

Barrie, L. A. and Vet, R. J. (1984): The concentration and deposition of acidity, major ions and trace metals in the snowpack of the Eastern Canadian Shield during the winter of 1980-1981. Atmospheric Environment, 18, 1459-1469.

Eriksson, E. (1960): The yearly circulation of chloride and sulfur in nature; meteorological, geochemical and pedological implications- II. Tellus, 12, 63-109.

Forland, E. J. and Gjessing, Y. T. (1975): Snow contamination from washout/rainout and dry deposition. Atmospheric Environment, 9, 339-352.

Kellogg, W. W., Cadle, R. D., Allen, E. R., Lazrus, A. L. and Martell, E. A. (1972): The sulfur cycle. Science, 175, 587-596.

Kimura, T. (1981): A study of the scavenging effect of precipitation. Master's thesis, Hokkaido University, 61pp.

Landsberger, S. and Jervis, R. E. (1985): Sulphur and heavy metal pollution in urban snow: multi-elemental analytical techniques and interpretations. Annals of Glaciology, 7, 175-180.

Landsberger, S., Jervis, R. E., Kajrys, G., Monaro, S. and Lecomte, R. (1983): Total soluble and insoluble sulfur concentrations in urban snow. Environmental Science and Technology, 17, 542-546.

Magono, C., Kikuchi, K., Kimura, T., Tazawa, S. and Kasai, T. (1966): A study on the snowfall in the winter monsoon season in Hokkaido with special reference to low land snowfall. Journal of the Faculty of Science, Hokkaido University, Ser. VII, 3, 287-308.

Magono, C., Kikuchi, K., Lee, S., Endo, T. and Kasai, T. (1965): An observation of snow crystals and their mother cloud. Journal of the Faculty of Science, Hokkaido University, Ser. VII, 2, 123-148.

McBean, G. A. and Nikleva, S. (1986): Composition of snow in pacific coastal mountains. Atmospheric Environment, 20, 1161-1164.

Naruse, R., Aburakawa, H. and Ishikawa, N. (1978): Characteristics on the distribution of snow accumulation in Sapporo, Hokkaido. Low Temperature Science, Ser. A, 36, 139-153.

Oliver, B. G., Milne, J. B. and LaBarre, N. (1974): Chloride and lead in urban snow. Journal of the Water Pollution Control Federation, 46, 766-771.

Suzuki, K. (1982): Chemical changes of snow cover by melting. Japanese Journal of Limnology, 43, 102-112.

Suzuki, K. (1987): Spatial distribution of chloride and sulfate in the snow cover in Sapporo, Japan. Atmospheric Environment, 21, 1773-1778.

Winchester, J. W., Zoller, W. H., Duce, R. A. and Benson, C. S. (1967): Lead and halogens in pollution aerosols and snow from Fairbanks, Alaska. Atmospheric Environment, 1, 105-119.

DISCUSSION ON "INFLUENCE OF URBAN AREAS ON THE CHEMISTRY OF REGIONAL SNOWCOVER"

T. D. Davies
School of Environmental Sciences
University of East Anglia
Norwich NR4 7TJ, UK

Although a relatively small proportion of the total area of seasonal snowcover is within urban areas, most seasonal snowcover is affected, to some degree, by anthropogenic emissions. A case-study of the composition of snowcover within a city therefore gives us useful information near the extreme end of the range of pollution impacts on the local snowpack.

The study by Suzuki illustrates some of the difficulty of attempting to assign causal influences on snowcover composition in complex situations. Local topography and geography, especially the proximity of the coast, make it difficult to confidently isolate the influence of the urban area.

Suzuki attributes the N-S gradients in snowcover density across the urban area to the greater deposition of hail and rimed crystals near the coast, but presented no direct evidence to support this explanation.At least two other causes may be operating. The observed N-S gradient in wind speed could lead to greater wind-packing/armouring near the coast, thereby imposing a spatial variation in snowcover density. Additionally, the moderating maritime influence could produce enhanced (partial) melting in coastal locations. The consequence would be greater snow densities.

The spatial distribution of Cl^- concentrations is explained in terms of the greater deposition of hail and rimed crystals near the shore. This may be the case (see above), but a likely major contributor is greater deposition of sea-spray near the coast.

Suzuki suggests that the convergent circulation induced by the urban "heat island" effect may lead to the low concentrations of nss SO_4^{2-} in the centre of Sapporo, but it is difficult to see, from the information presented, how this can be the case. The concentrations of SO_2 in the city centre are high compared to the concentrations elsewhere and, as Suzuki points out, urban circulations are most prevalent under conditions of light regional winds and clear skies. Snowfall is not likely to occur under these conditions.

A number of other possible explanations for the zone of high nss SO_4^{2-} concentrations to the SE of the city centre could be explored. The depth of the snowcover is relatively small to the south of the city. If this reflects precipitation amount, the usually-observed inverse relationship between

NATO ASI Series, Vol. G 28
Seasonal Snowpacks
Edited by T. D. Davies et al.

precipitation amount and ion concentration may be an important factor. Another possibility is a feeder-seeder mechanism operating in the lee of the hills to the west of the city, under the prevailing precipitation-bearing NW wind direction. Yet another possible explanation is dry deposition to the existing snowcover from large sources upwind.

Much of Suzuki's argument presupposes that the bulk composition of the urban snowcover reflects long-term (weeks to months) processes, but we know that wet deposition is generally very episodic. Consequently, the bulk composition is probably strongly influenced by a small number of events. The interpretive scale is, correspondingly, likely to be more appropriately focussed on events.

Suzuki's contribution, therefore, is a useful addition to knowledge of the chemical composition of an urban area via a case-study in Sapporo. There have been many other studies, predominantly in North America, some of which Suzuki makes passing reference to, but does not review in any detail. There is clearly scope for further observational work to isolate the factors which are controlling snowcover composition in Sapporo. Suzuki speculates on some of the possible mechanisms, but their establishment must await further research.

ORGANIC MICROPOLLUTANTS IN SEASONAL SNOWCOVER AND FIRN

D. J. Gregor
Lakes Research Branch
National Water Research Institute
Environment Canada
P.O. Box 5050
Burlington, Ontario, Canada
L7R 4A6

1. Introduction

Large areas of many temperate countries receive significant proportions of their total annual precipitation as snowfall and it makes up the primary source of precipitation in polar areas. Importantly, spring snow-melt is commonly the largest annual hydrologic event in temperate and polar areas and as such may be responsible for transporting pollutants rapidly over long distances. Lake sediments have been used to provide a crude temporal record of organic contaminant accumulation in the environment from the atmosphere and elsewhere. However, the chronicle of micropollutant contamination frozen in polar and alpine glaciers has, to date, received only limited attention. This has been the case despite the much improved temporal resolution possible in glaciers over that of most lake sediments, especially in the polar extremes where lake sedimentation rates tend to be very low.

NATO ASI Series, Vol. G 28
Seasonal Snowpacks
Edited by T. D. Davies et al.

In this context, it is surprising that research concerned with organic micropollutants in snow has been limited in geographic extent and sporadic in time. The purpose of this chapter is to review existing information on the role of snow in scavenging organic contaminants from the atmosphere, in temporarily storing contaminants and through snow-melt, in introducing these contaminants to the aquatic environment. In order to focus the content, only organochlorine pesticides, polychlorinated biphenyls and polycyclic aromatic hydrocarbons are considered. These classes of compounds are important environmentally as they tend to be long lived in the environment, may bioaccumulate and are frequently toxic and/or carcinogenic at elevated concentrations. Additionally, information regarding temporal trends of the levels of these compounds revealed from glacial studies and the apparent problems in utilizing these data will be considered. At the outset, the chapter will summarily review the physicochemical properties of typical organic micropollutants and the processes controlling their emission to and transport in the atmosphere from which they become incorporated into the falling snow.

2 Characteristics of Trace Organic Micropollutants

2.1 Potential sources

Two classes of chemicals, here called chlorinated industrial organics and pesticides (OCs) and polycyclic aromatic hydrocarbons (PAHs) are considered. These groups have distinct characteristics and sources and will be dealt with individually.

2.1.1 Chlorinated industrial organics and pesticides

Relevant OCs are those persistent compounds produced in large quantitities and released in part to the environment where they are widely dispersed. These include polychlorinated biphenyls (PCBs), hexachlorocyclohexanes (HCHs), chlordane, dichlorodiphenyl-trichloroethane (DDT), dieldrin, heptachlor and heptachlor epoxide, endosulfan, methoxychlor, hexachlorobenzene (HCB) and polychlorinated camphenes (PCCs) or toxaphene. Production and use statistics for the OCs are difficult to acquire but a recent summary of available production estimates for some of the compounds is provided in Barrie et al. (in press) using information from a variety of sources. Current use regulations regarding these compounds have also been summarized for selected countries by these authors while more detailed information is maintained by the International Registry of Potentially Toxic Chemicals (IRPTC) of the United Nations Environmental Programme (UNEP).

PCBs are a concern as most of the 209 congeners, especially those with 5 or more chlorines, are biomagnified (Jensen et al., 1969; Reijnders, 1980; Muir et al., 1988). PCBs have been used world-wide for a variety of purposes with as much as 370 tonnes (t) estimated to have been released into the environment (Tanabe, 1988).

Organochlorine pesticides are a concern because of their long half lives and tendency to bioaccumulate in aquatic and terrestrial food chains. Because pesticides are applied intentionally over large areas of land, their sources and pathways may differ from PCBs, other industrial organics and PAHs.

HCH can theoretically exist as eight different stero isomers of which only five are found in the technical product. The approximate abundance of the dominant isomers in technical HCH are α-HCH at 70% and lindane at 13% of the total (Deo et al., 1980). The insecticidal isomer lindane is currently used in pure form in North America and western Europe (Voldner and Ellenton, 1987; IRPTC, 1988). In contrast, technical HCH is widely used in southern Asia and China with estimated annual use in the late 1970's of about $6x10^4$ t (Tanabe et al., 1982; Agarwal et al., 1987; Rapaport and Eisenreich, 1988).

Technical chlordane, a multi component product, was used as an insecticide but has more recently been used only for subterranean use in Canada and the US (Candian Council of Resource and Environment Ministers [CCREM], 1987). Large quantities of chlordane have been used in the US (approximately $25x10^4$ t) (von Rumker et al., 1974). Within the last decade, agricultural use in the US and many other countries has been curtailled. Since heptachlor is more insecticidal than chlordane, efforts have been made to purify the technical product to more than 70% heptachlor. Heptachlor is mainly used as a soil insecticide and for seed dressing and has been reported to bioaccumulate. Heptachlor epoxide, a metabolite, has not been used as a pesticide but is commonly identified in the aquatic environment at concentrations greater than heptachlor. Heptachlor epoxide is reported to be more resistant to chemical and biological changes in the aquatic environment than heptachlor and has a predicted half life of several years (CCREM, 1987).

Dichlorodiphenyl-trichlorethane (DDT) has been banned or restricted in Canada, US and western Europe since the early 1970's; however, it continues to be manufactured and used in southern Asia, Africa and in Central and South America (Voldner and Ellenton, 1987). Global production is estimated at about $15x10^5$ t (Rapaport and Eisenreich, 1986). Although presently restricted in many countries, world-wide use of DDT remains extensive.

Dieldrin is considered to be persistent, volatile, and bioaccumulative (CCREM, 1987). This product has also been restricted in Canada and the US since the mid 1970's although it is still produced and used in a number of western European countries (IRPTC, 1988). Endrin has properties similar to dieldrin and has been applied widely to cotton and rice (Buchel, 1983). Its use is presently restricted or it is banned in many developed countries (IRPTC, 1988).

Endosulfan, consisting primarily of the α- isomer (70%) photolyzes in the gaseous state and although it is taken up by biota, it appears to be rapidly broken down in the organism or is eliminated and there is little danger of bio-accumulation (CCREM, 1987; Buchel, 1983). It is widely used on cotton and orchards, the latter because it is not toxic toward bees (CCREM, 1987).

Methoxychlor is not stored in fatty tissues of higher organisms (CCREM, 1987; Buchel, 1983) and is still approved for agricultural use in Canada, particularly for the control of blackfly larvae. Use has decreased considerably since the 1970's. It is also approved for use in the US, USSR and many European countries (CCREM, 1987; IRPTC, 1988).

Toxaphene was the most heavily used insecticide in the US during the 1960's and 1970's with over $5x10^5$ t produced. There it was used mainly for controlling weevil on cotton (Rapaport and Eisenreich, 1986; Voldner and Ellenton, 1987) while it has been used only in a very restricted way in Canada (CCREM, 1987). Use in Mexico was estimated to be about $2.2x10^3$ t in 1978 and projected as $2x10^3$ t in 1988 (Voldner and Ellenton, 1987). The USSR produces toxaphene derived from turpentine rather than camphene, for use as an agricultural pesticide (IRPTC, 1983) but production and use figures are not available.

2.1.2 Polycyclic aromatic hydrocarbons (PAHs)

PAHs, a number of which are carcinogenic, occur naturally in the environment as a result of forest and prairie grass fires. However, the quantity of PAHs in the environment due to incomplete combustion has increased as a result of the increased use of fossil fuels (CCREM, 1987). Studies of PAHs in sediments of lakes in the Adirondacks of the eastern US show that atmospheric emissions there peaked in the 1950's corresponding to the decline in the use of coal and the increased use of oil for space heating (Heit et al., 1981). Since the combustion of fossil fuels for space heating increases during the winter, it is to be expected that PAHs will be present in snowfall, especially in the vicinity of human settlements.

2.2 Emission and atmospheric exchange

The sources of organic micropollutants are potentially global in nature and may be emitted to the atmosphere either directly through use (e.g., agricultural pesticides) or indirectly and unintentionally as part of industrial, combustion and other processes (e.g., PCBs and PAHs). Once introduced into the atmosphere, these compounds may circulate in the atmosphere and hydrosphere for long periods of time prior to going to permanent sink or being broken down or bound up such that they are effectively removed from the system. Snow, therefore may contain "fresh" compounds recently emitted to the atmosphere and "aged" compounds which have evaded permanent deposition to a sink.

A semi-volatile organic compound that is chemically and biologically stable with respect to, for example, photo-degradation, biodegradation and/or chemical transformation, will tend to migrate over long distances through repeated deposition and volatilization. Many organic compounds are likely to be transported in this way (Bidleman, 1988). The vapor pressure, organic content and the concentration, size and surface area of total suspended particles control the distribution or partitioning between vapor and particles. Vapor-particle partitioning, in turn, will influence the distance travelled by these contaminants by means of determining whether the compound is scavenged by rain and snow scavenging of particles and vapors, by dry particle deposition or by direct vapor exchange across the air-water interface. Overviews of these atmospheric processes are available (Slinn et al., 1978; Eisenriech et al., 1981; Murphy, 1984; Bidleman, 1988). It is sufficient here to note that compounds with saturation vapor pressures greater than 1.3×10^{-2} Pa should exist almost entirely in the vapor phase while those having vapor pressures less than 1.3×10^{-6} Pa should exist almost entirely in the particulate phase (Eisenriech et al., 1981). Most high molecular weight organics have vapor pressures between these extremes, as shown in Table 1, and their atmospheric lifetimes depend on the nature and concentration of particles in the air mass.

Table 1 Physicochemical properties of selected organochlorine compounds and PAHs[a]

Compound	Henry's Law Constant[a] Pa(mol m^{-3}) freshwater(°C)		seawater (-2°C)	Vapor pressure[b] over liquid (25°C)(KPa)	water solubility[c] (ug L^{-1} @25°C)
organochlorines					
lindane	0.32	(25)	0.03	$6.5x10^{-2}$	2000-12000
α-HCH	0.55	(22-25)	0.10	$2.3x10^{-1}$	1200-2000
toxaphene	0.61	(20)	0.09	$1.7x10^{-3}$	300
hexachloro-benzene	83.0	(25)	7.8	$1.3x10^{-1}$	<20
α-endosulfan	1.09	(25)	0.10	$6.1x10^{-3}$	150-600
dieldrin	1.12	(20)	0.125	$1.0x10^{-2}$	186-200
chlordane	4.92	(25)	0.47	$5.1x10^{-3}$	56-1850
DDT	5.3	(25)	0.5	$5.1x10^{-4}$	<1-25
DDE				$3.3x10^{-3}$	1-140
2,4,5,2',4',5'-PCB	42.9	(25)	4.1	$5.0x10^{-4}$	
PAHs					
pyrene	1.2	(25)	0.11	$1.4x10^{-2}$	
phenanthrene	4.0	(25)	0.38	$1.0x10^{-1}$	1000-1800[d]
naphthalene	43	(25)	4.1	$3.8x10^{1}$	26000[d]
fluoranthene	220	(25)[e]		$6.4x10^{-3}$	≈200[e]

[a] modified from Barrie et al. (in press)
[b] Hinkley et al., 1990
[c] from Arctic Laboratories (1988)
[d] from CCREM (1987)
[e] from Mckay and Shiu (1981)

PCBs have been shown to be between 80 and 100% in the vapor phase in urban atmospheres and in the area of the Laurentian Great Lakes (Murphy and Rzeszutko, 1977; Doskey and Andren, 1981; Eisenreich et al., 1981). Duinker and Bouchertall (1989) investigated the partitioning of several PCB congeners and concluded that while only a very small fraction of total PCB was found in aerosols, this compartment accounted for more than 99% of the PCBs in rain. More highly chlorinated congeners were particle-associated while the lower chlorinated congeners remained essentially in the vapor phase. Thus the method of deposition and form of precipitation can affect the toxicity of the PCBs and can influence how readily they can be revolatilized from the depositional medium.

Particle deposition processes in polar regions have been reviewed by Barrie (1986). They depend mainly on physical factors including particle size distribution, surface roughness and wind speed. Vapor phase exchange with the surface - which may be water (fresh or saline), snow or soil - depends on the physicochemical properties of the compound, many of which are temperature dependent (Patton et al., 1989).

Exchange processes and rates between water or soil and the atmosphere have been researched but not at sub-freezing temperatures (Doskey and Andren, 1981; Atlas et al., 1982; Larsson, 1983; Larsson and Okla, 1987; Baker and Eisenreich, 1990). Only recently, have exchange processes between surface snow and the arctic atmosphere been considered for semi-volatile organic compounds. Results of replicate sampling of the 1985/86 snow layer on the Agassiz Ice Cap, Ellesmere Island, Canada (80° 40'N 73° 30'W) in May of both 1986 and 1987 are presented in Table 2. Except for PCBs, the concentration of the OCs measured in this layer in 1986 was an order of magnitude greater than the concentration measured in the same layer in the following year. Similarity between the two years of sampling for firn layers in which sampling was repeated in the two years indicated that this was not an analytical artifact. Comparison of the snow accumulation layer for the winter of 1986/87, represented only by a single sample in May of 1987 and again in May of 1989, shows residual concentrations of OC pesticides of about 1% in the glacier relative to the depositional concentration measured in 1987 (Gregor, unpublished data). In contrast to the 1985/86 situation, the PCBs also disappearred from the 1986/87 snow layer. Since the 1986/87 layer was characterized by only a single sample, as compared to the multiple samples for 1985/86, it is possible that this loss in 1986/87 is an analytical artifact and only future investigation of this phenomenon will confirm the fate of contaminants in the glacial snow during the first summer season.

Since melting on the glacier was limited and the snow layers below showed consistent low concentrations (i.e. no downward transport of contaminants with the percolation of meltwater through the annual layer) it must be concluded that large portions of the OCs, with the possible exception of PCBs, revolatilized to the atmosphere during the summer season as the air and snow temperatures rose. It is conceivable, because of the range of water solubilities and vapor pressures that characterize PCBs (Mackay and Wolkoff, 1973; Shiu and Mackay, 1986) and the different gas/particle partitioning coefficients of PCB congeners (Duinker and Bouchertall, 1989) that some portion of PCB will remain in the firn as was the case in 1986. This hypothesis needs to be confirmed. Thus, utilization of the glacial residues for temporal trend determinations must be approached cautiously until this re-volatilization phenomenon is better understood and quantified.

Table 2 Comparison of 1985/86 season concentrations for selected chlorohydrocarbons in Agassiz Ice Cap as measured in 1986 and again in 1987

compound	concentration ($pg\ L^{-1}$) (mean ± standard deviation)		
	1986 (n=5)	1987 (n=3)	residue (%)
lindane	4080 ± 909	128 ± 43	3
α-HCH	6576 ± 1441	497 ± 90	8
heptachlor epoxide	358 ± 55	47 ± 15	13
chlordane	760 ± 89	40 ± 26	5
dieldrin	1346 ± 238	213 ± 75	16
α-endosulfan	1094 ± 156	ND	nil
PCB	972 ± 344	648 ± 118	67

Supporting evidence for revolatilization from the snow in the Arctic comes from Hargrave et al. (1988) who detected HCB in snow collected in May-June, 1986 (33 $pg\ L^{-1}$) but not in snow from the same area in August-September. HCB has the highest vapor pressures and one of the lowest water solubilities of the compounds listed in Table 1 and thus it is readily revolatilized from the snow. At 0°C, more than 99% of HCB was found in the vapor phase in air by Bidleman et al. (1987). Similarly, chlordane was found at 91 $pg\ L^{-1}$ in the snow sampled in May-June while it was undetectable (<1 pg L^{-1}) in August (Hargrave et al., 1988). The quantity of DDT and lindane in surface runoff waters in agricultural river systems and plots in the Soviet Union were less than 1 % of the quantity present or seeded to the snowpack (Bobovnikova et al., 1978). Although it was not considered in the original work, volatilization from the snow may have been an important loss term.

Since the combustion of fossil fuels in the northern hemisphere increases during the winter season, it is to be expected that the products of combustion would be common in snow of even remote regions. The Arctic can be affected by this increase in fossil fuel combustion as a result of long range atmospheric transport. In winter, driven by the Siberian High pressure zone, air flows mainly from the Eurasian continent into the Arctic and then out over the North American continent or into major cyclonic regions in the Aleutians and off southern Greenland. Eastern North American and European emissions commonly reach the Arctic only in late winter and spring Barrie et al., 1981).

The sampling and analysis of PAHs is considerably more difficult than for OCs especially in the Arctic where local conditions and the equipment associated with sampling (e.g., aircraft) can indeed be sources. Edwards (1983) notes that the extraction and recovery of PAHs is problematic, resulting in underestimates and considerable sample variability.

Tuominen et al. (1988) demonstrated that the largest proportion of PAHs in the atmosphere in Finland was associated with particulate matter and that precipitation, specifically snowfall, was an efficient scavenger of particulate associated PAH. Czuczwa et al. (1988) identified naphthalene in precipitation in Switzerland with higher concentrations in snow and winter rain than rain events of other seasons. Concentrations of the higher molecular weight PAHs (phenanthrene, pyrene and benzo(a)pyrene) were higher in snow than in winter rain which in turn had higher concentrations than spring and summer rain. The distribution of odd and even numbered n-alkanes in precipitation in Switzerland indicate that fuel combustion sources are dominant in the winter while natural biogenic sources are dominant in the summer (Leuenberger et al., 1988). These authors also noted that snow samples contained five times as much PAH in the particulate phase as winter rain samples. This difference was thought to be explained by the higher particle scavenging efficiencies of snow compared to rain.

Leuenberger et al. (1988) indicate that gas scavenging of PAHs is temperature dependent and decreases by a factor of approximately 2 for every 10°C increase in temperature. This explains the relatively lower gas scavenging ratios of summer rain compared to winter rain but temperature dependency of snow scavenging has apparently not been investigated. At polar and sub-polar latitudes, the development of ice fog could be an important deposition mechanism of PAHs generated locally as well as those transported over long distances in the atmosphere. Bidleman (1988) argued that the proportion of aerosol bound PAHs increased inversely with temperature and thus both wet and dry deposition were likely to be important at ambient northern temperatures. The air/snow/air exchange of PAHs, as discussed above for OCs, has apparently not been addressed in the published literature.

Fresh snow in temperate and polar regions is evidently a scavenger of large quantities of organic micropollutants from the atmosphere. However, it appears that relatively little is known about the factors controlling scavenging efficiency and the fate of these compounds upon deposition.

2.3 Fate During Snow-Melt

There has been little study of the fate of organic micropollutants during snow-melt, probably in part a result of the few measurements in snow accumulations. The only laboratory investigation of the dynamics of trace organics during snow-melt found by this author in the published literature is that of Schondorf and Herrmann (1987). These authors considered the preferential elution of compounds with a range of physico-chemical properties using metamorphosed snow from Bavaria.

Theoretically, the metamorphosis of snow will change and reduce the surface of snow flakes resulting in the transfer of pollutants in either the dissolved or particulate phase to the surface. Subsequently, the first melt is able to carry off most of the dissolved pollutants. Organic compounds with high water solubility will readily be taken with the meltwater front whereas those that are adsorbed stay behind in the snowpack. Although solubility increases with temperature, the temperature range of meltwater will be small and this effect should be minor.

Schondorf and Herrmann (1987) confirmed that high concentrations of HCH isomers, with relatively high water solubilities (see Table 1), occurred in the first and final meltwater fractions. The melt water in between the first and final fractions contained nearly no HCH. More than 90 % of the mass of PAH compounds remained in the snow column and was only released with the final melt water, suggesting particle adsorption. The cumulative elution of ions and organic micropollutants is illustrated in Figure 1.

Schondorf and Herrmann (1987) also observed that a deeper snowpack shows a greater difference in concentrations with meltwater fractions, especially when the melt is rain induced. Polar snowpacks are likely comparable to the deeper, high altitude snowpack considered by these authors with frequent freeze and thaw cycles but this apparently has not been studied.

Simmleit and Herrmann (1987) considered empirically the transport of organic micropollutants in melting snow in karst water systems in Bavaria. During February, the 24 cm snowpack that had accumulated in January began to melt. Measurements included bulk snow samples on a daily basis, precipitation at weekly intervals and water samples from the karst springs. Typically for karst systems, the spring water was particulate free, except for a few hours during peak flows. This increase of particulate matter coincided with a sudden increase in benzo(a)pyrene and benzo(ghi)perylene concentrations. The major water soluble compounds such as HCH and fluoranthene responded more directly to discharge during the first melt fraction and were concluded to be transported in a dissolved state.

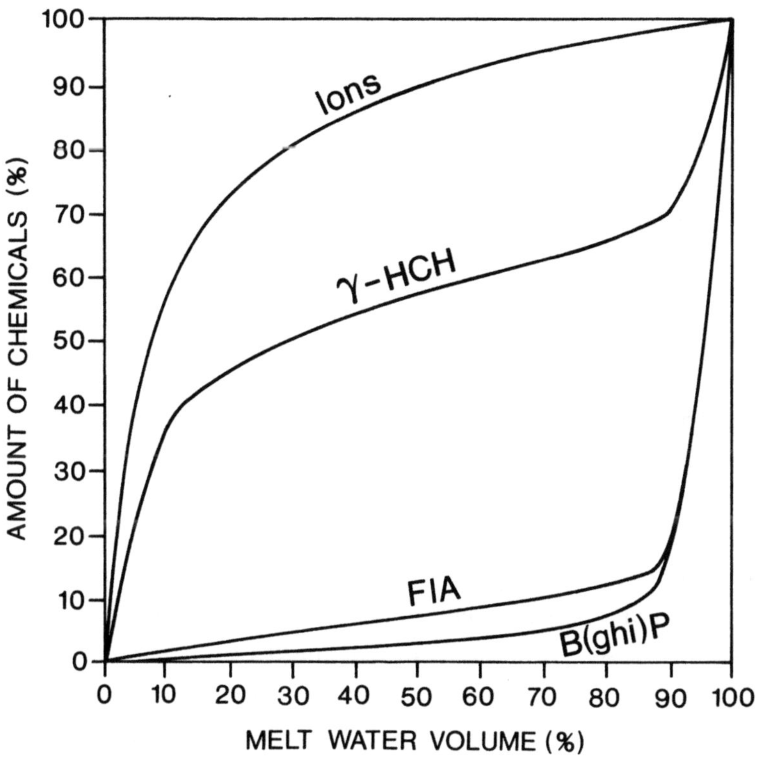

Figure 1 Cumulative elution of ions and organic micropollutants from snow-melt (redrawn from Schondorf and Herrmann, 1987)

The work of Simmleit and Herrmann (1987) also showed that 88 to 96% of the PAHs and HCHs were retained in the karst system. These calculations were based on complete retention of the compounds in the meltwater and conceivably, a portion of the high vapour pressure HCH compounds were revolatilized (see above) as opposed to being lost in the karst system. The interaction of the melting and revolatilization of at least high vapour pressure compounds in snow, especially in regions where sublimation may also be an important term in snow wasting, is a subject that needs careful investigation.

3. Measurements in Snow

3.1 Temperate and polar regions

3.1.1 Chlorinated industrial organics and pesticides

Pioneering work was undertaken in the Antarctic by George and Frear (1966) on snow samples collected at the South Pole and at Byrd Station (80°S 120°W) during November and December, 1964 and January, 1965. DDT was not detected in any of the 1 L water equivalent surface rime frost and snow samples with a detection limit of 1.0 $\mu g\ L^{-1}$. The British Antarctic Survey (Peel, 1975) reported DDT residues of the order of 100 to 2,000 pg L^{-1} in antarctic snow for samples representing the period of snowfall between 1965 and 1969. These results questioned those of Peterle (1969) which were considerably higher. Although twenty years have past, this work has apparaently not been pursued in the Antarctic until very recently.

The earliest known reported work on organochlorine compounds in the Canadian Arctic was undertaken in 1970 as part of the Ice Field Ranges Research Project (IRRP) in the Yukon Territory (Stengle et al., 1973). Sampling for subsequent analyses of DDT was undertaken at an elevation of 5,364 m on Mount Logan. Nineteen samples were taken in snow at depths of 1 to 15 m with DDT not detected in any of these samples at a detection limit of 5 ng L^{-1}.

The earliest atmospheric deposition measurements for OCs in northern Sweden were conducted in 1972 and 1973 using a silicone oil impregnated nylon mesh (Larsson and Okla, 1989). These data (see Table 3) were compared to measured depositions from a similar study conducted in 1984-1985 and it was concluded that the overall fallout of ΣDDT (p,p'-DDT plus p,p'-DDE) decreased during the intervening period. The ratios of DDT to DDE were found to have decreased from 4.10 (n=17) in 1972-1973 to 2.64 (n=52) in the more recent study. Since DDT degrades to DDE in ultraviolet light in the atmosphere and by the metabolism of organisms, it was concluded that the restrictions on DDT use in

Sweden in 1971 had been effective and that the DDT reaching Sweden presently has "aged" during long range transport from distant source regions. The trend of PCB concentrations over this time period was more complicated having decreased slightly in the northern part of Sweden. Other research in Scandinavia includes that of Lunde et al., (1977) and Haglund et al. (1987).

Concentrations in snow in temperate regions (not including Norway and Sweden) are limited. Probably best known of this work is that conducted on pesticides and PAHs in Bavaria (Schrimpff et al., 1979; Herrmann, 1981; Simmleit and Herrmann, 1987 and Schondorf and Herrmann, 1987). Some of these studies have been discussed above.

Strachan and Huneault (1979) collected accumulated snow from 17 remote sites throughout the Canadian Laurentian Great Lakes drainage basin in February, 1976. With the exception of PCBs and methoxychlor, all of the substances examined were at lower concentrations than in rainfall, collected as part of the same study as event samples between May and November. These concentrations are compared to Canadian arctic snow chemistry data in Table 7. The most noticeable differences between the results of these two studies are the much higher concentrations of PCBs and DDT in the snow of the Great Lakes region. Alpha-endosulfan and dieldrin, while present in the Arctic, were not detected by Strachan and Huneault (1979) although this is likely an artifact of detection limits.

An investigation of organic micropollutants at 28 sites in the Great Lakes region of Canada (International Environmental Consultants Ltd. [IECL], 1979) revealed only a limited number of samples with concentrations of PCBs and OC pesticides greater than detection limit (10 ng L^{-1} and 1 ng L^{-1}, respectively). Murphy and Schinsky (1983) have considered PCBs in snow at selected sites in the US portion of the Great Lakes basin. Trace organic pollutants in snowfall have also been studied in southern California and Oahu, Hawaii by Kawamura and Kaplan (1986) and Bevenue et al. (1972). Pesticide concentrations in the snow from Mauna Kea Summit (about 4,200 m) were 15, 2 and 3 ng L^{-1} for p,p'-DDT, dieldrin and lindane, respectively (Bevenue et al., 1972).

Table 3 Contaminant deposition data for northern Sweden and Moscow compared with Canadian arctic data ($\mu g\ m^{-2}\ yr^{-1}$)

compound	Moscow[a]	Sweden			Arctic[d]
	1978-79	1972-73[b]	1974-76[c]	1984-85[b]	1986-87
alpha-HCH	14				0.02-9.1
lindane	12				0.01-2.0
DDT	52	≈1.2-13.2		≈0.1-8.79	0.01-3.0[e]
PCB		≈1.2-16.6	24-60	≈24-154	0.04-0.6

[a] from the Moscow area as reported by Bobobnikova and Dibtseva (1980)

[b] estimated from figures published by Larsson and Okla (1989) for stations north of 60°N only. Actual data were reported for a monthly time period and these data have been multiplied by 12 to provide a comparable annual value.

[c] from Lunde et al. (1977) only for stations north of 60°N

[d] from Barrie et al. (in press) and determined from annual snowpack samples representing September to May and here assumed to represent the annual deposition

[e] includes only 1987 data

Although still limited, recent work involving snow has focussed on the Canadian Arctic. The Finnish Huure Expedition to the North Pole in 1984 collected a series of snow samples of 1 litre of water equivalent for analyses of a broad range of chlorinated pesticides and chlorophenols (Paasivirta and Paukku, 1987). None of the target compounds, which included OC pesticides, were quantified above the detection limit of 0.5 ng L^{-1} in these snow samples.

In 1979, McNeely and Gummer (1984) undertook snow sampling in the area of east-central Ellesmere Island. Sampling included snowpack at 5 sites, fresh snow at 3 sites and rime frost from 2 sites between May 24 and June 19, 1979, yielding 19 samples with replicates and repeat analyses for a total of 28 analyses. The results for organochlorine pesticides are summarized and compared to selected data from the much more extensive studies of Gregor and Gummer (1989) and Gregor et al. (in press) in

Table 4. The two stations selected for comparison were Alexandra Fiord sampled in 1986, the only station overlapping between the two studies, and the surface snow for both 1986 and 1987 from the Agassiz Ice Cap, approximately 175 km to the north. Although the results for Alexandra Fiord for 1986 were below the level of detection available to McNeely and Gummer, the concentrations observed at the Agassiz Ice Cap support concentrations of the magnitude found in the earlier study, at least for HCHs.

Table 4 Summary of pesticide snow chemistry data collected in 1979 by McNeely and Gummer (1984) from east-central Ellesmere Island compared with selected data from Gregor and Gummer (1989) for Alexandra Fiord (1986), and the surface snow layers from Agassiz Ice Cap in 1986 and 1987

pesticide	1979 range ($ng\ L^{-1}$)	confirmed detections[a]	unconfirmed detections[b]	Alexandra Fiord[c] ($ng\ L^{-1}$)	Agassiz L-1[d] ($ng\ L^{-1}$)	Agassiz L+1[c] ($ng\ L^{-1}$)
α-HCH	<1 - 18	21	5	0.43	6.7	19.1
lindane	<1 - 8	10	5	0.22	4.4	6.8
chlordane	<1 - 4	3	7	0.17	0.74	0.23
DDT	<1 - 6	4	10	0.07	0.01	0.13
α-endo sulfan	<1 - 2	1	2	0.24	1.1	1.1
dieldrin	<2 - 4	5	12	0.83	1.4	1.0

[a] Includes both detections above the detection limit and below the detection limit but confirmed by MID-GC/MS
[b] Indicate instrument responses below the detection limit but ofinsufficient concentration to permit confirmation by MID-GC/MS
[c] Single sample
[d] Represents the mean of 5 samples from this layer

Concurrent with the 1986 and 1987 snow chemistry survey of Gregor and Gummer, Arctic Laboratories Limited (1988) undertook work at the Canadian Ice Island during 1986 and 1987. Good agreement was found between the results of these two studies (Table 5) given the very low concentrations of these samples.

Table 5 Comparison of 1987 Ice Island samples for organochlorin residues as reported by Arctic Laboratories (1988) wit results from Gregor and Gummer (1989) based on 3 replicates for each laboratory

	concentration ($pg\ L^{-1}$) in melted snow	
compound	Gregor & Gummer (mean ± 1SD)	Arctic Laboratories (mean ± 1SD)
Aroclor (1254)[a]	129 ± 37	185 ± 160
hexachlorobenzene	11 ± 6	12 ± 3
heptachlor	<3[b]	6 ± 3
aldrin	Int	0.5 ± 0.5
o,p'-DDE	ND	2.8 ± 2.4
p,p'-DDE	ND	20 ± 14
Mirex	ND	<4.0
trans-chlordane	8 ± 1	20 ± 15
cis-chlordane	<12[c]	25 ± 18
p,p'-DDT	23 ± 19	<47 ± 26
heptachlor epoxide	30 ± 11	46 ± 26
alpha-endosulfan	21 ± 1	97 ± 58
dieldrin	54 ± 5	<140 ± 55
endrin	spike	27 ± 11
methoxychlor	55 ± 35	37 ± 26

Abbreviations: SD = standard deviation
Int= interference on chromatograms
ND = non-detectable

[a] For this comparison, the PCB data from Gregor and Gummer are reported as Aroclor 1254 whereas in all other Gregor and Gummer references in this report PCBs are reported as the 1:1:1 mixture of Aroclor 1242, 1254 and 1260.

[b] One of the three samples was ND. Therefore the concentration is reported as less than the maximum concentration observed with no standard deviation.

[c] Two of the three samples were ND. Therefore the concentration is reported as less than the concentration of the single value with no standard deviation.

HCH data for 1987 Ice Island snow samples are compared fo three different laboratories using three different sampling method in Table 6. The University of South Carolina (USC) results ar relatively high. The dominance of lindane in the USC results als does not agree with the other analyses and is not supported by HC ratios observed in the arctic atmosphere (Patton et al., 1989) The very low concentrations in the snow, and the inevitable desir or need to compare spatially and temporally among different studie underlie the obvious need for more extensive inter-laborator evaluations as part of future research.

Toxaphene measurements in the arctic snowpack are restricted to four samples. Two of these were 1985/86 snowpack accumulations collected from the Canadian Ice Island in May 1986 and two others were collected from the Devon Ice Cap on Devon Island (75° 6'N, 82° 45'W) and from the vicinity of Resolute Bay on Cornwallis Island (74° 42'N, 94° 54'W) in May, 1987. Concentrations for the two Ice Island samples were 85 and 307 pg L^{-1} while the Resolute Bay and Devon Ice Cap samples were 600 and 1700 pg L^{-1}, respectively (Bidleman et al., 1989). The wide range in concentration between the three stations was not surprising given the distance between these stations, the difference in elevation (the Ice Island and Resolute Bay locations are essentially at sea level while the Devon Ice Cap site is nearly 1,700 m above sea level) and the different snow accumulation rates.

Table 6 Comparison of 1987 Ice Island snow samples for hexachlorcyclohexane (HCH) residues among Gregor and Gummer (1989), Arctic Laboratories (1988) and University of South Carolina (USC) as reported by Arctic Laboratories

	concentration (pg L^{-1}) in melted snow		
	Gregor & Gummer[a] (mean ± SD)	Arctic Labs[b] (mean ± SD)	USC[c] (mean ± SD)
α-HCH	143 ± 13	290 ± 130	315 ± 25
lindane	83 ± 9	192 ± 96	365 ± 15

[a] Based on 3 samples collected June 6, 1988
[b] Based on 3 samples collected June 8,9 and 15, 1988
[c] Based on 2 samples collected June 7 and 9, 1988

The results of precipitation samples collected in the Moscow region of the USSR from January 1978 to September 1979 have been reported by Bobovnikova and Dibtseva (1980) for each month.

Although the authors have not indicated specifically which monthly samples were snow, it was reasonable to assume for this latitude that samples from December through March were primarily snowfall. Based on these four months, it can be estimated that ΣDDT ranged from 76 to 110 ng L^{-1}; α-HCH ranged from "trace" to 25 ng L^{-1}; and lindane ranged from "trace" to 30 ng L^{-1}. ΣDDT was taken here as the sum of p,p'-DDT, o,p'-DDT, p,p'-DDD and p,p'-DDE as reported by Bobovnikova and Dibtseva (1980) of which p,p'-DDT and p,p'-DDE were the dominant compounds. The detection limit for any of these analyses was not stated. The concentrations of α-HCH were similar to those noted for the Canadian Arctic (Gregor and Gummer, 1989) while lindane tends to be somewhat higher, likely reflecting relative proximity to sources. ΣDDT was two orders of magnitude higher in the Moscow snow samples than the arctic snow samples (Gregor and Gummer, 1989).

Lindane, in daylight, can be transformed to the α-HCH isomer with a half life, dependent upon temperature and other conditions, of the order of one year (Oehme and Ottar, 1984). Although degradation rates are likely substantially different in a polar climate than a temperate one, this ratio may be informative with respect to possible age or length of time of HCH transport. Also, chemical formulations differ. North America and western Europe use pure lindane. Aerosol ratios are reported at 1:7 (α-HCH:lindane) in the Great Lakes region (Eisenreich et al., 1981) while European ratios tend to be 1:3-4 (Oehme and Mano, 1984).

HCH ratios (α-HCH:lindane) in the air at the Ice Island, which appear to be indicative of the Canadian North, were 6:1 and 15:1 in the spring and summer, respectively, of 1986 (Hargrave et al., 1988). Based on Table 2, the ratio for surface snow from the Agassiz Ice Cap was 2:1 in 1986 and 4:1 in 1987. The average ratio for all surface snow samples reported by Gregor and colleagues (Gregor and Gummer, 1989; Gregor, 1990) in 1986 and 1987 was 2.5:1. These low ratios in the snow relative to the atmosphere, are thought to be related to the higher vapor pressure of α-HCH relative to lindane. Lindane, having a higher vapor pressure, will be more likely to be particle associated and thus will be preferentially scavenged by snowfall. Also, α-HCH may be preferentially revolatilized from the snowpack resulting in a lower isomeric ratio than observed in fresh snow and the atmosphere.

3.2.2 Polycyclic aromatic hydrocarbons

Rural, mountainous area snow samples from California collected in 1982 showed that of the total organic substances identified, more than 80% originated from natural plant waxes of biogenic origin. Accordingly, anthropogenic organic compounds were minor constituents of the rural snow samples (Kawamura and Kaplan, 1986).

PAHs have been measured in fresh snow samples in Bavaria, F.R.G. (Schrimpff et al., 1979; Schrimpff, 1980; Herrmann, 1981; Schondorf and Herrmann, 1987 and Simmleit and Herrmann, 1987) and in Switzerland (Giger et al., 1987). In a comparative, seasonal study of snow and rain near an urbanized area of Switzerland, Leuenberger et al. (1988) and Czuczwa et al. (1988) found that alkylbenzenes were the dominant volatile organic component with snow having higher concentrations (up to 10 $\mu g\ L^{-1}$) than rain at any time throughout the year. Based on the mean of five snow samples, the rank order of total (dissolved plus particulate) concentrations of the major compounds were - fluoranthene > phenanthrene > pyrene > Σmethylphenanthrene > chrysene > benzo(b+j+k)fluoranthenes. These six compounds comprised more than 80% of the mean total PAHs in the snow. Naphthalene concentrations were higher in winter than other seasons with winter rain concentrations higher than snow concentrations. Several factors are believed responsible for the higher winter concentrations. First and quite obvious is the higher hydrocarbon emissions due to domestic heating, Second, snow has a scavenging efficiency of particulates an order of magnitude greater than rain. Finally, there are more opportunities for dilution in the atmosphere in the summer time due to the increased atmospheric mixing heights in summer, the increased volatility of the compounds with summer temperatures and the increased photoreactivity or hydrolosis in the summer (Leuenberger et al.,1988; Czuczwa et al., 1988).

Arctic snow samples collected in 1987 have been analyzed by this author (unpublished data) for 17 PAHs by capillary gas chromatography with electron impact, multiple ion monitoring mass spectrometric detection (MIM-GC/MSD). ΣPAH concentrations were highly variable ranging from non-detectable to 125 ng L^{-1}. The seven PAHs detected consistently were phenanthrene, 1-methylnaphthalene, 2-methylnaphthalene, 1,2,3,4-tetrahydro-naphthalene and indene with generally small quantities of fluorene and/or acenaphthene. All seven of these PAHs are coal tar derivatives, practically insoluble in water (although having higher solubilities than many of the OCs) and of relatively low molecular weights for PAHs (Table 1). These PAHs are mildly toxic with some having a tendency to bioaccumulate, to be slightly carcinogenic and even to be mutagenic. Some, such as acenaphthene and 1,2,3,4-tetrahydro-naphthalene have commercial uses either alone or in conjunction with other products (Windholz, 1983). Compounds commonly associated with internal combustion exhausts (e.g., fluoranthene and pyrene which were dominant in the snow samples from Switerland [see above]) were either not detected or were detected only occassionally and were believed to be sample contamination. Phenanthrene, also a combustion product, was present in the Arctic. Relative to other PAHs, it appears that it is the lower molecular weight, high vapor pressure, more water soluble PAHs that are being found in the arctic snow, but the long range transport of PAHs to the Arctic requires a great deal more research.

3.3 Contaminant flux and trends

The comparison of contaminant concentrations in precipitation is frequently trivial without consideration of the total quantity of precipitation as required in determining flux. Median winter deposition rates for selected organochlorines based on concentrations in Canadian arctic snowpack in the spring of 1986 and 1987 are presented in Table 7. Although these are shown as winter deposition rates, it is reasonable to assume that these represent for all intents and purposes, annual deposition in the Arctic. There are two reasons supporting this assumption. First, approximately 80% of the total precipitation in the high Arctic arrives as snowfall (Woo et al., 1983). Second, the period of snow accumulation at these arctic stations generally ranges from September to May and this encompasses the period during which the arctic aerosol has been shown to be most heavily contaminated with anthropogenic pollutants (Barrie and Hoff, 1985; Barrie et al., 1985; Barrie et al., 1989) due to the tropospheric circulation patterns typical of the winter season.

These flux estimates are limited by the fact that the snowpack samples are commonly collected from remote sites and total annual snowfall at the specific sampling site has to be assumed to be equal to that recorded at the nearest weather station which may be several hundreds of kilometres distant. In addition, total snow storage in a basin generally exceeds weather station records by as much as from 130 to 300% (Woo and Marsh, 1977; Woo et al., 1983). It must also be remembered that deposition to the snowpack does not necessarily mean a direct transfer of this quantity of contaminant to the aquatic system following snow-melt as revolatilization may occur (see section 2.2).

Published data from the USSR appear to be limited to the annual fluxes in rainfall and snowfall provided by Bobovnikova and Dibtseva (1980) for the Moscow area. Fluxes there were estimated to be about 52 $\mu g\ m^{-2}\ yr^{-1}$ for ΣDDT (p,p'-DDT + o,p'-DDT + p,p'-DDD + p,p'-DDE) and 14 and 12 $\mu g\ m^{-2}\ yr^{-1}$ for α-HCH and lindane, respectively, for the 1978 calendar year. The flux of α-HCH is similar to the maximum of the range for the Arctic while lindane is about 6 times greater and ΣDDT is about 10 times greater in the Moscow region than in the Canadian Arctic. As noted in the discussion of concentrations, the Moscow data likely show much closer proximity to source regions. However, the Moscow data precede the arctic data by nearly 10 years and conditions could have changed considerably during this time due to different pesticide formulations, application procedures and quantities used.

Table 7 Median and range of concentrations and corresponding winter season deposition of trace organic compounds from Canadian arctic snow chemistry surveys in 1986 and 1987 compared with 1976 snow chemistry data from the Laurentian Great Lakes

compound	1986 (n=12) con-centration (pg L^{-1})	1986 (n=12) winter deposition (ng m^{-2})	1987 (n=15) con-centration (pg L^{-1})	1987 (n=15) winter deposition (ng m^{-2})	Great Lakes snow concentration (pg L^{-1})[a]
PCB	745 (20-1760)	73	907 (257-1773)	95	29000
alpha-HCH	3185 (430-8120)	285	910 (143-42700)	95	100
lindane	1160 (220-4446)	104	700 (83-10500)	73	900
heptachlor epoxide	97 (ND-362)	10	117 (ND-1600)	16	NA
alpha-endosulfan	345 (100-1094)	37	175 (ND-4880)	24	ND
dieldrin	540 (205-1402)	58	400 (ND-4420)	56	ND
HCB	28 (10-70)	3	16 (ND-104)	2	ND
ΣDDT	18 (10-50)	2	93 (ND-1380)	8	100
ΣCBZ	NA	NA	240 (ND-915)	30	NA
ΣPAH	NA	NA	19000 (190-149000)	4246	NA

NA = not available
ND = not detectable
HCB = hexachlorobenzene
ΣDDT = sum of all detected DDT compounds and metabolites
ΣCBZ = sum of all detected chlorobenzene compounds targetted
[a]from Strachan and Huneault (1979) reported as a mean of 17 samples

Rapaport <u>et al.</u> (1985) analysed snow and rain samples collected at Minneapolis, Minnesota, USA from 1981 through 1983 for p,p'-DDT. Concentrations in the snow ranged from 100 to 550 pg L^{-1} in 1981/82; 400 to 1,000 pg L^{-1} in 1982/83 and from 100 to 210 pg L^{-1} in 1983/84. Mean annual fluxes of p,p'-DDT for snow reported by these authors averaged 0.2, 0.4 and 0.13 $\mu g\ m^{-2}\ yr^{-1}$ for 1981/82, 1982/83 and 1983/84, respectively. The time interval and the variability present in these data preclude any assessment of trends.

To date there has been limited conclusive research regarding trends of trace organic contaminant deposition in the Arctic despite the extensive work conducted on glacial records for other anthropogenic contaminants. Periodic measurements in air and fresh snow have been too restricted temporally to provide any time series information with the exception of the Swedish study of Larsson and Okla (1987) noted in section 3.2. Stengle <u>et al.</u> (1973) sampled at depth on Mount Logan but could not quantify DDT. The extensive research on Greenland ice sheets has apparently not been extended to include trace organic contaminants. Thus the only research that has been directly concerned with determining temporal trends of OC deposition is that on the Agassiz Ice Cap on north-central Ellesmere Island.

The Agassiz Ice Cap has been sampled at a remote site at an elevation of approximately 2,000 m above sea level in both 1986 and 1987 to include 17 years of samples between 1969/70 and 1986/87 (Barrie <u>et al.</u>, in press; Gregor <u>et al.</u>, in press). Intuitively, this cold glacier should provide a credible historical record of anthropogenic pollutant deposition because (i) average summer melt affects only about 3% of the winter snow layer resulting in negligible redistribution of ions between adjacent snow layers (Barrie <u>et al.</u>, 1985) and (ii) annual snow accumulation at this latitude and elevation represents, on average, more than 75% of the total annual precipitation (Woo <u>et al.</u>, 1983). The seasonal specific conductance record of the firn (Figure 2) is preserved and is consistent with that reported by Barrie <u>et al.</u> (1985). As noted by these authors, this profile reflects the trace constituent composition of the atmospheric aerosol which undergoes a strong seasonal variation with maximum ion concentrations and therefore specific conductance during the arctic winter (Hoff and Barrie, 1986; Rahn and Shaw, 1982). In general, it appears that annual inventories of chlorohydrocarbons in the firn at the Agassiz Ice Cap have decreased over the period of record as follows:

- α-HCH	from >300 to 70 ng m^{-2};
- lindane	from 90 to <20 ng m^{-2};
- heptachlor epoxide	from 20 to 5 ng m^{-2};
- dieldrin	from >50 to 20 ng m^{-2}; and,
- PCB	from >300 to 100 ng m^{-2}.

The consistent decline among all compounds is quite remarkable. Support for a decrease in the supply of chlordane and α-endosulfan to the Agassiz Ice Cap is provided by the less frequent detection of these compounds in recent years.

However, as noted in section 2.2, the firn does not seem to be an effective permanant trap of the semi-volatile trace organic compounds, with the possible exception of PCBs. The residues in the firn therefore are not quantitatively representative of the annual flux of these pesticides to the glacier. Thus the residue trends seen in the glacier for lindane, α-HCH, heptachlor epoxide and dieldrin are not solely related to a trend in the annual loadings, but rather are being controlled to an unknown degree by meteorological conditions influencing the snow/atmosphere exchange. Notwithstanding the absence of confirmatory data regarding the retention of PCBs in the glacial firn, Gregor et al. (in press) conclude that, at the present time, only the trend for PCBs has any relevance to a real trend in annual deposition to the ice cap. Although the residue masses of PCBs in the glacier will be influenced somewhat by the annual variability of summertime losses, it is conceivable that the actual annual deposition is the dominant factor controlling the trend for PCBs, as was the case in 1985/86. Nonetheless, meteorologic conditions that control atmospheric transport to the Arctic and deposition to the snow vary from year to year. Thus, it is necessary to consider this variability and to attempt to adjust for it.

Gregor et al. (in press) normalized the PCB trend for annual meteorologic variability with ^{210}Pb flux measurements which were also determined for the glacier. The radioactive isotope ^{210}Pb is derived from the decay of ^{222}Rn which emanates from the earth's surface where it is produced as a member of the ^{238}U decay series. ^{222}Rn decays in the atmosphere to ^{210}Pb with a half life of 3.8 days. The charged daughter ions produced are chemically reactive and soon become irreversibly associated with aerosol particles (Robbins, 1978).

Since the source of ^{210}Pb (i.e., the earth's surface) remains more or less constant with time, it can be hypothesized that the deposition of ^{210}Pb to the glacier would be uniform from year to year if the air mass movements and the complex interaction of factors controlling transport and deposition to the glacier remain constant. Thus, the differences of measured decay corrected ^{210}Pb inventories in the snow layers can be assumed to represent, simplistically, the year to year differences in atmospheric transport and deposition of terrestrially derived anthropogenic pollutants. Under these assumptions, any trends in these adjusted trace organic substance residue data should represent real changes in the flux of contaminants to the glacier.

The ^{210}Pb normalized PCB inventories in the annual layers of the Agassiz Ice Cap are compared to measured fluxes in Figure 3. It is evident that the years having the greatest measured flux to the snow of PCBs generally coincided with high ^{210}Pb flux resulting in a much smoother normalized or adjusted trend line. These ^{210}Pb corrected inventories imply that after variable meteorological conditions have been considered, the annual normalized inventory of PCBs in the glacial layers have remained essentially constant since 1976/77 through to 1985/86 at a level approximately one half of that which prevailed between 1970/71 and 1975/76.

This decline in ^{210}Pb normalized PCB inventories in the glacier agrees in general with trends in PCB body burdens in ringed seals from the western Arctic between 1972 and 1981 (Addison and Zinck, 1986) and for the Lancaster Sound area between 1975-76 and 1984 (Muir et al., 1988). However, correspondence among these limited data may be merely fortuitous and a great deal more work is required to thoroughly evaluate trends in contaminant deposition in the Arctic.

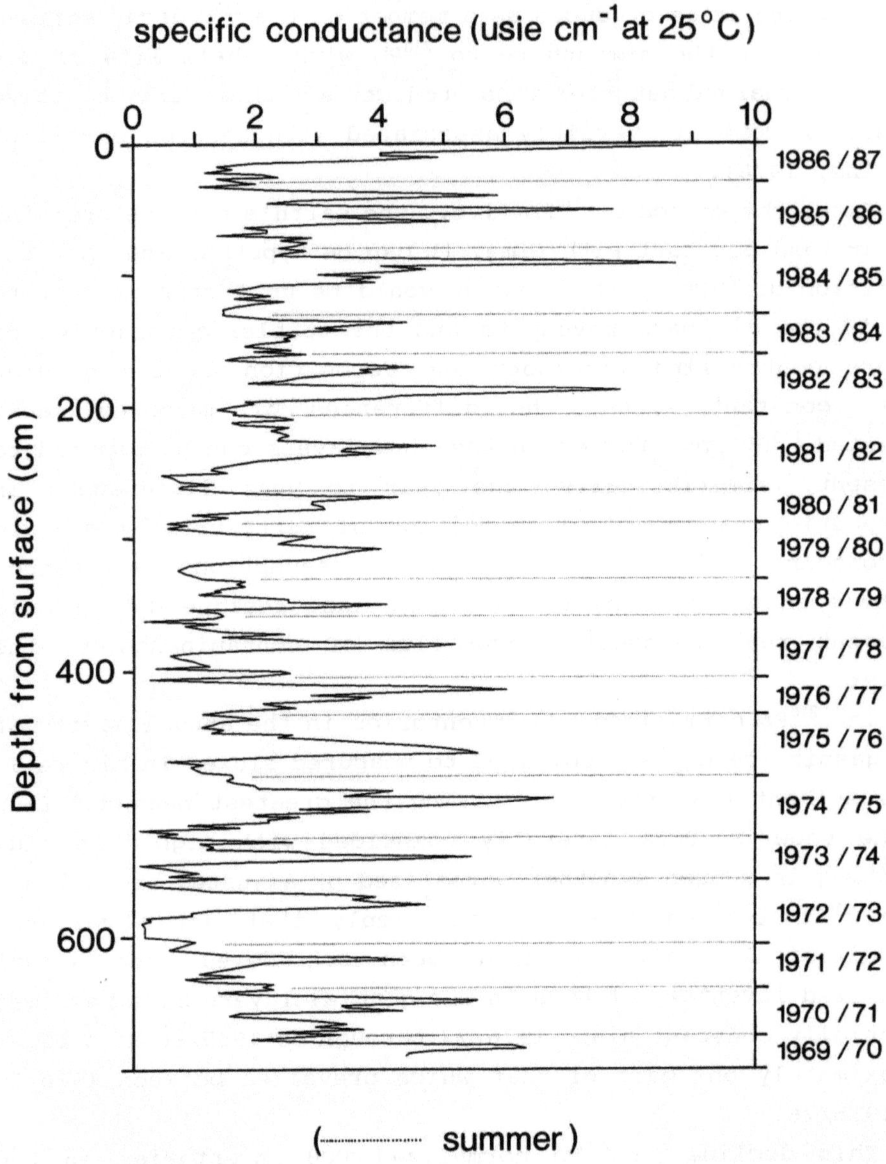

Figure 2 Plot of specific conductance versus depth for the Agassiz Ice Cap from the snow surface in the spring of 1987 to a depth of 6.9 m.

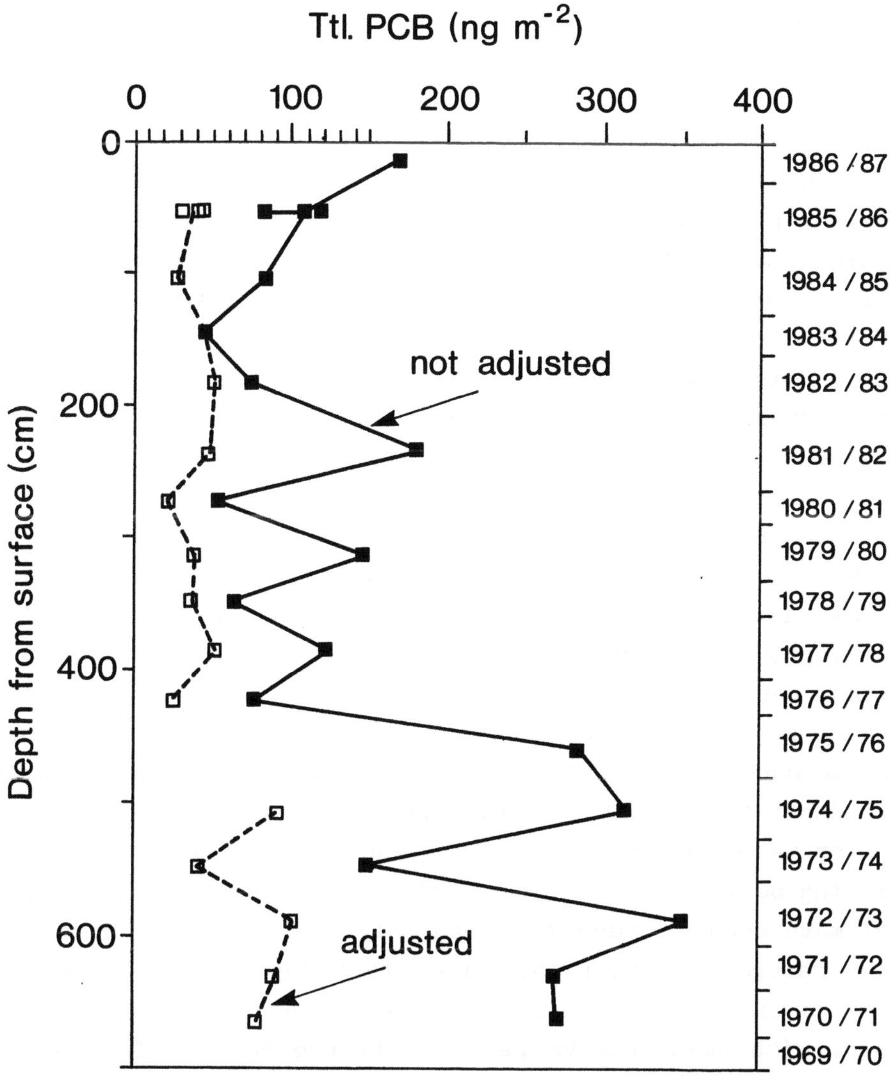

Figure 3 PCB inventories in annual layers of the Agassiz Ice Cap from the winter of 1970/71 to the winter of 1986/87 compared to the ^{210}Pb normalized inventories which are interpreted as the trend compensated for variable annual delivery resulting from variable meteorologic conditions

4. Summary and Conclusions

The aquatic environment and the atmosphere have received considerable research over the past several decades for the purpose of measuring the extent of environmental contamination by organic micropollutants. However, one of the major sources of precipitation in temperate and polar regions and a major atmospheric scavenger, snowfall, has to a large extent been ignored. Information on the sources, atmospheric pathways, present deposition rates, deposition trends and processes controlling the fate of organic micropollutants in snow is sparse.

Actual processes of deposition and re-volatilization from snow has received only limited attention. While deposition has been confirmed in a variety of concentration and flux measurements throughout the world, only recently has evidence been put forward that large percentages of the semi-volatile compounds may be revolatilized to the atmosphere from the snow as temperatures rise. These processes warrant more research under temperate and polar conditions.

Despite extensive work utilizing glacial ice cores for monitoring pollutant trends for such concerns as acid deposition and climate change, there has been only limited research on the glacial record of organic micropollutant deposition. The only glacier in the northern hemisphere that has been studied for this purpose, with quantitative results, is the Agassiz Ice Cap on Ellesmere Island. This research, although limited to 17 years of record at this time, indicates first that the glacial organochlorine pesticide residues are not directly related to actual deposition but are largely a function of variable meteorologic conditions. Only the PCB record appears relatively intact and these data, after correction for annual meteorologic differences, suggest that PCB deposition has decreased by a factor of approximately 2 since 1970 and has remained nearly constant over the past decade.

Clearly, more comprehensive research is required to expand the information base for organic micropollutants in snow. This research must seek a broader geographical as well as temporal coverage. Equally important though, is the research directed at understanding the processes and fate of these contaminants in temperate and polar environments and their effect on the cycling and storage of micropollutants. This must come from both laboratory and field based research. Only with this information, combined with ecological effects assessments, will an understanding of the role of snow in transporting and cycling organic contaminants be attained.

Acknowledgements

I wish to thank Dr. W. Giger of the Swiss Federal Institute for Water Resources and Water Pollution Control, Dr. M. Tranter of the Department of Oceanography, University of Southampton and Dr. Wm. Strachan of the National Water Research Institute, Environment Canada for their very helpful and knowledgeable comments. I also want to thank NATO for sponsoring this Advanced Research Workshop on Processes of Chemical Change in Snowpacks and congratulate T. Davies, P. Brimblecombe, M. Tranter, G. Jones and all others who were instrumental in making this workshop a success.

References

Addison RF, Zinck ME (1986) PCBs have declined more than DDT group residues in arctic ringed seals (Phoca hispida) between 1972 and 1981. Environ Sci Technol 20:253-256

Agarwal HC, Kaushik CP, Pillai MKK (1987) Organochlorine insecticide residues in the rainwater in Delhi, India. Water Air Soil Pollut 32:293-300

Arctic Laboratories Ltd. (1988) A study of the distribution of chlorinated hydrocarbon pesticides and PCBs in the Arctic Ocean using a Canadian Ice Island station. An unpublished report prepared for the Department of Supply and Services, Department of Indian and Northern Affairs and Department of Fisheries and Oceans by Arctic Laboratories Ltd., 2045 Mills Road, Sidney, B.C., V8L 3S1, Canada

Atlas E, Foster R, Giam CS (1982) Air-sea exchange of high molecular weight organic pollutants: laboratory studies. Environ Sci Tech 16:283-286

Baker JE, Eisenreich SJ (1990) Concentration and fluxes of polycyclic aromatic hydrocarbons and polychlorinated biphenyls across the air-water interface of Lake Superior. Environ Sci Tech 24:342-352

Barrie LA, Fisher D, Koerner RM (1985) Twentieth century trends in arctic air pollution revealed by conductivity and acidity observations in snow and ice in the Canadian High Arctic. Atmos Environ 19:2055-2063

Barrie LA, Hoff RM, Daggupaty SM (1981) The influence of mid-latitudinal pollution sources in the Canadian Arctic. Atmos Environ 15:1407-1419

Barrie LA, Hoff RM (1985) Five years of air chemistry observations in the Canadian Arctic. Atmos Environ 19:1995-2010

Barrie LA, (1986) Arctic air pollution: an overview of current knowledge. Atmos Environ 20:643-663

Barrie LA, Olson MP, Oikawa KK (1989) The flux of anthropogenic sulphur into the Arctic from mid-latitudes in 1979/80. Atmos Environ 23:2505-2512

Barrie LA, Gregor D, Hargrave B, Lake R, Muir D, Shearer R, Tracey B, Bidleman T (in press) Arctic contaminants:sources, occurrence and pathways. Sci Total Environ

Bevenue A, Ogata JN, Hylin JW (1972) Organochlorine pesticides in rainwater, Oahu, Hawaii, 1971-1972. Bull Environ Contam Toxicol 8:238-241

Bidleman T, Widequist U, Jasson B, Soderlund R (1987) Organochlorine pesticides and polychlorinated biphenyls in the atmosphere of Southern Sweden. Atmos Environ 21:641-654

Bidleman TF (1988) Atmospheric processes, wet and dry deposition of organic compounds are controlled by their vapor-particle partitioning. Environ Sci Tech 22:361-367

Bidleman TF, Patton GW, Walla MD, Hargrave BT, Vass WP, Erickson P, Fowler B, Scott V, Gregor DJ (1989) Toxaphene and other organochlorines in Arctic Ocean fauna: evidence for atmospheric delivery. Arctic 42:307-313

Bobovnikova TsI, Virchenko EP, Morozova GK, Sinitsyna ZA, Cherkhanov YuP (1978) Estimation of organochlorine pesticide loss in surface runoff waters. In: Symposium on Environmental Transport and Transformation of Pesticides. EPA-600/9-78-003, Washington, DC, p 103-107

Bobovnikova TsI, Dibtseva AV (1980) Organochlorine in precipitation. Meteorologiya i Gidrologiya 8:46-51

Buchel KH (ed) (translated by Holmwood G) (1983) Chemistry of Pesticides. Wiley Interscience Publications, New York, NY.

Canadian Council of Resource and Environment Ministers (CCREM) (1987) Canadian water quality guidelines. Environment Canada, Ottawa, Canada.

Czuczwa J, Leuenberger C, Giger W (1988) Seasonal and temporal changes of organic compounds in rain and snow. Atmos Environ 22:907-916

Deo PG, Hasan SB, Majumder SK (1980) Isomerization of ß-HCH in aqueous solution. J Environ Sci Health B15:147-164

Doskey P, Andren A (1981) Modelling the flux of atmospheric polychlorinated biphenyls across the air/water interface. Environ Sci Tech 15:705-711

Duinker JC, Bouchertall F (1989) On the distribution of atmospheric polychlorinated biphenyl congeners between vapor phase, aerosols, and rain. Environ Sci Technol 23:57-62

Edwards NT (1983) Polycyclic aromatic hydrocarbons (PAHs) in the terrestrial environment - a review. J Environ Qual 12:427-441

Eisenreich SJ, Looney BB, Thornton JD (1981) Airborne organic contaminants in the Great Lakes ecosystem. Environ Sci Tech 15:30-58

George JL, Frear DEH (1966) Pesticides in the Antarctic. J Applied Ecology 3(suppl):155-167

Giger W, Leuenberger C, Czuczwa J, Tremp J (1987) Organic micropollutants in the atmosphere: determination, origins and behavior in rain, snow and fog. EAWAG-NEWS 22/23:13-15

Gregor D, Gummer Wm (1989) Evidence of atmospheric transport and deposition of organochlorine pesticides and PCBs in Canadian arctic snow. Environ Sci Tech 23:561-565

Gregor DJ (1990) An overview of water quality research in the Canadian north. In: Ommanney S, Prowse T (eds) Proceedings of the Northern Hydrology Symposium. NHRI Science Report 1, National Water Research Institute, Environment Canada, 11 Innovation Blvd, Saskatoon, Sask, Canada, p 163-186

Gregor DJ, Dominik J, Vernet J-P (in press) Recent trends of selected chlorohydrocarbons in the Agassiz Ice Cap, Ellesmere Island, Canada. In: Shearer R, Bartonova A (eds) Proceedings of the 8th International Conference of the Comité Arctique International on Global Significance of the Transport and Accumulation of Polychlorinated Hydrocarbons in the Arctic. Oslo, Norway

Haglund P, Alsberg T, Bergman A, Jansson B (1987) Analysis of halogenated polycyclic aromatic hydrocarbons in urban air, snow and automobile exhaust. Chemosphere 16(10-12):2441-2450

Hargrave BT, Vass WP, Erickson PE, Fowler BR (1988) Atmospheric transport of organochlorines to the Arctic Ocean. Tellus 40B:480-493

Heit M, Tan YL, Klusek CS, Burke JC (1981) Anthropogenic trace elements and polycyclic aromatic hydrocarbons in sediment cores from 2 lakes in the Adirondack acid lake region. Water Soil Air Pollut 15:441-464

Herrmann R (1981) Transport of polycyclic aromatic hydrocarbons through a partly urbanized river basin. Water Air and Soil Pollution 16:445-467

Hinckley DA, Bidleman TF, Foreman WT, Tuschall JR (1990) Determination of vapor pressures for nonpolar and semipolar organic compounds from gas chromatographic retention data. J Chem Eng Data 35:232-237
Hoff RM, Barrie LA (1986) Air chemistry observations in the Canadian Arctic. Water Sci Tech 18:97-107
International Environmental Consultants Ltd (IECL) (1979) Toxic substances in winter snowpack in the Great Lakes Region. A report for Environment Canada, Burlington, Ontario
International Registry of Potentially Toxic Chemicals (IRPTC) (1983) Toxaphene. no 32 in a series "Scientific Reviews of Soviet Literature on Toxicity and Hazards of Chemicals", Izmerov NF (ed) USSR Academy of Medical Sciences, Centre for International Projects, GKNI, Moscow
International Registry of Potentially Toxic Chemicals (IRPTC) (1988) Unpublished report based on a computerized search of Registry files, United Nations Environmental Programme, Geneva, Switzerland
Jensen S, Johnels G, Olsson M, Otterlind G (1969) DDT and PCB in marine animals from Swedish waters. Nature 224:247-250
Kawamura K, Kaplan IR (1986) Biogenic and anthropogenic organic compounds in rain and snow samples collected in southern California. Atmos Environ 20:115-124
Larsson P (1983) Transport of ^{14}C-labelled PCB-compounds from sediment to water and from water to air in laboratory model systems. Water Res 17:1317-1326
Larsson P, Okla L (1987) An attempt to measure the flow of chlorinated hydrocarbons, such as PCBs, from water to air in the field. Environ Poll 44:219-225
Larson P, Okla L (1989) Atmospheric transport of chlorinated hydrocarbons to Sweden in 1985 compared to 1973. Atmos Environ 23:1699-1711
Leuenberger C, Czuczwa J, Heyerdahl E, Giger W (1988) Aliphatic and polycyclic aromatic hydrocarbons in urban rain, snow and fog. Atmos Environ 22:695-705
Lunde G, Gether J, Gjos N, May-Berit SL (1977) Organic micropollutants in precipitation in Norway. Atmos Environ 11:1007-1014
Mackay D, Shiu WY (1981) A critical review of Henry's Law constants for chemicals of environmental interest. J Phys Chem Ref Data 10:1175-1199
Mackay D, Wolkoff AW (1973) Rate of evaporation of low solubility contaminants from water bodies to atmosphere. Environ Sci Tech 7:611-614
McNeely R, Gummer WD (1984) A reconnaissance survey of the environmental chemistry in east-central Ellesmere Island, N.W.T.. Arctic 37:210-223
Muir DCG, Norstrom RJ, Simon M (1988) Organochlorine contaminants in arctic marine food chains: accumulation of specific polychlorinated biphenyls and chlordane-related compounds. Environ Sci Tech 22:1071-1079
Murphy TJ (1984) Atmospheric inputs of chlorinated hydrocarbons to the Great Lakes. In: Nriagu JO, Simmons MS (eds) Toxic Contaminants in the Great Lakes. Wiley Interscience, Toronto, Canada, p 53-80
Murphy TJ, Rzeszutko CP (1977) Precipitation inputs of PCBs to Lake Michigan. J Great Lakes Res 3:305-312

Murphy TJ, Schinsky AW (1983) Net atmospheric inputs of PCBs to the ice cover on Lake Huron. J Great Lakes Res 9:92-96

Oehme M, Mano S (1984) The long range transport of organic pollutants to the Arctic. Fresenius Z Anal Chem 319:141-146

Oehme M, Ottar B (1984) The long range transport of polychlorinated hydrocarbons to the Arctic. Geophys Res Letters 11:1133-1136

Paasivirta J, Paukku R (1987) Study of organochlorine pollutants in snow at North Pole. In: Nordlund OP (ed) The Huure Expedition to the North Pole at 1984. Technical Research Centre of Finland, Research Notes # 685, Espoo, Finland, p 24-31

Patton GW, Hinckley DA, Walla MD, Bidleman TF (1989) Airborne organochlorines in the Canadian high Arctic. Tellus 41B:243-255

Peel DA (1975) Organochlorine residues in antarctic snow. Nature 254:324-325

Peterle TJ (1969) DDT in antarctic snow. Nature 224:620

Rahn KA, Shaw GE (1982) Sources and transport of arctic pollution aerosol: A chronicle of six years of ONR research. Naval Research Reviews

Rapaport RA, Urban NR, Capel PD, Baker JE, Looney BB, Eisenrich SJ, Gorham E (1985) "New" DDT inputs to North America. Chemosphere 14:1167-1173

Rapaport RA, Eisenreich SJ (1986) Atmospheric deposition of toxaphene to eastern North America derived from peat accumulation. Atmos Environ 20:2367-2379

Rapaport RA, Eisenreich SJ (1988) Historical atmospheric inputs of high molecular weight chlorinated hydrocarbons to Eastern North America. Environ Sci Tech 22:931-941

Reijnders PJH (1980) Organochlorine and heavy metal residues in harbour seals from the Wadden Sea and their possible effects on reproduction. Neth J Sea Res 14:30-65

Robbins JA (1978) Geochemical and geophysical applications of radioactive lead. In: Nriagu JO (ed) The Biogeochemistry of Lead in the Environment. Elsevier, Amsterdam, p 285-393

Schondorf T, Herrmann R (1987) Transport and chemodynamics of organic micropollutants and ions during snowmelt. Nordic Hydrology 18:259-278

Schrimpff E (1980) The relationship between relief and deposition of some organic and inorganic contaminants in snow of northern Bavaria, F.R.G. In: Proceedings International Conference Ecological Impact Acid Precipitation. Norway, p 130-131

Schrimpff E, Thomas W, Herrmann R (1979) Regional patterns of contaminants (PAH, pesticides and trace metals) in snow of northeast Bavaria and their relationship to human influence and orographic effects. Water Air Soil Poll 11:481-497

Shiu WY, Mackay D (1986) A critical review of aqueous solubilities, vapor pressures, Henry's Law constants, and octanol-water partition coefficients of the polychlorinated biphenyls. J Phys Chem Ref Data 15:911-929

Simmleit N, Herrmann R (1987) The behavior of hydrophobic organic micropollutants in different karst water systems. Water Air Soil Poll 34:79-95

Slinn WGN, Hasse L, Hicks BB, Hogan AW, Lal D, Liss PS, Munnich KO, Sehmel GA, Vittori O (1978) Some aspects of the transfer of atmospheric trace consitituents past the air-sea interface. Atmos Environ 12:2055-2087

Stengle TR, Lichtenberg JA, Houston CS (1973) Sampling of glacial snow for pesticide analysis on the high plateau glacier of Mount Logan. Arctic 26:335-336

Strachan WMJ, Huneault H (1979) Polychlorinated biphenyls and organochlorine pesticides in Great Lakes precipitation. J Great Lakes Res 5:61-6888

Tanabe S, Tatsukawa R, Kawano M, Hidaka H (1982) Global distribution and atmospheric transport of chlorinated hydrocarbons: HCH (BHC) isomers and DDT compounds in the western Pacific, eastern Indian and Antarctic Oceans. J Oceanogr Soc Japan 38:137-148

Tanabe S (1988) PCB problems in the future: foresight from current knowledge. Environ Pollut 50:5-28

Tuominen J, Salomaa S, Pyysalo H, Skytta E, Tikkanen L, Nurmela T, Sorsa M, Pohjola V, Sauri M, Himberg K (1988) Polynuclear aromatic compounds and genotoxicity in particulate and vapor phases of ambient air: effect of traffic season and meteorological conditions. Env Sci Tech 22:1228-1234

Voldner EC, Ellenton G (1987) Production, usage and atmospheric emissions of priority toxic chemicals with emphasis on North America. Report ARD-88-4 prepared for the Canada/US International Joint Commission by Atmospheric Environment Service, Environment Canada, Downsview, Ontario, Canada

Von Rumker RE, Lawless EW, Meiners AF (1974) Production, distribution, use and environmental impact potential of selected pesticides. US Environmental Protection Agency, EPA/540/1-74-001

Windholz M (ed) (1983) The Merck Index: an encyclopedia of chemicals, drugs and biologicals. Merck & Co., New Jersy, 10th edition

Woo M, Marsh P (1977) Determination of snow storage for small eastern High Arctic basins. In: Proceedings of the Thirty-Fourth Annual Eastern Snow Conference. Belleville, Ontario, p 147-162

Woo M, Heron R, Marsh P, Steer P (1983) Comparison of weather station snowfall with winter snow accumulation in High Arctic basins. Atmos-Ocean 21:312-325

CHEMISTRY OF SNOW FROM HIGH ALTITUDE, MID/LOW LATITUDE GLACIERS

W. Berry Lyons *

Cameron Wake
and
Paul A. Mayewski
Glacier Research Group
Institute of the Study of Earth, Oceans, Space
University of New Hampshire
Durham, NH 03824 USA

INTRODUCTION

1.1 Snow chemistry records from high altitude, mid/low latitude (HAMLL) glaciers

Glaciochemical records describing the spatial and temporal variation in the chemical content of snow and ice in HAM/LL glaciers allows us to improve our understanding of the distribution of chemical species in the atmosphere at these latitudes, and their variation over space and time. The resulting data base provides the framework for describing and understanding aspects of atmospheric chemistry and circulation, biogeochemical cycling, climate change, anthropogenic emissions to the atmosphere, glacial hydrology and evidence of volcanic events. Furthermore, records from glaciochemical investigations are especially important when direct observations and measurements of the atmosphere are either spatially and/or temporally lacking.

Glaciochemical records recovered from high altitude/mid-low latitude sites provide a basis for determining *regional* climatic response (i.e., relative temperature, humidity, atmospheric circulation) to past *global* environmental change (as indicated by other paleoenvironmental records). These types of records could provide valuable regional constraints in the development of global climate models

Types of glaciochemical records include: 1) The collection of fresh snow, 2) snow pits in accumulation zones of glaciers that represent a time series of individual snow events or seasonal chemical signatures for a number of years, 3) and longer term records such as ice cores. These glaciochemical records undoubtedly represent the "best" paleoenvironmental records from these locations. Areas where records of these kinds have been or could be developed are shown in Figure 1.

---

*Present address
Hydrology/Hydrogeology Program
Mackay School of Mines
University of Nevada, Reno
Reno, Nevada 89557-0138 USA

NATO ASI Series, Vol. G 28
Seasonal Snowpacks
Edited by T. D. Davies et al.

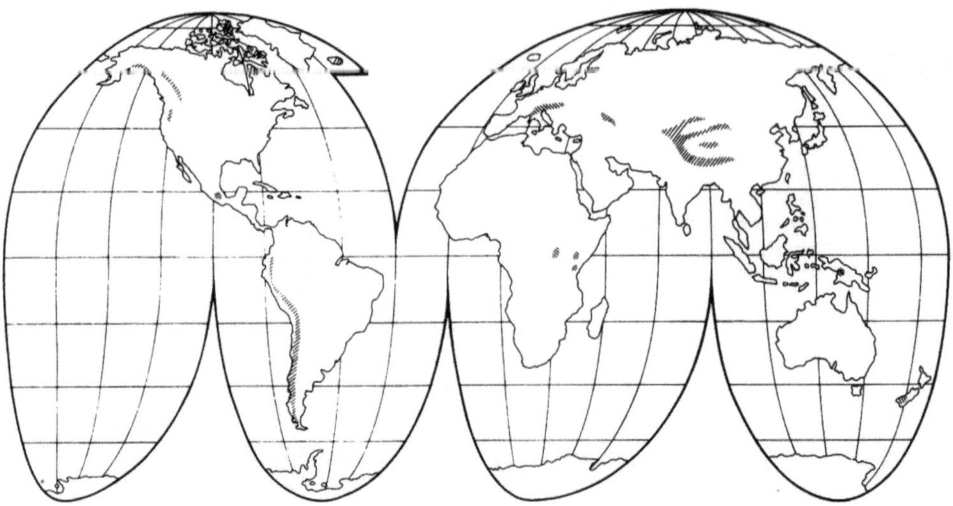

Figure 1. Location map for high altitude/mid-low latitude regions (outlined with hatching).

1.2 Site Selection, Sample Collection, Dating and Analyses

Site Selection

Glaciochemical samples recovered for paleoenvironmental interpretations should be collected from sites where the snow has experienced minimal postdepositional alteration, especially from the effects of wind, avalanches and meltwater percolation. Alpine glaciers in temperate regions are affected by strong vertical temperature and precipitation gradients (Barry 1981; Price 1981). Higher elevations within these glacier basins commonly experience cooler mean annual temperatures and higher rates of net annual snow accumulation, both important factors in preserving the chemical record within the snowpack. Cooler temperatures serve to limit the production of meltwater and thereby reduce the effects of postdepositional alteration due to percolating meltwater (Mayewski, et al. 1981). Collection sites should therefore be chosen at elevations as high as possible and in the dry snow zone. High rates of annual snow accumulation also help to preserve seasonal stratigraphy as meltwater must travel a greater vertical distance in order to significantly disturb the seasonal distribution of chemical species in the snowpack. However, glaciochemical records recovered from areas of high annual snow accumulation will show limited temporal coverage. There exists a balance between the quality of record preservation, sampling resolution and temporal extent of records recovered from HAM/LL glaciers. In alpine glacier systems there commonly exists strong local variations in the rate of snow accumulation due to aspect, local relief and the redistribution of snow by avalanches and the wind. An understanding of the physical processes that govern snow accumulation in mountainous terrain provides the basis for identifying study sites that are representative of the region in general. Sites should be located in areas removed from the effects of avalanching and local redistribution of snow by the wind. The physical character of the surface snow, such as the degree of rounding in snow grains and the buildup of sastrugi, provide a qualitative assessment of surface wind conditions. Relatively broad, flat areas within the accumulation zones of the glacier basins are commonly far removed from the effects of avalanching and the gentle surface topography acts to minimize the eddying effects of the wind. Sample locations should also be selected in areas removed as far as possible and/or upwind from local sources of chemical species which could potentially swamp regional chemical signals.

For ice cores deeper than ~20 meters (although this will vary depending on the velocity of the glacier, and rate of snow accumulation), the drilling site should be chosen in an area with relatively simple flow geometry, such as an ice flow divide on an ice cap or high col (Thompson, et al. 1986; Holdsworth, et al. 1989; Thompson, et al. 1989).

Spatial Sampling

The investigation of the variation of snow chemistry on two different spatial scales - local and regional - is essential to identify local and regional sources of chemical species in polar and HAM/LL regions (Mayewski, et al. 1987). Both scales of study are fundamental in order to extract regional signals from the data and confidently define regional trends. Strong local sources of chemical species (eg., dust from extensive deserts and evaporite deposits, anthropogenic emissions and marine salts) can significantly influence the chemistry of snow and mask regional trends (Thompson, et al. 1989; Lyons, et al. 1990; Wake and Mayewski 1990). The local influence must be identified before one can begin to interpret glaciochemical data on a regional scale.

Short term records of snow chemistry (single events to months) sampled over a wide spatial range (lateral and elevational) provide the basis for identifying discrete source areas and understanding the underlying processes responsible for the spatial variation in the chemical content of snow. For instance, the chemical analysis of snow samples collected over an elevational range of 5125 to 5512 m from a fresh snowfall event in the Ladakh Himalaya enabled Mayewski et al. (1983) to identify precipitation originating from two different air masses. The higher more pristine air mass was characteristic of high level east-west transport and the lower level air mass indicative of north-south flow. From the analysis of snow samples collected upwind and downwind of Urumqi, the major industrial center in north west China, Wake and Mayewski (1990) were able to identify the influence of anthropogenic activity on sulfate concentrations in the fresh snow.

Sample Collection

Extreme care must be exercised at all times during sample collection, handling and transport to assure samples remain uncontaminated. Non-particulating clean suits and hoods, plastic gloves and particle masks must be worn during all sampling procedures. For many types of chemical analysis samples can be melted in the field and transferred to precleaned leakproof sample containers. This is best done on the glacier itself as the environment in general is cleaner than anything down valley. Snow and ice can be returned to the laboratories frozen, but this generally involves considerable logistical hurdles and cost. However, while most major ions and stable isotopes can be analyzed reliably on samples that have been melted in the field under clean conditions, it is desirable to return at least some samples to the lab in a frozen state in order to analyze the samples for other important constituents such as nitrate.

Dating Snowpits and Ice Cores

A common problem in all snowpit/ice core investigations is the development of a reliable time scale. Several dating techniques have been used to establish chronologies for ice cores recovered from HAM/LL sites (Table 1). The most common dating techniques employed in the study of HAM/LL snowpits and ice cores are: (1) counting annual layers by observing the seasonal variation in stable isotopes, microparticle concentrations and ice and dirt layers; and (2) identifying periods of enhanced fallout (i.e., reference horizons) from atmospheric thermonuclear weapons testing and from volcanic eruptions. Confidence in dating snowpits and ice cores will improve with an increase in the number of different parameters used to refine the established chronology. Ice core dating is reviewed in detail elsewhere (Hammer, et al. 1978; Herron 1982; Oeschger and Langway 1989).

Seasonal Variations

As a result of relatively high snow accumulation rates (greater than 0.3 meters water equivalent per year-Table 1), seasonal variation of both chemical and physical parameters has proven extremely useful for establishing a chronology for snowpits and firn/ice cores. Seasonal variation in stable isotopes of hydrogen and oxygen, originally developed to date polar ice cores (Dansgaard, et al. 1973), has been used to date cores from Mt. Blanc (Jouzel, et al. 1977), Vemagtferner (Stichler, et al. 1982) and Mount Logan (Holdsworth, et al. 1984) and snowpits in the Karakoram (Wake 1989). Seasonal variations of stable isotopes and microparticle concentrations have been used to establish chronologies for dating ice cores recovered from Quelccaya (Thompson, 1986) and Dunde; Thompson, et al. 1989). The seasonal variation in the pollen content of snow was used by (Ambach and Eisner 1966) to date a 20 m snowpit on Kesselwandferner. In addition, ice and dirt layers and snow density have been used to identify summer and winter layers (Ambach and Eisner 1966; Han, et al. 1989; Wake 1989). However, in some locations physical stratigraphy has also produced ambiguous dating results when compared with other stratigraphic methods (Mayewski, et al. 1984; Miller, et al. 1965).

The seasonal variation in major ion concentrations, used extensively to refine chronologies in snowpits and ice cores from polar regions (Herron 1982; Legrand and Delmas 1988; Mayewski, et al. 1990), has seen only limited application in snowpits and ice cores from HAM/LL sites (Holdsworth, et al. 1989; Wake 1989). However, seasonal variation of major ion concentrations represents a potentially powerful technique for establishing chronologies. As the analysis of major ion content becomes a standard measurement in snow and ice recovered from HAM/LL sites, the use of seasonal variation in these constituents for dating purposes will most likely increase.

Reference Horizons

Thermonuclear weapons testing in the 1950's, 60's and 70's created a series of reference horizons in glaciers recognizable by significant increases in total beta activity (Crozaz, et al. 1966; Picciotto and Wilgain 1963) and tritium content (Miller, et al. 1965). Discrete peaks in total beta activity and tritium occur in 1952, 1954 and 1963 layers due to the Ivy, Castle and 1961-62 Soviet bomb tests, respectively. The identification of tritium and beta activity peaks in samples collected from snow pits and ice cores in HAM/LL regions provides a very reliable and accurate (+ 0.5 years) time horizon (Holdsworth, et al. 1984; Jouzel, et al. 1977; Mayewski, et al. 1984; Oerter and Rauert 1982). In addition, radioactive fallout originating from the Chernobyl accident has created a recent reference horizon in HAM/LL glaciers (Ambach, et al. 1988; Haeberli, et al. 1988; Pourchet, et al. 1988)

The positive identification of volcanic ash emitted during volcanic eruptions has been used to date ice cores from Peru (Thompson, et al. 1986) and western China (Han, et al. 1989). High concentrations of acidic anions have been used as indicators of soluble volcanic debris, resulting from volcanic eruptions, to help establish a chronology for ice cores recovered from Mt. Logan (Holdsworth, et al. 1989).

Holdsworth et al. (1988) used a well known forest fire event, identified in the firn core record by the occurrence of a strong peak in the nitrate profile and an ash layer consisting of forest fire charcoal, to help date a firn core from Snow Dome in the Columbia Icefields, Canada. In summary, reference horizons provide discrete time markers with which to date firn/ice cores and also serve to refine chronologies developed through the observation of seasonal variations in other physical or chemical constituents.

Radioactive Decay and Flow Models

The radioactive decay of ^{210}Pb has been used to date cores from the Austrian and Swiss Alps (Gaggeler, et al. 1983; Gunten, et al. 1982). The half life of ^{210}Pb (22.3 years) limits this method to firn/ice cores younger than ~100 years.

Time scales based on flow models (Hammer, et al. 1978; Reeh 1989) are commonly used to provide first approximation chronologies for deeper (i.e., > ~50m) ice cores. However, flow models are only used as the primary dating tool for the lower sections of deep ice cores where there exists no other applicable dating technique (Thompson, et al. 1986; Thompson, et al. 1989).

Types of Chemical Measurements

The vast majority of investigators on the glaciochemistry of high altitude snow reviewed in this manuscript have determined stable isotopes (i.e. $\delta^{18}O$) and/or radioactive isotopes, (i.e. beta activity) and/or major cations and anions. The potential now exists for other chemical constituents to be measured in this type of milieu. These other constituents include chemical species currently measured in polar ice cores such as H_2O_2, trace metals, MSA and cosmogenic radioactive isotopes.

2. GEOGRAPHICAL DATA BASE

2.1 TEMPERATE NORTHERN HEMISPHERE

European Alps

The best glaciochemical records form the European Alps come from Colle Gnifetti and to a lesser extent from Mt. Blanc. Because the Colle Gnifetti data have recently been extensively reviewed by both

Wagenbach (1989) and Mayewski et al., (in press), we will not detail it again here. The sulfate data indicate over a 4 fold increase in sulfate from the turn of the century to present day, while nitrate demonstrates over a 4.5 x increase over the same period (Wagenbach et al., 1988). The record from Colle Gnifetti also preserves an outstanding regional climate record (from the $\delta^{18}O$ measurements) as well as a detailed record of Saharan dust influx to Europe over the same time period (Haeberli et al., 1983; Wagenbach, 1989).

The Mt. Blanc record also shows the increase of anthropologically introduced strong acids in the regional troposphere from 1940 to 1976 (Delmas and Aristarain, 1978). Over a 25 year time span there was also a ~ 2 fold increase in the trace metals, lead, cadmium and vanadium in Mt. Blanc snow (Briat, 1978). Batifol and Boutron (1984) have argued that the vast majority of trace metal and metalloid data collected from alpine snow and ice prior to 1984 are incorrect due to sample contamination. They present their own data from Col du Gouter, Col du Midi, and Mt. Blanc, France indicating a possible 2 fold increase in Pb over the period ~ 1950 to ~ 1975. They also observed a 2 fold increase in Cd but no clear change in Zn or Cu concentrations over the same time period. Their mean Pb concentrations in surface snow were between 2.5 and 4.0 ug per kg, while for Cd, Zn, Cu and Ag and values were 20-200 ng per kg, ~ 1 to ~ 9 ug per kg, 0.1 to 2.5 ug per kg and 10 to 33 ng per kg, respectively.

Wagenbach (1989) points out that the interpretation of glaciochemical records from these glaciers can be difficult due in part to irregularities in snow deposition rates. However, Mayewski et al., (in press) have been able to relate the increasing trends of sulfate and nitrate in Colle Gnifetti snow with those in southern Greenland suggesting that the longer term records from the high Alps can indeed provide very useful historic records of regional atmospheric chemistry.

Western Cordillera

Glaciochemical investigations in the Western Cordillera have been dominated by the recovery and chemical analysis of ice cores and snowpit samples from the Mt. Logan region, St. Elias Range, Yukon Territory (Delmas, et al. 1985; Holdsworth 1986; Holdsworth, et al. 1989; Holdsworth, et al. 1988; Holdsworth and Peake 1985; Holdsworth, et al. 1984). In 1980, a 103 m firn/ice core was recovered from the icefield plateau at 5430 m. The core covers a time span of almost 300 years. Impurities in snow and ice from this site are predominantly the mineral acids H_2SO_4 and HNO_3, both of which show average concentrations of about 0.9 ueq/l. The sea salt component is extremely low (average sodium concentration in 0.06 ueq/l (Delmas, et al. 1985). There exists no significant long term trend in the background content of mineral acids over the last one hundred years indicating that this site is relatively unaffected by anthropogenic emissions, unlike sites in the Canadian Arctic (Barrie, et al. 1985), Greenland (Mayewski, et al. 1986) and Europe (Delmas and Aristarian 1978). This site may, therefore, be representative of the background aerosol in the free troposphere of this latitude and for air masses originating in the north Pacific.

Several peaks in the sulfate and chloride profiles have proven useful in the identification of volcanic events dating back to the 1690's. Spectral analysis of net snow accumulation time series revealed significant periodicities of 3.8 and 11.3 years. These periodicities are similar to those for ENSO events and the sunspot cycles, respectively.

Holdsworth, et al. (1988) also investigated the chemical concentration of shallow snowpits and ice cores at several locations in western Canada. Sample sites in the Seward (1800 m) and Kaskawulsh (2600 m) Glaciers, Yukon Territories, Flint Icefield, Northwest Territories (2290 m), Mt. Edziza, British Columbia (2787 m) and Snow Dome, Alberta all showed homogenization of seasonal chemical signals due to meltwater percolation. Particulate fallout from forest fires, associated with elevated nitrate concentrations, were identified in an ice core from Snow Dome, Alberta and snowpit samples from the Eclipse site near Mt. Logan. Butler et al. (1980) analyzed snow collected from a 7 m snowpit excavated at 2900 m on Athabasca Glacier. They conclude that nitrate, phosphate and reactive silica enters this region in much lower concentrations compared to Edmonton.

Central Asia

The vast extent of glacierized regions in the high mountains of central Asia (Figure 2) provide several suitable locations from which to recover glaciochemical records. A review of glaciochemical programs that have investigated the major ion content of snow and ice in central Asia appears in Table 2. Perhaps the single most obvious fact that emerges for this vast region is that the existing data set is limited in its spatial extent and, aside from the 140 m ice cores recovered from the Dunde Ice Cap, is also limited in its temporal coverage. The glaciochemical record contained in central Asian glaciers represents an extremely valuable, yet virtually untapped, resource. Furthermore, this resource will last only as long as climatic conditions in the region permit. Increases in the temperature of the troposphere could potentially destroy these records due to changes in the local thermal balance resulting in melting of the snowpack. In fact, global climate models developed by the Goddard Institute for Space Studies indicate that regions of unambiguous warming will appear earliest the interior of Asia (Hansen et al., 1988). From

Table 1 Dating techniques used to establish chronology in ice cores recovered from high altitude/mid-low latitude sites.

Location	depth (m)	length of record (yrs)	acc rate (m w.e $\cdot a^{-1}$)	Dating Techniques (RH - reference horizon; SV - seasonal variation)	Reference
Kesselwandferner, Austria	20	14	1.0	SV - pollen; physical stratigraphy (ice and dirt layers)	1
Mt Blanc, France	16	3	2.8	RH - ß activity, tritium; SV - deuterium	2
Colle Gnifetti, Switzerland	65	~150	0.3	Pb-210 (~1890-1977)	3
				RH - high concentration of ice layers (1947, 1950-hot summers) - Sahara dust (1936 and 1937 events); tritium (1953-1977)	4
Vernagtferner, Austria	81	~80	0.8	Pb-210 (~1900-1979)	5
				RH - tritium (1952-1979); Sahara dust (1936 and 1937 events)	6
				SV - excess deuterium (~1900 - 1979)	7
Mt. Logan, Yukon	103	290	0.4	RH - ß activity, tritium (1951-1980); volcanic ash (several)	8;9;10
				SV - oxygen isotopes; nitrate and sodium (1720-1980)	8;10
Sentik Glacier, India	17	17	0.6	RH - ß activity (1963-1980)	11
Chongce Ice Cap, China	32	60(?)	0.6	RH - Ashikule volcanic ash (1951); SV - physical stratigraphy	12
Yala Glacier, Nepal	60	70(?)	0.7	RH - tritium (1953-1982); SV - visible dirt layers (1912-1982)	13
Dunde Ice Cap, China	140	>11000	0.4	RH - ß activity (1963-1987)	14
				SV - oxygen isotopes, microparticle conc. (0-117 m - 4550 y bp) flow model (117 m to 140 m)	
Quelccaya Ice Cap, Peru	164	1500	1.0	RH - Huaynputina volcanic ash (1600 AD)	15
				SV - visible dirt layers, microparticle conc., cond., oxygen isotopes	

1. Ambach and Eisner, 1966
2. Jouzel et al., 1977
3.Gaggeler et al., 1983
4.Oeschger et al., 1977
5.Von Gunten et al., 1982
6.Oerter and Rauert, 1982
7.Stichler et al., 1982
8. Holdsworth et al., 1984
9.Holdsworth and Peake, 1985
10. Holdsworth et al., 1989
11. Mayewski et al., 1984
12. Han et al., 1989
13. Watanabe et al., 1984
14. Thompson et al., 1989
15. Thompson et al., 1986

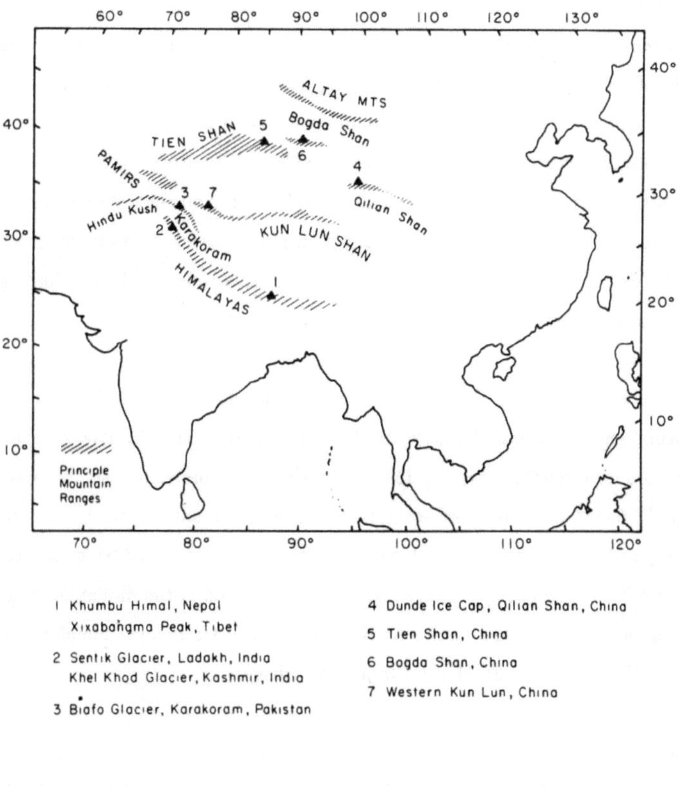

Figure 2. Location map for sites in the mountains of central Asia from which glaciochemical records have already been collected.

an analysis of oxygen isotope values in an ice core from central China, Thompson et al., (1989) suggests that the last sixty years have been one of the warmest periods in this region since the glacial-interglacial transition.

Glaciochemical data is only presently available for sites on the southern margin of the mountain ranges of central Asia (sites 1 through 4, Figure 2) and from the Tien Shan, Bogda Shan and Dunde Ice Cap along the northern margin (sites 5 through 7, Figure 2). Additionally, only the records from the Biafo, Sentik, Khel Khod, Yala and Dunde Glaciers extend for more than one year. Snow chemistry data recovered from the other southern margin areas (i.e., Kumjung Village, Mt. Everest, Xixabangma Peak, Tien Shan and Bogda Shan) represent samples of an individual, or a series of individual snowfall events covering a rather short period of time (i.e., less than one month). In addition, the types of chemical species analyzed differs between studies (Table 2), limiting comparison among the investigations.

The glaciochemical sample locations listed in Table 2 fall into three broad climatic zones. The Biafo, Khel Khod and Sentik Glaciers are characterized by precipitation sources from both the west and the south. In this region, snow accumulation increases toward the south-western margin as a result of increasing influence of precipitation derived from westerly moisture sources. The upper Biafo Glacier consists mainly of extensive open and gently sloping accumulation areas covering over 500 km^2. The percentage of ice free area surrounding the accumulation zone is minimal. Glaciochemical samples were collected from snowpits excavated from 4650 to 5520 m asl. The aspect of these collection sites varied from east to south to west. The northward facing Sentik Glacier is one of several glaciers emanating from the Nun Kun Plateau (325 km^2). While the plateau is largely ice covered, extensive areas surrounding the plateau are ice free. Khel Khod Glacier is one of several relatively small (<10 km^2) glacier basins in Kashmir. It faces southwestward and is surrounded by ice free areas.

Sample collection sites in the central Himalaya, including Yala Glacier, Xixabangma Peak, Mt. Everest and Kumjung village are characterized by a monsoonal maritime type of climate (Li and Xu, 1980). This region experiences heavy precipitation during the summer monsoonal period. Yala glacier covers an area of 2.6 km^2, flows in a north-easterly direction, and is surrounded by extensive ice free areas. Fresh snow samples from Xixabangma peak and Mt. Everest were collected from relatively high elevation sites (5850 m and 5600-7100 m asl, respectively) on the north side of the mountain, in an elevation band where perennial snow and ice dominate. The Kumjung village fresh snow samples were collected at a much lower elevation (4900 m asl) adjacent to a small village.

The Tien Shan and Bogda Shan mountains in north-west China experience an inland continental climate. Moisture is derived primarily from the west and the north (Li and Xu, 1980). Both Glacier No. 1 at the headwaters of the Urumqi River and Bogda Feng glacier are small (<2 km^2), and are surrounded by extensive ice free areas. The Dunde ice cap lies on the south west margin of the Qilian Shan mountains in central China and is also classified as an inland continental glacier (Li and Xu, 1980). The ice cap covers an area of 57 km^2 (LIGG 1980) with an average ice thickness of 140 m (Thompson et al., 1988). The southern margin of the mountain range is bordered by the Qaidam basin, a large desert with extensive salt deposits (Chen and Bowler 1986). As reviewed in the opening section, the preservation of chemical records in snow and ice deposited in HAM/LL glaciers depends upon glacier basin morphology and local meteorological conditions. A qualitative assessment of the degree of post depositional alteration of snow and ice samples collected in central Asia appears in Table 3.

The longest paleoenvironmental record obtained from glaciers in central Asia comes from three ~140 m deep ice cores drilled in the Dunde Ice Cap (Thompson et al., 1989). The glaciochemical record recovered from the Dunde ice cores extends back to the last glacial period. The transition from glacial to interglacial conditions is marked by a rapid decrease (~40 years) in microparticle concentrations and a slower increase in $\delta^{18}O$ (towards less negative values), chloride, nitrate and sulfate. Thompson et al., (1989) suggest that: 1) the rapid decrease in dust at the glacial-interglacial transition reflects a change in the prevailing climatic conditions which limited the capacity of dust transport in the atmosphere; 2) the trend towards less negative values of delta 18O indicates a gradual warming at the boundary; and 3) the increase in anion concentrations reflects the desiccation of freshwater lakes in the Qaidam Basin and the subsequent formation of evaporite deposits which acted as a strong local source of salt to the Dunde ice cap. In this respect the major anion content of ice cores from Dunde reflects a strong local source which tends to swamp any regional signals. In summary, the glaciochemical records from the Dunde ice cores suggest that late stage glacial conditions were colder, wetter and dustier than Holocene conditions. In addition, the $\delta^{18}O$ record suggests that the last sixty years have been one of the warmest periods in the entire record.

Although the glaciochemical investigations undertaken to date are few in number and limited in terms of spatial coverage and length of record, some preliminary observations can be made concerning regional and temporal trends in snow chemistry and their importance in helping to understand regional atmospheric circulation and atmospheric chemistry in central Asia (Wake et al., 1990). Table 2 provides a summary of major ion content of central Asian snow. Figure 3 illustrates the mean concentrations of sodium, chloride, nitrate and sulfate in snow collected from the mountains in central Asia. The columns in the histograms, from left to right, are roughly arranged west to east for sites along the southern margin of the mountains in central Asia (Biafo Glacier through Kumjung Village) and the northern margin (Tien Shan to Dunde Ice Cap).

Both sodium and chloride show strong regional trends with the lowest values in the Karakoram, and progressively higher values toward the south-east. This regional trend clearly demonstrates the

Table 2 Mean Concentrations of Major Ions in Snow and Ice Recovered from Central Asia

Location	LAT	Elev. (m)	Type of Sample*	No. of samples	Na (concentration in μ eq/kg)	Cl	NO3	SO4	Other	Ref
Biafo Glacier		4650-5520	SP	198	1.1	1.1	2.9	2.2		1
Sentik Glac		4908	IC	104	3.0	3.0	4.4		SiO_2-7.1	2,3
Khel Khod		4695	IC	19	15.6					4
Yala Glac		5400	IC	55	10.9	9.9	4.8			5
Xixabangma		5850	FSS	6	12.0	5.3	6.3	7.7		6
Mt Everest		5600-7100	FSS	120	7.5	6.1	4.5	5.7	Ca-32.1;SiO_2-0.5	7
Kumjung		3900	FSS	6	3.9		4.7	2.9	Ca-19.0	8
Chongce		6130/6366	IC	??						9,10
Tian Shan		3450-4110	FSS	37	27.9	9.3	3.7	23.3	Ca-60.7	11
Bogda Shan		1940-3550	FSS	38	44.0	9.0	6.0	51.1	Ca-66.7	11
Dunde		5450	IC	1200		32.4	4.2	18.9	cation data not reported	12

*IC-ice core; SP-snow pit; FSS-fresh surface snow

References:

1. Wake 1987; 1989
2. Lyons and Mayewski, 1983
3. Mayewski et al., 1984
4. Mayewski et al., 1981
5. Watanbe et al., 1982
6. Mayewski et al., 1986
7. Jenkins et al., 1987
8. Davidson et al., 1986
9. Han et al., 1989
10. Nakawo et al., 1990
11. Wake and Mayewski, in prep.
12. Thompson et al., 1990a,b

Table 3. Physical parameters affecting quality of glaciochemical samples collected in central Asia.

Location	morphology of collection site	elev. (m)	mean annual temp (C)*	annual acc. m w.e.	relative influence of wind	avalanches	meltwater infiltration	Type of Sample
Biafo	acc. basin	4650-5520	-4.5**	0.9-2.5	low	none	low	snowpits
K2 North	acc. basin	5320	------	------	low	high	high	ice core
Sentik	acc. basin	4908	-3.0	0.62	low	none	low	ice core
Khel Khod	circ glac	4695	------	0.48	?	?	low-med	ice core
Dunagiri	ridge	4400-6050	------	------	low	none	low	fresh snow only+
Yala Glac.	acc. basin	5400	0.0	0.60	?	?	high	ice core
Xixabnagma	acc. zone	5850	------	------	low	none	low	fresh snow only+
Mt. Everest	acc. zone	5600-7100	------	------	low	low	low	fresh snow only+
Kumjung	near village	3900	------	------	low	low	low	fresh snow only+
Kongma	ice cliffs	5650	------	0.58	?	?	med-high	ice cliff
Tian Shan	acc. basin	3450-4110	-5.4***	0.43	low	low	low	winter snow only
Dunde	ice cap	5250-5400	-5.4	0.20	?	low	low-med	snowpits/ice core

* from ~10 m borehole temperature measurements

**measured at 5450 m

*** from meteorological records at 3590 m

+ collected within 24 hours of snowfall

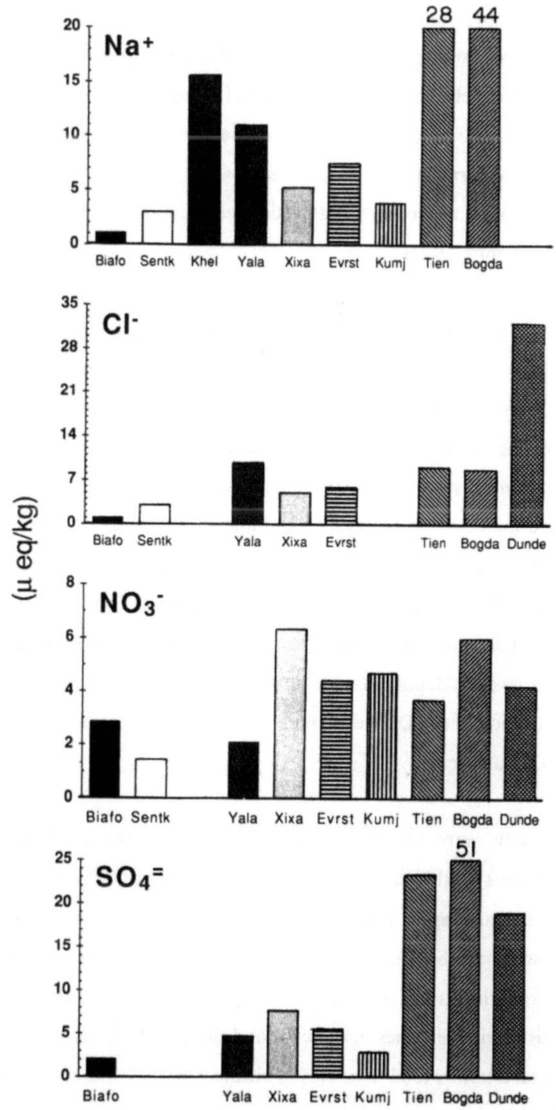

Figure 3. Comparison of mean ion concentration of glaciochemical records recovered from the mountains of central Asia. The seven sites on the left of the figure (i.e., Biafo to Kumjung) lie along the main crest of the Himalaya and are arranged from west to east.

increased influence of monsoonal derived moisture in the Himalayas when compared with the Karakoram. Seasonal signals apparent in records from the Karakoram (Wake, 1989) and western Himalaya (Mayewski et al., 1984) are characterized by relatively higher concentrations of sodium and chloride in summer strata. This is indicative of the influx of moisture derived from the Arabian Sea which is then transported to the mountains by summer monsoonal circulation. Winter strata are characterized by overall lower concentrations of sodium and chloride, reflecting moisture derived from distant marine sources such as the Mediterranean Sea and/or Atlantic Ocean.

The sodium:chloride ratio for samples collected from the Biafo and Sentik glaciers, Xixabangma Peak and Mt. Everest approach the ratio found in sea water (i.e. 0.86). Thus, most of the sodium and chloride entering the region is derived from marine sources. The slightly higher sodium:chloride ratio for the Sentik Glacier samples implies an additional source of sodium. Snow from Sentik Glacier is also characterized by high concentrations of reactive silicate. The elevated sodium and reactive silicate values suggest inputs of crustal derived material. The very high sodium values observed in samples from Khel Khod Glacier cannot be easily interpreted in terms of a regional trend as they most likely reflect a local crustal source from surrounding ice free areas. The high sodium and chloride concentrations in snow from the Tien Shan and the Bogda Shan most likely represent an input of sodium and chloride rich dust originating from the extensive evaporite deposits to the north and west of these mountain ranges. The extremely high chloride concentration in the Dunde ice core is indicative of a strong local source of "salt" chloride originating from the Qaidam desert.

Low nitrate concentrations were found in snow collected from the southern margin of the Himalayas (Sentik and Yala Glaciers) and high concentrations in snow deposited on the northern margin of the Himalayas (Biafo Glacier, Xixabangma Peak, and Mt. Everest) and in central and western China (Dunde Ice Cap, Tien Shan and Bogda Shan). This strong regional trend in the spatial distribution of nitrate suggests the influx of terrestrial dust, rich in nitrate, originating from the extensive arid regions of central Asia. (From an analysis of nitrate values in snow around the world, Lyons et al., 1990 found that aerosols derived from continental interiors tend to be rich in nitrate.) In addition, snow samples from Mt. Everest, Tian Shan and Bogda Shan show extremely high calcium concentrations (Table 2), also indicative of dust derived from the arid regions of central Asia.

Sulfate values are approximately 2 to 3 times higher in samples collected from Yala Glacier, Kumjung Village, Xixabangma Peak, Mt. Everest compared to samples from Biafo Glacier. Excess sulfate (non sea-salt) accounts for 90-95% of the total sulfate measured at all five sites. The elevated sulfate concentration in these samples could be due to an influx of anthropogenic emissions from the Indian subcontinent, transported to the Himalayas with monsoonal circulation. However, Mayewski et al., (1983) suggest that monsoonal air masses tend to dominate at lower elevations with more pristine air aloft. The calcium:sulfate ratio in snow from Mt. Everest is 0.17 suggesting that about one-fifth of the calcium and sulfate could enter the region in the form of gypsum rich dust from the arid regions of central Asia. The high sulfate values in the Dunde ice cores reflect the input of sulfate rich dust originating from the Qaidam basin. The very high sulfate concentrations found in snow from the Tien Shan and the Bogda Shan most likely reflect local dust and anthropogenic sources or perhaps a natural biogenic source.

While limited in spatial and temporal distribution, the glaciochemical data thus far collected from central Asia reveals local and regional trends, including a marine influence (sodium and chloride), a local crustal dust influence (sodium and reactive silica), a regional dust influence (nitrate and calcium) and an anthropogenic influence (sulfate) on snow deposited in central Asia.

The glaciochemical record contained in central Asian glaciers represents an extremely valuable resource. However, depending upon the sample location, glaciochemical records recovered from central Asian glaciers could potentially be swamped by high levels of marine, continental dust and/or anthropogenic inputs; a problem less prevalent in polar regions. Therefore, understanding the processes which control the chemical content of snow, the local to regional scale complexities, and the seasonal variability are fundamental steps necessary in order to assess the potential for recovering representative long-term glaciochemical records from central Asia.

Middle East

There are some recent data for the chemistry of snow in Mt. Hermon region, 35° 50' E and 33° 25' N (Gil'ad and Bonne, 1990). The snow with pH close to neutral (6.2-6.7) has Ca^{2+} as the major cation and the HCO_3^- and SO_4^{2-} as the major anions. The Cl^- and Na^+ concentrations increase with elevation. Due to the lack of concentration profiles with depth, little else can be inferred from these data.

2.2 Temperate Southern Hemisphere

Andes

A series of papers have been published in the physical and climatological aspects of the Quelccaya Ice Cap in the Peruvian Andes (14° S, 71° W) by L. Thompson and his colleagues. However, with the exception of $\delta^{18}O$ (Grootes et al., 1989) and microparticle measurements, little glaciochemical data are available from this location. Recently Wagenbach (1989) has reviewed the research at Quelccaya and, therefore, it will not be detailed here. Using $\delta^{18}O$ and microparticle measurements a detailed chronology of snow accumulation has been established (Thompson et al., 1979, 1984, 1985, 1986). The majority of snow accumulation occurs in the wet season. The $\delta^{18}O$ record is rather complex with seasonal changes in snow evaporation on the surface of the glacier greatly effecting the seasonal signal (Grootes et al., 1989).

The ice core records from Quelccaya have been utilized to develop El-Nino- Southern Oscillation (ENSO) records (Thomason et al., 1984) as well as establish the occurrence of the "Little Ice Age" in the tropics (Thompson et al., 1986). The Little Ice Age event at Quelccaya was identified as the wettest period of the past 1500 years.

Two snow pits on Quelccaya were analyzed for major element chemistry (Lyons et al., 1985). Crustal elements such as Si, Fe and Na correlate with the microparticle profiles. Sulfate and nitrate concentrations were relatively low and similar to values from snow from other Andean regions (Lyons et al., 1985). It was postulated that the sources of the sulfate and nitrate were probably the Amazon basin to the east. (The east is also thought to be the major source of moisture to the ice cap). The NO_3^- precipitated fluxes to Quelccaya are, however, much lower than those from tropical rain forest regions (Lyons et al., 1990) indicating a probable loss of NO_3^- on transport from the Amazon Basin.

Thompson et al., (1984) have suggested a number of other potential sites in the Andes where glaciochemical records could be obtained. These records could be extremely useful in evaluating the atmospheric response to ENSO's as well as determining the regional atmospheric impact of biomass burning in Amazonia. Efforts should be made to obtain detailed glaciochemical records from this important geographic region as soon as possible.

Africa

In the 1970's, a number of studies were conducted on the permanent ice fields on Mt. Kenya (Thompson, 1979; Vincent et al., 1979; Thompson and Hastenrath, 1981). In addition to these investigations, Thompson et al., (1984) presented $\delta^{18}O$ data for Kilimanjaro (Fonfiantini, 1970). The data from these ice fields include total beta activity, microparticles, density and $\delta^{18}O$ values. In addition, there are some major cation data (Na, Mg and Ca) for the Lewis Glacier (0° 09'S, 37° 18' E) on Mt. Kenya (Vincent et al., 1979).

The oxygen isotope record from Mt. Kenya show little variation with depth although there is a relationship between total beta and microparticle concentrations, with the highest concentrations in the ice layers, (Thompson and Hastenrath, 1981). There is no doubt that Lewis Glacier has retreated considerable since the beginning on the 20th century and that melting has greatly affected the glaciochemical record there (Vincent et al., 1979). The down hole record demonstrates very high concentrations of cations, especially Na^+, with no significant correlations between cation concentrations and particle amounts. Core sections with extremely high concentrations of Na^+ are thought to have been produced via melting and refreezing processes (Vincent et al., 1979). The daily melting and refreezing remobilize the chemical record so that it cannot be utilized for paleoreconstructions.

Fresh surface snow samples from these areas, however, can provide useful information regarding atmospheric chemical/transport processes. The $\delta^{18}O$ values become more negative with elevation at Kilimanjaro (Gonfiantini, 1970), indicating, as expected, more fractionization with altitude. The mean Cl^-, NO_3^- and SO_4^{2-} of snow on Kilimanjaro are 965, 1710 and 2707 ug/kg, respectively, (Spencer and Mayewski; unpublished data; Lyons et al., 1990). These values are very high compared to other high elevation snow and suggest an extremely enriched aerosol source. This high nitrate value indicates a probable arid to semi-arid, particle-rich source, (Lyons et al., 1990).

2.3 Sub Arctic / Sub Antarctic

Heard Island

Heard Island is a 44 km by 19 km island located south of the Antarctic Convergence (53° 05' S, 73°, 30' E). The island is 80% covered by snow and ice with glaciers that apparently respond very rapidly to climatic fluctuations (Lambeth, 1951; Allison, 1980). A 3 meter snow pit was collected at 2450m and analyzed for various chemical species (Spencer et al., 1985). Allison (1980) calculated that the accumulation rate at 2400m is 4.2 ma^{-1} water equivalent and ablation is zero. The $\delta^{18}O$ values from the snow at the sampling location was - 12 per mil. The Cl^- concentrations vary from 2 - 105 ueq $^{-1}$ L^{-1}, while the Na^+ values are much less (2 to ~ 50 ueq L^{-1}. Both the Na^+ and Cl^- concentrations have maxima and minima at the same depths with a correlation coefficient of $r = 0.93$ (Spencer et al., 1985). The sulfate concentrations are lower (0.3 - 9.0 ueg L^{-1}) but correlate quite well with both Cl^- and Na^+. It is probable that a portion of both the Cl^- and SO_4^{2-} have a local, volcanic source as an active volcano, Big Ben is the dominant geomorphic feature of the island. A strong volcanic source function is also supported by the very high F^- concentrations observed in the snow pit samples (mean $F^- = 1.6$ ueq L^{-1}). The nitrate concentrations are very low having a mean of ~ 0.3 ueq L^{-1} (Lyons et al, 1990).

The major variation in Cl^-, Na^+ and sea salt SO_4^{2-} are undoubtedly due to seasonal variation in sea - ice extent with maxima concentrations occurring in the spring-summer period (Spencer et al., 1985). The highest Fe concentrations also occur in the "summer" suggesting increased input of crustal material at this time as well. Using these seasonal indicators the summer portion of the mass balance at this location is at least 55% of the total (Spencer et al., 1985). These data suggest that Heard Island would be a very useful location to obtain paleo-records of several decades to hundreds of years through ice coring activities.

Marion Island

Smith (1987) has reported on the chemical composition of precipitation collected over a 1 year period on Marion Island (46° 54' S, 37° 45' E). Because the precipitation collected and analyzed was predominantly rain, we will not detail the results. However, the ionic chemistry decreased in order: $Cl > Na > SO_4 > Mg > K >= Ca$. This is the order of the ionic concentration of seawater and indicates the overwhelming influence of marine aerosol on the precipitation chemistry. Nitrate concentrations varied from 0.1 to 6.4 ueq L^{-1} suggesting other than a marine source. Smith (1987) suggest that the major source of nitrate to the precipitation is from the penguin rookeries in the coastal zone of the island.

Iceland

Iceland has numerous glaciers which cover ~ 11% of the island (Bjornsson, 1979). Because of its unique climatological location located in a region of high cyclonic activity between cold polar air masses to the north and warm, temperate air masses to the south, theoretically, Iceland is an ideal place to obtain paleo-atmospheric records (Bjornsson, 1979). However, because of the temperate nature of these glaciers, the original chemical signatures in the accumulating snow in Icelandic glaciers are subject to melt-out and redistribution (Mayewski and Lyons, personal observation). In general, ~ 25% of the snow accumulation of Icelandic glaciers is lost through melting inside the accumulation zone (Bjornsson, 1979). Therefore, little glaciochemical data are available from Iceland.

The UNH group has analyzed a series of snow pit samples and shallow cores from Icelandic glaciers. Two, 2 meter snow pits were collected from the central potion of Hofsjokull in May, 1985. Hofsjokull is located in central Iceland. The average precipitation rate in this region is 24-32 cm a^{-1} and the firn line is at about 1300m (Bjornsson, 1979). The mean nitrate concentration of these samples were reported in Lyons et al. (1990) and is 0.7 ueq L^{-1}. The Cl^- and Na^+ concentration varied from less than 100 ug L^{-1} to greater than 1000 ug L^{-1} with both profiles having coincidental maxima and minima. The

SO_4^{2-} concentration varies between ~100 ug L^{-1} to less than 1000 ug L^{-1} with little correlation to Cl^-. This lack of similarity between the Cl^- and SO_4^{2-} emphasized the importance of other local sources of SO_4^{2-} besides sea-salt to these glaciers. These other sources include marine biogenic sources via MSA production as well as local volcanic input. Unfortunately, much of the paleo-record was "melted out" of these samples.

3. Potential Environmental Information and the Future

It is not in the scope of this manuscript to discuss in detail the usefulness of snow/ice chemistry to provide environmental information. However, the study of HAM/LL snow can provide important information regarding regional and global atmospheric phenomenon. The brief summaries of studies tabulated above demonstrate this. Information regarding atmospheric circulation, the source of moisture and precipitation patterns have been documented. For example, ENSO events and monsoonal variations have been well established in HAM/LL ice core research. Sources of aerosols and turbidity in the atmosphere can be discerned and the importance of volcanic and biogenic sources of particular salts can also be documented. Spatial studies such as the on-going one in Asia by the UNH group help to develop a regional picture of atmospheric circulation and aerosol source strengths. The influence of anthropogenic activities such as fossil fuel and biomass burning can also be established by these studies. Maybe even more importantly, where long term records can be retrieved such as in Qualccaya, detailed paleoclimate records can be developed so that mid to low latitude climate events can be correlated to those obtained from ice cores in the polar regions.

As our knowledge of global snow chemistry evolves, it is apparent that new programs ought to incorporate analyses of a wider variety of chemical species in order to determine the relative importance of various sources (eg., marine, crustal, biogenic, anthropogenic, volcanic). For example, marine sulfate could be quantified by measuring methyl sulfonic acid (MSA). Inputs from the arid regions of central Asia and local crustal debris could be identified by measuring the aluminum, sodium, calcium, magnesium, reactive silica and iron content of the samples. A potential volcanic source of sulfate could be identified by looking at the microparticle, chloride and fluoride concentrations, and possibly even sulfur isotope ratios. Anthropogenic influence could be quantified by analyzing the snow for such constituents as ammonium, nitrogen and sulfur isotopes, and various trace metals. Stratospheric versus tropospheric inputs can be distinguished using ^7Be and ^{210}Pb as a tracer (Dibb, 1990).

Glaciochemical records can provide a wide range of "global change" records over varying time frames. In the future, both time series and regional survey studies should be expanded in order to better establish the chemistry of the remote atmosphere and how it has changed through time. Full realization of the potential climatic information recorded in HAM/LL ice cores has been established by the work of Thompson et al., (1986) and Mayewski et al., (1984). Other locations for such studies do exist in South America, Asia and possibly Iryan Jaya. Proper selection of HAM/LL core sites with respect to elevation and latitude is necessary if samples, unaffected by post-depositional effects, are to be recovered. For example, our work indicates that southern and central Iceland are not appropriate sites for coring activities due to the "melt-out" of the glaciochemical record. Information for accumulation rates and glacier dynamics also must be available before coring programs can be developed (Wagenbach, 1989). Regional records from the Pamirs, Caucuses, the southern Andes, Iryan Jaya as well as the Canadian Arctic would be extremely useful and informative. More detailed ice coring from any of these areas should also be encouraged.

Acknowledgements

Some of the glaciochemical work summarized above was collected and analyzed on NSF grants to P. A. Mayewski and W. B. Lyons. This support is gratefully acknowledged. We greatly appreciate the typing efforts of Ms. Elizabeth Gadsden and Ms. Therese Hyatt.

REFERENCES

Allison, I.F. 1980. A preliminary investigation of the physical characteristic of the Yashel Glacier, Heard Island. ANARE Scientific Reports, Ser. A (IV). Glaciology Publication No. 128.

Ambach, W. and H. Eisner. 1966. Analysis of a 20 m firn pit on the Kesselwandferner (Oetztal Alps). *J. Glaciol. 6,* 223-231.

Ambach, W., W. Rehwald, M. Blumthaler, M. Eisner and P. Brunner. 1988. Chernobyl fall-out on glaciers in the Austrian Alps. *J. Glaciol. 34,* 255-256.

Barrie, L. A., D. Fisher and R. M. Koerner. 1985. Twentieth century trends in Arctic air pollution revealed byconductivity and acidity observations in snow and ice in the Canadian high Arctic. *Atmos. Environ. 19,* 2055-2063.

Barry, R. G. 1981*Mountain Weather and Climate.* New York, Methuen pp. 313.

Batifol, F.M. and Boutrol, C.F., 1984. Toxic metals and metalloids in high altitude alpine glaciers snow and ice. Physico-Chemical Behavior of Atmospheric Pollutants, 3rd European Symp. 395-403.

Bjornsson, H. 1979. Glaciers in Iceland. Jokull 29, 74-80.

Briat, M. 1978, Evaluation of levels of Pb, V, Cd, Zn and Cu in the snow of Mt. Blanc during the last 25 years. Atmospheric Pollution Proceedings of the 13th International Colloquium, Paris, France, 225-228.

Butler, D., W. B. Lyons, P. A. Mayewski and J. Hassinger. 1980. Shallow core snow chemistry of Athabasca Glacier, Alberta. *CJES* 17, 273-281.

Chen, K. and J. M. Bowler. 1986. Late Pleistoscene evolution of salt lakes in the Qaidam Basin, Qinghai Province, China. *Palaeogeo. Palaeoclim. Palaeoeco. 54,* 87-104.

Crozaz, G., C. C. J. Langway and E. Picciotto. 1966. Artificial radioactivity reference horizons in Greenland firn. *Earth Planet. Sci. Lett. 1, 42-48.*

Dansgaard, W., S. J. Johnsen, H. B. Clausen and N. Gunderstrup. 1973. Stable isotopes in glaciology. *Medd. Groenl.* 197, 1-53.

Delmas, R. and A. Aristarian 1978. Recent evolution of strong acidity of snow at Mt. Blanc. *Atmospheric Pollution - Proceedings of the 13th International Colloquium,* Paris, France 233-237.

Delmas, R. J., M. Legrand and G. Holdsworth. 1985. Snow chemistry on Mount Logan, Yukon Territory, Canada. *Ann. Glaciol.* 7, 213.

Dibb, J.E., 1990. Recent deposition of ^{210}Pb on the Greenland ice sheet: Variations in space and time. Ann Glaciol. 14, 51-54.

Gaggeler, H., H. R. v. Gunten, E. Rossler, H. Oeschger and U. Schotterer. 1983. 210Pb-Dating of cold alpine firn/ice cores from Colle Gnifetti, Switzerland. J. *Glaciol.* 29,165-177.

Gil'ad, D. and J. Bonne, 1990. The snowmelt of Mt. Hermon and it's contribution to the sources of the Jordan River. J. Hydro. 114, 1-15.

Gonfiantini, R. 1970, In: Isotope Hydrology, IAEA, p. 56.

Grootes, P.M., M. Stuiver, L.G. Thompson and E. Mosley-Thompson, 1989. Oxygen isotope changes in tropical ice, Quelccaya, Peru. J. Geophys. Res. 94, 1187-1194.

Gunten, H. R., V. E. Rossler and H. Gaggeler. 1982. Dating of ice cores from Vernagtferner (Austria) with Fission products and Lead-210. Z. *Gletscherkd. Glazialgeol 18,* 37-45.

Haeberli, W., H. Gaggeler, U. Baltensperger, D. Jost and U. Schotterer. 1988. The signal from the Chernobyl accident in high-altitude firn areas of the Swiss Alps. *Ann. Glaciol.* 10, 48-51.

Haeberli, W., Schotterer, D. Wagenbach, H. Haeberli-Schwitter and S. Bortenschlager, 1983. Accumulation characteristics on a cold, high-alpine from saddle from a snow-pit study on Colle Gnifetti, Monte Rosa, Swiss Alps. J. Glaciol. 29, 260-271.

Hammer, C. U., H. B. Clausen, W. Dansgaard, N. Gundestrup, S. J. Johnsen and N. Reeh. 1978. Dating of Greenland ice cores by flow models, isotopes, volcanic debris and continental dust. J. *Glaciol. 20,* 3-26.

Han, J., T. Zhou and M. Nakawo. 1989. Stratigraphic and structural features of ice cores from Chonce Ice Cap, West Kunlun mountains. *Bull. Glac. Res. 7,* 21-28.

Hansen, J., I. Fung, A. Lacis, D. Rind, S. Lebedeff, R. Ruedy and G. Russel. 1988. Global climate changes forecast by Goddard Institute for Space Studies three-dimensional model. J. *Geophys. Res. 87,* 3052-3060.

Herron, M. M. 1982. Glaciochemical dating techniques. *Nuclear and Chemical Dating Techniques: Interpreting the Environmental Record, ACS Symposium Series No. 176,* 303-318.

Holdsworth, G. 1986. Evidence for a link between atmospheric thermonuclear detonations and nitric acid. *Nature 324,* 551-553.

Holdsworth, G., H. R. Krouse, M. Nosal, M. J. Spencer and P. A. Mayewski 1989. Analysis of a 290-year net accumulation time series from Mt. Logan, Yukon. *Snow Cover and Glacier Variations,* 71-79.

Holdsworth, G. and E. Peake. 1985. Acid content of snow from a mid-troposphere sampling site on Mount Logan, Yukon Territory, Canada. *Ann. Glaciol. 7,* 153-160.

Holdsworth, G., M. Pourchet, F. A. Prant. and D. P. Meyerhof. 1984. Radioactivity levels in a firn core from the Yukon Territory, Canada. *Atmos. Environ.* 18, 461-466.

Jouzel, J., L. Merlivat and M. Pourchet. 1977. Deuterium, tritium and B activity in a snow core taken on the summit of Mont Blanc (French Alps). Determination of the accumulation rate. J. *Glaciol.* 18, 465-470.

Lambeth, A.J., 1951. Heard Island. Geography and glaciology. J. Proceed. Royal Soc. NSW 84, 92-98.

Legrand, M. R. and R. J. Delmas. 1988. Formation of HCl in the Antarctic atmosphere. J. *Geophys. Res 93,* 7153-7168.

Li, J. and S. Xu. 1980. The distribution of glaciers on the Qinghai-Xizang Plateau and its relationship to atmospheric circulation. *In* Miller, K., *ed. International Karakoram Project*, Vol. I, Cambridge, Cambridge University Press, pp. 84-93.

LIGG. 1980. *Glacier Inventory of China I - Qilian Mountains*. Lanzhou, PRC, Lanzhou Institute of Glaciology and Geocryology, Academia Sinica pp. 249.

Lyons, W.B., P.A. Mayewski, L.G. Thompson and B. Allen III, 1985. The glaciochemistry of snow-pits from Quelccaya Ice Cap, Peru, 1982. Ann Glaciol. 7, 84-88.

Lyons, W. B., P. A. Mayewski, M. J. Spencer and M. S. Twickler. 1990. Nitrate concentrations in snow from remote areas: implications for the global NO_x flux. *Biogeochem. 9*, 211-222.

Mayewski, P. A., W. B. Lyons and N. Ahmad 1981. Reconnaissance glacio-chemical studies in the Indian Himalayans. *38th Annual Eastern Snow Conference, Syracuse, New York*, 45-48.

Mayewski, P. A., W. B. Lyons and N. Ahmad. 1983. Chemical composition of a high altitude fresh snowfall in the Ladakh Himalayas. *GRL* 10, 105-108.

Mayewski, P. A., W. B. Lyons, G. Smith, N. Ahmad and M. Pourchet. 1984. Interpretation of the chemical and physical time-series retrieved form Sentik Glacier, Ladakh, Himalayas. J. *Glaciol.* 30, 66-76.

Mayewski, P. A., W. B. Lyons, M. J. Spencer, M. Twickler, W. Dansgaard, B. Koci, C. I. Davidson and R. E. Honrath. 1986. Sulfate and nitrate concentrations from a south Greenland ice core. *Science* 232, 975-977.

Mayewski, P. A., M. J. Spencer, W. B. Lyons and M. S. Twickler. 1987. Seasonal and spatial trends in South Greenland snow chemistry. *Atmos. Environ.* 21, 863-869.

Mayewski, P. A., M. J. Spencer, M. S. Twickler and S. Whitlow. 1990. A glaciochemical survey of the Summit region, Greenland. *Ann. Glaciol.* 14, 186-190.

Mayewski, P.A., M.J. Spencer and W. B. Lyons in press. A review of glaciochemistry with particular emphasis on the recent record of sulfate and nitrate. Proc. of the 1988 Global Change Inst. Snowmass, Co.

Miller, M. M., J. S. Leventhal and W. F. Libby. 1965. Tritium in Mt. Everest Ice-Annual glacier accumulation and climatology at equatorial altitudes. J. *Geophys. Res 70.*

Oerter, H. and W. Rauert. 1982. Core drilling on Veragtferner (Oetztal Alps, Austria) in 1979: Tritum contents. Z. *Gletscherkd. Glazialgeol.* 18,13-23.

Oeschger, H. and C. C. J. Langway 1989. *The Environmental Record in Glacier and Ice Sheets.* New York, John Wiley & Sons, pp. 401.

Picciotto, E. and S. Wilgain. 1963. Fission products in Antarctic snow: A reference level for measuring accumulation. J. *Geophys. Res.* 68, 5965-5972.

Pourchet, M., J. F. Pinglot, L. Reynaud and G. Holdsworth. 1988. Identification of Chernobyl fall-out as a new reference level in Northern Hemisphere glaciers. J. *Glaciol.* 34, 183-187.

Price, L. W. 1981 .*Mountains and Man: A Study of Processes and Environment.* Berkeley, University of California Press pp. 508.

Reeh, N. 1989. Dating by Ice Flow Modeling: A useful tool or an exercise in applied mathematics. *In* Oeschger, H. and C. C. J. Langway, *ed. The Environmental Record in Glacier and Ice Sheets,* New York, John Wiley & Sons, pp. 141-160.

Smith, V.R. 1987. Chemical composition of precipitation at Marion Island (Sub-Antarctic) Atmos. Environ. 21, 1159-1165.

Spencer, M.J., P.A. Mayewski and W.B. Lyons, and M.R. Hendy, 1985. A preliminary assessment of the potential application of glaciochemical investigations on Heard Island, South Indian Ocean, J. Glaciol. 31, 233-236.

Stichler, W., D. Baker, H. Oerter and P. Trimborn. 1982. Core drilling on Vernagtferner (Oetztal Alps, Austria) in 1979: Deuterium and Oxygen-18 contents. Z. *Ietscherkd. Glazialgeol. 18,* 23-35.

Thompson, L.G. 1979. Ice core studies from Mt. Kenya, Africa and their relationship to other tropical ice core studies. In: Sea Level, Ice and Climate Change, IAHS Pub. No. 131, 55-62.

Thompson, L.G., S. Hastenrath and B.M. Arnao, 1979. Climatic ice core records from the tropical Quelccaya Ice Cap, Science 203, 1240-1243.

Thompson, L.G., E. Mosley-Thompson, P.M. Grootes, M. Pourchet and S. Hastenrath, 1984. Tropical glaciers: Potential for ice core paleoclimatic reconstructions. J. Geophys. Res. 89, 4638-4646.

Thompson, L.G., E. Mosley-Thompson, J.F. Bolzan and B.R. Koci, 1985. A 1500-year record of tropical precipitation in ice cores from the Quelccaya Ice Cap, Peru, Science.

Thompson, L. G., E. Mosley-Thompson, W. Dansgaard and P. M. Grootes. 1986. The Little Ice Age as recorded in the stratigraphy of the tropical Quelccaya ice cap. *Science* 234, 361-364.

Thompson, L. G., E. Mosley-Thompson, M. E. Davis, J. F. Bolzan, J. Dai, T. Yao, N. Gundestrup, X. Wu, L. Klein and Z. Xie. 1989. Holocene-Late Pleistocene climatic ice core records from Qinghai-Tibetan plateau. *Science* 246, 474-477.

Thompson, L. G., X. Wu, E. Mosley-Thompson and Z. Xie. 1988. Climatic records from the Dunde ice cap, China. *Ann. Glaciol.* 10, 178-182.

Vincent, C.E., T.D. Davies and P. Brimblecombe, 1979. The Lewis Glacier (Mt. Kenya) and possible links with tropical climate. In: Sea Level, Ice and Climatic Change, IAHS Publ. No. 131, 63-78.

Wagenbach, D., K.O. Munnich, U. Schotterer and H. Oeschger, 1988. The anthropogenic impact on snow chemistry at Colle Gnifetti, Swiss Alps. Ann Glaciol. 10, 183-187.

Wagenbach, D. 1989. Environmental Records in Alpine Glaciers. *In* Oeschger, H. and C. C. J. Langway, *ed. The Environmental Record in Glaciers and Ice Sheets,* New York, John Wiley & Sons, pp. 69-84.

Wake, C. 1989. Glaciochemistry as a tool to delineate the spatial variation of snow accumulation in the Karakoram Himalaya, northern Pakistan. *Ann. Glaciol.* 13, 279-284.

Wake, C., P. A. Mayewski and M. J. Spencer. 1990. A review of central Asian glaciochemical data. *Ann. Glaciol. 14.*, 301-306.

Wake, C. and P. A. Mayewski. in prep. The chemical content of snow from the Tien Shan and Bogda Shan, north-west China

USE OF SNOW AND FIRN ANALYSIS TO RECONSTRUCT PAST ATMOSPHERIC COMPOSITION

Albrecht Neftel
Physics Institute, University of Bern
Abteilung für Klima und Umweltphysik
Sidlerstrasse 5
CH - 3012 Bern, Switzerland

1. Introduction

There is no doubt that the knowledge about temporal and spatial variability of trace constituents is crucial for understanding the interaction between climate and atmospheric chemistry as well as the effects of pollutants upon vegetation.

These days, a large amount of money is spent on characterizing a whole zoo of different atmospheric trace constituents and to learn how they interact with each other. This detailed knowledge about the spatial and temporal abundance of chemical trace species in the atmosphere is limited to the most recent few years, with very few records extending more than 20 years into the past (Meteorological measurements are avaiable further back in time, but still, no records date back to more than 400 to 500 years ago). The recent global changes of atmospheric constituents due to anthropogenic emissions are often discussed as excursions of deviations from "natural" conditions, which must be derived indirectly.

Alpine glaciers and polar ice sheets have accumulated quietly and free of charge the precipitation of the past. These stores of precipitation have incorporated information about atmospheric composition at the time when the snow was formed and deposited.

These cold archives offer a unique possibility tracing back in time the variations of different atmospheric trace constituents and of learning from them the interplay between atmospheric chemistry and climate. Polar and alpine archives do not provide direct information about atmospheric composition. Measured parameters in ice cores (PIC's) have to be translated into "atmospheric variables".

It is the scope of this contribution to discuss how PIC's can be quantitatively related to atmospheric variables, to determine which components can be reconstructed and to identify where the obstacles are. Of course, this cannot be done for all PIC's ever measured. Because this contribution is part of a workshop entitled " Processes of

NATO ASI Series, Vol. G 28
Seasonal Snowpacks
Edited by T. D. Davies et al.

chemical change in snowpack ", the discussion will focus upon chemical constituents and possible chemical interactions in snow and ice. I restrict myself to the discussion of snow and ice archives in remote locations with temperatures well below the freezing point, "eternal" snow fields as contrast with seasonal snow fields, where such "ugly" things as meltwater percolate through the snow, making the interpretation much more complicated.

2. How do the cold archives work?

Figure 1 shows a cross-section of an ideal ice sheet. Greenland comes close to this model. Ice is a highly viscous fluid which flows under its own pressure from the inside to the outside. Snow is accumulated in the interior higher elevation and is lost at the edges by calving of ice bergs and due to melting in the summer period. By drilling a core in the middle of such an ice sheet, a complete chronology of past precipitation is recovered. Due to ice flow, yearly layers are thinner going back in time.

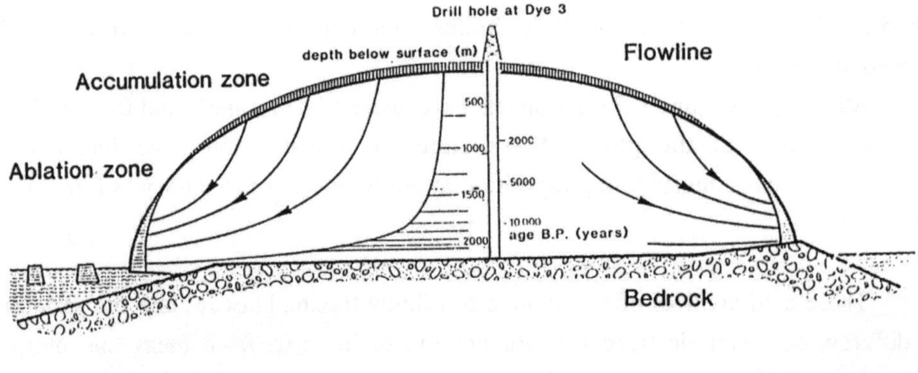

Figure 1: Cross section through an idealized ice sheet with flow lines indicated.

In reality, the stratigraphy of recovered ice cores is not as simple as that shown in Fig. 1. Drill holes are usually not located on the crest of an icesheet. The farther a hole is drilled from the middle line, the more the origin of the ice in the core deviates from the drilling location with depth. Bedrock topography influences ice flow and can disturb the stratigraphy. Fig. 2 shows an examples: the bedrock topography of the Dye 3 core

(Overgaard and Gundestrup, 1985). The whole section of the Dye 3 core originating from the last glaciation comes from depths were influences from the bedrock can be expected.

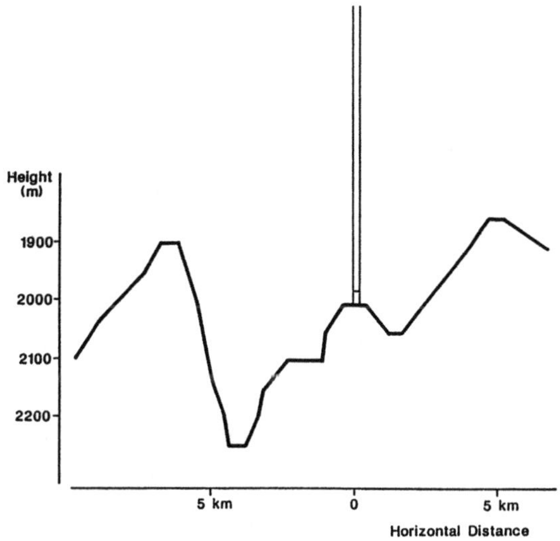

Fig. 2: Bedrock topography of the Dye 3 site.

Compared to the large ice sheets, alpine glaciers are much more complicated. They are smaller, faster flowing and thus far from being ideal ice sheets. Nevertheless, they provide useful information, especially about the deposition of chemical trace species and their recent changes due to anthropogenic emissions. Because much less ice is stored in such alpine glaciers, they cannot provide information reaching as far back in time as the polar ice sheets.

3. Criteria of drill site selection

The selection of a specific drill site depends upon the scientific question asked. The lower the temperature, the better conserved are the PIC's. But lower temperatures generally mean lower precipitation and consequently lower temporal resolution. The 2000 m core drilled at Vostok, the permanent station with the coldest mean temperature (T_{mean} = -55°C) reaches 160'000 years back in time (Barnola et al., 1987). Evidently, seasonal variations cannot be resolved in such a core. If one is interested in information on a shorter timescale, e.g. the changes since the industrial revolution, high accumulation cores are best suited, because they allow seasonal variation to be resolved . The main disadvantage that these sites experience is higher temperatures, which can lead to melting of summer snow layers.

4. The transformation of snow to ice

Fig 3 shows a scheme of the transformation of snow to ice. During the early lifetime of the snow, the sharp edges of the flakes are rounded off, because the water vapor pressure is inversely proportional to the radius of curvature (Kelvin effect). This leads to a fast reduction of the specific surface. Any species that are only absorbed on the surface and have vapor pressures similar to or larger than water vapor pressure over ice, will be transferred, proportionately to surface reduction into the gas phase. Deeper down, temperature gradients lead to evaporation and recondensation of water molecules. The molecules diffuse into the open pore space, so that the initial distribution is spread out. The mean displacement length of water molecules during the whole firnification process was determined from the smoothing of the seasonal $\delta^{18}O$ variations to be 8-10 cm (Johnson, 1977).

The snow is compressed by rearrangements of the snow grains. Above a density of about 550 kg/m^3, further rearrangement of the ice grains no longer leads to an increase in the density; instead sintering and plastic deformation become the dominant process leading to increased density. At a density of a about 800 kg/m^3, the pores are gradually pinched off to form individual bubbles (Schwander and Stauffer, 1984). These bubbles contain atmospheric air and are the best air sample container that nature could have invented.

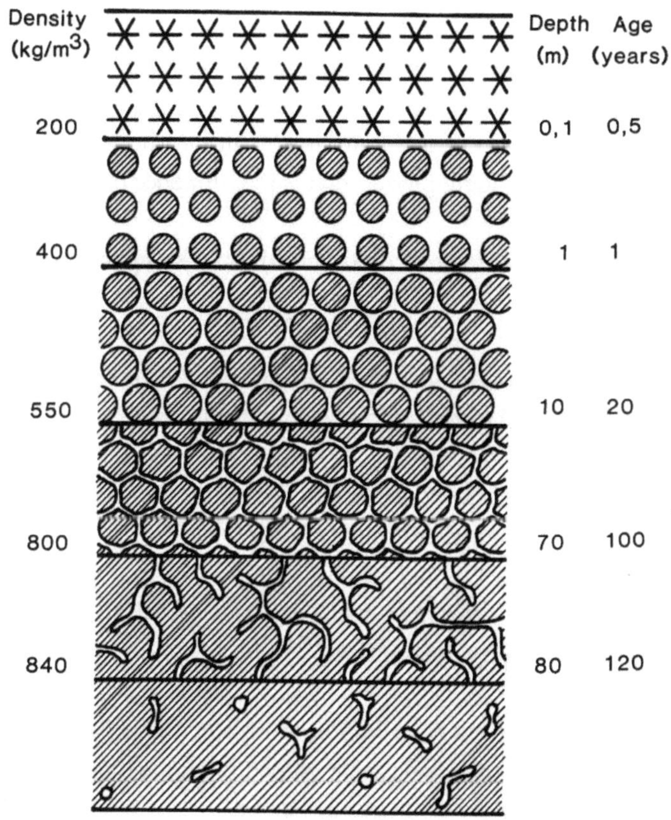

Fig. 3: Schematic transformation from snow to ice.

5. Impurities in snow and ice samples - Relationships with corresponding atmospheric concentrations

Everything that can be measured today in liquid water samples can, theoretically, also be measured in snow and ice samples. Such a list would be long, thus I restrict myself to parameters which are of interest for this workshop and where I feel able to contribute to the understanding of the relation between PIC's and corresponding atmospheric concentrations.

Additional general information about PIC's can e.g. be found in the Dahlem Workshop report "The Environmental record in Glaciers and Ice Sheets (Oeschger & Langway, 1989).

In order to trace back a measured parameter in snow or ice to its atmospheric concentration, the following steps are involved and will be discussed individually:

1) Do the reported results correspond to the real concentration originally present in the sample? If measured in the liquid phase, what is the influence of melting?

2) Are there any changes in concentration during sampling or storage and transportation of samples?

3) Does the firnification process alter the concentration? If it does, how; and is the altered concentration still linked to the atmospheric concentration?

4) What percentage of the snow deposited subsequently is blown away? What fraction of the snow deposited is retained by the surface?

5) How are the impurities incorporated into the snow, both in clouds and during deposition? What is the role of dry deposition?

It is beyond the scope of my contribution to answer these five questions for all PIC's. I restrict the discussion to a few species which illustrate some of the problems and characterize what kind of information can be gained.

Each species discussed is representative of a certain category of parameters and associated difficulties of relating ice concentration to atmospheric concentrations.

They are:

CO_2: an important, apparently chemically boring component. But, is CO_2 really as unreactive as it seems?

NO_3^- and SO_4^{2-}: Potential to learn about past S-and N-cycles from polar ice.

Peroxides: How stable is an unstable compound in the ice?

Formaldehyde (HCHO): Is this a key parameter for a better understanding of atmospheric methane cycle in the the past ?

6. The historical record of atmospheric greenhouse gases revealed from polar ice cores

It is well known that the course of changes in the atmospheric greenhouse gases CO_2 and CH_4 for the last 160'000 years can be reconstructed based on measurements from air extracted from the bubbles of polar ice (e.g. Barnola et al., 1987 , Neftel et al., 1982, 1988, Chappelaz et al., 1990). There is good evidence that the extracted air does represent the unchanged ancient atmosphere, as long as melting of the surface layer does not occur during summertime (Stauffer et al., 1985). This is also true, when the gases incorporated in the ice are transformed into clathrates due to the pressure of ice masses lying above. After pressure relaxation, the original gas composition is reestablished in newly formed bubbles (Neftel et al., 1983).

The CO_2 results reported by the three laboratories performing such measurements (Laboratoire de Glaciologie, Grenoble, CSIRO, Australia and Physics Institute, Bern) differ systematically by up to 3% on parallel measured samples (Barnola et al., 1983). The reasons for these discrepancies are not fully understood, but must be related to differences in the extracting and analytical procedures, thus to inequalities on the laboratory level. More intercalibration work is needed to overcome these problems.

There are measured CO_2 signatures in polar ice cores, especially in the Dye 3 core, which suggest that a reaction occurs between CO_2 incorporated as gas and that in the ice matrix . Changes in CO_2 concentrations of up to 50 ppm have been measured over distances of only 2.5 cm in an ice core section from the last glaciation, where the occurence of melt layers is unlikely. This CO_2 step parallels similar, sharp changes in other PIC's (Figure 4). The ice with the low CO_2 concentrations has all of the signatures of ice originating from cold climate conditions; the ice with the high CO_2 concentrations has the signature of an interglacial origin (Staffelbach et al., 1988). The cold signature is characterized by a higher load of alkaline dust, mainly calcium carbonate. This was measured recently in our laboratory and revealed a calcium concentration change, perfectly parallel to the CO_2 step going from 0.2 μM to 3 μM from the "warm" side to the "cold" side. Hammer (pers. comm.) suspected that CO_2 might be trapped by alkaline spots in the ice and that, consequently, the composition of the extracted air from such samples no longer represents the true atmospheric one.

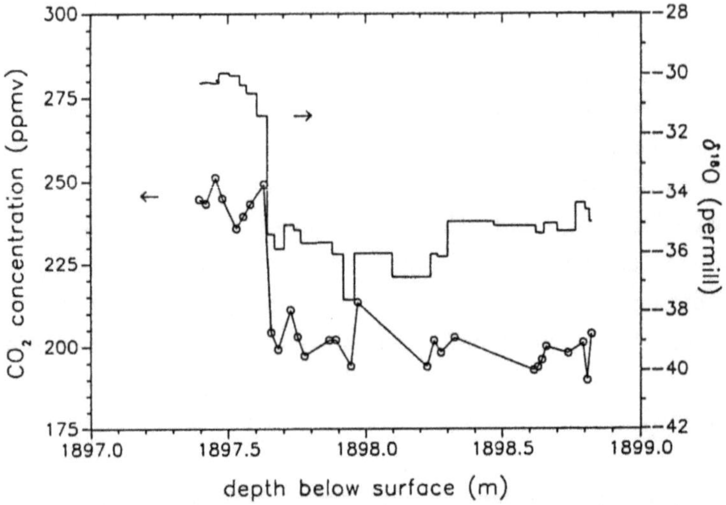

Fig. 4: Rapid transition of different PIC's at a depth of 1897 m.b. surface of the Dye 3 deep core (after Staffelbach et al., 1988).

There are a number of arguments against this hypothesis. If a kind of chemical equilibrium is installed between the CO_2 in the gas or clathrate phase and the ice, the pCO_2 would depend upon the alkalinity in the ice. During the glacial maxima 30'000-20'000 years B.P., alkalinity values show a large variability in cores from different locations, whereas the CO_2 concentrations ar all around 200 ppm. Greenland ice cores were alkaline, whereas the ice from the Byrd core was acidic. Alkalinity is mainly controlled by carbonate equilibria. Episodes with an elevated acidity level, e.g. after volcanic eruption, should lead to an additional release of CO_2 from carbonate incorporated in the ice. Such a response has not yet been observed.

For Holocene conditions, the potential contribution from carbonate to the total CO_2 content is up to 50 ppm, and it is much higher for ice from the last glaciation.

Another possible source for CO2 in ice core is the oxidation of different organic compounds, such as formic acid or formaldehyde. The total level of such compounds is of the order of 0.3 μmol/kg ice. If all of these organic compounds were quantitatively oxidized, 50 ppm additional CO_2 would be produced. The potential is there, but a systematic increase of CO_2 with increasing age, accompanied by a parallel decrease of organic compounds, has not been measured.

In the Summer 1990 two parallel deep drilling operations at the highest elevation in Central Greenland began, the American GISP 2 (Greenland Ice Sheet Project) and the European GRIP (Greenland Research Ice Project). The drill sites are 30 km apart.

At these drill sites, the transition to the last glaciation will occur far from the bedrock, which excludes many of the uncertainities encountered at Dye 3. The mean air temperature of -30°C also guarantees the exclusion of a meltwater influence on these future cores. Hopefully, the competition between the two drilling projects to be the first to reach the bedrock does not disable most of the scientific action.

Summary: The statement that CO_2 is a "chemically boring" substance in snow and ice can be safely made. The same is true of N_2O and CH_4. This statement underlines the importance of the polar ice cores in reconstructing the past atmospheric composition of these important Greenhouse gases. The main difficulty in interpreting "greenhouse data" from polar ice cores is the exact determination of the age of the enclosed gas.

7. Nitrate and sulfate concentrations in snow and ice samples

Nitrate and sulfate are the oxidized forms of nitrogen and sulfur species in the atmosphere. Precursors of nitrate are mainly NO and NO_2, whereas sulfate has SO_2 and organic sulfur compounds, such as DMS as precurors. Other forms of nitrogen oxides besides NO_3^- or HNO_3 have not been found to date in polar samples. Sulfur is present in the highest oxidation state as sulfate. Only small amounts of sulfur have been found, in the form of methane sulfonate acid (MSA) (Legrand and Sage, 1988).

Let us first take a look at sulfate: Theoretically sulfate has two sources: a) deposition of aerosols, either sulfate particles or sulfuric acid droplets and b) deposition of S(4) with successive oxidation to S(6). Valdez et al., (1987) showed that the latter process might occur on snow surfaces, if enough SO_2 is present in the gas phase. The corresponding deposition velocities are in the range of 0.05 - 0.1 cm/sec for temperatures near the melting point and clearly lower for colder temperatures.

Field data from alpine and polar snow (thus from remote regions) show that the contribution of S(4) to the total sulfur deposition can be neglected (Neftel et al., 1987). In polar regions, gas phase concentrations of SO_2 are far below 1 μg/cm3 and dry deposition velocities are of the order of 0.05 cm/sec or less. The potential contribution to the total sulfate deposition does not exceed a few percent. The situation is different for the alpine region in Europe, where mean SO_2 air concentration of 1-3 μg/m3 between 2000 and 3000 m.a.s.l. can be expected (Baltensberger et al, 1989). There, dry deposition of S(4) amounts at best to 10% of the total deposited sulfate.

What happens if such an SO_2 molecule hits the surface? (Best way to answer this question: Ask Roger Bales.) There is a good chance that nothing at all happens and that the SO_2 molecules leave the snow surface again. This is the impression that we get from the following findings (Sigg et al., 1987):

1) The pattern of sulfate and acidity in the snow does not change with storage time in the snow or ice field, as long as the temperature stays well below 0°C. Sulfate always shows the typical aerosol signal, with a very noisy pattern. No smoothing of sulfate occurs during the firnification process.

2) Snow layers with higher alkalinity should have a higher uptake rate of SO_2, which we did not observe.

3) The observed decrease of H_2O_2 in alpine snow and ice with storage time is not linked with a corresponding increase of sulfate, so that the oxidation of S(4) to S(6) by means of H_2O_2 cannot be resonsible for the observed decrease.

The situation for nitrate is different. Nitrate comes from aerosols and gaseous HNO_3. Fig. 5 shows a high-resolution pattern of sulfate and nitrate from a study near Dye 3 (South Greenland). Nitrate shows a smoother pattern than sulfate, because gaseous HNO_3 is displaced in the firn in a similar manner to water molecules. This implies that HNO_3 molecules have a sticking coefficient, that is not too far away from that of water molecules. It is difficult to guess whether dry deposition of HNO_3 contributes to total nitrate deposition. In a cold snow cover an equilibrium between HNO_3 in the air and /or on the snow surface might be established. Each snow layer has therefore a corresponding HNO_3 concentration, which will depend among other physical parameters on the alkalinity. The sign of the HNO_3 flux from such a layer will depend upon the difference between the initial concentration in the snow and the "hypothetical" equilibrium concentration. So it is likely that a net flux from the snow surface to the air exists. Recently published measurements of NO_3^- on low accumulation cores in the Antarctic support this hypothesis (Mayewski and Legrand, 1990). Deeper down in a snow cover, escaped HNO_3 molecules will be trapped on another ice surface.

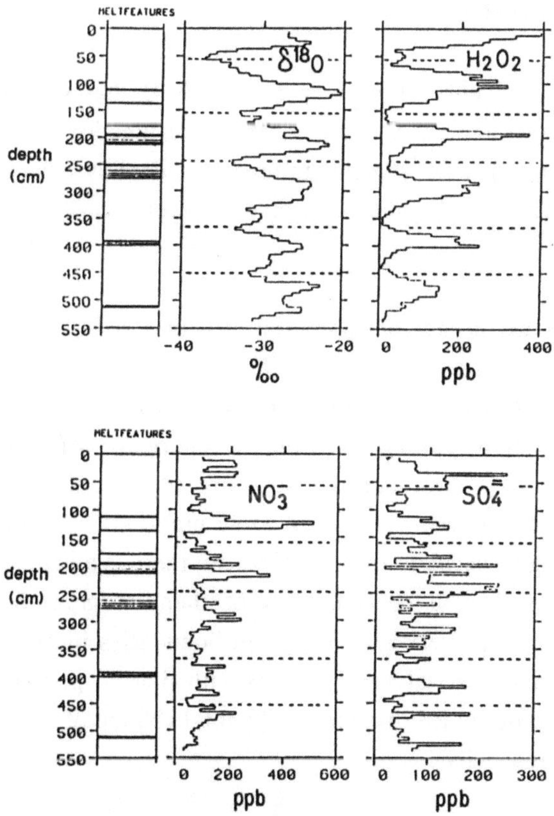

Fig. 5: $\delta^{18}O$, H_2O_2, SO_4^{2-} and NO_3- concentration in a 5 meter pit near Dye 3 (South Greenland). The time resolution is approx. 20 samples/year (after Beer et al., 1990).

Summary:

The large scatter found in measurements relating snow concentration directly to air concentration of aerosol based parameters (Pourchet et al, 1983) is due to complicated washout processes taking place above the snow layer. Sulfate and nitrate concentration behave as conservative tracers once they are within a snow or ice cover. If one is using mean concentrations as being representative for larger time periods (a few years or more) the stochastic fluctuation of the washout processes will be averaged out, and the concentrations will be proportional to mean atmospheric concentrations, as long as no systematic shift in the precipitation pattern occurred (Stauffer and Neftel, 1988).

8. Where are the impurities located in the snow?

There is experimental evidence (Mulvaney et al., 1989) that sulfuric acid is located at the junctions where three snow grains meet (triple junctions). There, the concentration can be so high, that even at temperatures considerably below zero, the liquid phase is maintained. These triple junctions would form a network of veins in the polycrystalline ice.

At the surface of an ice crystal a disordered structure exists, the thickness depends in a non-linear way on the temperature. The most direct evidence of the existence of such a layer was published by Golecki and Jaccard in 1978. They measured the distribution of back-scattered protons at different energies, corresponding to different penetration depth. To a disordered ice structure corresponds a disordered back scatter pattern. The thickness was determined to 90 nm at -1°C, 30 nm at -10°C and 10 nm at -30°C. One might speculate that, in this layer, there is a higher solubility of many impurities and, therefore, a concentration of them at grain boundaries. Whether such a liquid structure also exists at intergrain boundaries in polycrystalline natural ice is uncertain. Because crystal size increases with time in permanent snow fields, the relative ratio of disordered ice structure to the total volume of ice decreases, as well as the total "chemical reactivity".

The electrical conductivity measurement (ECM), a kind of solid state pH-measurement, was introduced by Hammer back in the late 70ies (Hammer, 1980). In its simplest form, a pair of copper electrodes (area 1 mm^2, distance 10 mm) with an applied tension of 1000 Volt is drawn over a fresh cleaned flat surface of an ice core. The resulting transient current is related to the acidity in the ice. The relation between measured current and acidity in the ice is given by the equation

$$[H+] = 0.045 * I^{1.73} \ (T = -15°C) \ mEq/kg$$

I: measured current (mA)

The method only works for acidic ice, alkaline ice does not show any measurable current. What are the implications from ECM for the distribution of acidic impurities in the ice? A simple mechanistic model to explain qualitatively the ECM current assumes homogenous distribution of proton defects in the ice (Neftel et al, 1985). This is supported by ECM measurements using very thin electrodes. Measurements along grain boundaries and within individual crystals have been made and no difference in the ECM signal was observed (Hammer, pers. comm.).

The ECM signal shows another feature, which is related to chemical processes in the ice: The aging effect. Briefly, if an ice sample following a longer storage period is remeasured the signal dies out in the surface layer. If this layer is removed, the original signal returns, as shown in Fig. 6. Schwander et al., 1983, showed that NH_3 penetrating the ice leads to a similar die-out of the ECM signal. This penetration proceeds with an astonishingly high diffusion constant. Because NH_3 is always present if scientists are around (they are also human beings) and a piece of an ice sheet is only converted to a sample by the action of a person, NH_3 penetrating into the surface layer could be responsible for the observed loss of ECM signals.

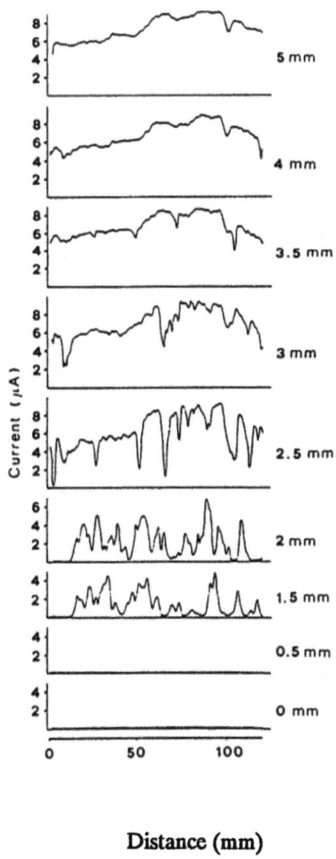

Fig. 6: Aging effect related to ECM signal. Recovery of the signal as function of the removed surface (after Schwander et al., 1983).

This is contradictory to the picture of a network of liquid veins in the ice. The recovery of the signal was linear to the amount of removed surface layers. If the neutralisation process took place in a liquid vein system, it would spread out deeper and more inhomogenously in the ice.

NH_3 is one of the interesting molecules in ice. If it is incorporated in this ionic form, mainly as ammoniumsulfate, it behaves as a conservative tracer. In the gaseous form, it is apparently able to penetrate into the ice matrix and to neutralize acidity, where it is then fixed. Ammonium levels in polar are generally very low. After recovery of the cores, they are exposed to relatively high atmospheric NH_3 concentrations. As a consequence, ammonium measurements should be carried out directly in the field, immediately after core drilling to avoid contamination. Such measurements were performed by Thomas Staffelbach last summer during the Eurocore operation in summit, Greenland. These continuous, high resolution NH_4^+ measurements along some core sections, revealed an interesting insight into the behaviour of ammonium concentrations in the ice, as can be seen from Figure 7. The sharpness of the ammonium peak suggests that ammonium was incorporated as ammoniumsalt and not as NH_3. At this stage, it is too early to give an interpretation of the causes of the the ammonium peaks. In his thesis, Staffelbach consideres forest fires (biomass burning) and NH_3 emissions from large bird colonies. NH_3 emissions from both sources are reported in the literature. Currently, the evaluation of a 600 meter ammonium record from the GRIP core from Summit is in progress.

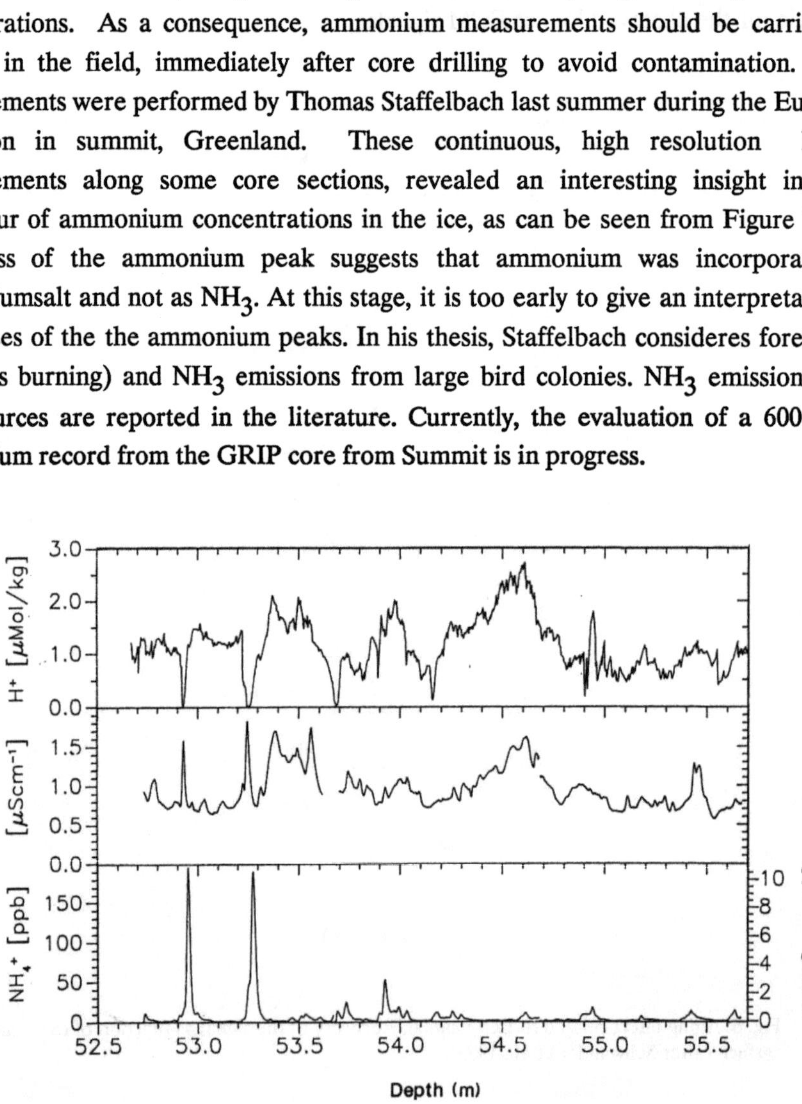

Fig. 7: Ammonium concentration, liquid conductivity and ECM signal in a core section from Summit Greenland (52-55 m.b.surface)

9. H_2O_2 in snow and ice

It was discovered 1984 that H_2O_2 is one of the dominant trace species in polar ice (Neftel et al., 1984). As with other parameters that have been measured, the question arose, how the ice H_2O_2 concentration could be translated into atmospheric concentrations. Atmospheric H_2O_2 is, in many respects, an interesting substance. It is the most important oxidant for the conversion of S(4) to S(6) in clouds, under acidic conditions (Penkett, 1979). It is produced by recombination of two HO_2 radicals. Photolytical decay yields two OH radicals. H_2O_2 is a sort of odd hydrogen radical reservoir. Figure 8 shows a simplified scheme of the processes in the atmosphere, where the family of odd hydrogen (OH, HO_2 and H_2O_2) is involved (after Logan et al., 1981). The relative stability of H_2O_2 and the possibility of measuring atmospheric concentrations of this substance make H_2O_2 a potential parameter for verifing atmospheric chemistry models. The question is on the table: Do the H_2O_2 concentrations in polar ice sheets provide an easy way to trace back the atmospheric oxidation capacity?

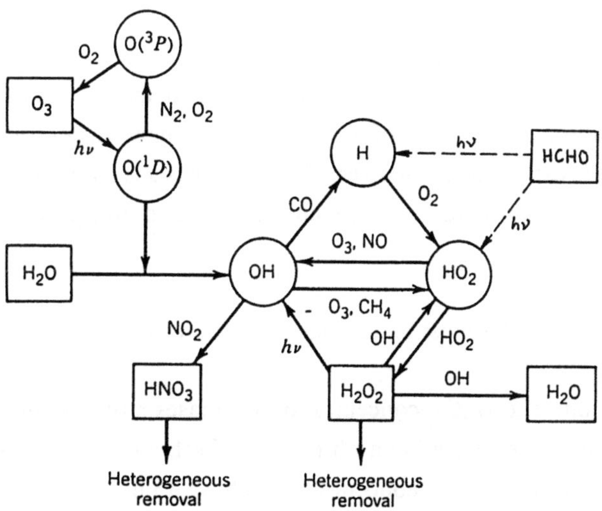

Fig. 8: Reaction scheme of odd hydrogen in the atmosphere (after Logan et al., 1981).

There is good evidence that during snow formation, H_2O_2 is incorporated by a so called co-condenation process. In snowflakes, the same ratio between H_2O_2 and H_2O molecules is established as in the gas phase (Jacob, 1987). Thus knowing the absolute water content and the H_2O_2 concentration of the falling snow one can calculate the atmospheric H_2O_2 concentration. This model was verified in the field by measuring freshly taken snow and the air concentration with a cryogenic sampling technique (Sigg and Neftel, 1988). What is the meaning of such a calculated air concentration? Unfortunately not too much. Because of the extremely high solubility of H_2O_2 in a cloud, over 90% of the total H_2O_2 will be dissolved in the droplets, whereas the larger part of water molecules remain in the gas phase.

At temperatures where supercooled water drops coexist with ice particles, the latter grow at the expense of the former. This might occur by water vapor diffusion to ice particles (deposition) and/or by drops colliding with and freezing on the ice particles (rimming). The H_2O_2 concentration in freshly formed snow will depend upon the complicated process of snow formation. Especially in rimmed ice particles, the fast freezing process of the impacted water will trap H_2O_2 in an unpredicible way, so that higher concentrations, as predicted by the co-condensation model will be present. In the case where all liquid water in a cloud is transformed into ice, the co-condensation model will be valid.

But fortunately, all of these complicated in-cloud processes are not so important for the H_2O_2 recorded in a snow cover. The deposited snow loses its memory for the original concentration by a number of processes , and a new concentration that depends on various factors is established . Andreas Sigg PhD thesis is a perfect compilation of all that is known today about H_2O_2 in snow and ice. The following statements are based on his work.

Seasonal snow covers must be distinguished from ice cores polar ice sheets.

a) Alpine region: The H_2O_2 concentration decreases strongly with the aging of the snow. Levels up to 1 ppm have been observed in fresh snow, whereas firn cores from cold alpine glaciers rarely exhibit concentrations above 15 ppb with concentrations most below 5 ppb.

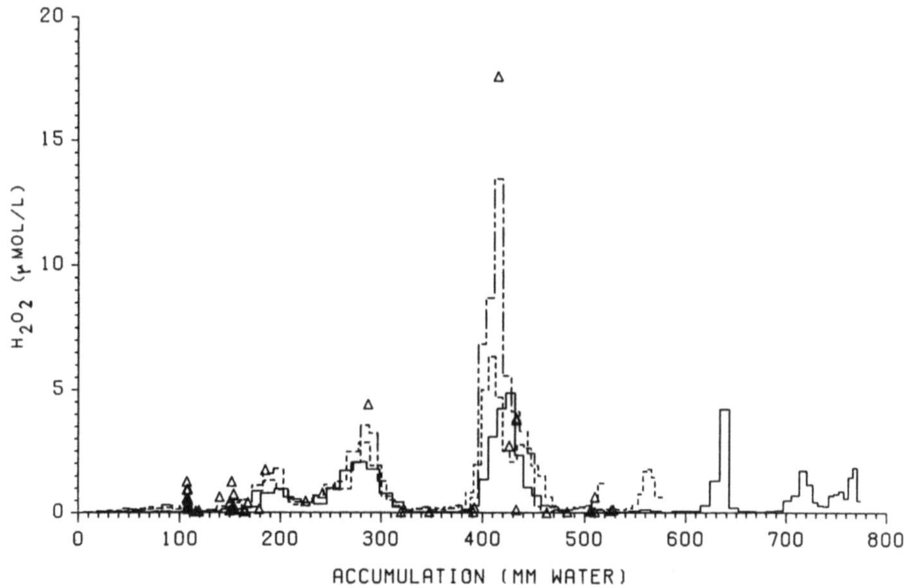

Fig. 9: H_2O_2 record of three consecutive profiles. Surface snow samples, freshly taken and stored in the dark are plotted with triangles at the corresponding accumulation.

Extensive studies were conducted in the experimental snow field at the Swiss Avalanche and Snow Research Institute at the Weissfluhjoch in the Swiss alps (Sigg et al., 1987). Fig. 9 illustrates the observed loss of H_2O_2 in consecutive profiles. The mechanism causing this loss is still not understood, most probably it is a light-enhanced catalytic destruction on dust particles. A closer inspection of the Weissfluhjoch data shows, that the loss rate is not dependent on snow temperature, but decreases if light is shed by snow masses lying above. As already mentioned , the loss of H_2O_2 is not accompanied by a corresponding molar increase of sulfate, thus oxidation of SO_2 cannot explain the loss. It is known that H_2O_2 can be produced or destroyed catalytically on semi conductor surfaces (Kormann et al., 1988). Dust particles are always present in alpine snow and could serve as destruction, that is to say oxidation centers. To summarize , it is evident that in the alpine regions peroxides are not preserved in snow and ice. There is absolutely no possibility to reconstruct atmospheric concentrations from such samples .

b) Polar regions: In cores from the Greenland and Antarctic ice sheets, H_2O_2 is found throghout the Holocene, but virtually no H_2O_2 in ice from the glaciation was found (Neftel et al., 1986). It is evident that H_2O_2 is conserved far better in polar snow and ice than in alpine snow masses. Andreas Sigg measured continuous H_2O_2 profiles reaching over 100 years back in time on two Greenland ice cores Summit and Dye 3, and of the Siple core, Antarctica. Table 1 gives the main characteristics of these drill sites.

Table 1

Site	Location	Mean annual air temperature [°C]	Mean annual accumulation [meter water equivalent]
Siple	75°55'S 83°55'W	- 24.0	0.5
Dye 3	65°11'N	- 19.6	0.5
Summit	72° N 39° N	- 31.8	0.23

In polar regions snow falls mostly as fine dendrites with a thickness around 20 - 100 μm (Pruppacher and Klett, 1978). H_2O_2 will diffuse out, as long as the H_2O_2 partial pressure outside the snow crystals is smaller, as the partial pressure, which correponds to an equilibrium concentration. Schwander calculated in his Masters thesis the time needed for 90% of a gas trapped in ice to diffuse out to be:

$0.046\ a^2/D$ for a sphere and $0.212\ a^2/D$ for a plate.

(a: radius or thickness; D: Diffusion constant)

The diffusion constant of H_2O_2 in ice is not known, as a minimum value, we can assume the self diffusion coefficient of water in the ice, which was determined to be 0.5 . 10^{-15} m^2/sec at -20°C (Ramseier, 1967). It takes 24 hours for 90% of the gas to diffuse out of a 20 mm plate and 600 hours in case of a 100 mm plate.

Thus the potential exists for a large part of trapped H_2O_2 to leave the snow. In addition evaporation of snow support the reentering of trapped H_2O_2 in the gas phase. This leads to a new distribution of H_2O_2 in snow. As a consequence, the original structure becomes smoothed as is beautifully demonstrated by the continuous record from the Summit core from Central Greenland (Figure 10). If the yearly accumulation is below 25 cm water equivalent, the very pronounced seasonal variations with high values in summer snow and low values in winter snow is equalized.

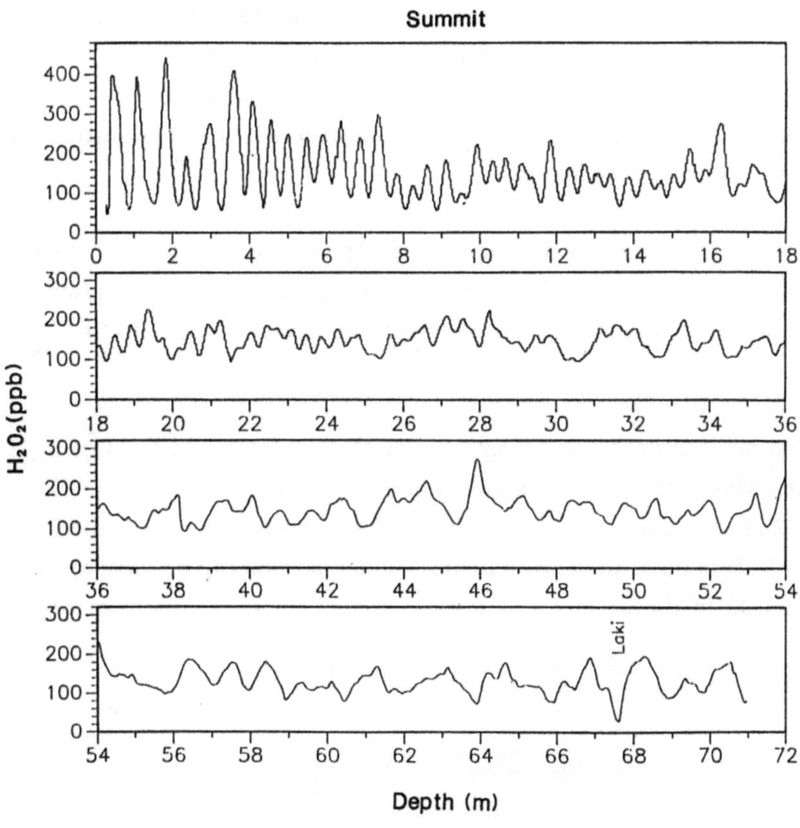

Fig. 10: Continuous record of the H_2O_2 profile of the Summit core of the first 72 meters.

This smoothing process can mathematically be described by a low pass filter. The amplitude of a seasonal variation will be reduced by a factor R given as:

$R = \exp(-2\pi 2L^2/A^2)$ A: yearly accumulation (ice/years)

L corresponds to the mean displacement length of an H_2O_2 molecule in the firn. This is exactly the same concept as was applied by Johnsen (1977) to describe the attenuation of the seasonal $\delta^{18}O$ variation in the firn. Sigg calculated an L to be 12-13 cm. from three different H_2O_2 records (Siple, Antarctica and Dye 3 and Summit, Greenland). This length is 60% higher than the calculated displacement length for the water molecules in the firn, based on the smoothing of the $\delta^{18}O$ seasonals. What causes this larger diffusion length? There are two main possibilities. First, the sticking probability of H_2O_2 to an ice surface is lower than the corresponding value for water molecules. As a consequence, H_2O_2 reentering open pore spaces in the firn will remain longer than water molecules, and will therefore, diffuse over longer distances. The sticking probability cannot be too low, otherwise the snow would loose its H_2O_2 content. Secondly, a higher diffusion constant of H_2O_2 in the ice than the self diffusion of water molecules would lead to a loss of H_2O_2 molecules in excess of a "mean equilibrium" concentration. This H_2O_2 again would be redistributed in the open pore space. It is important to quantify these two proposed mechanisms by experimental data and mathemaical models in the future.

A distinct increase in the yearly mean H_2O_2 concentration during the last 20 years is visible (Fig. 11). Does this reflect an increase in atmospheric H_2O_2 concentration at high latitudes? At this stage, the answer is yes. Two main arguments are support a corresponding atmospheric increase:

- The mean displacement length of H_2O_2 molecules in the firn is 12 cm ice equivalent, which coresponds to a layer thickness of roughly half a year. H_2O_2 from a depth of 2 m is no longer in contact with the free atmosphere. From the fourth to the sixth year back in time, an increase in the summer maxima can be seen, which could not be maintained in the case of a systematic loss of H_2O_2 from greater depths to the surface.

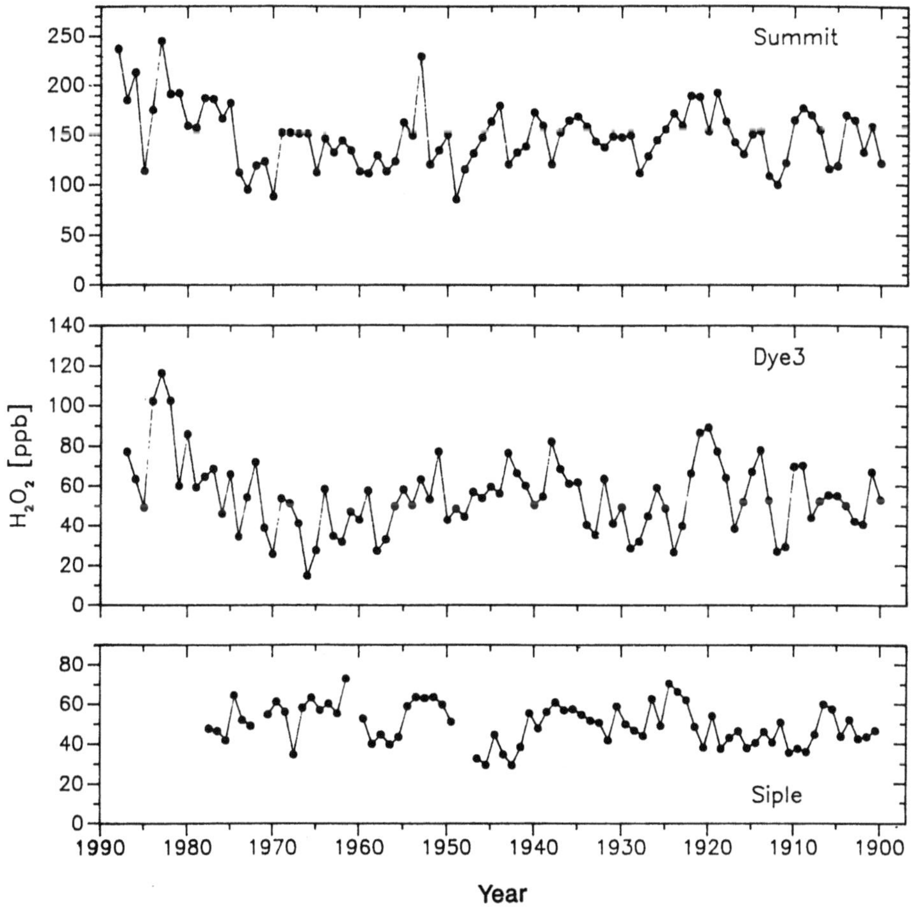

Fig. 11: Yearly mean H_2O_2 concentrations of the last 100 years from Dye 3, Summit and Siple.

- The continuous record from the Dye 3 station shows a similar general trend, but the higher accumulation rate (50 cm water equivalent) yields another depth/age relation. A restriction regarding the Dye 3 core: A detailed interpretation of the results from the Dye 3 core is complicated by the regular ocurence of melt layers due to summer temperature around 0°C.

An increase in the atmospheric concentration of H_2O_2 is compatible with model calculations and is ascribed to increased hydrocarbons emissions (Hov and Isaaksen, 1987, Thompson et al., 1989).

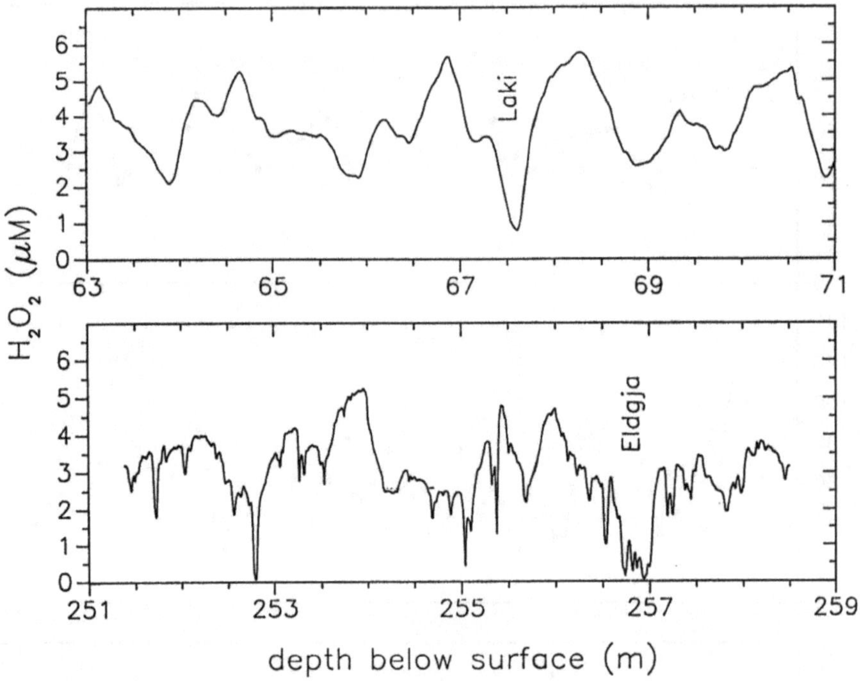

Fig. 12: H_2O_2 profile between 251 and 259 m.b.s. of the Summit core.

The continuous H_2O_2 measurement on the 300 m Eurocore from Summit also revealed a clear disintegration of H_2O_2 in narrow ice layers. Fig. 12 shows the depth interval 251 m to 259 m below the surface, where sharp "H_2O_2 holes" are visible and an H_2O_2 depression linked with the eruption of the volcano Eldgjia are visible. The continuous, high resolution H_2O_2 analysis makes it possible to distinguish between "post deposition" disintegration, corresponding to the sharp holes, and "predeposition" H_2O_2 depression in the snow, due to a large amount of emitted SO_2 from the volcanic eruption. The sharp holes in the H_2O_2 profile cannot have been initially there, because diffusion during the firnification processes discussed previously would have flattened out the edges. The holes are associated with higher Ca^{2+} concentrations (Daniel Zurmühle pers. comm), an indication of elevated alkaline dust concentration. This is clear evidence of a slow H_2O_2 disintegration process in ice in the dark, most probably a catalytic decay on small particles incorporated in the ice. In order to decay, the H_2O_2 would have to meet these particles in the ice. The mean displacement in the ice is given as d = $(2Dt)^{0.5}$. Assuming for the diffusion constant of H_2O_2 molecules in ice a lower limit equal to the self diffusion coefficient of water molecules in ice, yields a displacement of roughly 1 mm in 100 years, much more than the distance to the next particle.

The second type of "H_2O_2 holes" present in Fig. 12 is linked to volcanic eruptions. The high ECM signal indicates elevated levels of sulfuric acid, the oxidation product of SO_2. This oxidation process consumes H_2O_2 molecules in the air, so that the initial the H_2O_2 concentration in the firn is already low. The signatures of these holes are perfectly compatible with the smoothing process taking place in the firn.

Summary:

- *H_2O_2 concentration in snow and ice that can be related to atmospheric concentreations are only preserved in locations with low dust concentrations such as the polar ice caps.*

- *The H_2O_2 signatures of individual snowfalls are lost during the firnification process. Depending on the accumulation rate, a temporal mean value is conserved in the ice. The transfer function atmosphere - ice is well described by a co-condensation model.*

- *The mean displacement length of H_2O_2 molecules in the firn is 12 - 13 cm ice equivalent, integrated over the whole firnification process.*

- *H_2O_2 disintegration occurs on dust particles by a slow process in the dark, or much faster, if sunlight is present.*

- *In order to extract atmospheric information of the past, continuous measuring series must be available.*

10. The formaldehyde (HCHO) record in ice cores

The reason that we began measuring HCHO in polar ice samples is simple: HCHO is an oxidation product of hydrocarbons, mainly methane. The methane record of the last 160'000 years is well known (Chappellaz, 1990). If the formaldehyde record is also known, it may be possible to establish a record of the atmospheric oxidation capacity.

The atmospheric lifetime of HCHO against photolysis and reaction with OH is less than 1 day. Deposition is an additional sink for HCHO but should be relatively unimportant in polar regions; wet deposition is limited by the scarcity of precipitation, and dry deposition to cold ice is inefficient even for water-soluble species (Bales et al, 1989). The HCHO concentration in polar air can be expressed as a steady state between chemical production and loss rates:

$$[HCHO] = \frac{k_1 \cdot [OH] \cdot [CH_4]}{k_p + k_2 \cdot [OH]}$$

where $k_1 = 2.4 \cdot 10^{-12} \cdot e^{-1710/T} cm^3 molecule^{-1} s^{-1}$ and $k_2 = 1.0 \cdot 10^{-11}$ cm^3 $molecule^{-1}$ s^{-1} are the reaction constants for oxidation of CH_4 and HCHO by OH, respectively (Brasseur and Solomon, 1986); and k_p is the rate of HCHO photolysis.

The eye of the needle to determine from the HCHO concentration the atmospheric concentration is the characterization of the transfer function. Let us first take a look at a summary of the experimental results. All measurements were performed by Thomas Staffelbach and are compiled in his thesis. He passes again a summer freezing on the Greenland ice sheet. Fig. 13 summarizes different pit studies performed in Greenland in the summer 88 and 89 at the Dye 3 and Summit respectively. They show a surprising behaviour of the course of the formaldehyde concentration in shallow depth. It starts with a sharp decrease in concentration in the first 10 cm, then passes a minimum and then recovers to a more or less constant concentration for the next 100 meters. We explain this behaviour qualitatively as follows: Freshly deposited snow looses HCHO molecules by two mechanisms. First a reduction of specific surface and evaporation leads to a loss of absorbed HCHO molecules on the snow surfaces. Secondly - we believe that this is the more important process - decay by photolysis of "fixed" HCHO molecules in the snow. The photolytical lifetime of HCHO molecules in the gas phase is under summer conditions around 1.2 days. If for HCHO in or at the surface of snow flakes, a similar photolytical rate holds, this is an effective way to decrease the HCHO concentration. Because the light flux is stronlgy attenuated by snow (A 1/e attenuation depth of ca. 20 cm for blue light, which has the weakest attenuation coefficient), this destruction mechanism becomes very inefficient below 20 cm depth.

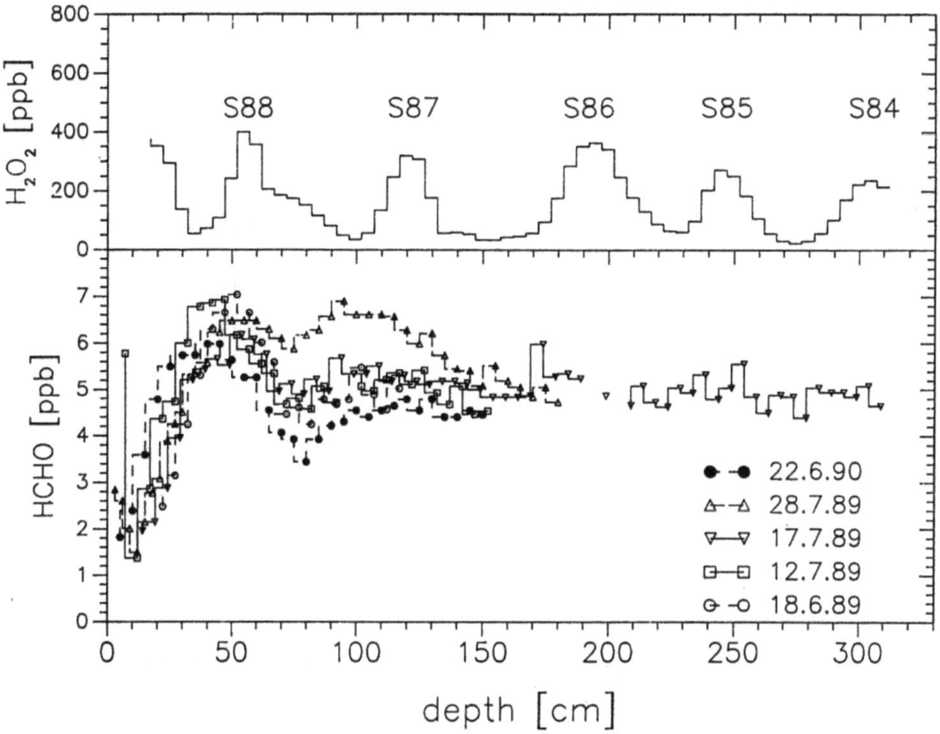

Fig. 13 : HCHO concentrations in different pit studies from Summit.

At this depth a new equilibrium between snow and open pore space HCHO concentrations is slowly established. Because of the rapid exchange between the firn and the air in the first meter the photolytical decay of HCHO in the open pore space of the firn will be small. The adjustment to a new HCHO concentration in the snow is independent of the initial snow concentration, of the Greenland sites examined. The transfer coefficent defined as C_{ice} (ppb(w))/C_{air} (ppb(v)) amounts to 40 for Dye 3 and 50 for Summit assuming a constant air concentration of 0.1 ppb for Greenland (Lowe and Schmidt, 1983). This is clearly lower than the transfer coefficient calculated based on the co-condensation model, which is roughly one order of magnitude larger. It must, therefore, be assumed that the sticking probability of HCHO on a snow surface is much lower than that of water molecules. As a consequence, the HCHO concentration of the firn is proportional to a multiyear average of the air concentration. This hypothesis has to be verified in a field experiment, where the HCHO concentration is measured in the air of the firn at different depths. So far, we assume that the relative variations of the HCHO concentrations measured along a core reflect corresponding variations in the atmosphere (Staffelbach et al., 1990). Figure 14 shows the long-term trend of HCHO concentration in the ice core, that clearly reveals lower values during the last glaciation, which points towards a reduced oxidation capacity in the atmosphere at that time.

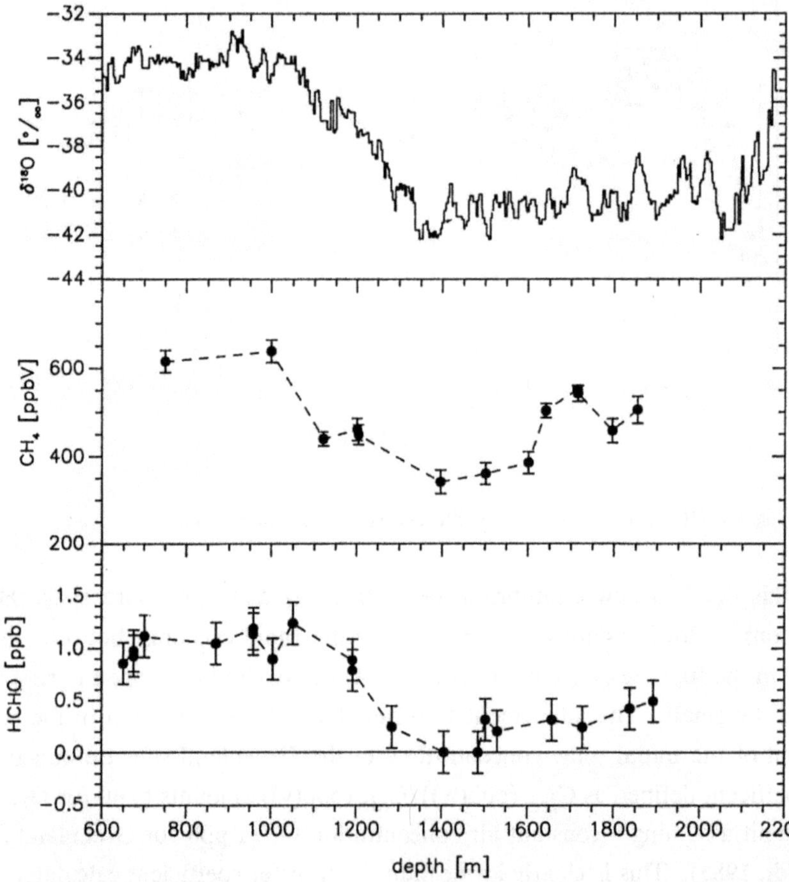

Fig. 14a: $\delta^{18}O$, **methane and formaldehyde results over the deep ice core from Byrd, Antarctica .**

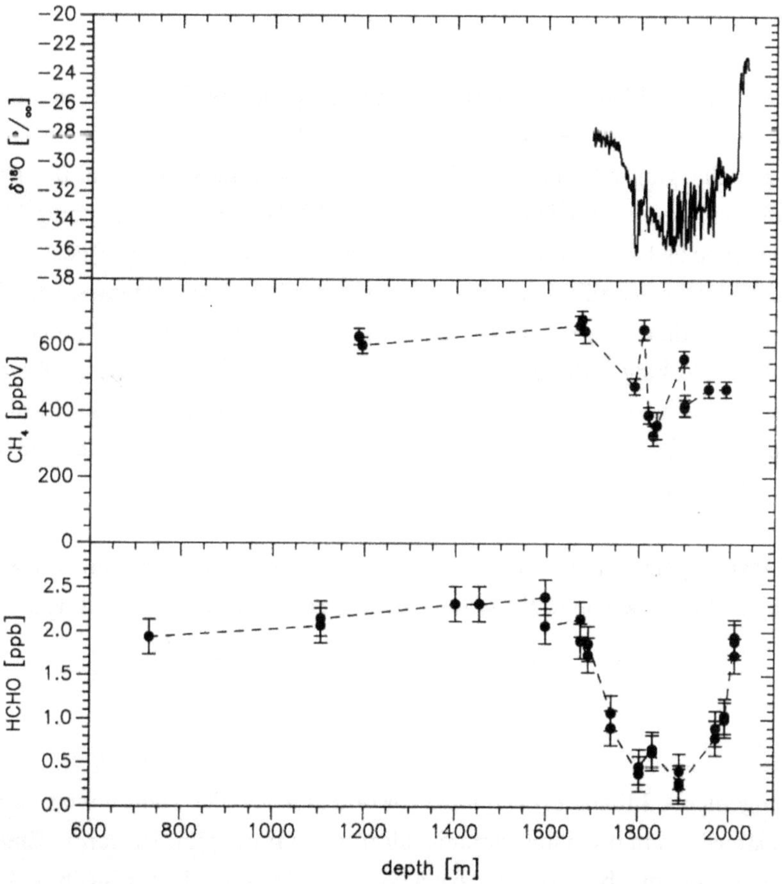

Fig. 14b: $\delta^{18}O$, methane and formaldehyde results from the deep ice core from Dye 3, South Greenland.

Summary of the HCHO chapter:

- *Initial HCHO concentrations are not preserved in polar snow.*

- *A new equilibrium is established between 20 to 50 cm below surface, which can be related to a mean multiyear atmospheric concentration.*

- *The transfer coefficient points towards a sticking coefficient far below that of water molecules.*

Closing remarks

The analysis of ice cores provides a time-lapse picture of changes in atmospheric chemistry processes. It demonstrates at rather low cost, how fast anthropogenic activities are changing different chemical cycles in the atmosphere and it makes clear how blindly we are progressing in an uncertain future. We are spending billions in different currencies for technological progress and important and (un)necessary things like more highways, satellites for more (commercial) TV programmes, etc. and very little money and time is spent to step back and reflect upon the consequences of what we are doing. Working with ice cores, spending months in the cold white solitude of the ice sheets, recovering information from the past, helps to get a needed critical distance to the hectic state of our modern society of consumption, also consumption of scientific discoveries and ideas.

In this respect, I hope that I have succeeded in awakening the interest of the participant of this workshop to the stories that the cold, white snow is telling us.

Acknowledgements: I thank my two main coworkers Andy Sigg and Thomas Stafelbach for their enthusiasm in providing climatic information from polar ice cores. That kind of work was only possible, because we have a good working team in our institute. I am gratefull to Dr. Ann Arquit. She carefully reviewed the manuscript and removed (hopefully) all of the many language errrors.
The Swiss office for science and education provided financial support through the COST action 611.

References

Baltensberger U., H.W. Gäggeler, M. Gloor, D.T. Jost, U. Siegenthaler, A. Neftel, D. Wagenbach, K. Geis and J. Beer. (1989)
Preliminary results of a transfer study on Weissfluhjoch-Davos.
Contribution to 4th Alptrac Meeting, Grenoble, January 1989

Barnola J.M., D. Raynaud, A. Neftel and H. Oeschger (1983)
Comparison of CO_2 measurements by two laboratories on air from bubbles in polar ice.
Nature, **303**, 410-413

Barnola J.M., D. Raynaud, Y.S: Korotkevich and C. Lorius (1987)
Vostok ice core provides 160'000-year record of atmospheric CO_2
Nature, **329**, 408-414

Beer J., R.C. Finkel, G. Bonani, H. Gäggeler, U. Görlach, P. Jacob, D. Klockow, C.C. Langway Jr., A. Neftel, H. Oeschger, U. Schotterer, J. Schwander, U. Siegenthaler, M. Suter, D. Wagenbach and W. Wölfli (1990)
Seasonal variations in the concentration of ^{10}Be, Cl^-, NO_3^-, SO_4^{2-}, H_2O_2, ^{210}Pb, ^3H, Mineral Dust and $\delta^{18}O$ in Greenland Snow
(in press Atmospheric Environment)

Chappellaz J., J.M. Barnola, D. Raynaud, Y.S. Korotkevich and C. Lorius (1990)
Ice-core record of atmospheric methane over the past 160'000 years
Nature, **345**, 127-131

Golecki I.and C. Jaccard (1978)
Intrinsic surface disorder in ice near the melting point.
Journal of Physics C: Solid state physics, **11**, 4229-4237

Hammer C.U. (1980)
Acidity of polar ice cores in relation to absolute dating, past volcanism, and radioechoes.
J. of Glaciology, **25**, 359-371

Jacob, P. (1987)
Wasserstoffperoxid in der atmosphärischen Gas- und Flüssigphasenchemie: Laborstudien und Felduntersuchungen. **Dissertation** am Fachbereich Chemie der Universität Dortmund

Johnsen, S.J. (1977);
Stable Isotope Homogenization of Polar Firn and Ice. International Association of Hydrological Sciences Publication **118**, Isotopes and Impurities in Snow and Ice, Grenoble, 210-214

Kormann C., D. W. Bahnemann and M.R. Hoffmann (1988)
Photocatalytic production of H_2O_2 and organic peroxides in aqueous suspensions of TiO_2, ZnO and desert sand.
Env. Science and Technology, **22**, 798-806

Legrand M. and C. Saigne (1988)
Formate, acetate and methanesulfonate measurements in antarctic ice: some geochemical implications
Atmospheric Environment, **22**, 1011-1017

Logan J.A., M.J. Prather, S.C.Wofsy, and M.B. McElroy (1981); Tropospheric Chemistry: A Global Perspective.
J.Geophys. Res., **86 (C8)**, 7210

Lowe, D. C., U. Schmidt, and D. H. Ehalt (1981)
The Tropospheric Distribution of Formmaldehyde, Kernforschungsanlage Jülich, Jülich, 1981

Mulvaney R., E.W. Wolff and K. Oates (1989)
Sulphuric acid at grain boundaries in Antarctic ice.
Nature, **331**, 247-249

Neftel A., H. Oeschger, J. Schwander and B. Stauffer (1983)
CO_2-concentration in bubbles of natural cold ice.
Journal of Physical Chemistry, **87**, 4116-4120

Neftel A., P. Jacob and D. Klockow (1984)
Measurements of hydrogen peroxide in polar ice core samples.
Nature, **311**, 43-45

Neftel A., M. Andrée, J. Schwander, B. Stauffer and C.U. Hammer (1985)
Measurements of a kind of DC-conductivity on cores from Dye 3
In Langway C. C. Jr, H. Oeschger, W. Dansgaard (eds) Greenland ice core: geophysics, geochemistry and the environment. Washington DC, American Geophysical Union: 32-38 (Geophysical Monograph **33**)

Neftel, A., P. Jacob, and D. Klockow; (1986)
Long-Term Record of H_2O_2 in Polar Ice Cores. Tellus, **38 B**, 262-267

Neftel, A., A. Sigg, and F. Zürcher. (1987)
Acid Deposition in a Snow Field at 2500 m.a.s.l. in Switzerland. COST Air Pollution Research Report **2**, 'Physico-Chemical Behaviour of Atmospheric Pollutants', Stresa, D. Reidel Publishing Company, Dordrecht, 500

Neftel A., H. Oeschger, T. Staffelbach and B. Stauffer (1988)
CO_2 record in the Byrd ice core 50'000 - 5'000 years BP
Nature, **331**, 609-611

Oeschger H. and C.C. Langway Jr (1989)
The Environmental Record in Glaciers and Ice Sheets.
Physical, Chemical and Earth Science Research Report **8**
John Wiley & Son.

Overgaard S. and N. Gundestrup (1985)
Bedrock topography of the Greenland ice sheet in the Dye 3 area.
In Langway C. C. Jr, H. Oeschger, W. Dansgaard (eds) Greenland ice core: geophysics, geochemistry and the environment. Washington DC, American Geophysical Union: 49-56 (Geophysical Monograph **33**)

Penkett, S.A., B.M.R. Jones, K.A. Brice, and A.E.J. Eggleton; (1979)
The Importance of Atmospheric Ozone and Hydrogen Peroxide in Oxidizing Sulfur Dioxide in Cloud and Rainwater. Atmos. Environ., **13**, 123-137

Pourchet M., F. Pingolet and C. Lorius (1983)
Some meteorological applications of radioactive fallout measurements in Antarctic snows.
Journal Geophys. Res., **88**, 6013-6020

Pruppacher H.R.and J.D. Klett (1978)
Microphysics of clouds and precipitation.
D. Reidel Publishing company 714p

Ramseier R.O. (1967)
Self diffusion in ice monocrystals.
CRREL Reserach Report **232**, Hanover, New Hampshire..

Schwander J. (1980)
CO_2-Gehalt von Schnee- Firn- und Eisproben.
Master thesis, Physical Institute University of Bern (unpublished)

Schwander J., A. Neftel, H. Oeschger and B. Stauffer (1983)
Measurements of Direct Currwnt Conductivity on Ice Samples for Climatological Applications.
J. of Physical Chemistry, **87**, 4157-4160

Schwander J. and B. Stauffer (1984)
Age difference between polar ice and the air trapped in its bubbles.
Nature, **311**, 45-47

Sigg A. and A. Neftel (1988)
Seasonal variations of hydrogen peroxide in polar ice cores.
Annals Glaciol., **10**, 157-162

Sigg, A. (1990)
Wasserstoffperoxid-Messungen an Eisbohrkernen aus Grönland und der Antarktis und ihre atmosphärenchemische Bedeutung,
Dissertation an der Universität Bern,

Staffelbach T., B. Stauffer and H. Oeschger (1988)
A detailed analysis of the rapid changes in ice-core paramters during the last ice age
Annals of Glaciology **10**,

Staffelbach T., A. Neftel, B. Stauffer and D. Jacob (1990)
Formaldehyde in polar ice cores; a possibilty to characterize the atmospheric sink of methane in the past.
(in press Nature)

Stauffer B., A. Neftel, H. Oeschger and J. Schwander (1985)
CO_2 concentration in air extracted from greenland ice samples.
In Langway C. C. Jr, H. Oeschger, W. Dansgaard (eds) Greenland ice core: geophysics, geochemistry and the environment. Washington DC, American Geophysical Union: 85-89 (Geophysical Monograph **33**)

Stauffer B., G. Fischer, A. Neftel and H. Oeschger (1985)
Increase of atmospheric methane recorded in Antarctic ice core
Science, **229**, 1386-1388

Stauffer B., E. Lochbrunner, H. Oeschger and J. Schwander (1988)
Methane concentration in the glacial atmosphere was only half that of the preindustrial Holocene.
Nature, **332**, 812-814

Stauffer B.and A. Neftel (1988)
What have we laerned from ice cores about atmospheric changes in the concentrations of nitrous oxide, hydrogen peroxide, and other trace species?
In Rowland F.S. and I.S.A. Isaksen (eds) 63-77
John Wiley & Sons Ltd, Dahlem Konferenzen 1988

Valdez M.P., R. Bales, D.A: Stanley and G. Dawson (1987)
Gaseous deposition to snow 1. Experimental study of SO_2 and NO_2 deposition
JGR, **92**, 9779-9787

USE OF SNOW AND FIRN ANALYSIS TO RECONSTRUCT PAST ATMOSPHERIC COMPOSITION - DISCUSSION

E.W. Wolff
British Antarctic Survey
Natural Environment Research Council
High Cross
Madingley Road
Cambridge CB3 0ET
UK

Polar ice cores (and cores from certain high-altitude lower latitude sites) have proved to be a rich resource of information on past atmospheres and climate. The ideal environmental archive would faithfully record the composition of the atmosphere with no post-depositional changes. Of course, the picture is not this simple even for polar ice, but it is worth recalling a few of the many successes of ice coring.

The Vostok (Antarctica) core has given the most detailed and important record of climate over a complete glacial cycle, and has shown the close linkage between climate and trace gases such as CO_2 and CH_4. It is hoped that a northern hemisphere record of still greater length will be obtained by drilling operations now underway at Summit in Greenland. Existing cores from Greenland have revealed detailed climatic and chemical changes, and have shown the extent to which signals are global rather than regional. Shorter records have provided the clearest indication we have of increasing concentrations of pollutants in the atmosphere. These and other developments are detailed elsewhere (eg Oeschger and Langway 1989).

Ice cores are being used for studies of increasing sophistication. For instance, they have the potential to reveal changes related to biospheric productivity, and details of past atmospheric chemistry (eg oxidative capacity). The importance of obtaining century-scale records of climate is also recognised. All these studies require an improved ability to interpret records in ice correctly.

A number of factors currently obscure our capacity to use ice core records to the full. These include gaps in our understanding of sources, transport and deposition processes, and dating and analytical problems. An area that has been neglected is that of changes occurring after snowfall, and this is the interest of this meeting. In general, the processes in cold snow are more limited and better understood than those in wet snow (assuming that cores are taken at sites with no melt, or at worst restricted melt).

Some of the processes that can occur in cold ice sheets are listed below:

1. Re-emission of vapour to the atmosphere (possibly following decomposition).
2. Vapour movement in firn, leading to smoothing of records.

NATO ASI Series, Vol. G 28
Seasonal Snowpacks
Edited by T. D. Davies et al.

3. Uptake of vapour by sub-surface firn.
4. Chemical reaction in snow and ice (on dust particles, at grain boundaries, in melt layers).
5. Percolation of liquid (in surface melt features, or near warm beds).
6. Changes in composition of air bubbles; interaction with the ice.
7. Loss or overturning of annual layers.

The paper by Neftel in this volume has concentrated particularly on vapour-phase problems. For species such as H_2O_2, and HCHO, the interpretation of ice core records appears to depend crucially on understanding incorporation, smoothing and loss mechanisms. HNO_3 is another species where it has been suggested that initial concentrations may be reduced by photochemical decomposition and vapour phase loss after deposition (Neubauer and Heumann 1988). The time is certainly ripe for detailed field and laboratory experiments to establish the importance of these processes, and the factors that control the eventual concentrations found in deep ice.

Apart from discussion of the possible decay of H_2O_2 on dust particles, there has been little consideration of reactions in the ice. Although in many cases, the records from ice make it clear that only minor changes can have occurred, it will be valuable to consider the possibilities further. This will require a better understanding of possible reactions on small particles, of the concentration effect of some impurities to grain boundaries, and of the nature of such boundaries. The most direct experimental attack on these problems is to make microchemical analyses using techniques such as scanning electron microscopy. There is still debate as to how general is the finding that in some ice, acid is concentrated in veins (believed to be liquid) at triple junctions (Mulvaney et al. 1988). Under what circumstances does this localisation occur? Studies are underway to try and answer this question. A particular interest is to see if localisation occurs both in acidic Holocene ice and in generally alkaline Wisconsin-age ice from Greenland.

Where some impurities are localised to grain boundaries or more specifically to triple junctions, what will be the consequences for the ice core record? There has been no discussion in the literature of possible reactions at the boundaries. These could occur if other impurities (such as dust) remain in contact with the boundary liquid for any time. The concentration of acids in liquid veins in ice is determined only by the temperature of the ice, so that there is no concentration gradient driving diffusion. However, physical movement of the liquid may be possible, particularly near a warm bed, where the veins may become large enough to drain. If the localisation of impurity is found to be widespread, then these ideas will need some theoretical and field study.

Many of the processes that are vital in cold snow may be obscured in seasonal snowcover by larger effects due to the presence and movement of water. Nonetheless, there are certainly common interests between the two fields in areas such as vapour uptake by snow, and the nature of material at grain boundaries. Above the snow surface, there is a common need to understand incorporation mechanisms of impurities from the atmosphere to the snow.

Although work on interpreting the chemistry of ice core records is seen as less glamorous than retrieving long cores, it must increasingly be seen as a necessary part of getting full benefit from costly logistic efforts. Much valuable information remains untapped at present. Some future work in this field is already planned under the auspices of the IGAC (International Global Atmospheric Chemistry) Programme, and other studies are underway in Antarctica and Greenland.

REFERENCES

Mulvaney R, Wolff EW, Oates K (1988) Sulphuric acid at grain boundaries in Antarctic ice. Nature 331:247-249

Neubauer J, Heumann KG (1988) Determination of nitrate at the ng/g level in Antarctic snow samples with ion chromatography and isotope dilution mass spectrometry. Fres Z Anal Chem 331:170-173

Oeschger H, Langway CC Jr (eds) (1989) The environmental record in glaciers and ice sheets. Wiley-Interscience, Chichester

CLIMATIC CHANGE AND SEASONAL SNOWCOVERS: A REVIEW OF THE FACTORS REGULATING THE CHEMICAL EVOLUTION OF SNOWCOVER AND A PREDICTIVE CASE STUDY FOR NORTH-EASTERN NORTH AMERICA

T. D. Davies and S. J. Vavrus
School of Environmental Sciences
University of East Anglia
Norwich NR4 7TJ, UK

Introduction

There have been a large number of "what-if" studies of the possible future impact of climatic change on systems within the biogeosphere. It may be argued that such studies are presently futile, given the current poor knowledge of likely future regional climates (Rowntree, 1990). Nevertheless, it is more responsible to assess the type, or nature, of likely impacts based on those predicted elements of climatic change about which we have a greater relative degree of confidence. Such studies may give us a notion of the direction and scale of a future impact and, hence, enhance our preparedness. Although the range of possible future perturbations to the chemistry of seasonal snowcover is large - and we engage in some speculation on these at the qualitative level - our quantitative example is constrained and, we contend, sensibly restricted to what we regard, at the level of current knowledge, to be amongst the least uncertain of climatic perturbations and effects.

We briefly discuss the physical links between snowcover and the atmosphere since these are of direct relevance to deposition processes. We then examine some relationships between transport and deposition processes and snowpack chemistry, since these are strongly climate-dependent. Then follows a review of factors influencing the chemical evolution of an existing snowcover. The last component of the review element of this paper is a discussion of climate variability and climate predictions/projections. These discussions set the scene and context, and establish the justification for the approach we have taken in a first attempt to assess the response of the composition of seasonal snowcover to climate projections. Since our methodology concerns the acid content of meltwater leaving the snowcover, there is useful incidental information on future potential impact on aquatic biota.

Physical interactions of snowcover and atmosphere

The areal extent of snowcover in the northern hemisphere is large (maximum seasonal extent ~ $43\text{-}52 \times 10^6$ km^2) (Ropelewski,

NATO ASI Series, Vol. G 28
Seasonal Snowpacks
Edited by T. D. Davies et al.

1986). Hence, there are important two-way global-scale links with climate (e.g. Barnett et al., 1988). Berry (1981) has summarised the range of interactions between snowcover and the physical state of the atmosphere. In winters with more extensive snowcover over North America, depressions take a more southerly track, becoming less frequent over eastern Canada (Dewey, 1986). It is likely that such changes in atmospheric circulation patterns have a strong influence on pollutant wet deposition (e.g. Davies et al., 1986), as well as on melt regimes. On a local scale there are also important effects; for example, the presence or absence of snowcover influences the dispersion potential for air pollutants. Dispersion efficiency is lower over snow (Berry,1981); a factor which might be of some importance close to pollutant sources.

Surface albedo-temperature feedbacks are an important component of the global climate system and snowcover is one of the most important elements, both in respect of "present-day" climate and future projections (e.g. Hansen et al., 1984).The relationships are complex; for example, higher temperature influences not only the extent of snowcover, but also the albedo of the snow surface because of accelerated snow metamorphism (Kuhn, 1989). Changes in seasonal snowcover also affect the regional and global hydrological cycles (Kuhn, 1989), with implications for precipitation, deposition and snowpack evolution regimes.

These few examples serve to illustrate the pervasive, and complex, character of snowcover-atmosphere relationships and warn of the problems to be encountered in any study of future climate influences on snowcover composition.

The composition of the unmodified snowcover

Snowcover is a long-term integrator of wet and dry deposition (Johannes, 1987), but its composition can be modified by many factors, such as melting, forest cover and wind redistribution (Jeffries, 1989). Here we disregard in-pack modification, and consider the initial controlling factors.

(a) Atmospheric transport, chemistry and wet deposition

The initial controlling factor is atmospheric transport. Wright and Dovland (1978) related the north-south variation in Norwegian snowcover composition to the long-range transport of pollution from European sources. There is seasonal and interannual variability in long-range transport (e.g. Davies et al., 1986; Farmer et al., 1989; UKRGAR, 1987, 1990). Mean annual airborne sulphur concentrations in Europe can be as dependent on meteorology as on emissions

(Mylona, 1989). Wet deposition can be very episodic (Skartveit, 1982; Fowler and Cape, 1984; Davies et al., in review). It is not surprising, therefore, that the bulk composition of seasonal snow cover can also exhibit considerable variability.

Less obvious is the fact that there might be trends over time (as opposed to "random" fluctuations) in snowcover composition, because of observed trends in atmospheric circulation patterns which affect air quality (UKRGAR, 1990) and/or wet deposition (Davies et al., 1986; UKRGAR, 1990). It is necessary to consider past trends in climate when attempting to define emission-deposition relationships (Davies et al., 1986), let alone any future climatology (Shannon et al., 1985).

Reactions and transformations in the atmosphere are also dependent on the prevailing climate. Changes in transit time from pollutant source to deposition zone may alter NO_3^-/SO_4^{2-} ratios in precipitation (Brimblecombe et al., 1988). Warmer winters may lead to faster SO_2 oxidation than currently exists, modifying the currently observed NO_3^-/SO_4^{2-} ratios in snowfall (Summers and Barrie, 1986).

The scavenging process may be altered because of differing future distributions of atmospheric liquid water, or because of the dependence of different ice-nucleii activations on temperature (Schemenauer, 1981). Different temperature distributions may also alter snow crystal riming regimes, which could also affect the composition of snow crystals. Any change in the relative quantities of rainfall versus snowfall may have an impact on snowcover composition because of the known differences in composition, part of which relates to the NO_3^-/SO_4^{2-} temperature dependence, but part of which is to do with relative scavenging efficiencies of gases and particles (Knutson et al., 1976). Rainfall impinging on the snowcover can be regarded as representing a deposition process: rain events can directly alter the chemical composition (e.g. Cadle et al., 1984, 1987; Dasch, 1987; DeWalle, 1987). In this respect rainfall can be considered as one of the initial inputs to the "unmodified" snowcover , but it is also a major modifying agent in terms of initiating melt or modifying melt patterns (see below).

Deposition in dew, fog, mist and cloud droplets is an important part of the pollutant flux to snowcover and is frequently ignored (Johannes, 1987). These deposition mechanisms could all have significant effects on snowcover composition, and their occurrences could alter in future climate states.

There is an established inverse relationship between ion concentration in precipitation and precipitation amount (e.g. Smith, forthcoming). Consequently, any change in snowfall amount (individual events, or over a longer term) may have an impact on snowcover composition. Smith has considered this relationship in respect of model predictions of increased future winter/spring precipitation amounts over the UK. He suggested that the wet deposition of sulphur in southern Scotland may increase in future decades (assuming no change in emissions); the predicted increase in precipitation amount being more important than the concentration-amount relationship.

(b) Dry deposition

The dry deposition rates of many chemical species are substantially different over snow when compared with the uncovered surface. A change in the climate which led to a change in the (time-dependent) areal distribution of seasonal snowcover would significantly change dry deposition regimes (Davies, 1989) and, consequently, influence atmospheric transport. Even neglecting the effect of changes in the geographical distributions of seasonal snowcover, climate-induced changes in the physical state of the snowcover will have an important influence (Sirois and Barrie, 1988). These influences will, themselves, vary spatially in terms of the net effect on snowpack composition, since the dry deposition component is generally less important than wet deposition (Jeffries, 1989), although its relative importance increases close to sources (Dasch and Cadle, 1986; Cadle and Dasch,1987).

Particle dry deposition to a snowcover is greatest immediately after snowfall, and decreases as metamorphism proceeds (Johannes, 1987). The dry deposition of various gases depends on snowcover surface and on climate (Sirois and Barrie, 1988, Jeffries, 1989). For example, SO_2 and HNO_3 deposition is greater over snowcover with a high liquid water content (Bales et al., 1987). The implications of a warmer climate are clear.

The composition of the modified snowcover

Jeffries (1989) points out that there is conflicting evidence concerning the "pre-melt" stability of snowcover composition. He reviews the literature and indicates that, in many cases' there is unrealised (or unobserved) melt because of heat flux from the underlying ground and, hence, the gradual removal of solute (e.g. Motoyama et al., 1986; English et al., 1987). In many locations, underlying soils rarely freeze because of snowcover insulation and microbial decomposition (Peters, 1984). Even if meltwater does not leave the pack, solute can be redistributed in the snowpack, leaving the bulk composition intact, but ultimately affecting the chemical composition of the meltwater when it does leave the pack (Davies et al., 1987). The net result is that the bulk composition of the remaining snowcover is impacted. Hence, conditions antecedent to the measurable melt-events (i.e. preceding climatic influences) may modify the composition of the snowcover.

There are other processes which may lead to in-pack (i.e. involving no removal of water) transformations. It is possible that there are upward fluxes of gases from the sub-strate (Johannes, 1987). Chemical interactions with organic debris can have significant effects on snowcover composition, especially in forest environments (Jones and Sochanska, 1985; Jones and Deblois, 1986; Jones, 1987). A wide range of microbiological processes may modify, or mediate in, snowcover chemistry; these are going to be strongly influenced by meteorological and climatological conditions (e.g. transport of gases and water, diurnal freeze-thaw cycles, and episodic events such as rainfall: ICAIR, 1987).

Nitrogen dynamics in seasonal snowcover might be influenced by non-biological processes; Delmas and Jones (1987) speculated on the production of NO_3 at the surface of the snowcover through the oxidation of dry-deposited NO_2 by O_3 or the OH radical. Palmer et al. (1975) reported the production of O_3 in snowpacks during periods of bright sunshine. In a study on an Alpine snowpack, not affected by organic detritus, Sigg et al. (1987) found that the vertical chemical concentration profiles did not change over time, as long as the snowpack temperature stayed $<0^oC$, except for H_2O_2, which exhibited considerable disintegration upon the incidence of sunlight.

Redistribution of recently-fallen snow, or pack snow, can influence the composition of snowcover. Transport of snow particles can occur at windspeeds as low as 0.15 m s^{-1} (Kind, 1981). Delmas and Jones (1987) suggested that there may be resuspension of impacted aerosols from snow crystals to the atmosphere during transport, but Pomeroy et al. (this Volume) have established that sublimation of resuspended snow particles during transport increases solute concentrations leading to higher concentrations in depositional zones of the snowcover.

All of the processes discussed above in this section are meteorology- or climate-dependent. Changes in these assemblages of causal factors are likely to have an impact on the resultant snowpack chemistry.

The impact of snowmelt

The impact of snowmelt on the remaining snowcover is dramatic DeWalle (1987), because of the "fractionation effect" which occurs at melt (Johannessen et al., 1975), when the solute concentrations in the initial fractions of melt can be much higher than the pre-melt average concentrations in the snowpack (e.g. Davies et al., 1987). The net radiative flux is generally most responsible for melt (Jeffries,1989), although rain can be an effective initiator (Jones, 1987) and can be particularly effective in removing solute (Jones et al., 1989). The mode of melting (rate, intermittency, etc.) can have a marked effect on drainage pathways in the snowpack (Jeffries, 1989) and consequently, and additionally (see Davies et al., 1987; Jones et al., 1989), on the composition of the meltwater. The rate of impurity removal by meltwater is also partially governed by earlier conditions, such as depositional (physical and chemical) characteristics, and type and degree of snow metamorphism (Colbeck, 1981). Snowpack depth is also important (Davies et al., 1987; Jones et al., 1989). The overwhelming importance of the climate in influencing the precise physical and chemical nature of snowmelt has been stressed by Wigington et al. (1990), who also pinpoint the necessity of considering temporal trends due to climatic change.

Besides having the dramatic effect of depleting solute concentrations in the remaining snowcover because of the "fractionation effect", the melting process, or the scavenging process of percolating rainwater, may also change the relative concentrations of individual chemical species because of "preferential elution" (Davies et al., 1982; Tranter et al., 1986). This also depends on predominating mode of melting (i.e. climate). Particles in the snowcover tend to remain in the pack during at least the early phases of melt; the composition will be changed in this respect too.

Future climate

It is clear, from the above discussion, that the composition of snowcover is strongly conditioned and governed by atmospheric conditions, which we can loosely describe as "climate". Any future change, or trend, is, therefore, likely to alter the future compositional state of future seasonal snowcovers (which are also likely to be physically altered). A major future influence, which we shall ignore (except for the "greenhouse gases"!) because of the large uncertainties, are changes in pollutant emissions. These may be a result of future control strategies, or they may, in turn, be a consequence of a changing climate.

We have two problems in trying to assess the effects of climate change. The first is our lamentable lack of quantitative measures for many of the relationships discussed above. This deficiency immediately proscribes our study. The second problem is the great uncertainty inherent in predictions of future climate (e.g. Rowntree, 1990), especially at the regional scale (Wilson and Mitchell, 1987). The most suitable models for predicting the atmospheric response to increasing greenhouse gas concentrations are 3-D Global Climate Models (GCMs) (Cess and Potter, 1988). The first layer of uncertainty is the prediction of future greenhouse gas emissions (Warrick and Russell, 1988). This is compounded by the fact that CO_2 is not the only radiative forcing gas; however, the combined radiative forcing of the gases is expressed in terms of a "CO_2 equivalent". Current best estimates indicate that CO_2 equivalent concentrations will double ($2xCO_2$), compared to pre-industrial concentrations, by ~2030.

Ideally, for the purposes of this study, we would require regional-scale GCM output for $2xCO_2$ conditions. Such ideal predictions would include changes in mean temperature and precipitation; diurnal temperature range; trajectory climatologies; variability on daily, seasonal, and interannual scales; and frequency and magnitude of extreme events. Unfortunately, numerical studies do not show consistent changes at the regional scale, even for major climatic factors such as temperature and precipitation (Wilson and Mitchell, 1987; Schlesinger and Mitchell, 1987; Wigley, 1989). Furthermore, predicted changes in other climatic factors such as variability and diurnal temperature range are often unrealistically represented by GCMs, or are unreliable due to lack of comparison with the current climate simulation (Karl et al., 1984; Rind et al., 1987). The absence of variability information is serious, since relatively small shifts in the mean of a variable can lead to large changes in the frequency of extreme values (e.g. Wigley, 1988).

The assumptions we make about future climate, and our justifications, are presented later.

An indicator of the effect of future climate change on seasonal snowcover

(a) Preamble

Melt has been seen to dramatically modify the nature of seasonal snowcover. Melt models have been developed which perform well in their simulation of the physical characteristics of snowmelt in certain regions, whilst requiring only temperature and precipitation as input variables (see, for example, Marmorek et al., 1984). Such models have been adapted (see Marmorek et al., 1984; Jeffries, 1989) to give predictive information on the release of acid from the snowcover as melting occurs. We have adapted such approaches to our first approach to assessing the impact of future climatic change on the composition of seasonal snowcover in a region of eastern

Canada and the north-eastern United States. The output is in terms of what departs the snowcover, rather than what remains; the impact projection is indirect in that respect, but it does have an additional environmental interest. Episodic acidification, induced by snowmelt, may impact aquatic biota (Wigington et al.,1990).

Several hydrological or hydrometeorological snowmelt models have been applied regionally to estimate the potential acidic "insult" or "shock" to watersheds due to snowmelt and/or precipitation episodes (Wilson and Barrie, 1981; Eshleman,1988; Schnoor et al., 1989). All of these studies considered acidification under current climatic conditions. Goldstein et. al. (1984) suggested that because climatic factors affect the acidity of surface waters, long term climatic trends and multi-year climatic fluctuations could potentially produce major effects. These authors proposed that a stochastic model be developed which would simulate an altered climate by changing annual or seasonal means or higher moments within the model.

In this study, we have developed an adaptation of the hydrometeorological snowmelt shock potential model used by Wilson and Barrie (1981) to simulate the life cycle of acids in snowpacks under altered climatic conditions. Because of the gross disagreement of GCM estimates of regional scale climatic changes due to increasing greenhouse gases, this study uses the best estimates of global scale climatic changes as input into the model (itemised below). Climatological data for the past thirty snow seasons (1959-60 - 1988-89) are used to represent the current climate. Twenty-eight stations in southeastern Canada and the northeastern United States are chosen to represent conditions in the area 40°-50°N, 70°-80°W, which is one of the most highly acidified areas in North America and an area that experiences snowcover during the winter months (Figure 1). The results of this study will indicate the areas where lakes and streams are most susceptible to significant changes in melt frequency and snowmelt shock potential and, consequently, give some pointers to an assessment of the seasonal and spatial perturbations to the physical and chemical characteristics of snowcover as a result of the assumed future climate.

(b) Methodology, model and assumptions

The model simulates snowpack accumulation and ablation, while calculating the acid budget of the pack. Fluxes of wet and dry deposition at each station are estimated using data from the National Atmospheric Deposition Program and the Canadian Atmospheric Environment Service from 1979-1989 (Barrie and Sirois, 1982; Barrie, 1982). The values of the H^+ concentration in precipitation and the H^+ dry deposition (Table 1) are assumed constant (unlike precipitation amount). The required climatic inputs are daily maximum and minimum temperature, daily snowfall, and daily rainfall.

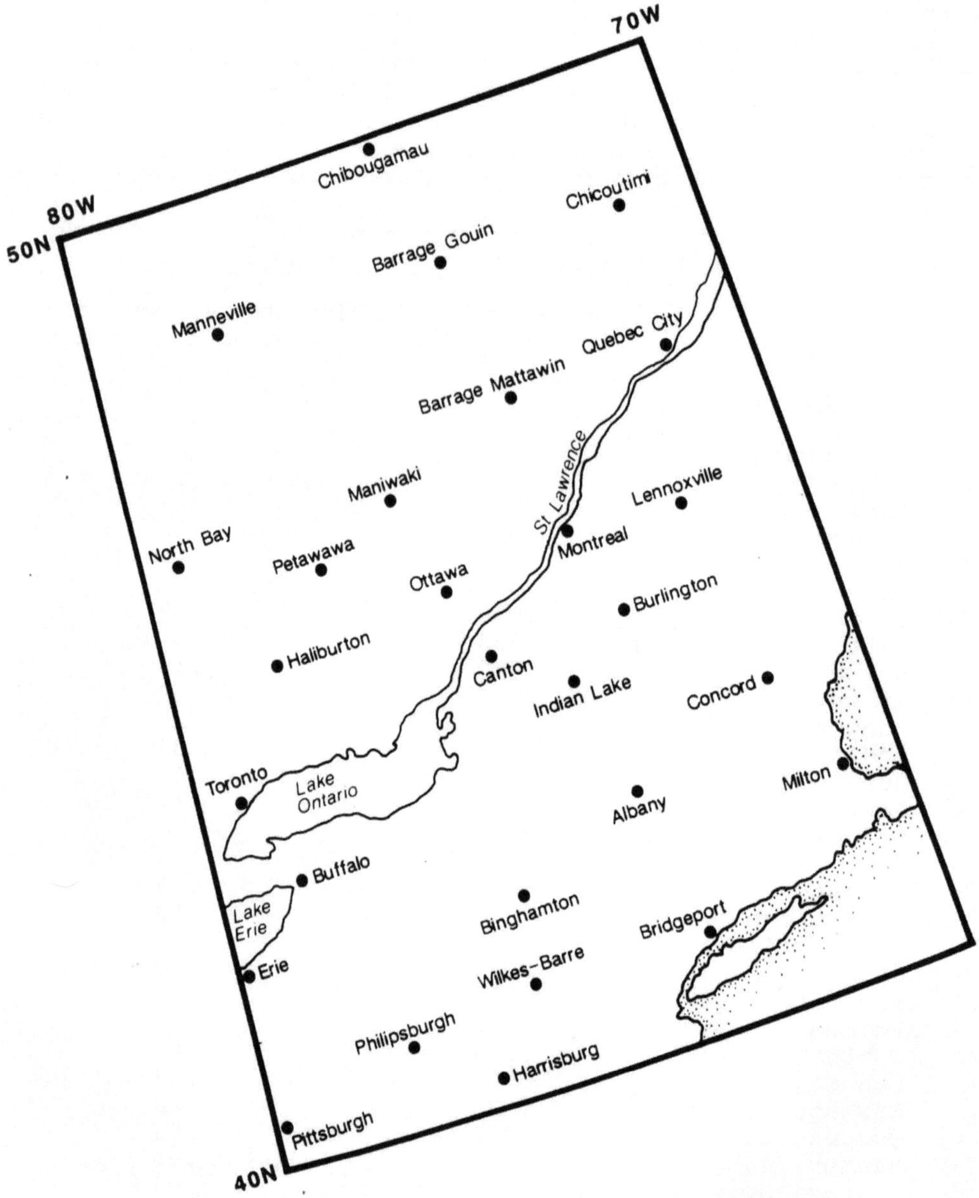

Figure 1. Station network used in study

TABLE 1

Mean H^+ concentration in precipitation and dry deposition of H^+ during the winter period for each station (based on data for the period 1979-1989, National Atmospheric Deposition Program (USA) and Atmospheric Environment Service (Canada)).

Station	H^+ concentration (m mole l^{-1})	Dry deposition of H^+ (m mole m^{-2}-day)
(United States)		
1. Harrisburg	.052	.13
2. Pittsburgh	.054	.13
3. Philipsburg	.054	.13
4. Bridgeport	.040	.10
5. Wilkes-Barre	.053	.13
6. Erie	.056	.14
7. Binghamton	.050	.12
8. Buffalo	.055	.14
9. Milton	.040	.10
10. Albany	.041	.10
11. Concord	.041	.10
12. Indian Lake	.036	.09
13. Burlington	.038	.09
14. Canton	.040	.10
(Canada)		
15. Toronto	.041	.12
16. Haliburton	.040	.10
17. Lennoxville	.032	.10
18. Ottawa	.039	.12
19. Montreal	.034	.11
20. Petawawa	.040	.12
21. North Bay	.029	.07
22. Maniwaki	.033	.09
23. Barrage Mattawin	.030	.08
24. Quebec City	.032	.08
25. Manneville	.035	.07
26. Barrage Gouin	.025	.07
27. Chicoutimi	.020	.05
28. Chibougamau	.015	.04

These data were obtained from the National Climatic Data Center and the Canadian Climate Centre. For each day of the snow season, the model computes the amount of meltwater that leaves the pack, based on the snowmelt equation given by Bruce and Clark (1966)

$$M = 1.8T_{ave}(1.88 + 0.007R) + 12.7 \qquad (1)$$

and Yoshida (1962)

$$M = M_f(T_i - T_b) \qquad (2)$$

where M = melt produced (millimetres per day)
T_{ave} = average daily temperature (^oC)
R = rainwater (millimetres)
M_f = melt factor (mm/(^oC*day))
T_i = mean air temperature for that portion of the day above 0^oC
T_b = base temperature (0^oC)

Equation (1) has produced excellent results in studies of snowmelt runoff from sizeable drainage basins (Bruce, 1961; Louie and Pugsley,1980) and is applicable to mainly forested areas, as evidenced by its lack of a wind speed term. Equation (2) was derived from measurements in the Tadami River Basin in Japan, for which the M_f values were found to range from 4 to 8 mm/(^oC*day), depending on location, time of year and meteorological conditions. This equation's advantage is its implicit inclusion of a diurnal temperature range term and a range of acceptable values of M_f. On days in which the minimum temperature is below 0^oC and the maximum temperature is above 0^oC, Yoshida's results indicate the following approximate relationship between the daily maximum temperature and the melt produced

$$M = 0.43M_fT_{max} \qquad (3)$$

Thus by setting equations (1) and (3) equal to one another, a relationship can be derived between the daily maximum and daily mean temperature

$$T_{max} = (1/M_f)(7.87T_{ave} + 0.03RT_{ave} + 29.5)$$
$$[4 \le M_f \le 8] \qquad (4)$$

We propose modifying Equation (1) to account for a diurnal temperature range by altering this equation's last term. This last term is included to account for non-longwave heat transfer effects such as short wave solar radiation, which is related to the diurnal temperature range. Our new snowmelt equation must meet three criteria: its melt production must equal that of Equation (1) over long time periods, it must remain linear, and it must be valid over acceptable ranges of M_f (4 to 8 mm/(^oC*day). The following equation was derived, based on these conditions

$$M = 1.8T_{ave}(1.88 + 0.007R) + 12.7(d/D) \quad (5)$$

where d = a station's diurnal temperature range for a specific melt day (^{o}C)

D = a station's mean diurnal temperature range for all melt days (^{o}C)

Equation (5) is linear and will equal Equation (1) over long periods. The diurnal temperature range terms can then be expressed in terms of daily mean and daily maximum temperatures (range = $2(T_{max} - T_{ave})$). If Equation (4) is then substituted for T_{max}, the range of applicability of Equation (5) can be determined. It can be shown that Equation (5) is valid over a wide range of daily mean temperatures.

It is assumed that the maximum retention of interstitial water by a snowpack is 5% (Bruce and Clark, 1966). Thus, an "effective melt" occurred if, during a melt, the sum of daily meltwater and rainfall was greater than 5% of the snowpack water content at the beginning of the melt. If a melt did not qualify as an effective melt, then no runoff was assumed to occur.

Because the greatest proportion of pollutants is released during the early stages of a melt, a concentration factor was needed to calculate the acidic depletion from the snowpack. Based on field and laboratory work by Johannessen and Henriksen (1978), it was assumed that 63% of the snowpack acid is removed by the first 30% of meltwater (and this is a big assumption: see Marmorek et al., 1984). Thus the following equation was used in the model

$$C_f = (1 - (1 - F_s)^{2.78}) / F_s \quad (6)$$

where C_f = concentration factor (ratio of meltwater acidity to snowpack acidity)

F_s = fraction of snowpack that melts during a thaw

An effective melt was defined to continue until melting ceased or the snowpack disappeared. A shock period was defined as the number of consecutive days during an effective melt when the daily meltwater acid concentration was greater than 10% of the meltwater acid concentration on the first day of the melt. Two shock indices were calculated for each effective melt. The stream shock index, I_{st}, given by

$$I_{st} = Cm * Ts \quad (7)$$

where Cm = the arithmetic mean meltwater acid concentration for a shock period (m mol l^{-1})

Ts = duration of shock period (days)

and the lake shock index, I_{lk}, given by

$$I_{lk} = \sum_{i=1}^{Ts} Cm_i * (M + R)_i \qquad (8)$$

where Cm_i = daily meltwater acid concentration on the ith day (m mol l^{-1})
$(M + R)_i$ = daily runoff on the ith day (l)

I_{st} represents the maximum possible pollutant dosage to a point in a stream during the shock period, and I_{lk} represents the total acid released per square meter during a shock period. [Although it may be more appropriate to refer to these two indices as a time-concentration/dosage index and a load index, respectively (D. Jeffries, pers. comm.), we retain the terminology of Wilson and Barrie, for continuity and ease of comparison with the original work]. It must be stressed that this model does not consider the possible modifications of the pollutant discharge from the snowpack by groundwater and soil interaction processes. Thus the model calculates the potential shock to streams and lakes. In fact, in many cases, sub-snowcover hydrology and biogeochemistry are of overwhelming importance in controlling, modifying or moderating the chemical composition of surface waters (Rascher et al., 1987; Jeffries et al., 1989; Wigington et al., 1990), although its influence is probably less for lakes. This, however, is no disadvantage for the purposes of this study; since we are employing the model as an indicator of future snowpack behaviour, and the potential shock indices are useful additional environmental information.

So, the model described above is developed from that used by Wilson and Barrie (1981) with some modifications. They used Equation (1) only and therefore did not consider the effect of the diurnal temperature range on snowmelt. In this study the dry deposition fluxes given in Table 1 were reduced by 25% on the first day of a snowfall and by 50% on the last day of a snowpack, to account for the decreased opportunity for deposition on these days. On the first day of a snowfall, it was assumed that snow fell continuously for half of the day, centered around midday. In such a scenario, there exists only 18 of 24 hours available for dry deposition into the snowpack, requiring a reduced flux of 25%. On the last day of a snowpack, it was assumed that the pack melted entirely by midday, allowing only 12 of 24 hours in which dry deposition could occur. In addition, as proposed by Christophersen et al. (1984), a "sleet interval" was used to determine precipitation type at temperatures around $0^{o}C$. Precipitation was assumed to change linearly from snow to rain as the mean daily temperature increased from $0^{o}C$ to $1.6^{o}C$. It has been shown that relatively simple snowpack accumulation and ablation models of this type have been successful in describing these processes and perhaps perform as well as more complicated models (Lundquist, 1980;

Marmorek et al., 1984). Rainfall was added simply to the water content of the snowpack, and no account was taken of the possible "rain-on-snow" effects on solute removal from the pack (Tranter, this volume).

Due to the lack of suitable regional climatic predictions (see above), we use the best estimates of future global climatic change from the results of five major GCMs as input parameters into the model. The best estimate of global mean temperature change and precipitation change for a doubling of CO_2 is 1.5 - 4.5°C and positive 10% respectively (Wigley, 1989). Such a temperature estimate for our study region is probably conservative, since the largest temperature increases are expected at high latitudes in the winter hemisphere, due to the positive feedbacks of reduced albedo and reduced static stability resulting from diminished snowcover (Schlesinger and Mitchell, 1987). The authors know of no global scale predictions of diurnal temperature changes due to a doubling of CO_2, though Rind et al. (1989) modeled regional scale effects in the United States and found that the diurnal temperature range decreased in summer. The effect of increased concentrations of greenhouse gases on the diurnal temperature range in calm weather should be similar to that caused by increased atmospheric moisture, which produces a rise in nocturnal temperature under these conditions (Elsasser, 1942; Saunders, 1952). The model of Gall and Herman (1980) confirmed that minimum surface air temperature is highly sensitive to changes of atmospheric water vapor. Based on radiation theory, the nocturnal temperature should be sensitive to CO_2 variations as well (Haltiner and Martin, 1957). In this study we use a reduced diurnal temperature range of 10% at all stations for a doubled CO_2 climate. This figure is not based on GCM output and is not intended as a prediction. Rather, the sign of this value follows expectations that the diurnal temperature range will decrease in a doubled CO_2 climate, and its magnitude merely allows a sensitivity study to be made on the impact of such a climatic change on snowmelt shock potential.

(c) Shock indices in the current climate

Melt Frequency

The distribution of seasonal mean effective melt frequency for the current climate is presented in Figure 2. The values range from a minimum at Bridgeport in the southeast to a maximum at Buffalo in the west. The main characteristics are the southeast to northwest distribution and the peak which occurs in the middle of the study region, around the Great Lakes. The lowest frequencies occur in the southeast and the northwest, apparently resulting from two very different types of wintertime climates. Stations in the southeast are

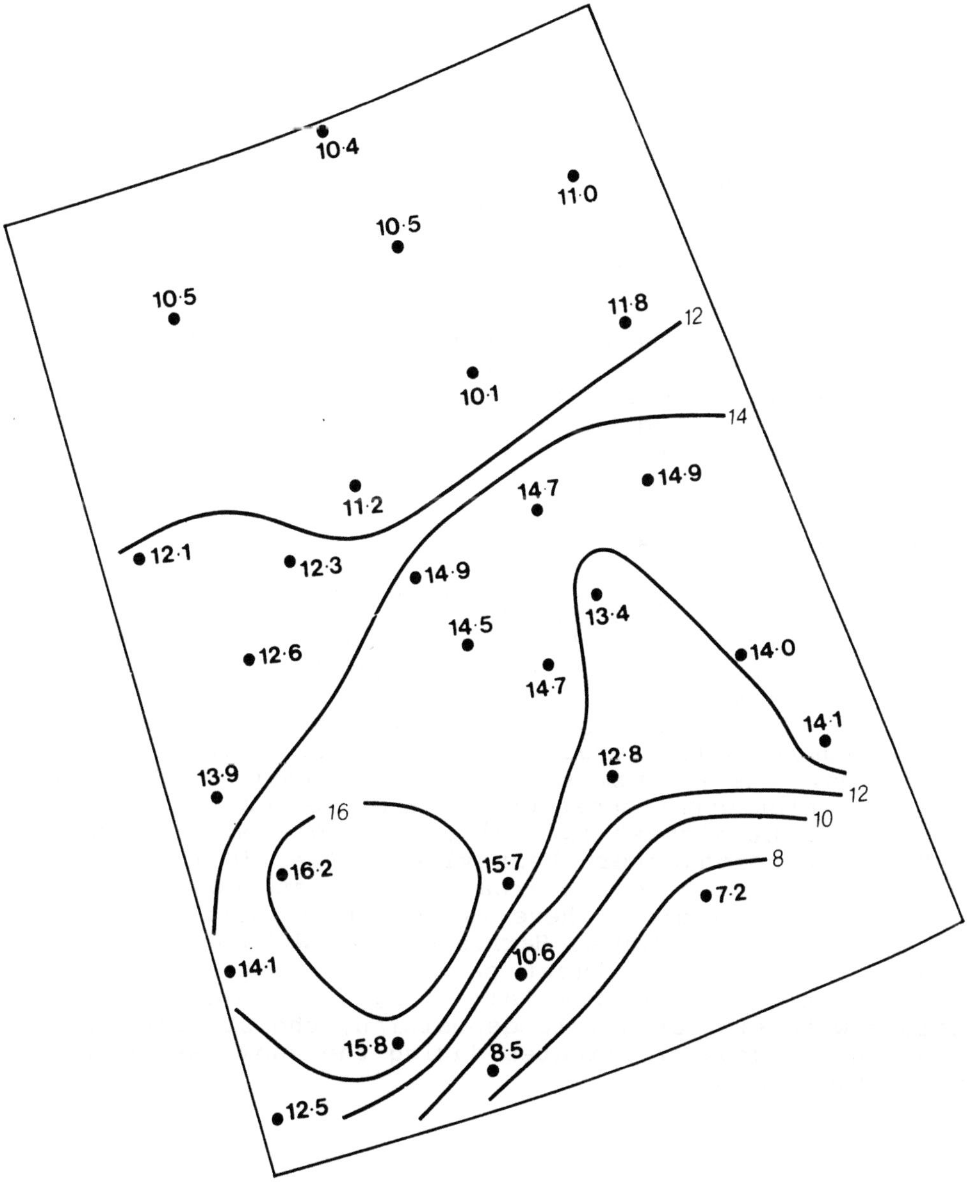

Figure 2. Distribution of the average number of melts (see text) throughout a winter season.

susceptible to mild, maritime air masses during this time of year, particularly those near the ocean, such as Bridgeport. This type of air mass inhibits the occurrence of snowfall and thereby lowers the frequency of melts in the south and southeast regions. Northwestern stations such as Manneville and Chibougamau often experience Continental Polar and Continental Arctic air masses during the snow season. These air masses are often very cold, and they tend to inhibit the northward migration of cyclones, reducing the climatic variability in this region. The frequent presence of these air masses during the snow season in northern sections of the study region explains the low melt frequencies there. The band in between these two contrasting climates--extreme southern Canada, western New York, and northern Pennsylvania--is subject to frequent incursions of both types of air masses and therefore experiences quite variable weather during these months. This high degree of variability causes frequent melts to occur in this area.

Stream Shock Index (I_{st})

The seasonal mean I_{st} pattern (Figure 3) generally follows an east to west distribution. The I_{st} is closely correlated with the amount of acidic deposition that a station receives. This is to be expected, since the I_{st} (Equation 7) is simply a measure of the acidic concentration of the meltwater. The minimum I_{st} is located at Chibougamau (0.10 m mol l^{-1}-day), which receives the least acid deposition of any station. The maximum occurs at Wilkes-Barre (0.35 m mol l^{-1}-day), which experiences one of the greatest amounts of deposition. The highest I_{st} values are located in western New York and northwestern Pennsylvania (>0.30 m mol l^{-1}-day), corresponding to the area of greatest deposition in the entire study region. There is also a climatic influence on this index. For example, Harrisburg and Wilkes-Barre have nearly identical acid input parameters, yet the latter station has a noticeably larger I_{st}. In general, for stations with similar acidic deposition, the stations with colder and/or snowier climates during the snow season will have a higher I_{st}.

Lake Shock Index (I_{lk})

Because the I_{lk} accounts for both the concentration of the meltwater and the runoff (Equation 8), one would expect this index to show a greater correlation with climatic conditions than the I_{st} does. Indeed, the pattern of I_{lk} (Figure 4) is more climate-dependent than that of the I_{st}. The values range from a minimum of 1.09 m mol m^{-2} at Pittsburgh in the extreme southwest to a maximum of 2.44 m mol m^{-2} at Quebec City in the east. The lowest values (<1.20 m mol m^{-2}) are located in the southeast and the southwest: in southeastern New York, northeastern Pennsylvania, western Connecticut, and western Pennsylvania. The maximum values (>2.00 m mol m^{-2}) occur in the northcentral region, from Haliburton and Manneville to Quebec City. The acidic input parameters have

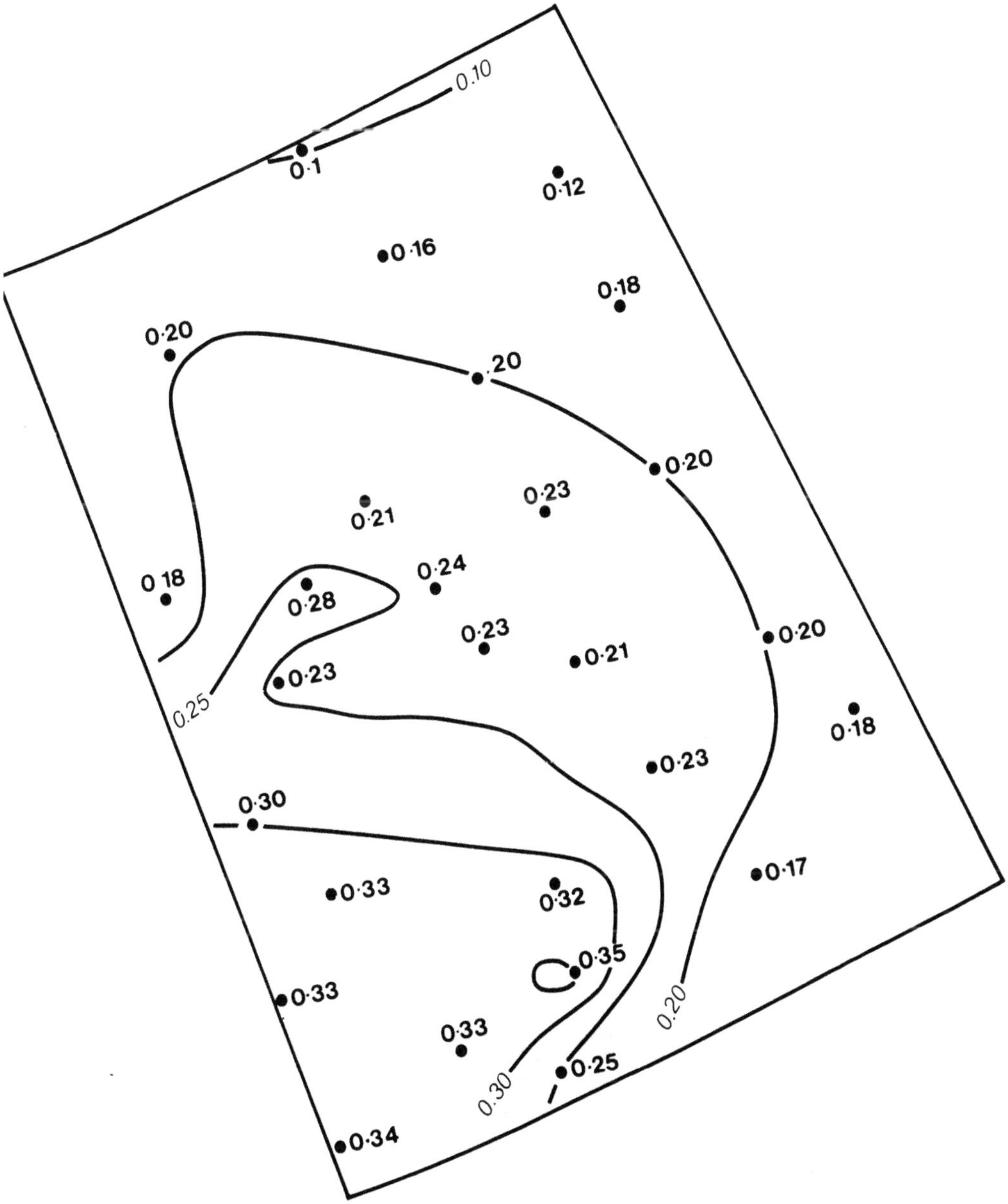

Figure 3. Seasonal mean "stream shock" index (1_{st}) value (m mol 1^{-1}-day).

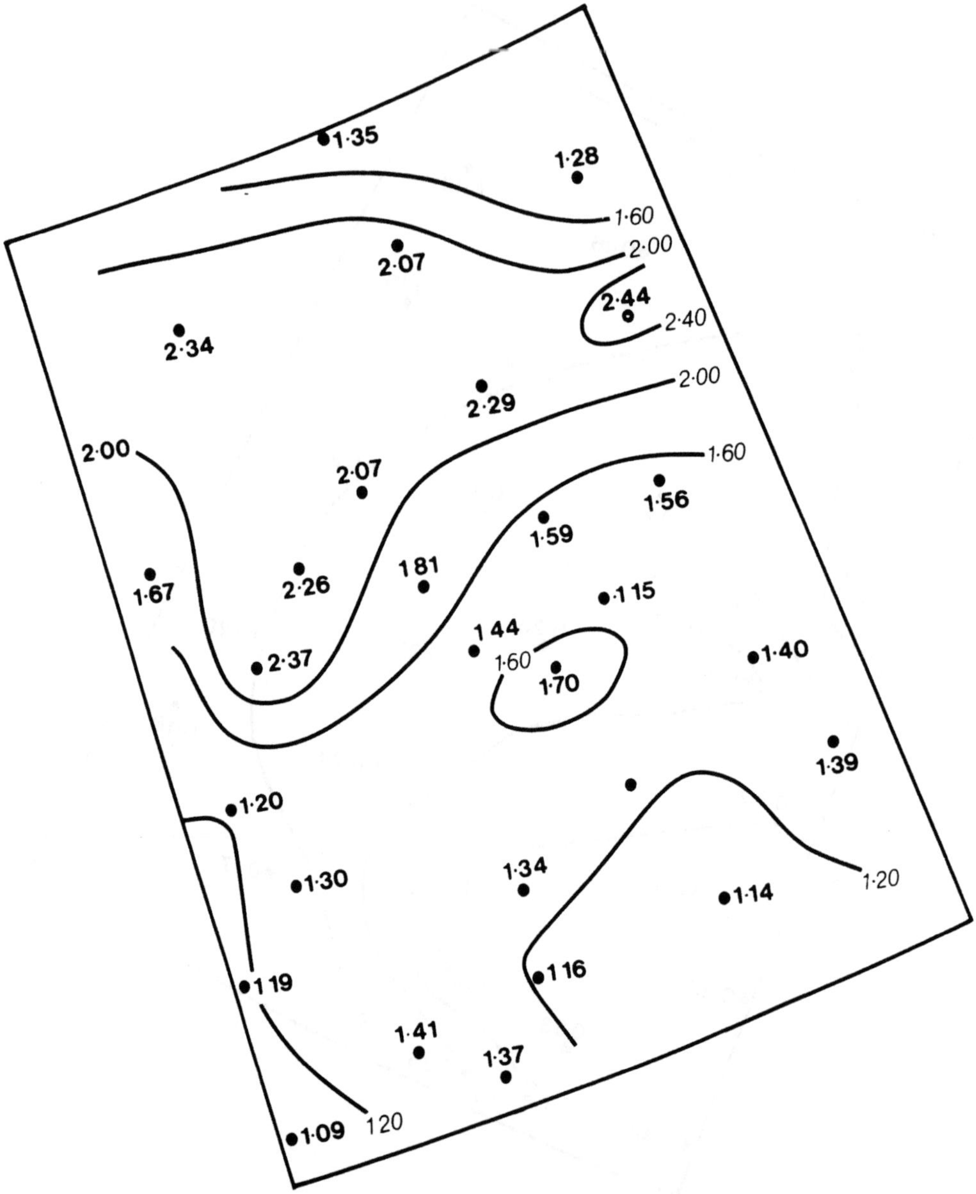

Figure 4. Seasonal mean "lake shock" index (I_{lk}) value (m mol m^{-2})

very little effect on the I_{lk}, as evidenced by the high values in central Canada and the low values in western Pennsylvania.

(d) Projected indices

The current climate time series was altered for each station to account for projected climatic changes. The model was run for five different scenarios: 1.5°C, 3.0°C, and 4.5°C temperature increases with no change in precipitation or diurnal temperature range; 3.0°C temperature increase, 10% increase in precipitation, and no diurnal temperature range change; and 3.0°C temperature increase, a 10% increase in precipitation, and a 10% reduction in the diurnal temperature range. Temperature increases were simulated by simply increasing both the maximum and minimum temperature for each day of the current climate by either 1.5, 3.0 or 4.5°C. The precipitation increase was simulated similarly, by increasing precipitation for each day by 10%. The reduced diurnal temperature range was simulated by adding (subtracting) 5% of the mean October - April diurnal temperature range of the current climate to the minimum (maximum) for each day. The results of these model runs are shown in Tables 2 - 4. In addition, the changes in melt frequency and shock indices for the 3.0°C temperature increase case are spatially represented in Figures 5 - 7.

Melt Frequency Changes

Changes in melt frequency are determined almost solely by changes in temperature. Table 2 reveals the dominant effect of this climatic variable on the seasonal melt frequency changes for each station. In general, melt frequencies are expected to decrease in a warmer climate. For all stations, decreases of 7%, 14% and 22% are projected for the three temperature increase scenarios. The effect of a precipitation increase and a reduction in the diurnal temperature range are negligible, as each account for only 1% increases in the melt frequency.

Perhaps as important as the sign and magnitude of these projected changes is their spatial distribution. As shown in Table 2, the seasonal melt frequency for US stations is expected to drop considerably (between 17% and 48%), whereas the Canadian stations are to increase slightly (between 3% and 6%). The largest reductions are expected at stations which currently have the highest melt frequencies (Bridgeport and Harrisburg). The model predicts very large reductions of nearly 80% at these stations for a 4.5°C temperature increase. Farther north, the reduction of melt frequencies decreases until north of about 46°N, where increases are predicted (Figure 5). The largest increases in melt frequency (up to 28%) for a 3.0°C temperature rise are expected in a band between 46°N and 47°N. These results support the logical expectation that the band of greatest

TABLE 2

Percentage changes in melt frequency from the current climate for the five climate scenarios (ppn.; precipitation: diur.; diurnal temperature range). The figures in brackets represent the percentage change due to those scenarios when scenario 4 is compared with scenario 2, and scenario 5 is compared with scenario 4.

	Scenario	1	2	3	4	5
	Station	$+1.5^{o}$	$+3.0^{o}$	$+4.5^{o}$	$+3.0^{o}$ +10% ppn.	$+3.0^{o}$ +10% ppn. -10% diur
(United States)						
1.	Harrisburg	-35	-56	-75	-55 (+3)	-54 (+3)
2.	Pittsburgh	-16	-35	-55	-33 (+4)	-30 (+5)
3.	Philipsburg	-13	-30	-48	-29 (+2)	-30 (-1)
4.	Bridgeport	-29	-56	-78	-54 (+3)	-54 (0)
5.	Wilkes-Barre	-26	-48	-61	-47 (+2)	-47 (0)
6.	Erie	-21	-33	-43	-31 (+3)	-30 (+2)
7.	Binghamton	-10	-25	-40	-24 (+2)	-22 (+2)
8.	Buffalo	-19	-32	-40	-31 (+1)	-31 (+1)
9.	Milton	-21	-48	-64	-48 (+1)	-47 (+1)
10.	Albany	-15	-27	-45	-26 (+2)	-25 (+1)
11.	Concord	-11	-29	-45	-29 (0)	-28 (+1)
12.	Indian Lake	- 1	-10	-16	- 9 (+1)	- 8 (+1)
13.	Burlington	- 9	-16	-29	-15 (+1)	-14 (+1)
14.	Canton	-13	-20	-30	-19 (+1)	-18 (+2)
	Average (US)	-17	-33	-48	-32 (+2)	-31 (+1)
(Canada)						
15.	Toronto	-21	-35	-42	-33 (+2)	-32 (+1)
16.	Haliburton	+ 7	+ 7	+ 4	+ 8 (+1	+ 7 (-1)
17.	Lennoxville	+ 1	-10	-17	- 9 (+1	- 8 (+2)
18.	Ottawa	- 4	- 9	-22	- 7 (+2	- 7 (-1)
19.	Montreal	- 5	-18	-26	-16 (+3)	-16 (0)
20.	Petawawa	+15	+11	+ 4	+11 (0)	+14 (+3)
21.	North Bay	+13	+21	+23	+23 (+1)	+25 (+1)
22.	Maniwaki	+19	+21	+14	+21 (+1)	+20 (-1)
23.	Barrage Mattawin	+10	+28	+32	+27 (-1)	+25 (-2)
24.	Quebec City	+14	+16	+14	+14 (-1)	+17 (+2)
25.	Manneville	- 7	+ 3	+12	+ 5 (+2)	+ 2 (-1)
26.	Barrage Gouin	- 1	+10	+20	+11 (+1)	+15 (+3)
27.	Chicoutimi	+ 9	+15	+16	+15 (0)	+15 (0)
28.	Chibougamau	- 1	+ 4	+16	0 (-4)	+ 5 (+1)
	Average (Canada)	+ 4	+ 5	+ 3	+ 5 (+1)	+ 6 (+1)
	Average (Total)	- 7	-14	-22	-14 (+1)	-13 (+1)

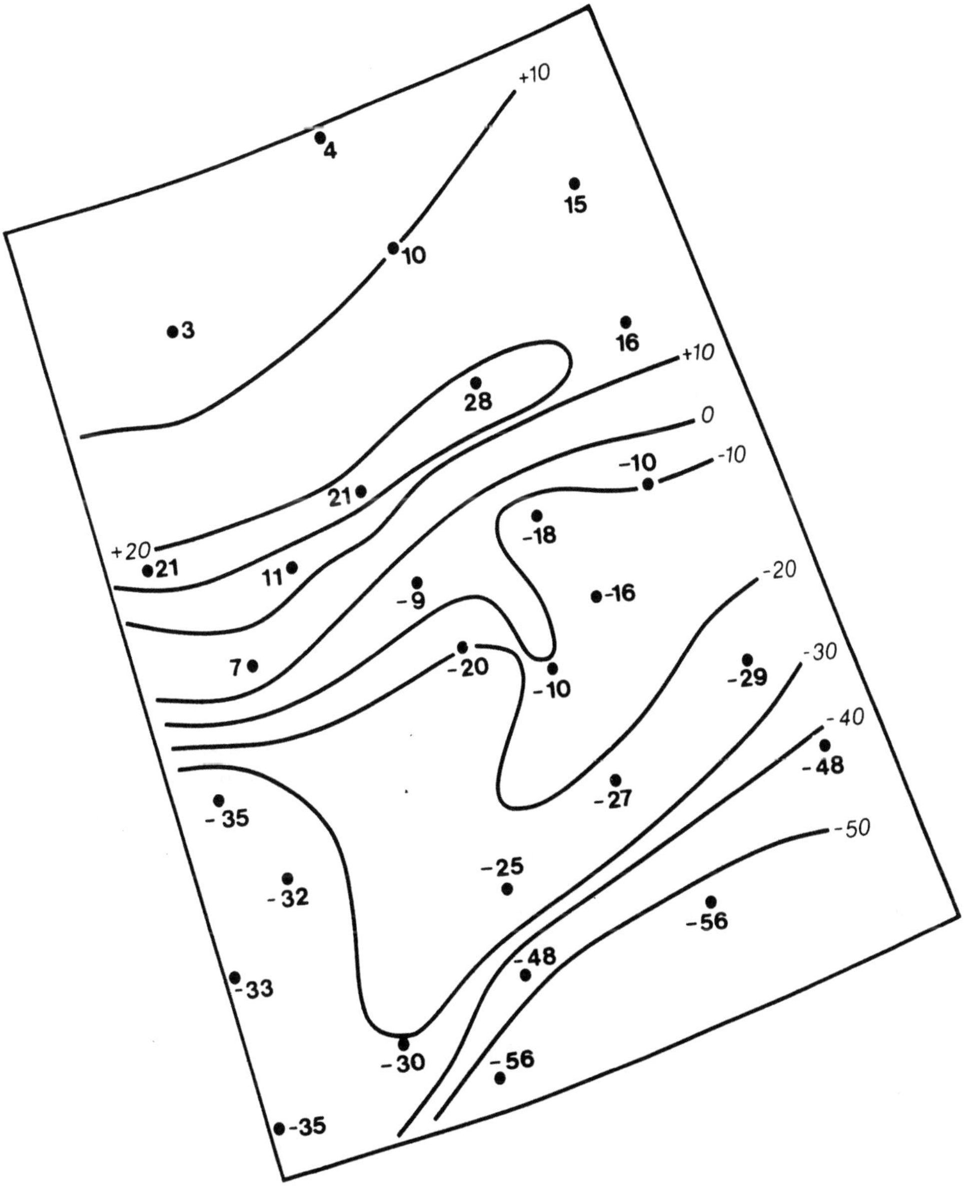

Figure 5. Percentage changes in melt frequency in the +3.0°C climate scenario case.

melt frequencies will shift northward in a warmer climate, regardless of changes in precipitation or diurnal temperature range.

Stream Shock Index Changes

The model predicts much smaller changes in the I_{st} for the altered climates (Table 3). Overall, reductions of less than 5% are forecast for temperature rises only, but with increases at US stations and decreases at Canadian stations. For this index the effect of increased precipitation is more important than higher temperatures. A 10% increase in precipitation accompanied by a 3.0°C temperature rise results in a 7% reduction of the I_{st} for all stations compared with the current climate and a 5% reduction compared with a 3.0°C warmer climate. This result reflects the fact that the potential for stream shock is a measure of the acidic meltwater concentration. The net effect of increased precipitation on this index appears to be dilutory, offsetting the enhanced contribution of wet deposition into the snowpack. Once again, the effect of a reduced diurnal temperature range is found to be negligible (1% increase).

The pattern of I_{st} changes for a 3.0°C temperature rise is presented in Figure 6. This pattern is less spatially coherent than that of melt frequency change. All but two stations in the US show no change or an increase in their I_{st}, contrasting the pattern of Canadian stations. The largest increases in the I_{st} (>10%) are expected at Pittsburgh and Canton in the US, whereas the greatest decreases (>10%) are found in the northeast of Canada, around Barrage Mattawin and Chicoutimi. In general, I_{st} decreases are forecast for northerly regions, while increases are to occur farther south. However, the magnitude of these changes for either country is small compared to melt frequency changes.

Lake Shock Index Changes

The I_{lk} is the index most sensitive to projected changes in climate. Model results suggest that I_{lk} decreases of 13% to 33% (all stations) will occur if estimates of climatic changes are accurate (Table 4). Canadian stations may expect the greatest decreases (nearly 50% for some stations for a 4.5°C temperature increase). The effect of a 10% precipitation increase on the I_{lk} is similar in magnitude and opposite in sign to its effect on the I_{st}. Enhanced precipitation raises the value of the I_{lk}, as the increased runoff which results from this climatic change offsets the extra dilution that occurs (Equation 4). The reduced diurnal temperature range is found to have no effect on the I_{lk}.

TABLE 3

Percentage changes in the stream "shock index", I_{st}, from the current climate for the five climate scenarios (ppn.; precipitation: diur.; diurnal temperature range). The figures in brackets represent the percentage change due to those scenarios when scenario 4 is compared with scenario 2, and when scenario 5 is compared with scenario 4.

Scenario	1	2	3	4	5
Station	+1.5°	+3.0°	+4.5°	+3.0° +10% ppn.	+3.0° +10% ppn. -10% diur.
(United States)					
1. Harrisburg	-12	+ 8	-32	- 8 (-15)	- 8 (0)
2. Pittsburgh	+ 3	+18	+21	+ 6 (-10)	+ 3 (- 3)
3. Philipsburg	+ 6	0	+ 3	- 3 (- 3)	- 3 (0)
4. Bridgeport	+12	0	- 6	-12 (-12)	0 (+13)
5. Wilkes-Barre	- 6	0	+14	- 6 (- 6)	+ 6 (+12)
6. Erie	+ 3	+ 6	+30	+ 6 (0)	+ 6 (0)
7. Binghamton	0	+ 6	+19	+ 3 (- 3)	- 3 (- 6)
8. Buffalo	- 6	+ 6	+21	0 (- 6)	- 3 (- 3)
9. Milton	-11	0	+11	- 6 (- 6)	- 6 (0)
10. Albany	- 4	- 9	- 9	-13 (- 5)	-13 (0)
11. Concord	- 5	+ 5	+15	0 (- 5)	0 (0)
12. Indian Lake	0	-10	-10	-10 (0)	- 5 (+ 5)
13. Burlington	+ 9	+ 9	+17	+ 9 (0)	+ 4 (- 4)
14. Canton	+ 4	+13	+17	+ 4 (- 8)	+ 9 (+ 4)
Average (US)	- 1	+ 4	+ 8	- 2 (- 6)	- 1 (+ 1)
(Canada)					
15. Toronto	+10	0	+13	- 3 (- 3)	- 7 (- 3)
16. Haliburton	- 9	-13	-22	-17 (- 5	-17 (0)
17. Lennoxville	- 5	- 5	0	-10 (- 5)	-10 (0)
18. Ottawa	- 4	- 4	0	- 8 (- 4)	- 8 (0)
19. Montreal	- 9	+ 4	0	+ 9 (+ 4)	+ 9 (0)
20. Petawawa	-11	- 4	- 4	-11 (- 7)	-14 (- 4)
21. North Bay	0	0	- 6	- 6 (- 6)	-11 (- 6)
22. Maniwaki	-10	-10	-14	-14 (- 5)	-14 (- 5)
23. Barrage Mattawin	-10	-15	-15	-20 (- 6)	-15 (+ 6)
24. Quebec City	-11	-11	-11	-11 (0)	-11 (0)
25. Manneville	0	- 5	-10	-15 (-11)	-10 (+ 6)
26. Barrage Gouin	- 6	- 6	-13	-19 (-13)	-19 (0)
27. Chicoutimi	- 8	-17	-17	-25 (-10)	-17 (+11)
28. Chibougamau	+10	0	-10	0 (0)	0 (0)
Average (Canada)	- 5	- 6	- 8	-11 (- 5)	-10 (0)
Average (Total)	- 3	- 1	0	- 7 (- 5)	- 5 (+ 1)

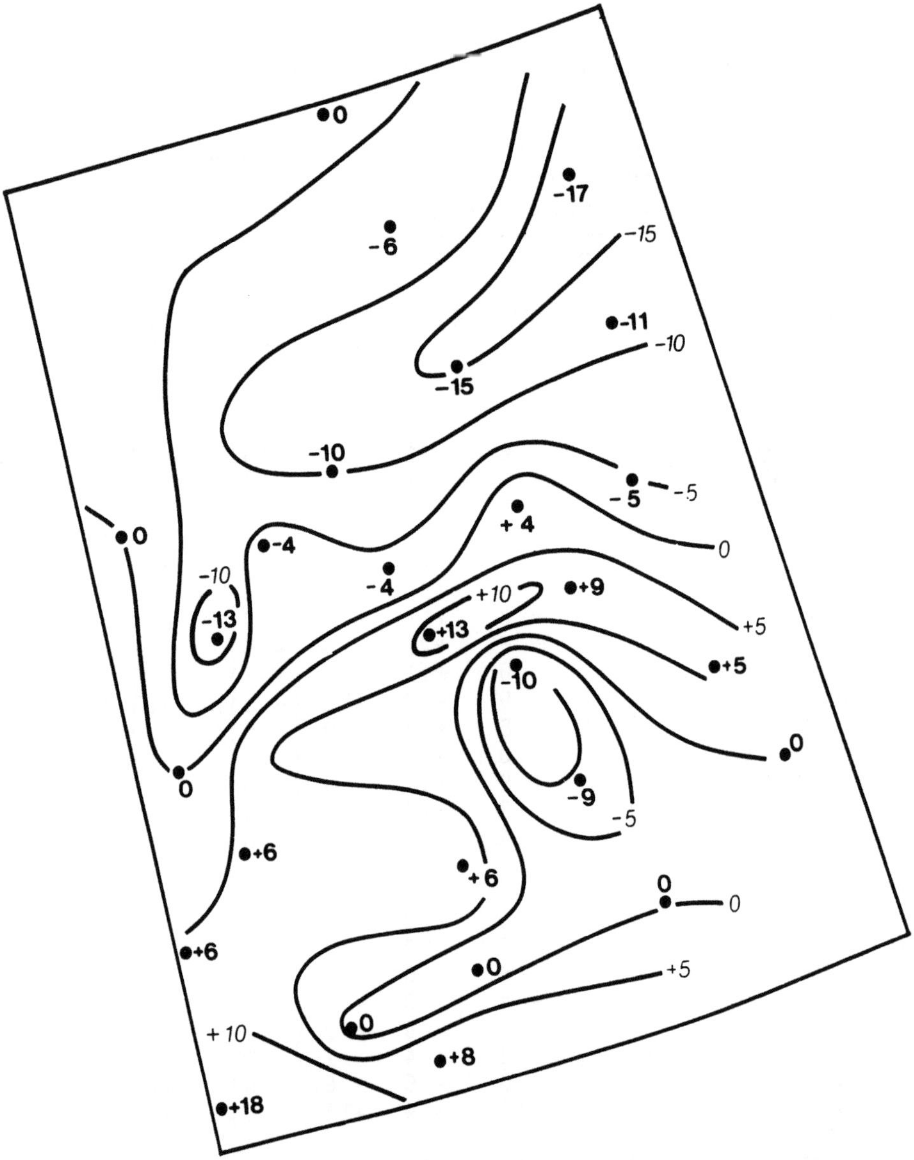

Figure 6. Percentage change in I_{st} values in the +3.0° climate scenario case.

TABLE 4

Percentage changes in the lake "shock index", I_{1k}, from the current climate for the five climate scenarios (ppn.; precipitation: diur.; diurnal temperature range). The figures in brackets represent the percentage change due to those scenarios when scenario 4 is compared with scenario 2, and scenario 5 is compared with scenario 4.

	Scenario / Station	1	2 +1.5°	3 +3.0°	4 +4.5° +10% ppn.	5 +3.0° +10% ppn. -10% diur.
(United States)						
1.	Harrisburg	- 7	-14	-44	-19 (- 6)	-22 (- 4)
2.	Pittsburgh	- 9	- 6	- 6	+ 1 (+ 8)	- 6 (- 7)
3.	Philipsburg	-16	-23	-31	-21 (+ 4)	-22 (- 2)
4.	Bridgeport	-18	-28	-44	-24 (+ 6)	-22 (+ 2)
5.	Wilkes-Barre	- 8	-12	-26	- 9 (+ 3)	- 2 (+ 9)
6.	Erie	- 4	-18	-31	-16 (+ 3)	-12 (+ 5)
7.	Binghamton	-10	-25	-37	-23 (+ 3)	-26 (- 4)
8.	Buffalo	- 2	-11	-25	- 6 (+ 5)	- 6 (0)
9.	Milton	- 6	- 2	-10	+ 1 (+ 4)	+ 4 (+ 2)
10.	Albany	-10	-13	- 3	- 7 (+ 7)	- 3 (+ 4)
11.	Concord	-11	-20	-24	-17 (+ 4)	-14 (+ 4)
12.	Indian Lake	-18	-27	-44	-24 (+ 5)	-24 (0)
13.	Burlington	-15	-27	-28	-23 (+ 5)	-23 (0)
14.	Canton	-10	-29	-26	-26 (+ 5)	-24 (+ 2)
	Average (US)	-10	-18	-27	-15 (+ 4)	-14 (+ 1)
(Canada)						
15.	Toronto	- 8	-13	-29	- 7 (+ 7)	- 8 (- 2)
16.	Haliburton	-20	-35	-48	-32 (+ 6)	-32 (0)
17.	Lennoxville	-15	-22	-28	-19 (+ 3)	-20 (+ 1)
18.	Ottawa	-14	-27	-34	-22 (+ 6)	-23 (- 1)
19.	Montreal	-16	-26	-36	-19 (+ 9)	-20 (- 2)
20.	Petawawa	-25	-29	-34	-26 (+ 5)	-27 (- 2)
21.	North Bay	-18	-37	-47	-34 (+ 4)	-34 (+ 1)
22.	Maniwaki	-27	-37	-42	-33 (+ 6)	-32 (+ 1)
23.	Barrage Mattawin	-16	-37	-47	-33 (+ 6)	-31 (+ 3)
24.	Quebec City	-24	-34	-44	-27 (+12)	-29 (- 3)
25.	Manneville	- 3	-22	-36	-19 (+ 4)	-17 (+ 3)
26.	Barrage Gouin	-10	-27	-36	-23 (+ 5)	-26 (- 4)
27.	Chicoutimi	-15	-27	-36	-21 (+ 7)	-23 (- 2)
28.	Chibougamau	- 4	-21	-37	-14 (+ 9)	-16 (- 2)
	Average (Canada)	-15	-28	-38	-24 (+ 6)	-24 (- 1)
	Average (Total)	-13	-23	-33	-19 (+ 5)	-19 (0)

There is generally a south to north increase in the reduction of the I_{lk} for a 3.0°C warming (Figure 7). All stations are expected to experience decreases in their I_{lk}, although Milton and Pittsburgh in the US show only slight (<10%) reductions. The most pronounced reductions (>30%) are to occur in a wide band in the central Canadian region, extending from Quebec City in the east to North Bay and Haliburton in the west. This region is approximately colocated with the area of greatest I_{lk} values in the current climate.

Intra-seasonal Changes

The results presented thus far have focused on the mean seasonal melt frequencies and shock indices for the current climate and those expected for various climatic change scenarios. However, within a snow season the effect on aquatic biota of the temporal distribution of acid shocks may be as important or more important than the average seasonal values. Goldstein et al. (1984) proposed that the time of occurrence of snowmelt could alter the effect of the major acidic pulse in springtime by changing the occurrence of this shock relative to spawning. In their Integrated Lake-Watershed Acidification Study (ILWAS), they modeled the effect of a 2°C temperature increase on the outlet pH of Panther Lake in the Adirondack Region of New York. This simulation produced pH depressions similar to those in the current climate, but the springtime pulse occurred earlier and did not last as long.

The median monthly distributions of changes in melt frequency, I_{st}, and I_{lk} for a 3.0°C temperature rise are presented in Figure 8. The overall temporal pattern of melt frequency change is bell-shaped, peaking at 17% as an average for all stations in February. The months of January and February show overall increases in melt frequency, whereas all other months show decreases. This result is consistent with expectations that the major acid pulse in springtime would occur earlier in the snow season in a warmer climate. The amplitude of the temporal distribution is greater for Canadian stations than for US stations. Canadian stations show positive melt frequency changes from December through March, but US stations show negative changes in all months.

Temporal variation in the change of the I_{st} is not as large. Decreases of 100% occur in October because many stations experience no snowfall in October in the warmer climate simulation. In general there is a slight downward trend to the data from November to April, but in none of these months does the projected reduction of I_{st} exceed 16%. There are much smaller differences between the monthly values of US and Canadian stations for changes in this index than for the melt frequency changes, although US stations show small non-negative values during most of the snow season. It appears from these results that the concentration of melt water varies only slightly during the snow season.

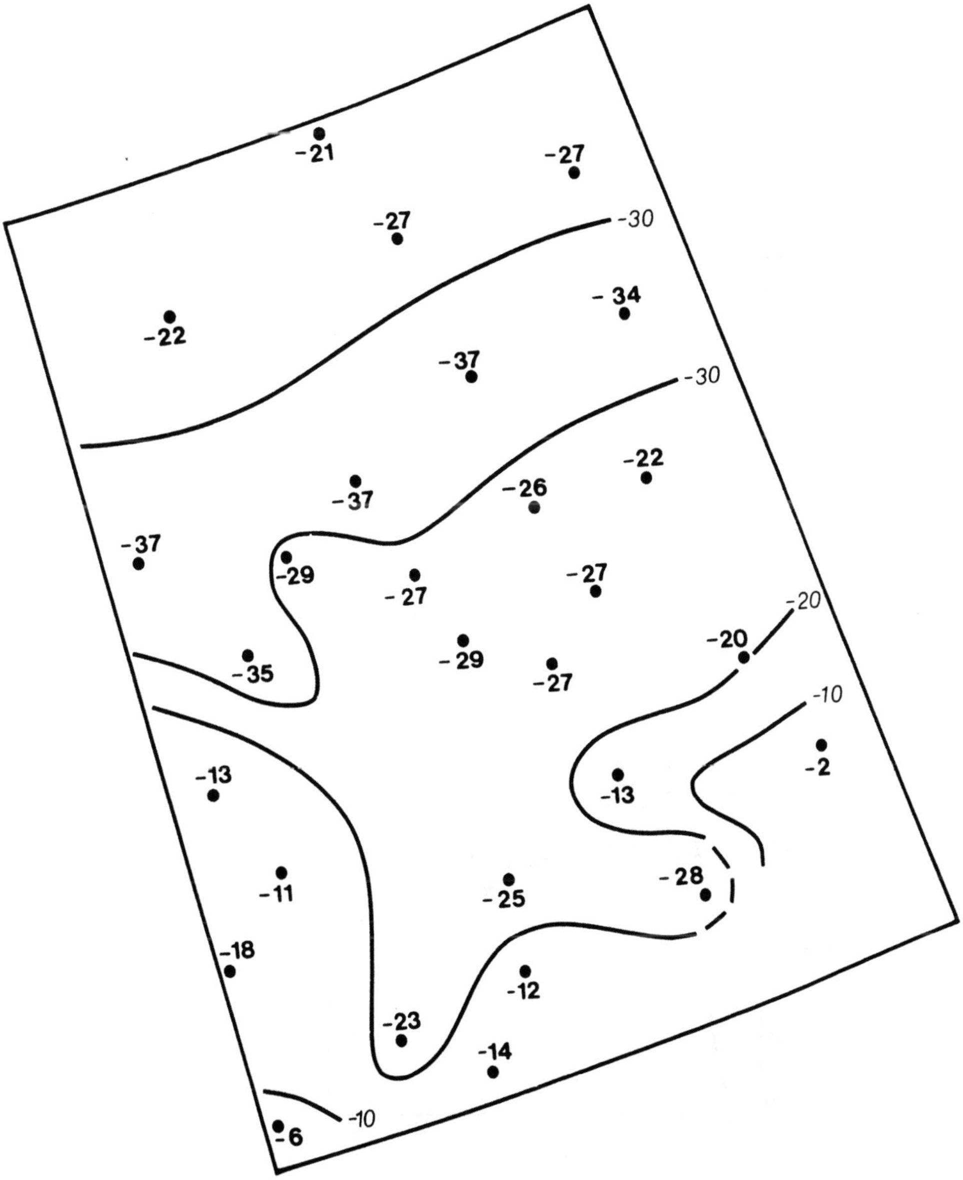

Figure 7. Percentage change in I_{1k} values in the + 3.0°C climate scenario case.

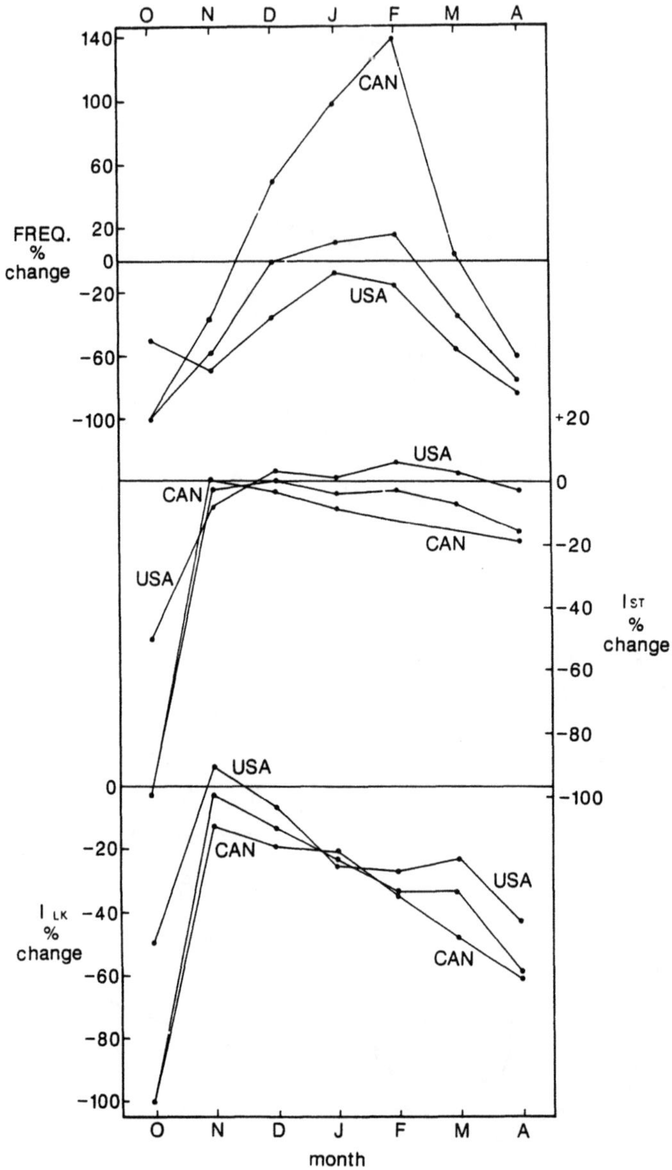

Figure 8. Percentage changes in melt frequency, I_{st}, I_{lk} by month for (1) the whole study region, (2) Canada, (3) USA. The values shown represent the medians of the station values.

There is a distinct downward trend in the temporal distribution of monthly I_{1k} changes in the warmer climate. With the exception of October, this index change decreases steadily in nearly every month from November to April. Similar trends exist between stations in the US and Canada, although the US stations show smaller decreases for almost every month. It is unclear why this downward trend in I_{1k} should occur. One would expect large reductions in March and April, due to a reduction in runoff caused by the shifting of the spring melt earlier into the season. However, November would also be expected to have reduced runoff, as the warmer climate, andconsequent less snowfall, would lead to fewer melt episodes during this month. Little of the trend in the I_{1k} changes can be attributed to the temporal variability of acidic meltwater concentration, since the previous graph showed only a small decrease of concentration throughout the snow season.

These results suggest that the greatest overall change in acid shock potential will occur late in the snow season, during March and April, if climatic change estimates are realised. Both of these months show relatively large decreases in the melt frequency and shock indices. Although melt episodes in January and February in the simulated climate show smaller shock indices, their frequency is expected to increase.

(e) Conclusion

In a network of 28 stations in the highly acidified region of southeastern Canada and the northeastern US, a hydrometeorological snowmelt model was used to predict changes in acid shock potential to streams and lakes due to projected climatic changes. Significant overall reductions are expected in the melt frequency, but the northward migration of the high melt frequency band is to cause increased melt episodes north of 46°N. General decreases in the potential for stream shock are expected to be smaller, with reductions in Canada and increases in the US. The potential for lake shock is most sensitive to warmer climate scenarios, as large decreases are predicted for most of the study region. The greatest reductions in the I_{1k} are expected from 45°N to 48°N, approximately co-located with the region of largest projected melt frequency increases. The reduced diurnal temperature range has a negligible effect on acid shock potential and melt frequency. Melt episodes during mid-winter are expected to increase in a warmer climate, corresponding to a shift in the springtime acid pulse toward earlier in the snow season. The most significant reductions in melt frequency and acid shock potential will occur during the months of March and April.

Overall conclusions

It is likely that future climatic change will exert a considerable influence on seasonal snowcover; in terms of its distribution, extent and physical nature. There will be important feedbacks to the atmosphere system. There may also be fundamental changes to the chemical composition of the snowpacks. There are severe difficulties in attempting to assess these effects: we are in no position to quantify many current climate\snowcover relationships; and we know little about likely relevant future climates. For these reasons, we have developed an existing model approach, with simple climate input, and used it to predict the removal of acid from seasonal snowcover. We have chosen to retain the outputs as potential "acid shock" indices, since they represent information on the melt status and acid load of the snowpack, and because of their possible interest from the point of view impact on aquatic biota. The changes, in a possible future climate, for melt frequency, stream shock index and lake shock index, give pointers to the likely magnitudes of the perturbations to the physical and chemical evolutions of seasonal snowcover. The combined effects of the predicted changes in melt frequency and acid removal (as indicated by the stream shock index, in particular) suggests that the seasonal climatology of the physical and chemical state of the snowcover will be considerably modified as a result of greenhouse gas induced climatic change. Thus far, we have adopted only the simplest approach, capitalising on existing methodology. It is possible that the impacts will be very pronounced when all causal relationships, outlined at the beginning of this chapter, are considered. But even within the scope of our analysis there are potentially-large uncertainties; the contribution of rain-on-snow events to acid removal from snowpacks is not quantified (Jeffries, 1989), and such events may become more common in a warmer world.

Our results give an indication of the potential perturbation to aquatic ecosystems. The model does not consider the effect that a warmer climate may have on spawning or feeding patterns, for example, nor does it simulate the potential for biological adaptation to sudden environmental stresses caused by greenhouse warming. Research into these areas would help to refine the model. The most serious limitations in applying this model, though, are its neglect of soil and groundwater interactions with meltwater and the uncertainty of future climatic change scenarios. The former problem could be rectified by transferring the output from this model into a more sophisticated one which incorporates realistic simulations of soil and groundwater effects on meltwater flowpaths. The limitation of reliable climate predictions should become less important in the future, as GCMs are improved to accurately assess the impact of greenhouse warming on regional scales.

Acknowledgements

We are extremely grateful to the following scientists for data, for very mind cooperation and for advice: Carol Simmons (National Atmospheric Deposition Program Colorado State University Fort Collins); Bob Vet, (Atmospheric Environment Service, Downsview, Ontario); Dean Jeffries (National Water Research Institute Burlington Ontario); Jack Shannon (Argonne National Laboratory Argonne Illinois), Jake Peters (United States GeologicalSurvey, Doraville, Georgia); Tony Marsh (National Power Technology and Environmental Centre, Leatherhead, UK); Rudy Boxleitner (Environmental Protection Agency, Raleigh, North Carolina).

References

Bales RC, Valdez MP, Dawson GA, Stanley DA (1987) Physical and chemical factors controlling gaseous deposition of SO2 to snow. In: Seasonal Snowcovers: Physics, Chemistry and Hydrology (Eds. Jones HG and Orville-Thomas WJ) NATO ASI Series, D Reidel Publ. Co., Dordrecht, 289-298.

Barnett TP, Dumenil L, Schlese U, Roekner E (1988) The effect of Eurasian snowcover on global climate. Science 239, 504-507.

Barrie, LA (1982) Environment Canada's Long Range Transport of Atmospheric Pollutants Program-Atmospheric Studies, Chapter 7 in Acid Precipitation, Effects on Ecological Systems (Ed. D'Itri FM), Ann Arbor Science Publishers.

Barrie LA and Sirois A (1982) An Analysis and Assessment of Precipitation Chemistry Measurements made by CANSAP (Canadian Network for Sampling Precipitation), 1977-1980, Internal Report AQRB-82-003-T, Atmospheric Environment Service, Downsview, Canada.

Berry MO (1981) Snow and climate. In: Handbook of snow; principles, processes, management and use (Eds. Gray DM and Male DH), Pergamon Press, Oxford, 776pp.

Brimblecombe P, Davies TD, Tranter M (1988) The composition of precipitation and chemical evolution of air masses. In: Proceedings of the WMO Conference on Air Pollution Modelling and its Application (Vol 3), WMO/TD No. 187, Geneva, Switzerland, 17-30.

Bruce JP (1961-2) Snowmelt contributions to maximum floods. In: Proceedings Eastern Snow Conference, 85-103.

Bruce JP and Clark RH (1966) Introduction to Hydrometeorology (Chapter 3, Melting of snow and ice), Pergamon Press, London, 319 pp.

Cadle SH, Dasch JM, Grossnickle NE (1984) Retention and release of chemical species by a northern Michigan snowpack. Wat. Air Soil Poll. 22, 303-319.

Cadle SH and Dasch JM (1987) The contribution of dry deposition to snowpack acidity in Michigan. In: Seasonal Snowcovers; Physics, Chemistry, Hydrology

(Eds. Jones HG and Orville-Thomas WF), NATO ASI Series, D. Reidel, Dordrecht, 299-320.
Cadle SH, Dasch JM, Kopple RV (1987) Composition of snowmelt and runoff in northern Michigan. Environ. Sci. Technol. 21, 295-299.
Cess RD and Potter GL (1988) A methodology for understanding and intercomparing atmospheric climate feedback processes in general circulation models. Journ. Geophys. Res., 93 (D7), 8305-8314.
Christophersen N, Rustad S, Seip HM (1984) Modelling streamwater chemistry with snowmelt. Phil. Trans. Roy. Soc. Lond., 305, 427-438.
Colbeck SC (1981) A simulation of enrichment of atmospheric pollutants in snow cover runoff. Wat. Resour. Res. 17, 1383-1388.
Dasch JM (1987) On the difference between SO_4^{2-} and NO_3^- in wintertime precipitation. Atmos. Environ. 21, 137-141.
Dasch JM and Cadle SH (1986) Dry deposition to snow in an urban area. Wat. Air Soil Poll. 29, 297-308.
Davies TD, Vincent CE, Brimblecombe P (1982) Preferential elution of strong acids from a Norwegian ice cap. Nature 300, 161-163.
Davies TD, Kelly PM, Brimblecombe P, Farmer G, Barthelmie RJ (1986) Acidity of Scottish rainfall influenced by climatic change. Nature 322, 359-361.
Davies TD, Brimblecombe P, Tranter M, Tsiouris S, Vincent CE, Abrahams PW, Blackwood IL (1987) The removal of soluble ions from melting snowpacks. In: Seasonal snowcovers; physics, chemistry, hydrology (Eds. Jones HG and Orville-Thomas WJ) NATO ASI Series, D. Reidel, 337-392. Dordrecht,
Davies TD (1989) As pure as the driven snow? New Scientist 122, 45-49.
Davies TD, Tranter M, Jickells TJ, Abrahams PW, Landsberger S, Jarvis K, Pierce CE (in review) Heavily-contaminated snowfalls in the remote Scottish Highlands: a consequence of regional-scale mixing and transport. Atmos. Environ.
Delmas V and Jones HG (1987) Wind as a factor in the direct measurement of the dry deposition of acid pollutants to snowcovers. In: Seasonal snowcovers; physics, chemistry, hydrology (Eds. Jones HG and Orville-Thomas WJ). NATO ASI Series, D. Reidel, Dordrecht, 321-336.
DeWalle DR (1987) Review of snowpack chemistry studies. In: Seasonal snowcovers; physics, chemistry, hydrology (Eds. Jones HG and Orville-Thomas WJ). NATO ASI Series, D. Reidel, Dordrecht, 255-268.
Dewey KF (1986) The relationship between snowcover and atmospheric thermal and circulation anomalies; Snow Watch 1985, Glaciological Data, Report GD-18, World Data Center A for Glaciology, Boulder, USA, 37-53.
Elsasser WM (1942) Heat transfer by infrared radiation in the atmosphere. Harvard Meteorological Studies No. 6, Harvard University Press, 494 pp.
English MC, Semkin RG, Jeffries DS, Hazlett PW, Foster NW (1987) Methodology for investigation of snowmelt

hydrology and chemistry within an undisturbed Canadian Shield watershed. In: Seasonal snowcovers; physics, chemistry, hydrology (Eds. Jones HG and Orville-Thomas WJ), NATO ASI Series, D. Reidel, Dordrecht, 467-500.

Eshleman KN (1988) Predicting regional episodic acidification of surface waters using empirical models. Wat. Resour. Res. 24, 1118-1126.

Farmer G, Davies TD, Barthelmie RJ, Kelly PM, Brimblecombe P (1989) The control by atmospheric pressure patterns of sulphate concentrations in precipitation at Eskdalemuir, Scotland. Int. Journ. Climatol. 9, 181-189.

Fowler D and Cape JN (1984) On the episodic nature of wet deposited sulphate and acidity. Atmos. Environ. 18, 1859-1866.

Gall RL and Herman BM (1980) The sensitivity of the nocturnal minimum temperature over dry terrain to temperature and humidity changes aloft. Mon. Wea. Rev. 108, 286-291.

Goldstein RA, Gherini SA, Chen CW, Mok L, Hudson RJM (1984) Integrated acidification study (ILWAS): a mechanistic ecosystem analysis. Phil. Trans. Roy. Soc. Lond., B 305, 409-425.

Haltiner GJ and Martin FL (1957) Dynamic and Physical Meteorology, McGraw Hill, 116 pp.

Hansen J, Lacis D, Rind D, Russel G, Stone P, Fung I, Ruedy R, Lerner J, (1984) Climate sensitivity: Analysis of feedback mechanisms. AGU Geophysical Monograph 29, Maurice Ewing Volume 5, 130-163.

ICAIR (1987) Snow pack/ground interface. Appendix 5 in: Workshop on acid deposition monitoring for snowfall and snowpack. ICAIR work assignment 081438, for USEPA, 401 M Street, SW, Washington DC 20460, p4-1.

Jeffries DS (1989) Snowpack storage of pollutants, release during melting, and impact on receiving waters. In: Acidic precipitation Vol. 4, Soils, aquatic processes, and lake acidification (Eds. Norton SA, Lindberg SE, Page AL), Springer-Verlag, New York 107-132.

Johannes AH (1987) Monitoring chemical changes in snow pack: state-of-the-science and proposed research. Discussion paper D-898-8-2 (1986) in: Workshop on acid deposition monitoring for snowfall and snowpack (summary report TR 898 24A), ICAIR, Life Systems Inc. submitted to USEPA, 401 M. Street SW, Washigton DC 20460. 25pp.

Johannessen M, Dale T, Gjessing ET, Henriksen A, Wright RF (1975) Acid precipitation in Norway: The regional distribution of contaminants in snow and the chemical concentration processes during snowmelt. Proc. Isotopes and Impurities in Snow and Ice Symp., Grenoble 1975. Int. Ass. Sci. Hydrol. Publ. 118, 116-120.

Johannessen M and Henriksen A (1978) Chemistry of snow meltwater: changes in concentration during melting. Wat. Resour. Res. 14, 615-619.

Jones HG (1987) Chemical dynamics of snowcover and snowmelt in a boreal forest. In: Seasonal snowcovers; physics, chemistry, hydrology (Eds. Jones HG and Orville-Thomas WJ), NATO ASI Series, D. Reidel, Dordrecht, 531-574.

Jones HG and Deblois C (1986) Chemical dynamics of N-containing ionic species in a boreal forest snowcover during the spring melt period. Hydrol. Processes 1, 271-282.

Jones HG and Sochanska W (1985) THe chemical characteristics of snowcover in a northern boreal forest during the spring run-off period. Ann. Glaciol. 7, 167-174.

Jones HG, Tranter M, Davies TD (1989) The leaching of strong acid anions from snow during rain-on-snow events: evidence for two component mixing. In: Atmospheric deposition, Proceedings of the Baltimore symposim, May 1989. IAHS Publ. no. 179, International Association Hydrological Sciences, 239-250.

Karl TR, Kukla G, Gavin J (1984) Decreasing diurnal temperature range in the United States and Canada from 1941 through 1980. Journ. Clim. Appl. Meteorol. 23, 1489-1504.

Kind RJ (1981) Snow drifting. In: Handbook of snow (Eds. Gray DM and Male DH). Pergamon Press, Oxford. 776pp.

Knutson EO, Sood SK, Stockham JD (1976) Aerosol collection by snow and ice crystals. Atmos. Environ., 10, 395-402.

Kuhn MH (1989) The role of land ice and snow in climate. In: Understanding climate change (Eds. Berger A, Dickinson RE, Kidson JW), AGU Geophysical Monograph 52. IUGG Vol 7, AGU, 2000 Florida Ave., Washington DC 20009, 17-30.

Louie PYT and Pugsley WI (1980) Application of near real-time water budgets in monitoring climate-related events. Proceedings of the Fifth Climate Diagnostics Workshop, Seattle, Washington, Oct. 22-24, 1980 (sponsored by NOAA).

Lundquist D (1980) Studies of snow models in Dyrdalen. Norwegian Hydrological Committee, Oslo, report no. 5, 27 pp. (in Norwegian). p73

Marmorek DR, Cunningham G, Jones ML, Bunnell P (1984) Snowmelt effects related to acidic precipitation: a structured review of existing knowledge and current research activities. LRTAP Workshop no. 3, Foret Montmorency, Quebec, Oct. 16-18, 1984. ESSA Env. Social Sys. Analysts Ltd., 678 W. Broadway, Vancouver, BC, 116pp.

Motoyama H, Kobayashi D, Kojima K (1986) Jap. Journ. Limnol. 47, 165-176.

Mylona SN (1989) Detection of sulphur emission reductions in Europe during the period 1979-1986. Report EMEP MSC-W 1/89, Norwegian Meteorol. Inst. Oslo.

Palmer YT, Smith LR, Neirink J (1975) Ozone in the valleys of the high Sierra Nevada, Paper 32-7, Int. Conf. Environ. Sensing Assessment, Sept. 14-19, 1975, Las Vegas, New IEEE, publ. N.Y.

Peters NE (1984) Comparison of air and soil temperatures at forested sites in the west-central Adirondack Mountains. Northeastern Environ. Sci., 3, 67-72.

Pomeroy JW, Davies TD, Tranter M (this vol.) The impact of blowing snow on snow chemistry.
Rascher CM, Driscoll CT, Peters NE (1987) Concentration and flux of solutes from snow and forest floor during snowmelt in the most central Adirondack region of New York. Biogeochemistry 3, 209-224.
Rind D (1987) The doubled CO_2 climate: impact of the sea surface temperature gradient. Journ. Atmos. Sci., 44, 3235-3268.
Rind D, Goldberg R, Ruedy R (1989) Change in climate variability in the 21st century. Climatic Change, 14, 5-37.
Ropelewski CF (1986) Snow cover in real time climate monitoring. Snow Watch 1985, Glaciological Data, Report GD-18, World Data Center A for Glaciology, Boulder, 105-108.
Rowntree PR (1990) Estimates of future climatic change over Britain. Part 1: mechanisms and models. Weather, 45(2), 38-42.
Saunders WE (1952) Some further aspects of night cooling under clear skies. Quart. Journ. Roy. Meteorol. Soc. 78, 603-612.
Schemenauer RS, Berry MO, Maxwell JB (1981) Snowfall formation. In: Handbook of Snow (Eds. Gray DM and Male DH), Pergamon Press, Oxford, 776pp, 129-152.
Schlesinger ME and Mitchell JFB (1987) Climate model simulations of the equilibrium climatic response to increased carbon dioxide. Rev. Geophys., 25, 760-798.
Seip HM (1980) Acid snow-snowpack chemistry and snowmelt. In: Effects of acid precipitation on terrestrial ecosystems (Eds.Hutchinson TC and Havas M), Plenum Press, New York, pp. 77-94.
Schnoor JL, Nikolaidis NP, Nikolaidis VS (1989) Assessment of episodic freshwater acidification in the Sierra Nevada, California. Draft final report, Contract A732-036, California Air Resources Board.
Shannon JD, Lesht BM, Samson PJ (1985) Variability in patterns of acid deposition resulting from climatological variability. Argonne National Lab.IL. USA, Report Conf-85 1028-2 E1.99:DE85 018392, 4pp.
Sigg A, Neftel A, Zurcher F (1987) Chemical transformations in a snowcover at Weissfluhjoch, Switzerland, situated 2500 masl. In: Seasonal snowcovers: physics, chemistry, hydrology (Eds. Jones HG and Orville-Thomas WJ), NATO ASI Series, D. Reidel, Dordrecht,269-280.
Sirois A and Barrie LA (1988) An estimate of the importance of dry deposition as a pathway of acidic substances from the atmosphere to the biosphere in Eastern Canada. Tellus 40(B), 59-80.
Skartveit A (1982) Wet scavenging of sea-salts and acid compounds in a rainy, coastal area. Atmos. Environ. 16, 2715-2724.
Smith FB (forthcoming) Possible future trends in acid rain in the UK. Royal Society (London) Discussion Meeting: Technology in the third millenium; approaches to the

handling and treatment of wastes (proceedings in press).
Summers PW and Barrie LA (1986) The spatial and temporal variation of the sulphate to nitrate ratio in precipitation in eastern North America. Wat. Air Soil Pollut., 30, 275-283.
Tranter M, Brimblecombe P, Davies TD, Vincent CE, Abrahams PW, Blackwood IL (1986) The composition of snowfall, snowpack and meltwater in the Scottish Highlands-evidence for preferential elution. Atmos. Environ., 20, 517-525.
Tranter M (this volume) Controls on the chemical composition of meltwater.
UKRGAR (United Kingdom Review Group on Acid Rain) (1987) Acid deposition in the United Kingdom 1981-1985. Warren Spring Lab., Stevenage, UK. 104pp.
UKRGAR (United Kingdom Review Group on Acid Rain) (1990) Acid deposition in the United Kingdom 1986-1988. Warren Spring Lab., Stevenage, UK. 124pp.
Warrick RA and Russell JE (1988) Estimations of future greenhouse warming and the probabilities of occurrence. Report to Cent. Elect. Gen. Board, Climatic Research Unit, Univ. East Anglia, UK. 12pp.
Wigington PJ, Davies TD, Tranter M, Eshleman, KN (1990) Episodic acidification of surface waters due to acidic deposition. State of Science/Technology Report 12, National Acid Precipitation Assessment Program, Washington DC.
Wilson EE and Barrie LA (1981) Climatological study of acidic snowmelt shock potential for eastern Canada. Proceedings of the Thirty-Eighth Annual Eastern Snow Conference, Syracuse, New York, June 4-5, 1981, 23-32.
Wilson CA and Mitchell JFB (1987) Simulated climate and CO_2-induced climate change over western Europe. Climatic Change, 10, 11-42.
Wigley TML (1989) Scientific assessment of climate change and its impacts. Seminar on Climate Change, London, England, April 26, 1989.
Wigley TML (1988) The effect of changing climate on the frequency of absolute extreme events. Climate Monitor, 17 (2), Climatic Research Unit, Univ. East Anglia UK, 44-55.
Wright RF and Dovland H (1977) Regional surveys of the chemistry of the snowpack in Norway late winter 1973, 1974, 1975 and 1976. Research report 12,SNSF Project, Box 61, 1432 Aas-NLH, Norway. 29pp.
Yoshida S (1962) Hydrometeorological study on snow melt. Journ. Meteorol. Res., 14, 879-899 (in Japanese).

CHEMICAL PROCESSES IN SNOW—WHERE SHOULD WE GO?

Jeff Dozier
University of California
Center for Remote Sensing and Environmental Optics
Santa Barbara, CA 93106 U.S.A.

Yngvar T. Gjessing
Universitetet i Bergen
Geofysisk Institutt
N-5000 Bergen, Norway

Arland J. Johannes
Oklahoma State University
Department of Chemical Engineering
Stillwater, OK 74078 U.S.A.

Elizabeth M. Morris
Ice and Climate Division
British Antarctic Survey
High Cross, Madingley Road
Cambridge CB3 0ET, U.K.

UNDERSTANDING OF PROCESSES

In this conference we have seen excellent work on conceptual understanding of processes. Naturally we need to improve our quantitative understanding, but investigations at the scale of the laboratory experiment or the field plot have revealed many important details about:

- incorporation of chemicals into snowfall or snowpacks
  — but wet deposition is better understood than is dry deposition;

- link between snow metamorphism and fractionation of elutants
  — but effects and importance of solid particles are not well studied, and meltwater movement and fractionation are not so well linked.

Here is our list (non-inclusive!) of problems, where our conceptual and quantitative models of the process fall short.

I. Formation of Snow in Clouds and Incorporation of Solutes and Solids

Len Barrie showed links between physics and chemistry for simple geometries, but knowledge is still lacking for incorporation of solutes and solids into the variety of crystal shapes that snow assumes. Snowfall chemistry is different from that of rain. Snow and rain transport ionic species from the atmosphere through two main processes, nucleation scavenging within the cloud and below-cloud scavenging by attachment of particles to existing hydrometeors.

NATO ASI Series, Vol. G 28
Seasonal Snowpacks
Edited by T. D. Davies et al.

II. Boundary Conditions at Air-Snow Interface and Their Effect on Dry Deposition and sublimation

Steve Cadle and John Pomeroy examined physical processes of wind stress and transport. We think, however, that the dry deposition work over snow has lagged behind useful findings by investigators of heat and mass transfer. Compared to other surfaces, snow is smooth, and air over snow is usually stable, whereas most of the equations used to estimate dry deposition are taken from general formulations for turbulent transfer, where the focus has usually been on rougher surfaces under neutral or unstable conditions. The dry-deposition investigators would benefit from examination of recent work in transfer of heat and water vapor from snow, ice, and other surfaces.

Sublimation of blowing snow depends on the temperature and humidity differences between the grains and the surrounding air. The approach used by Pomeroy calculates these variables. An alternative might be to directly measure them, and the resulting equations would be much simpler, similar to the Bowen-ratio method for calculating evaporation from the land surface.

III. Metamorphism and Chemical Fractionation in Snow

a. ''Dry'' snow (i.e. $T < 0°C$) and the role of the quasi-liquid film

Dry snow is a two-phase mixture of ice and air. A disordered layer on the surface of the ice influences adsorption and surface tension but is attached to the surface. It is not truly a liquid phase. Bales' experiments have shown that the layer is important, causing uptake of gas to depend on temperature, but additional experiments are needed to connect chemistry and physics. We want to know how the liquid layer behaves, learn how to measure its thickness, and understand its physical and chemical properties. The work so far has led to a reasonable picture of the following:

i. evaporation of HNO_3 from firn;

ii. diffusion of H_2O_2 and ^{18}O in ice;

iii. loss of pesticides by degassing;

iv. adsorption of SO_2.

b. Snow with low water content

Snow must pass through this stage, in which there are concentrated solutions—10 M or so—in the small liquid inclusions. We need to formulate the proper thermodynamics for the reactions in these thin layers. There are strong links to the physics, with effects caused by surface tension and viscosity (e.g. Brimblecombe). Chemical fractionation will take place in this ''brine.''

c. ''Wet'' snow, where gravity moves the liquid through the pack

There appear to be two causes of fractionation and preferential elution. On the microscale, formation and conditioning of snow produce inhomogeneous samples. In addition, physico-chemical processes within the snowpack have a chromatographic effect, involving hydrological dispersion, adsorption, and differential trapping in immobile water.

IV. Biological Processes

H. G. Jones presented some intriguing data and speculations. Clearly more work is needed on the roles of animals and animal wastes in snow chemistry.

V. Photochemistry

Sunlight penetrates up to 0.5 m into the near-surface layer, particularly in the spring when grain size is large. Photochemical reactions are important, and must be part of the interpretation of ice cores.

ROLE OF EXPERIMENTS

To refine our understanding of these processes, we think laboratory experiments need to be more adventurous and creative. In the history of investigations of snow chemistry in the laboratory and the field, many of the observations and experiments have been designed for reconnaissance, to learn enough to formulate hypotheses, rather than to test hypotheses. Future progress, though, will occur through experiments carefully designed to isolate the important unanswered questions.

In designing future experiments, then, we should keep the following questions in mind:

i. What do we hope to learn? Can we compare our results with those of previous studies?

ii. How do we couple laboratory experiments with models?

iii. What new insights do we need in laboratory experiments?

iv. What data (or models) do we trust? What important parameters and physical constants are not well enough known?

v. How might we experiment on biological processes?

INVESTIGATIONS AT LARGER SCALES

In the future we must move toward larger scales—catchment, regional, global. Among the publicized effects of global change, climate warming caused by increased atmospheric concentrations of radiatively active gases is the one most emphasized. Expected consequences include changes in precipitation patterns, increased frequency of severe storms, and a rise in sea level. Though not emphasized as much in the news media, changes in precipitation chemistry are certainly equally profound, so we must investigate the chemical and nutrient balances at these larger scales if we are to make our important contributions to the science of environmental change. The current international global change research program focuses on three needed improvements:

i. monitoring of changes;

ii. understanding of processes; and

iii. development of predictive models.

In this workshop we have seen mainly a focus on the second of these elements.

In the workshop we have also seen a reluctance to address the larger-scale questions. As scientists we have a natural tendency to be cautious, to proceed in small steps, to seek refinement of existing knowledge instead of bold steps. Compared to our present level of understanding of many Earth System processes, the global change questions are so complicated that addressing them with any rigor is impossible, but address them we must. If we seek to understand every detail at each step in increasing scale from laboratory to field plot to catchment to large drainage basin, we will progress slowly. It is permissible to use toy models where better ones are unavailable.

There are important variables that we can measure or model over large regions with a judicious combination of remote sensing, global climate models, and hydrologic models. Climatologists and hydrologists are developing methods of measuring the following list of variables, and we would expect that by the end of the 20th century they could be reliably ascertained at the needed spatial or temporal sampling:

atmospheric gases and aerosols;
precipitation;
snow accumulation;
snow melt rate;
runoff.

An ambitious but necessary undertaking, then, is to ask about the measurements needed for investigations in snow chemistry. Can we develop imaginative methods to obtain reliable estimates of the following concentrations, fluxes, and processes?

precipitation chemistry;
snow chemistry;
spatial / temporal integration of chemical hydrographs;
biological processes in and below the snowpack;
interaction between snow chemistry and soil chemistry;
stream and lake chemistry.

INDEX

NATO ASI Series G

NATO ASI Series G

Vol. 22: **Evolutionary Biogeography of the Marine Algae of the North Atlantic.** Edited by D. J. Garbary and G. R. South. 439 pages. 1990.

Vol. 23: **Metal Speciation in the Environment.** Edited by J. A. C. Broekaert, Ş. Güçer, and F. Adams. 655 pages. 1990.

Vol. 24: **Population Biology of Passerine Birds. An Integrated Approach.** Edited by J. Blondel, A. Gosler, J.-D. Lebreton, and R. McCleery. 513 pages. 1990.

Vol. 25: **Protozoa and Their Role in Marine Processes.** Edited by P. C. Reid, C. M. Turley, and P. H. Burkill. 516 pages. 1991.

Vol. 26: **Decision Support Systems.** Edited by D. P. Loucks and J. R. da Costa. 592 pages. 1991.

Vol. 27: **Particle Analysis in Oceanography.** Edited by S. Demers. 428 pages. 1991.

Vol. 28: **Seasonal Snowpacks. Processes of Compositional Change.** Edited by T. D. Davies, M. Tranter, and H. G. Jones. 484 pages. 1991.

MIX
Papier aus verantwortungsvollen Quellen
Paper from responsible sources
FSC® C105338

If you have any concerns about our products,
you can contact us on
ProductSafety@springernature.com

In case Publisher is established outside the EU,
the EU authorized representative is:
Springer Nature Customer Service Center GmbH
Europaplatz 3, 69115 Heidelberg, Germany

Printed by Libri Plureos GmbH
in Hamburg, Germany

develop processes that will be safe to operate and robust towards scale up whilst retaining economic application.

4.3 Work Package 3: Biocatalysis

The third work package is taking a holistic "Systems Biocatalysis" approach to developing biocatalytic sustainable manufacturing routes (Figure 3). The systems approach means that in addition to working on the biological aspects of biocatalysis there is consideration of reaction engineering principles and the overall sustainability of the process. The use of biocatalysts in non-aqueous solvents is well precedented; however, commonly used organic solvents may themselves have sustainability issues. Thus, work has been directed to characterising biocatalytic reactions in bio-derived solvents.[13] The advantages of renewable solvents and considerations for selection are described in Chapter 3: Renewable Solvent Selection in Medicinal Chemistry. This work package is also evaluating the utility of carbon dioxide as a solvent (as supercritical CO_2) and as an aid to the isolation and purification of amines produced by biocatalysis. The holistic systems approach extends to development of bio–bio and bio–chem cascades, in particular the intensification of processes by using biocatalytically produced reagents for

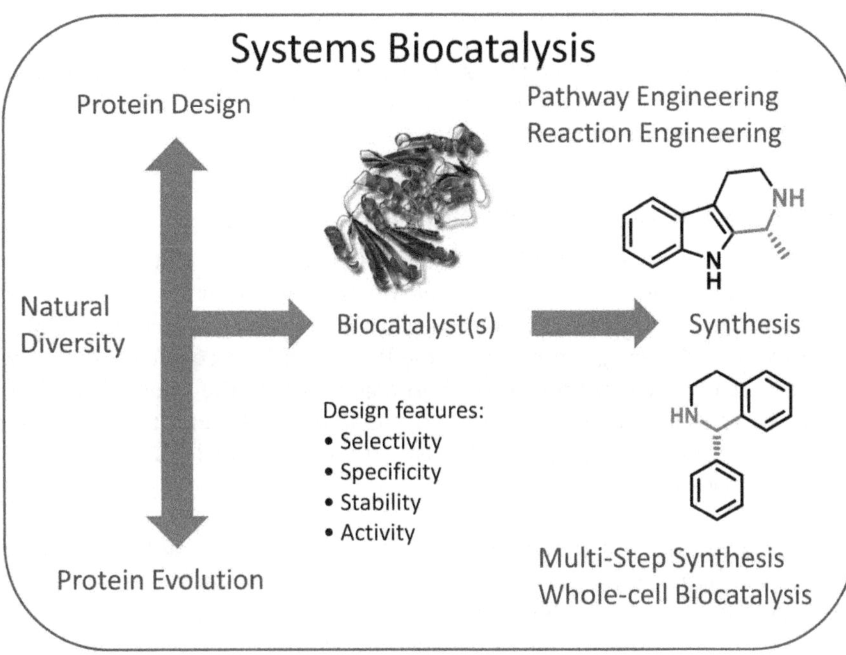

Figure 3 A systems approach to biocatalysis highlights the need to consider the whole process involved in developing a manufacturing route (figure kindly provided by Professor Nick Turner).

multi-component reactions (MCRs).[14] An overview of Biocatalysis for Medicinal Chemistry is given in Chapter 15.

Specific biocatalytic reaction classes have been chosen to target amide and chiral amine synthesis, stereo- and regio-specific hydroxylation of complex molecules as well as other redox reactions, the use of enzymes for stereo-controlled synthesis of fluorinated compounds, and production of homochiral quaternary centres.

4.4 Work Package 4: Synthetic Biology

There is no single agreed definition of Synthetic Biology, but all agree that it involves the engineering of biology for useful purposes and that it has great potential application in manufacturing the complex molecules found in modern medicines. A comprehensive overview of Synthetic Biology for Organic Synthesis is given in Chapter 14. The work within CHEM21 is divided into two phases. The first phase has been the generation of a toolbox of genetic technologies, parts, devices and chassis with clear freedom to operate in response to the complex and convoluted intellectual property landscape inherent to this emerging area of science. During this phase, the work package has also assembled an impressive arsenal of biocatalytic enzymes, in addition to the enzymes being worked on in work package 3, to put into synthetic pathways. The second phase is focussing on exemplification of the toolbox for producing compounds of interest including unnatural heterocycles, chiral amines and alcohols from cheap feedstocks using short cascades of the available enzymes. The technologies are also being challenged to provide improved routes to natural products, or close analogues of interest to the EFPIA partners.

4.5 Work Package 5: Education and Training

The fifth work package provides the control aspect of the DMAIC by monitoring sustainability improvements *via* the metrics toolkit and providing education and training materials to embed the principles of sustainability in the current and next generation of medicinal and process chemists, the European academic community and more generally to the scientific and lay public. This is being done *via* the development and dissemination of tailored educational and training material, case studies, and reviews. Specific objectives of work package 5 include:

- Formation of an expert network to drive forward the WP5 work plan, ensuring the longevity of WP5 outputs
- Identification of a preferred sustainability metrics toolkit
- Interconnect all work packages and embed green chemistry principles throughout *via* integration of the preferred sustainability metric toolkit
- Review and evaluation of the syntheses (gate to gate); environmental fate; and cradle-to-grave analysis of selected drugs

- Develop new, verifiably green synthetic routes to specific target molecules as defined by the EFPIA partners, as part of postgraduate and postdoctoral training
- Provision of support and guidance for the demonstration of the most promising transformations generated by the project at scale

The book you are reading is one element in this process. Other aspects beyond the usual scientific publications include running workshops for audiences beyond the CHEM21 project.[15] Within CHEM21, there is an enthusiastic and engaged Young Researchers Network encouraged to affiliate to NESSE (the Network of Early-Career Sustainable Scientists and Engineers).[16]

5 Concluding Remarks

The challenges in bringing sustainable technologies to a manufacturing environment should not be underestimated, particularly in a heavily regulated sector such as pharmaceuticals where, understandably, any change introduced has to be demonstrated to have a benefit to the patient. However, it is the duty of us all to be responsible stewards for our planet, to hand it over in good state to our children and grandchildren, and to treat its finite resources with care. The CHEM21 project is but one piece in a global effort to develop sustainable chemistry and biotechnological solutions. I hope that after reading this book you have a better understanding of what we are trying to do and feel better placed to approach you own career goals with a more sustainable outlook.

References

1. NHS Sustainable Development Strategy for the Health and Social Care System 2014–2020 http://www.sduhealth.org.uk/policy-strategy/engagement-resources.aspx (accessed Jun 2015).
2. http://www.sduhealth.org.uk/areas-of-focus/carbon-hotspots/pharmaceuticals.aspx (accessed Jun 2015).
3. Swedish Medical Products Agency Booklet on Sustainable Development and Pharmaceuticals https://lakemedelsverket.se/upload/eng-mpa-se/OF2009/Booklet%20Sustainable%20Development%20and%20Pharmaceuticals.pdf (accessed Jun 2015).
4. http://www2.epa.gov/green-chemistry/presidential-green-chemistry-challenge-award-recipients-industry-sector (accessed Jun 2015).
5. http://www.gsk.com/en-gb/responsibility/, http://www.pfizer.com/responsibility, http://en.sanofi.com/csr/planet/planet.aspx, http://www.bayer.com/en/sustainability.aspx, http://www.janssenbelgium.be/en/sustainability, http://www.orion.fi/en/Orion-group/Sustainability/.
6. http://www.gsk.com/en-gb/responsibility/our-planet/.

7. http://www.imi.europa.eu/content/documents#calls_for_proposals_-_imi_1_programme (accessed Jul 2015).
8. D. J. Constable, P. J. Dunn, J. D. Hayler, G. R. Humphrey, J. L. Leazer Jr, R. J. Linderman *et al.*, *Green Chem.*, 2007, **9**, 411.
9. A. Wells and H. P. Meyer, *ChemCatChem*, 2014, **6**, 918.
10. D. Prat, J. Hayler and A. Wells, *Green Chem.*, 2014, **16**, 4546.
11. C. P. Ashcroft, P. J. Dunn, J. D. Hayler and A. S. Wells, *Org. Process Res. Dev.*, 2015, **19**, 740.
12. C. R. McElroy, A. Constantinou, L. C. Jones, L. Summerton and J. H. Clark, *Green Chem.*, 2015, **17**, 3111.
13. G. Paggiola, A. J. Hunt, C. R. McElroy, J. Sherwood and J. H Clark, *Green Chem.*, 2014, **16**, 2107.
14. R. C. Cioc, E. Ruijter and R. V. Orru, *Green Chem.*, 2014, **16**, 2958.
15. https://www.uantwerpen.be/en/rg/orsy/sustainable-chemistry/ (accessed Jun 2015).
16. http://www.sustainablescientists.org/ (accessed Jun 2015).

About the Authors

John Blacker has been involved in the Fine Chemical/Pharmaceutical industry for 20 years and is currently Chair of Process Chemistry (www.chem.leeds.ac.uk/john-blacker) and Technical Director of the Institute of Process Research and Development (www.iprd.leeds.ac.uk) at the University of Leeds, part of both the School of Chemistry and the School of Chemical and Process Engineering. His interests are in sustainable chemical manufacture using catalysis, flow and efficient batch processes. He undertakes much work with industry to assist in the development of better production processes and improved products.

RSC Green Chemistry No. 46
Green and Sustainable Medicinal Chemistry: Methods, Tools and Strategies for the 21st Century Pharmaceutical Industry
Edited by Louise Summerton, Helen F. Sneddon, Leonie C. Jones and James H. Clark

Published by the Royal Society of Chemistry, www.rsc.org

Richard A. Bourne is currently a Tenure Track Fellow at the iPRD at the University of Leeds. He completed a PhD working on reactions in supercritical carbon dioxide under the supervision of Prof. Martyn Poliakoff, CBE, FRS at the University of Nottingham. He then worked as a research fellow with Prof. Martyn Poliakoff and Prof. Michael W. George looking at reactions of singlet oxygen and as part of the EU FP7 SYNFLOW project. He now investigates rapid process development and continuous flow chemistry using automated flow reactors and inline analysis.

Jessica Breen is currently a freelance journalist. She gained her PhD from the Durham University on fluorine chemistry in flow, under the supervision of Prof. Graham Sandford. After a year's postdoctoral fellowship in carbon capture chemistry with Prof. Christopher Rayner at the University of Leeds, she began working in the iPRD with Prof. John Blacker on the synthesis of chiral amines in flow.

Murray Brown joined SmithKline Beecham in 1997 (subsequently to become GSK) after completing a PhD and post-doctoral research at Cambridge University studying enzymes of the Shikimate pathway and polyketide synthases. He has had a variety of roles in early drug discovery in the hit ID, lead optimisation and early safety prediction arenas leading groups developing *in vitro* biochemical and cellular assays. Recognising that reagent generation, assay development and screening skills could equally be applied to the discovery and optimisation of novel enzymes, he led an Innovation group seeking biological alternatives to chemical synthesis. He is now part of the Synthetic Biochemistry team within the Advanced Manufacturing Technologies group developing platform capabilities for faster, better development of biocatalysts and Synthetic Biology approaches leading to robust biological processes for manufacturing drug substances. He is lead coordinator for CHEM21 (a €20M Innovative Medicines Initiative project), and represents GSK on various IB advisory boards, such as SynbiCITE (http://synbicite.com/), biocatnet (http://biocatnet.com/), SynBioCDT (http://www.synbio-cdt.ac.uk/), BrisSynBio (http://www.bristol.ac.uk/brissynbio/) and SynBERC (http://www.synberc.org/).

James Clark is Professor of Chemistry at York, Director of the Green Chemistry Centre of Excellence, and a Director of the Biorenewables Development Centre. He is also Chief Technical Officer for the University technology company Starbons Ltd. James has been at the forefront of green chemistry worldwide for about 20 years: he was founding scientific editor of the world-leading journal *Green Chemistry*, senior editor for the Royal Society of Chemistry Green Chemistry book series, and President of the Green Chemistry Network. His research and collaboration with industry has led to numerous awards including the 2011 RSC Environment Prize, the 2011 SCI Chemistry for Industry award, the RSC John Jeyes and SCI Environment medals, the Royal Academy of Engineering Clean Technology Fellowship, and distinctions from universities worldwide, including an honorary doctorate from Gent University in 2013. He was also research leader for projects that won EU and Royal Society of Arts Better Environment awards and the Prince of Wales award for innovation, as well as the 2012 Rushlight environment award. He has published over 400 original articles and written or edited over 20 books. He has given plenary lectures worldwide, and advises companies and governments across the globe on these topics.

Andri Constandinou is a 2nd year PhD student in the Green Chemistry Centre of Excellence (GCCE) at the University of York. She is also involved in the CHEM21 project and her research focusses on assessing the green credentials of current and improved synthetic routes to target APIs.

Ian Fairlamb (born 1975 in Crewe, England, UK) was appointed to a lectureship in Organic Chemistry at the University of York in October 2001, following a PhD under the guidance of Dr J. Dickinson investigating the synthesis of squalene synthase inhibitors (1996–1999) and a post-doctoral research position with Prof. G. C. Lloyd-Jones studying the mechanisms of various Pd-catalysed processes (2000–2001). In January 2010, he was promoted to Full Professor in Chemistry (Chair). His research is at the interface between Inorganic and Organic Chemistry. 15 PhD students have graduated from the Fairlamb research group over the past 13 years. The current group consists of MChem, MSc, PhD and PDRAs (*ca.* 16 members), many of whom are working on interdisciplinary collaborative projects. Key areas involve synthetic chemistry (*e.g.* cross-couplings, pericyclic processes, Pauson–Khand reactions), transition metal catalyst design (halide and pseudohalide effects, olefin ligands), mechanistic understanding specifically involving palladium nanoparticles and clusters, biological probes and therapeutic transition metal-containing complexes. The latter area involves the design of thermal and photochemical carbon monoxide-releasing molecules (CO-RMs) and anticancer compounds (both organic and transition metal-based), work that involves collaboration with chemists, biologists and biophysicists.

Farhana Ferdousi received her BSc in Chemistry in 2003 from the University of Dhaka, Bangladesh. She achieved her MS in Inorganic Chemistry in 2005 at the same university. Before joining as a Lecturer in the Department of Chemistry, University of Dhaka, Bangladesh in 2007, she spent a short period as a Lecturer in the Department of Natural Sciences, Stamford University, Bangladesh. She is currently pursuing a PhD under the direct supervision of Prof. Andy Whiting at Durham University. Her work mainly involves the development of new catalytic approaches for direct amide bond formation and their application in peptide synthesis as well as in some important drug syntheses.

Anton Glieder is Professor for Biotechnology at the Graz University of Technology, where he leads a research group for the engineering of protein and pathway expression systems by synthetic biology. After his studies in chemistry at the University Vienna and his PhD studies in Microbiology at the University Graz, he spent several years as a researcher and innovation manager in industry and as a postdoc in Biocatalysis at the University of Technology in Graz. During his research stage at Caltech and in Graz, he gained experience in protein engineering, which also led to close and successful collaborations with industry. Later he became a cofounder, CEO and CSO of ACIB in Austria. Today's major scientific interest is in innovative eukaryotic expression systems and their application for protein and pathway engineering.

John Hayler read chemistry at the University of Exeter and studied for a PhD in organic synthesis at the University of Bath. He joined SmithKline and French (subsequently SmithKline Beecham and GlaxoSmithKline) in 1987 and is currently a manager in the API Chemistry department, part of Product Development and Supply. His scientific interests include the application of green and sustainable principles to the manufacture of pharmaceutically active compounds.

Christopher Hone is currently studying for a PhD in Chemical Process Research and Development at the iPRD at the University of Leeds. He is working on the development of methodology for the design, optimization and scale-up of continuous flow processes. The project is part funded by AstraZeneca and Chris is working under the supervision of Prof. Frans Muller, Dr Richard Bourne and Prof. Steve Marsden.

Andrew J. Hunt gained a PhD in Chemistry from University of York (2006) focussed on "the extraction of high value chemicals from British upland plants". Dr Hunt now leads the Natural Solvents section of the Green Chemistry Centre of Excellence at the University of York. The use of bio-derived solvents and supercritical fluids in extractions, reactions and chromatography is a key aspect of his work. Research highlights include innovative work on the use of supercritical carbon dioxide for the extraction and recovery of liquid crystals from defunct display devices. This collaborative project led to a Rushlight award for innovation in recycling. Other related work on the recovery and expansion of waste polyvinyl alcohol from this waste stream has received significant press attention including an ASC press conference at the ASC green chemistry conference Washington DC, June 2010. More recently, he leads the EU SUNLIBB work package on high value products as part of an integrated biorefinery and now manages the PHYTOCAT project on metal recovery by plants (http://www.phytocat.org/). His research interests include elemental sustainability (metal recovery), materials science (utilisation of waste residues), waste treatment, catalysis, biorefineries and green solvents.

Klaus Kümmerer is Professor for Sustainable Chemistry and Material Resources and Director of the Institute for Sustainable and Environmental Chemistry at Leuphana University in Lüneburg, Germany. Before, he was professor for environmental chemistry and environmental hygiene at the Department for Environmental Health Science at the University Medical Centre Freiburg. He has been conducting research into sources, fate effects and risks connected to chemicals and pharmaceuticals in the aquatic environment for 15 years and is one of the worldwide leading scientists in the field. Only recently, he developed the framework of green and sustainable pharmacy and elaborated in detail the concept benign by design as a major building block of sustainable chemistry and pharmacy. He has published numerous articles in peer reviewed scientific journals and he (co)edited several books on the topic (latest: *Pharmaceuticals in the Environment*, Springer, 3rd edition, 2008; *Green and Sustainable Pharmacy*, Springer, 2010; *Water Reuse*, Springer forthcoming). In 2009, he won the Recipharm International Award for his work on pharmaceuticals in the environment and the conceptualization of green and sustainable pharmacy. Furthermore, in June 2015 he received the Water Resource Management Award from the Rüdiger Kurt Bode Foundation, Germany. He is founding editor of the peer-reviewed journals *Sustainable Chemistry* and *Pharmacy* as well as associate editor of *Chemosphere*. He is member of the management board of the European technology platform SusChem on behalf of the German Chemical Society. He is also a member of the Commission for Water Research of the Deutsche Forschungsgemeinschaft and of the board of the Division of Sustainable Chemistry within the German Chemical Society as well as a reviewer for national and international funding bodies.

Bert U. W. Maes was born in Belgium, graduated in 1997 from the University of Antwerp (Belgium) and obtained his PhD in 2001 from the same institute. He performed postdoctoral work in the group of Prof. Anny Jutand (CNRS) at the École Normale Supérieure in Paris (France) studying fundamental mechanisms in catalysis. In 2003, he was appointed as an Assistant Professor in the Department of Chemistry at UAntwerp. In 2008, he was promoted to Associate Professor and in 2014 to Full Professor. He currently holds a research position, which allows him to spend more time on research. In 2015, he became the chair of the Department of Chemistry. His research interests centre on heterocyclic chemistry, homogeneous catalysis, organometallic chemistry and sustainable chemistry. Bert Maes is a member of the Industrial Research Council (IOF) of UAntwerp and deputy member of the Committee on Innovation Policy (CIB) of the Flemish Council for Science and Innovation (VRWI), advising the Flemish government for science and innovation policy. He is a member of the board of the section Medicinal and Bioorganic Chemistry of the Royal Flemish Chemical Society (KVCV), the Belgian Organic Synthesis Symposium (BOSS) committee, the European Colloquium on Heterocyclic Chemistry (ECHC) committee, the Arkat/Arkivoc/Flohet Steering Committee, the advisory board of *Advances in Heterocyclic Chemistry* and the editorial board of the book series *Topics in Heterocyclic Chemistry*.

Jens Maes was born in Geel, Belgium and obtained his Master Degree in 2010 from the University of Antwerp (Belgium). He currently works as a PhD student in the group of Prof. Bert Maes at the University of Antwerp. His research is focused on the development of base metal-catalyzed direct amination reactions.

Josef Messinger was born in Duisburg (Germany). He started his career in chemistry as a chemical assistant in a research laboratory of Bayer AG (Uerdingen). He studied chemistry at the University of Duisburg (Germany) and McMaster University in Hamilton (Ontario, Canada) and in 1991 completed his doctoral thesis on "Conformational analysis of charge transfer complexes and benzodiazepine derivatives" in the group of Prof. Volker Buss (Theoretical Chemistry, University of Duisburg). From 1991, he worked for Solvay Pharmaceuticals GmbH (later Abbott Products GmbH) in Hannover as a medicinal chemist on different projects and in different indication areas. He extended his activities further as lecturer at the Gottfried Wilhelm Leibniz University Hannover. Since 2011, he has been working in the medicinal chemistry department of Orion Pharma (Espoo, Finland) as a senior advisor. He has over 45 patents and publications covering synthetic tools and techniques, biotransformations, new pharmacological active molecules, structure–activity relations, and modelling and modelling tools.

Emily A. Mitchell was born in Montréal, Canada and received her PhD in 2008 from Queen's University (Kingston, Ontario) under the supervision of Dr Michael C. Baird. Her research focused on fundamental studies in the optimization and nature of palladium(0) complexes for applications in cross-coupling reactions. She currently holds a postdoctoral position in the group of Prof. Bert Maes at the University of Antwerp. Her research focuses on direct functionalization through transition metal catalyzed sp^3 C–H activation.

Laurianne Moity is a chemist whose previous experience involves working in the fields of formulation and organic synthesis at ARD (Agro-industries Research and Development, Reims, France) in order to valorise bio-based succinic acid into valuable bio-surfactants, and solvent selection and bio-based products at the GCCE (Green Chemistry Centre of Excellence, York, UK). Laurianne recently completed a PhD at the University of Lille for Prof. J.-M. Aubry and Dr Valérie Molinier, where she worked on the design (Computer-Assisted Organic Synthesis), modelling (COSMO-RS and Hansen solubility parameters) and evaluation (synthesis, physico-chemical property measurements) of new bio-based solvents. She focused on isosorbide, glycerol and itaconic acid derivatives to replace hazardous solvents in agrochemical formulation.

Jonathan Moseley obtained his BSc in chemistry from Southampton University in 1986 and his PhD from Cambridge University in 1992. He worked as a medicinal chemist for Merck Sharp and Dohme for three years both before and after his PhD. In 1996, he moved to Zeneca (now AstraZeneca) as a process development chemist where he worked on pharmaceutical development projects from early route selection, kilo-lab and pilot plant scale-up, through to established manufacturing processes. His research interests include the application of new technologies to chemical synthesis, notably microwave chemistry, on which he has published extensively. At the beginning of 2011, he joined former colleagues to form their own company, CatSci Ltd, which delivers expertise in the development, optimisation and understanding of catalysis and other challenging reactions across the chemical research, development and manufacturing sectors. At CatSci, he is the Research Director, and leads project delivery and the scientific research programme. CatSci makes regular routine use of DoE on many of its projects.

Leena Otsomaa (Ph.Lic. eMBA) was born in Savonlinna (Finland). She is currently working as Head of Medicinal Chemistry at Orion, a globally operating Finnish pharma company developing pharmaceuticals and diagnostics. She studied chemistry at the University of Oulu (Finland) and completed her licentiate thesis on "Synthesis of 6-Deoxyamino Sugars" in the group of Prof. Ari M. P. Koskinen. She completed her eMBA studies in the Turku School of Economics. She started her career in chemistry as a research scientist in a synthetic chemistry laboratory at Orion in 1996. Since the beginning of 2010, she has been an active member (Orion representative) of the Scientific Committee of Chemical Industry Federation of Finland and later in 2012–2013 as vice chairperson. Currently Leena Otsomaa is acting as the chairperson of the Scientific Committee of Chemical Industry Federation of Finland.

Graham Pattison graduated with a first-class MChem degree from Durham University in 2005. He then remained at Durham for a PhD in organofluorine chemistry under the supervision of Prof. Graham Sandford, examining the use of highly fluorinated aromatics as scaffolds for the synthesis of highly substituted heterocycles. In 2009, he moved to the University of Edinburgh to work with Prof. Hon Wai Lam, working in the area of asymmetric catalysis. In late 2012, he moved to the University of Warwick as an IAS Global Research Fellow to commence independent research. His research interests are in the development of new sustainable catalytic reactions and in organofluorine chemistry.

Sirpa Rasku obtained her PhD in Organic Chemistry from the University of Helsinki (Finland) in 2000. In 2001, she joined the research function of Orion, a globally operating Finnish company developing pharmaceuticals. She started as a research scientist in Medicinal Chemistry working in early phase drug discovery projects. Currently she is working as a laboratory manager in one of the synthetic chemistry laboratories in Medicinal Chemistry.

James Sherwood is a research scientist at the Green Chemistry Centre of Excellence at the University of York. His previous experience includes working as a medicinal chemist at GlaxoSmithKline, before conducting postgraduate research under the supervision of Prof. Tom Welton and Dr Paul Lickiss at Imperial College London in the field of solvent effects. He has also worked for a company making bio-diesel from waste cooking oils. James recently completed a PhD at the Green Chemistry Centre of Excellence for Prof. James Clark and Dr Duncan Macquarrie, where he examined the synthesis and application of bio-based solvents in examples of organic transformations of relevance to the pharmaceutical industry. James' research interests now also include the analysis and certification of bio-based products, including solvents (http://www.biobasedeconomy.eu/research/kbbpps/).

Helen Sneddon read Natural Sciences at Christ's College, Cambridge University, and stayed at Cambridge to complete her PhD in organic chemistry, on β-keto-dithianes and their application to the synthesis of natural products, with Professor Steven V. Ley. After postdoctoral work on the asymmetric catalytic chemistry of palladium(II) with Professor Larry Overman at the University of California, Irvine, she joined GlaxoSmithKline in Stevenage, UK in 2007 as a medicinal chemist, working on respiratory medicines. In late 2011, she founded GSK's Green Chemistry Performance Unit, a team she still leads, looking at improving the environmental sustainability of research and development, and the routes arising from it. She has particular interests in solvent and reagent selection and the development of more efficient transformations, has published over 30 papers in scientific journals, and is honorary professor in Sustainable Chemistry at the University of Nottingham, School of Chemistry.

Louise Summerton is the Training, Education and Networks Manager in the Green Chemistry Centre of Excellence (GGCE) at the University of York and has over 10 years' experience in the preparation and delivery of green chemistry training and educational material. Louise is currently leading on the creation of an Education and Training package for the CHEM21 project. She has also been involved in securing the funding for, as well as the coordination and delivery of, numerous desk-top research projects on a national and European level, of which two included close collaboration with the pharmaceutical industry. Prior to joining the GCCE, Louise worked as a Medicinal Chemist for GlaxoSmithKline in Stevenage, UK and a Research Chemist for Croda in Hull, UK.

Andy Wells is an expert in chemical process research and development (R&D) and has experience across the pharmaceutical R&D lifecycle, especially in synthesis and active pharmaceutical ingredient (API) supply from discovery chemistry through to commercial operations. Andy spent 25 years working at Smith Kline and French, SmithKline Beecham and AstraZeneca before setting up Charnwood Technical Consulting Ltd (CTC Ltd) to promote the uptake of more sustainable manufacturing. Key areas of interest are in the application of homo/hetero-metal catalysis, organo-catalysis, industrial biotechnology, solvents and green chemistry solutions. Andy has experience across a broad range of scientific disciplines, and is a keen exponent of working at the interface of the physical and biological sciences to deliver more sustainable manufacturing. Andy Wells has worked on over 200 industry projects using biocatalysis as a key enabling technology across all scales from discovery to manufacturing. CTC Ltd is a member of the CHEM 21 consortium, and Andy Wells is a work package leader (http://www.chem21.eu/). Andy Wells is a visiting professor at the Chemistry Department at York University UK, a visiting industrial fellow at the Centre for Sustainable Chemical Technologies at Bath University, UK and is Chairman of the Board of the Centre for Biocatalysis, Biotransformations and Biocatalytic Manufacture at Manchester University, UK. Andy has ~150 publications, patents and invited lectures.

Andy Whiting carried out PhD studies with Professor R. J. Stoodley at Newcastle University, working on beta-lactam chemistry, before moving on to postdoctoral research at Boston College with Professor T. Ross Kelly working on natural product synthesis and the development of chiral Diels–Alder Lewis-acid catalysts. After a short period in industry with Ciba-Geigy Central Research, he moved to his first academic position as Lecturer in Chemistry at UMIST. In 2001, he moved to a Readership at Durham University and became Professor in 2009. He is currently Director of the Centre for Sustainable Chemical Processes at Durham, with research interests in organic synthesis, methodology and catalysis.

Birgit Wiltschi is the head of the Junior Group Synthetic Biology at the Austrian Center of Industrial Biotechnology in Graz, Austria. She earned her PhD from Graz University of Technology, stayed at the University of California in Berkeley for a research sabbatical, and carried out post-doctoral research at the Max Planck Institute of Biochemistry in Martinsried, Germany and the Albert Ludwigs University of Freiburg, Germany. Her research focuses on the engineering of proteins with unnatural amino acids. She is also interested in methods for the combinatorial assembly of multi-gene expression constructs, *e.g.* for metabolic engineering.

Acknowledgements

This book was written as part of the education and training aspects of the CHEM21 (Chemical Manufacturing Methods for the 21st Century Pharmaceutical Industries) project (http://www.chem21.eu). CHEM21 has received funding from the Innovative Medicines Initiative Joint Undertaking under grant agreement n°1115360, resources of which are composed of financial contribution from the European Union's Seventh Framework Programme (FP7/2007–2013) and EFPIA companies' in kind contribution (www.imi.europa.eu).

The editors would like to thank all the contributing authors; both those within the CHEM21 consortium and the external academics with complementary expertise to the CHEM21 members.

The views expressed in regards to education and training materials represent the aspiration of the CHEM21 consortium, although may not always be the view of each individual organisation.

RSC Green Chemistry No. 46
Green and Sustainable Medicinal Chemistry: Methods, Tools and Strategies for the 21st Century Pharmaceutical Industry
Edited by Louise Summerton, Helen F. Sneddon, Leonie C. Jones and James H. Clark

Published by the Royal Society of Chemistry, www.rsc.org

Glossary

Active pharmaceutical ingredient (API)	An active pharmaceutical ingredient (API) is the substance in a drug or a pesticide that is biologically active. Besides the API, several other substances are used as an excipient (*i.e.* the inactive or inert substances present inside a drug). These compounds are added as diluents (add volume/mass to the drug to allow a more precise dosing for patients), additives (*e.g.* flavors, colours) or for pharmacokinetic considerations (*e.g.* facilitating drug absorption or solubility).
Anti-angiogenesis agent	These substances suppress the formation of new blood vessels, which are vital to the growth of tumours.
Aquatic environment	Rivers, lakes, ground water, sea; a broader view also includes surface run-off, effluents and sewage.
Atom economy	Atom economy (atom efficiency) defines the conversion efficacy of a chemical operation in terms of all atoms concerned (desired products produced). Atom economy can be written as: Atom economy = [(molecular mass of all desired products)/(molecular mass of all reactants)] $\times$ 100

RSC Green Chemistry No. 46
Green and Sustainable Medicinal Chemistry: Methods, Tools and Strategies for the 21st Century Pharmaceutical Industry
Edited by Louise Summerton, Helen F. Sneddon, Leonie C. Jones and James H. Clark

Published by the Royal Society of Chemistry, www.rsc.org

Benign by design A concept that requires the targeted design of active pharmaceutical ingredients and adjuvants from the very beginning in order to render them readily and completely mineralized after their release into the environment.

Bio-based solvent A fluid derived from a biomass feedstock, which is used to dissolve other chemicals without undergoing a chemical change.

Biocatalysis The use of an enzyme or enzymes, either as a free protein(s) or as part of a whole cell system, to catalyse synthetic reactions.

Chemical substitution Seeking to replace currently used chemicals with more benign alternatives whilst ensuring that one hazard is not being exchanged for another.

Circular economy The application of closed loop or "cradle to cradle" strategies for the production, utilisation, recovery and reuse of products or resources.

Concerted metalation deprotonation (CMD) A term used to describe the arrangement of the organic and acetate/carbonate groups around a palladium(II) in the rate-determining step found in many C–H bond functionalization reactions.

Confounding Occurs when the main effects cannot be separated from the interaction effects due to the level or type of fractionation in the design (see also Interaction).

COSMO-RS: COnductor-like **S**creening **MO**del for **R**eal **S**olvents A quantum chemistry-based equilibrium thermodynamics method that is used to accurately predict a wide range of solvent properties (solubility, vapour pressure, partition coefficient *etc.*).

Critical elements Elements with significant supply risk issues that if restricted could harm a company's business or nations economy are considered to be "critical".

Cross-coupling reaction The reaction of two organic compounds leading to loss of two fragments and formation of one organic product. A leading example is the reaction of an organohalide (R–X) with an organoboronic acid $\{R'B(OH)_2\}$ to give a cross-coupled product R–R′ (containing two different organic groups—hence the term "cross-product".

Dean–Stark water removal	Used for the continuous removal of water that is produced during chemical reaction performed at reflux temperature, in combination with a reflux condenser and a batch reactor. It was invented by E. W. Dean and D. D. Stark in 1920 for determination of the water content in petroleum.
Design of Experiments (DoE)	A structured and efficient approach to experimentation that employs statistical techniques to investigate potentially significant factors and determine their cause-and-effect relationship on the outcome of an experiment.
Ecopharmacovigilance	Integrating an awareness of the adverse effects related to the presence of pharmaceuticals in the environment into the concept of pharmacovigilance.
Elemental sustainability	Elemental sustainability is a concept whereby each element within the periodic table is guaranteed for use by both current and future generations.
Equielutropic series	A schematic showing the equivalence in polarity of different solvent mixtures.
Factor	A controllable experimental parameter that can be varied to change the reaction conditions. Factors can be continuous (*i.e.* variable at many settings, such as concentration, pH, temperature) or discrete (*i.e.* discontinuous, such as solvent, ligand, base).
Fate, degradation and transformation	Fate is the whereabouts of a chemical in the environment; it can distribute between soil, air, water, sediment and organisms; there it can be fully degraded *i.e.* mineralized to inorganic chemicals or only incompletely degraded ("transformed") or accumulate in soil, water, sediment or organisms and may hereby enter the food chain.
Genotoxic impurities	DNA-reactive substances that have a potential to cause direct DNA damage and consequently require control to very tight limits in API.

Geopolitics	The field of geopolitics traditionally studies the links between political power and geographic space (*e.g.* natural resources) in international relations. Industries that are dependent on products coming from a limited number of regions (such as oil or precious metals) are quite vulnerable to political tension. The policy of a regional or federal government in a certain country can therefore have a major influence on business and economics worldwide.
Heterogeneous catalysis	Refers to the form of catalysis where the phase of the catalyst differs from that of the reactants.
Heterologous expression	The use of a carrier or host organism for the production of an enzyme that does not occur naturally in that organism.
Homogeneous catalysis	There are two types of catalysts, namely heterogeneous and homogeneous catalysts. In contrast to heterogeneous, the homogeneous catalyst is in the same phase as the reactants. In general, this allows better mixing and interaction of the catalyst with the reaction mixture.
Immobilized Pd catalysts	A catalyst that has been immobilized or encapsulated by a suitable support, *e.g.* polymer, containing appropriate palladaphilic groups (*e.g.* alkenes or amides).
Immobilized Pd nanoparticles (PdNPs)	A similar definition to the immobilized Pd catalysts; the nanoparticles describe a range of higher order Pd species that are present under the reaction conditions. For example, spherical PdNPs that are *ca.* 2 nm in size consist of *ca.* 250 Pd atoms, whereas PdNPs that are *ca.* 4 nm in size, consist of >4000 Pd atoms.
Inherently safe	Pharmaceuticals and chemicals that have little or no hazards.
Interaction	Occurs when two or more factors are dependent on each other, so that the overall result is different from what would be expected from the sum of their individual effects. Two factors interacting (A and B), results in a two factor interaction term (AB); three factors interacting gives a three factor term (ABC) *etc.* It is rare for three factor or higher interactions to be significant.

LD50 (lethal dose 50%) The median lethal dose, LD50, of a toxin is the dose required to kill half the members of a tested population after specified test duration in toxicology. LD50 figures are frequently used as a general indicator of a substance's acute toxicity.

Life Cycle Assessment A method for assessing the environmental impacts associated with all stages of a product's life, from raw material extraction right through to disposal (cradle-to-grave).

Mass-based metrics Calculations that assess the greenness of a reaction based solely on mass inputs and outputs.

Mass intensity A metric parameter indicating the amount of waste produced for a certain reaction/synthetic path. It is defined by the quotient of all the material that went into a certain reaction and the amount of material actually produced.

Micro-pollutant Pollutant, so-called due to the low (micro- to nano-gram per litre) concentrations in which they are found.

Mineralization Full degradation of a molecule to simple compounds, such as water, carbon dioxide, ammonia and sulfate, without the formation of stable transformation products, either by organisms or non-biotic processes.

Multi parameter space In drug discovery property space for optimisation is defined by potency, solubility, absorption, specificity, safety, metabolism and other project (indication, target) related properties.

Mutant An enzyme that has had one or more of the amino acids in the sequence changed by random or directed mutation to improve a particular catalytic parameter, such as enantiomeric excess, stability, tolerance to heat, and solvents.

One-variable-at-a-time (OVAT) An approach to practical experimentation in which one factor is varied at a time, keeping all other factors constant. Despite its apparent logic and simplicity, it often fails to uncover the best results for a process or reaction, usually because it fails to account for the dependency between factors, aside from other reasons given in the main text (see also interaction).

Optimisation cycle During optimisation of a compound series, researchers run through a process of hypothesis and trials. The process starts with the design and synthesis of a compound, which after biological testing in a multiparameter space leads to a verified hypothesis and design of compounds until the optimisation goal has been achieved.

Organocatalysis In contrast to metal catalysis, the rate of a chemical reaction is increased by small organic molecules. These organocatalysts are predominantly composed of C, H, O, N, S and P and contain no metal atoms. In general, they should preferentially be insensitive to moisture or oxygen, be readily available and possess low toxicity.

Oxidant An oxidant (oxidizing agent) accepts electrons from another species in an oxidation–reduction (redox) reaction. It is thus itself reduced, while the reactant in question is oxidized.

Oxidase The terminology oxidase and oxygenase finds its origin in enzyme catalysis using molecular oxygen as the oxidant. In the first case, the oxidant acts as an electron acceptor and is itself reduced. In the latter case, a substrate is oxidized by transferring the oxygen atom from molecular oxygen to the substrate.

Peptide Peptides are small chains of amino acid monomers joined by peptide (amide, –CO–NH) bonds. The covalent chemical bonds are formed when the carboxyl group of one amino acid reacts with the amino group of another.

Platform molecule Functionalised molecules (which are derived from renewable resources) used as chemical 'building blocks' to access a wide variety of fine chemicals.

Polarity The sum of the strength of all hydrogen bonding (specific) and other non-specific intermolecular interactions between a solute and its solvent.

Proof of concept (POC) Demonstration of a hypothesis in a limited patient population. Achievement of POC will inform investment in the larger clinical trials required to demonstrate safety and efficacy.

Quality by Design	An approach to drug product development using a combination of risk management and scientific knowledge to understand the relationship between process parameters and unit operations to API and/or drug product critical quality attributes (CQA).
REACH	Registration, Evaluation, Authorisation and restriction of CHemicals (REACH) is the system for controlling hazardous chemicals in the European Union. It aims to provide protection to human health and the environment, and makes manufacturers and importers responsible for understanding and managing the risks of chemicals.
Reagent selection guide	A comparative ranking of the sustainability credentials of different reagents for a given transformation. They may be used to help identify alternatives to reagents whose use may be becoming increasingly restricted by legislation or company targets around reducing the use of materials of concern. They may help identify reagents/procedures that have lesser EHS implications, are more atom efficient or with fewer operational complexities, or that can be run in a more benign solvent.
Resource efficiency	Maximising the supply of products, whilst minimising waste.
SELECT criteria	An acronym describing the criteria (Safety, Environmental, Legal, Economic, Control and Throughput) against which a synthetic route can be assessed.
SIN list	The SIN (Substitute It Now!) List is a list of hazardous chemicals based on the REACH Substances of Very High Concern criteria. The list is maintained by ChemSec, (the International Chemical Secretariat) and aims to speed up the process of phasing out hazardous chemicals.
Solvent selection guide	A comparative ranking of the sustainability credentials of different solvents. Multiple categories can be viewed in detail to derive the optimum solvent for a given use, and rules can be applied to combine categories to provide generalised guidance.

Supercritical fluid	A chemical substance that is above its critical temperature and critical pressure, resulting in properties between those of a liquid and a gas.
Sustainability	"Development that meets the needs of the present without compromising the ability of future generations to meet their own needs." The concept considers the impact of a process on ecological protection, society, and economic viability.
Sustainable management of water resources	Usage and protection of natural water bodies in order to not deplete quantity or quality by direct or non-direct human impact.
Toxicity/eco-toxicity	Unwanted effects of a chemical to humans/ organisms in the environment.
Transformation	Change of the chemical structure of a molecule without complete mineralization because of the formation of stable reaction products.
Waste valorization	The process of deriving value from waste, for example by extracting useful chemicals and energy.

Contents

RSC Green Chemistry No. 46
Green and Sustainable Medicinal Chemistry: Methods, Tools and Strategies for the 21st Century Pharmaceutical Industry
Edited by Louise Summerton, Helen F. Sneddon, Leonie C. Jones and James H. Clark

Published by the Royal Society of Chemistry, www.rsc.org

CHAPTER 1

Green and Sustainable Chemistry: An Introduction

JAMES H. CLARK

Green Chemistry Centre of Excellence, Department of Chemistry, University of York, York YO10 5DD, UK
Email: james.clark@york.ac.uk

1.1 What is Green Chemistry?

"Green Chemistry", "Green Engineering" and "sustainability" are often used interchangeably to describe the concept of making processes and products that have less environmental impact and are (ideally) based on renewable resources. However, if one examines these concepts more deeply, it will become clear that there are significant differences in their philosophy, which impacts the applicability of the methodologies and techniques in the development of an environmentally appropriate society.

Green Chemistry is quite well defined by the twelve principles of Anastas and Warner.[1] These principles focus mostly on how one should carry out chemical reactions and make chemical products, and describe the synthesis of chemicals in an environmentally preferable way. Thus, specific ideas, such as the use of benign auxiliaries including solvents for reactions and separations, reducing the number of steps, and the concept of atom economy, or incorporating all of your raw materials into your product, are focal points of Green Chemistry. The 12 Principles were written 20 years ago and don't fully reflect the modern way of thinking. Additional issues, such

RSC Green Chemistry No. 46
Green and Sustainable Medicinal Chemistry: Methods, Tools and Strategies for the 21st Century Pharmaceutical Industry
Edited by Louise Summerton, Helen F. Sneddon, Leonie C. Jones and James H. Clark

Published by the Royal Society of Chemistry, www.rsc.org

as toxicology and biodegradability, now play an important role in Green Chemistry as part of a greater emphasis on product safety as well as on renewable resources. In many respects, Green Chemistry can be considered as the scientific underpinning of environmentally preferable manufacturing.

Green Engineering, on the other hand is the design, commercialization and use of processes and products that are feasible and economical while minimizing generation of pollution at source, as well as minimizing risk to human health and the environment.[2] The green engineer uses the tools of recycling, process intensification and design optimization to maximize the efficiency of a process and reduce its burdens on the environment. Green engineering evaluates the manufacturing process as a system and seeks to optimize its design, and in the truest sense, incorporates the concepts of lifecycle analysis and environmental economics into an appropriate evaluation of the overall environmental impact. Green engineering requires the development of a set of metrics that appropriately evaluate the environmental parameters that we seek to control.

Sustainable design looks even more broadly to try to understand the relationships between the manufacturing system and the ecosystem. Sustainability focuses on the triple bottom line: the integration of ecological integrity, societal responsibility, and economic viability. Sustainability takes the broadest level systems approach, looking at the planet as the system of interest, but in order to optimize design at this scale, new ways of measuring human impacts on the environment will be required.

The chemical and allied industries now face as tough a challenge as they have ever faced. The 20th century saw enormous growth in chemicals manufacturing but this growth has come at a cost. Inefficient processes leading to unacceptable levels of pollution, hazardous operations resulting in a number of disasters, and a lack of knowledge of the human and environmental toxicity of most chemicals in widespread use, all leading to an exponential growth in chemicals legislation. The industry now needs to achieve environmental and social acceptability as well as economically viable manufacturing in the toughest-ever legislative framework. Recent chemicals legislation, such as REACH, is causing major changes in the chemical product supply chain.[3] Yet somehow this has to be done in a way that satisfies the demands of a growing population. Sustainable chemical production can only be realized through a re-assessment of the entire chemical product lifecycle from resources, to manufacturing and production, through to product use and ultimate fate (Figure 1.1).

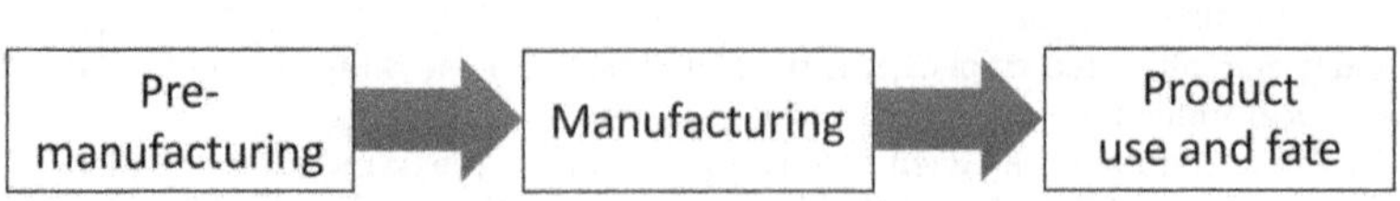

Figure 1.1 Critical stages in the lifecycle of chemical products.

1.2 Drivers for Change

1.2.1 Legislation

Pressure on chemical manufacturing, especially legislative and also from customers, continues to be applied and generally leads to cleaner and safer manufacturing. The chances of another disaster such as Bhopal are lower, at least in most regions of the world (*e.g.* as a result of penalties and restrictions over the storage of hazardous substances) although there is still considerable manufacturing in regions with less control and therefore higher risk.[4] The control over pollution and the severe penalties that can be imposed have discouraged significant emissions from plants in most locations.

REACH is the most talked about legislation affecting chemical products.[3] This and other chemical legislation directly affect chemical and allied manufacturing through restricted availability of an increasing number of common chemicals. While very hazardous substances such as organomercury and lead compounds have been the subject of rigorous scrutiny for many years, new restrictions on the use of others such as chromates and cobalt compounds may have considerable impact on some industrial chemistries, including oxidations. While REACH is slow in its progression (it will be the next decade before all the chemicals subject to REACH have been tested), unofficial lists of substances to substitute have appeared. Probably the most prominent of these is the so-called SIN ("substitute it now") list.[5] Several hundred chemicals appear on this list and it is influencing some end-users who do not want their products to contain any chemicals on such publically available "red lists".

Perhaps the biggest impact will be over the use of solvents since many of the more common organic solvents are under threat from REACH: these include *N*-methyl-2-pyrrolidone (NMP), dimethylformamide (DMF) and dimethylacetamide (DMAc).[6] (For more information on solvent substitution guides see Chapter 2, 'Tools for Facilitating more Sustainable Medicinal Chemistry', by Helen Sneddon and James Sherwood's Chapter 3 on renewable solvent selection.) The electronics industry has also been subject to chemical legislation that aims to replace especially hazardous substances. RoHS (restriction on hazardous substances) targets certain chemicals, including lead, mercury, cadmium chromates and polybrominated flame retardants.[7]

1.2.2 Elemental Sustainability

In addition to substances becoming restricted or unavailable due to changes in legislation, they may also be at risk due to issues with supply. Elements used by the chemicals industry, both in the manufacturing steps (*e.g.* as catalysts) and in the products themselves, include organohalogen compounds and numerous organic compounds containing heteroatoms, such as phosphorus, sulfur and boron. These are extracted from virgin ores and other natural sources, which, like petroleum, are finite and require a large

Table 1.1 Some important elements that have limited availability based on current practices.

Element	Some areas of use
Phosphorus	Detergents, agrochemicals
Germanium	Fiber optics, semiconductors
Indium	Solar cells, LCDs
Antimony	Batteries, catalysis
Neodymium	Hybrid cars, wind turbines

amount of energy for extraction. While renewable carbon has been a hot topic for the last decade (see the next section), it is only in the last couple of years that attention has been broadened to include other critical elements including phosphorus and many metals (see Chapter 5 on elemental sustainability by Andrew Hunt). Some of the elements of concern are listed in Table 1.1. Many elements are now considered to be under threat in terms of predicted usage rates and known reserves.[8]

Some may run out within 10 years (*e.g.* indium and germanium). While new reserves are being discovered, as with oil they are often of relatively poor quality and come at high economic and environmental costs. It is ironic that some shortages (*e.g.* lithium and some of the rare earths) are a result of increasing rates of use for low-carbon technologies. While we certainly need low *fossil* carbon technologies, we must introduce these with our eyes wide open to any consequential issues, such as high use of other critical elements.[9]

One way that researchers are responding to the criticality of some elements is by much smarter use of these elements, *i.e.* better catalyst design and recycling (discussed in Chapter 11), and by developing catalysts that avoid the use of critical elements by focusing on the use of more plentiful base metals (Chapter 16).

1.2.3 Renewable Resources

Manufacturing of chemicals is resource-dependent. Petroleum has dominated the industry as a carbon feedstock with a few exceptions, including a small percentage of naturally derived compounds (*e.g.* for use in personal care products and pharmaceuticals) and chemicals derived from coal in South Africa (developed to overcome trade barriers introduced in the apartheid era).

There is increasing pressure, especially from consumers, on manufacturers to produce bio-derived chemicals as replacements for fossil resources and substances now considered to be hazardous to us or to the environment. The move towards bio-based products is considered to have a number of advantages:[10]

- Use of renewable and expendable resources
- Less dependency on limited and increasingly expensive fossil resources
- The potential to reduce greenhouse gas emissions (carbon neutral/low carbon impact)